Proceedings of the Sixth International Congress on Catalysis

Imperial College, London, 12-16 July, 1976

Edited by

G. C. Bond
Professor of Industrial Chemistry, Brunel University

P. B. Wells
Reader in Physical Chemistry, University of Hull

F. C. Tompkins
Professor of Physical Chemistry, Imperial College

Volume 2

The Chemical Society
Burlington House, London, W1V 0BN

Text set in 9/10 pt Monotype Times New Roman, printed by letterpress,
and bound in Great Britain at The Pitman Press, Bath

Contents

iii

List of Contents

iv

List of Contents

v

List of Contents

List of Contents

vii

List of Contents

List of Contents

ix

List of Contents

List of Contents

Subject and Author Indexes are printed at the end of both Volumes

Dispersion and Uniformity of Supported Catalysts by X-Ray Photoelectron Spectroscopy

PHILIP J. ANGEVINE, JAMES C. VARTULI,† and W. NICHOLAS DELGASS*

School of Chemical Engineering, Purdue University, West Lafayette,
Indiana 47907 U.S.A. and Department of Engineering and Applied Science,
Yale University, New Haven, Connecticut 06520 U.S.A.

ABSTRACT   X-Ray photoelectron spectroscopy (XPS) has been used for quantitative characterization of supported metal and metal oxide catalysts. Active component/support XPS peak intensity ratios of 100% dispersed catalysts were found to be proportional to the surface density of active component atoms. Results for Na/Al_2O_3 and Eu/Al_2O_3 as a function of Na and Eu loading indicated high dispersion at low loading and agglomeration or occlusion of the Eu phase at $1\cdot2 \times 10^{19}$ Eu atoms m^{-2} Al_2O_3. Analysis of XPS intensity ratios also identified 100% dispersion of Pt on SiO_2 and suggested that the relative uniformity of deposition of active component atoms on a support surface can be indicated by XPS data. With further refinement of the analysis and calibration procedures, the method has particular promise for estimation of relative dispersions of elements in a multicomponent supported catalyst.

INTRODUCTION

Few techniques for measuring the dispersion of supported catalysts have the general ability to measure both metal and oxide particle sizes < 5 nm or to detect individual dispersions of several components on a support. X-Ray photoelectron spectroscopy (XPS), while still a developing technique,[1] is emerging rapidly as a formidable tool in surface chemistry. Its sensitivity to all elements and specificity to thin surface layers make it particularly attractive for this important task. Investigation of a series of Pt/SiO_2 catalysts by Sharpen[2] has shown an excellent linear correlation between μmoles of H_2 adsorbed/g catalyst and the XPS intensity ratio, $I_{Pt(4f)}/I_{Si(2p)}$, for five samples prepared by ion exchange. Two samples prepared by impregnation did not fit this correlation, however. In a second study of supported metals, Brinen et al.[3] showed a smooth variation of $I_{Rh(3d)}/I_{C(1s)}$ with particle size determined from X-ray diffraction line broadening for 12 wt % Rh on charcoal. In this paper we examine in more detail the quantitative relationship getween XPS peak intensity ratios and catalyst dispersion and uniformity.

Several guidelines and precautions are available in the literature on the use of XPS for quantitative analysis.[4-7] A simplified picture of a supported catalyst is employed here to ascertain the functional relationship between active component/support intensity ratios and nominal surface coverage and particle size of the active component. The catalyst is envisioned as a uniform distribution of cubes of the active component on a flat, semi-infinite support surface. If the electron take off angle is taken at $90°$ from the surface plane,

$$\frac{I_m}{I_s} = \frac{I_m^\infty f\{1 - \exp[-c/\lambda_m(E_m)]\}}{I_s^\infty[1 - f\{1 - \exp[-c/\lambda_m(E_s)]\}]},\tag{1}$$

where $I_{m,s}$ = peak area of a particular XPS line in m, the active component, or s, the support; $I_{m,s}^\infty = I_{m,s}$ for an infinitely thick sample of m or s; f = fraction of surface

† Present address: Texaco Inc., Beacon, New York 12508, U.S.A.

covered by m; c = length of cube edge; and $\lambda_x(E_y)$ = mean free path for inelastic scattering of an electron with kinetic energy E_y in material $\times$ (usually of order 1·5 nm). In the limit of monatomic dispersion, $c/\lambda \ll 1$ and $f \ll 1$, eqn. (1) becomes

$$I_m/I_s = KN_m/S_{BET},\qquad (2)$$

where $K = \sigma_m/[\sigma_s\lambda_s(E_s)n_s]$; $\sigma_{m,s}$ = effective cross-section for production of electrons that will reach the detector; n_s = number density of the support element chosen for reference; N_m = number of atoms of reference element m per g of catalyst; and S_{BET} = support surface area, taken as the BET area per g of finished catalyst. Finally, when $c/\lambda \gg 1$ and f $\ll 1$, *i.e.* large particles at low coverage,

$$I_m/I_s = K\lambda_m(E_m)N_m/(cS_{BET}) = I_m^\infty N_m/(I_s^\infty n_m cS_{BET}).\qquad (3)$$

In the following presentation we examine XPS data from supported catalysts with reference to eqn. (2) and (3). More accurate geometric models have been found to conserve the basic functional relationships.[8]

EXPERIMENTAL

Catalyst Preparation. Na/Al$_2$O$_3$(AC) was produced from a slurry of American Cyanamid PHF alumina in an aqueous solution of NaOH by evaporation to dryness at 353 K while stirring, and then heating in air at 393 K for 16 h. Eu/Al$_2$O$_3$(AC) and Eu/Al$_2$O$_3$(H) were prepared similarly from PHF alumina and Harshaw Al-0102 P alumina and aqueous solutions of Eu(NO$_3$)$_3$·6H$_2$O (Research Organic/Inorganic Chemical Corp., 99·9% pure).

The silica-supported platinum catalysts were obtained through the courtesy of M. Boudart and co-workers at Stanford University and R. L. Burwell, Jr. and co-workers at Northwestern University. The catalysts obtained from Professor Boudart were the same as those examined by Sharpen.[2] Five of these were prepared by ion exchange of tetra-ammineplatinum(II) chloride monohydrate[9] while two were prepared by impregnation.[10] Three of the catalysts from Professor Burwell were prepared by ion exchange while two were made by impregnation. All twelve platinum catalysts were prepared on Davison No. 62 silica gel and prereduced in H$_2$.

Surface area and dispersion. Total surface areas were measured for most of the catalysts by the standard nitrogen BET method[11] after outgassing at 773 K. Dispersions were measured at Stanford by the H$_2$/O$_2$ titration method[12] and at Northwestern by a pulsed H$_2$ chemisorption method similar to that of Freel.[13] Adsorption of CO$_2$ at 773 K and 5 $\times$ 10^4 Pa was used to monitor surface Na$^+$ in Na/Al$_2$O$_3$. A large uptake by the alumina support, however, obscured measurement of CO$_2$ chemisorption on surface Na$^+$ at 195 K.[14]

X-Ray photoelectron spectroscopy. XPS data for the alumina-supported samples were obtained on the Hewlett Packard model 5950 A spectrometer in the Department of Chemistry, Yale University. The Pt/SiO$_2$ samples from Professor Burwell were examined using a similar spectrometer in the Department of Chemistry at Purdue University. An electron flood gun set at 100 μA emission and zero kinetic energy minimized sample charging effects in all measurements.

Catalyst samples, which were used as prepared or received, were pressed into pellets and placed in a standard pellet holder. They were evacuated to 0·7 Pa in order to remove adsorbed water before being placed in the spectrometer. The spectra were recorded at a spectrometer pressure of 3—7 $\times$ 10^{-6} Pa. Total data accumulation times for Eu (3$d_{5/2}$), Al (2p), and Na (1s) spectra for Na/Al$_2$O$_3$ or Eu/Al$_2$O$_3$ were less than 1 h per sample. Because 12—14 h were required for the Pt (4$f_{7/2,5/2}$) lines, Si (2p) spectra were recorded before and after the Pt spectra and the spectral areas averaged to compensate for drift in spectrometer performance. The signal intensity, I, was obtained by

Table 1

Catalyst	Surface Area (m² g⁻¹)	I_m/I_s	Dispersion (%)	Catalyst	Surface Area (m² g⁻¹)	I_m/I_s
0·38% Pt/SiO₂[a,c]	279	0·022²	100[a]	0·93% Na/Al₂O₃	291	0·66
0·48% Pt/SiO₂[b,c]	325[e]	0·0085	63[b]	1·85% Na/Al₂O₃	290	1·08
0·49% Pt/SiO₂[b,c]	325[e]	0·0107	63·5[b]	2·75% Na/Al₂O₃	290	1·39
0·53% Pt/SiO₂[a,c]	245	0·019²	56[a]	3·64% Na/Al₂O₃	278	2·45
0·80% Pt/SiO₂[a,d]	518	0·047²	34[a]	6·0% Eu/Al₂O₃(AC)	204	5·86
1·17% Pt/SiO₂[b,d]	325[e]	0·026	40[b]	12·0% Eu/Al₂O₃(AC)	189	20·3
1·46% Pt/SiO₂[a,c]	376	0·067²	100[a]	24·0% Eu/Al₂O₃(AC)	174	39·6
1·48% Pt/SiO₂[b,c]	325[e]	0·015	27[b]	32·0% Eu/Al₂O₃(AC)	106	56·8
1·5% Pt/SiO₂[a,c]	273	0·092²	100[a]	6·0% Eu/Al₂O₃(H)	95	164
1·91% Pt/SiO₂[b,d]	325[e]	0·048	7[b]	12·0% Eu/Al₂O₃(H)	71	20·6
2·3% Pt/SiO₂[a,c]	298	0·076²	62[a]	12·0% Eu/Al₂O₃(AC) + 3·64% Na	175	8·75
3·7% Pt/SiO₂[a,d]	526	0·069²	14[a]			

[a] From the Laboratory of M. Boudart.
[b] From the Laboratory of R. L. Burwell, Jr.
[c] Prepared by ion exchange.
[d] Prepared by impregnation.
[e] Approximate value given by Davison Chemical Co.; measurement of 2 samples indicated slightly (< 20%) lower values.

measuring the area under the peaks and normalizing to unit attenuation and time. Reproducibility of measured intensity ratios was $\pm$ 2%. A multiplicity of Pt chemical states broadened the Pt (4f) lines and in many cases prevented clear separation of the Pt($4f_{7/2}$) and Pt ($4f_{5/2}$) components. Thus, I_{Pt} is reported as the total intensity under the Pt ($4f_{7/2,5/2}$) envelope. In order to compare intensity ratios measured in this work with those reported by Sharpen,[2] the 3·7 wt % Pt on SiO_2 catalyst from Professor Boudart was examined on the Purdue spectrometer. A relative spectrometer calibration constant, calculated by comparison of the measured intensity ratio for this sample with the corresponding ratio reported by Sharpen, was then used to normalize all measured $I_{Pt(4f)}/I_{Si(2p)}$ intensity ratios to the Sharpen data.

RESULTS

Values for BET areas, dispersions and XPS intensity ratios are reported in the Table. The behaviour of I_m/I_s as a function of catalyst loading was examined first for sodium on alumina, as a system expected to show high dispersion at submonolayer Na^+ coverage. The XPS intensity ratio, $I_{Na(1s)}/I_{Al(2p)}$, is plotted against nominal surface density, $[N_m/S_{BET}]/[10^{18}$ Na atoms $m^{-2}]$, as the open circles in Figure 1 (upper and left

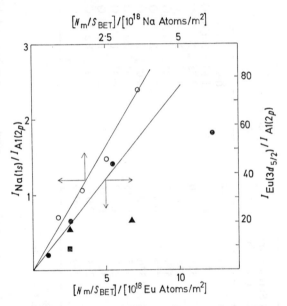

FIGURE 1 Relative XPS peak intensity against nominal surface density for Na/Al_2O_3 (upper and left axes) and Eu/Al_2O_3 (lower and right axes): $\bigcirc$–Na/Al_2O_3(AC); $\bullet$–Eu/Al_2O_3(AC); $\blacktriangle$–Eu/Al_2O_3(H); $\blacksquare$–Eu/Al_2O_3(AC) + Na.

axes). As shown by the upper solid line, the relation between these variables is linear with the intercept at the origin. A plot of CO_2 uptake at 773 K and 5×10^4 Pa against N_{Na}/S_{BET} was also linear. This linearity shows that the dispersion was constant, but the slope corresponds to a ratio of 1 CO_2 molecule to 10 Na atoms in the sample. Since the stoicheiometry of the adsorption of CO_2 on surface Na at these conditions could not be established by comparison with low temperature CO_2 chemisorption, a value for the dispersion could not be extracted from the data. The linearity of the uptake plot up to at least 0·25 monolayers of Na is taken as strong evidence for high dispersion of the surface Na, but it is possible that some Na was present in the bulk of the Al_2O_3.

The dependence of the XPS intensity ratio on nominal surface density was also studied for Eu/Al_2O_3. $I_{Eu(3d_{5/2})}/I_{Al(2p)}$ is plotted against $[N_m/S_{BET}]/[10^{18}$ Eu atoms m$^{-2}]$ as the filled points in Figure 1 (lower and right axes). The filled circles correspond to Eu on PHF alumina. The low coverage region again produced a linear relation with intercept at the origin, as indicated by the lower solid line in Figure 1. In this region, further evidence for high dispersion of Eu on Al_2O_3 is provided by both CO_2 chemisorption and Mössbauer data.[15] The point corresponding to the highest nominal surface density in this series falls below the correlation line as expected, since the coverage exceeds an effective monolayer and some agglomeration of the Eu phase is anticipated. Assuming that the linear correlation corresponds to 100% dispersion of Eu on Al_2O_3, analysis of the high coverage point in terms of eqn. (1) yields an average particle size of the order of 1—2 nm (depending on λ) for the Eu phase, but possible occlusion of Eu in the Al_2O_3, as suggested by the loss of surface area shown in the Table, cannot be ruled out.

Data for the two $Eu/Al_2O_3(H)$ catalysts are shown in Figure 1 as filled triangles. The point corresponding to the catalyst with lower nominal surface density falls on the linear correlation line, but the other point falls below it. This behaviour can be rationalized by observing that the Harshaw alumina contained sodium, which might facilitate agglomeration of the Eu phase. This hypothesis is supported by the finding that the addition of sodium to Eu on PHF alumina by NaOH impregnation led to the decrease in $I_{Eu(3d_{5/2})}/I_{Al(2p)}$ shown by the filled square in Figure 1.

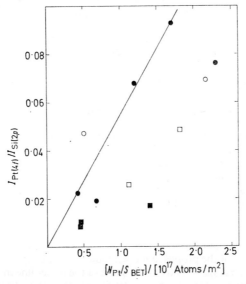

FIGURE 2   Relative XPS peak intensity against nominal surface density for Pt/SiO_2: ●–ion-exchanged Stanford catalysts; ○–impregnated Stanford catalysts; ■–ion-exchanged Northwestern catalysts; □–impregnated Northwestern catalysts.

Figure 2 is a plot of normalized $I_{Pt(4f)}/I_{Si(2p)}$ against the nominal surface density, $[N_{Pt}/S_{BET}]/[10^{17}$ atoms m$^{-2}]$ for all the Pt/SiO_2 samples. The filled symbols correspond to ion-exchanged catalysts while the open symbols correspond to catalyst preparation by impregnation. The circles represent the Stanford samples; the squares, those from Northwestern. The only three points corresponding to 100% Pt dispersion fall nicely on a line passing through the origin, as indicated in the Figure and expected on the basis of the results for the alumina-supported samples. The remaining points, corresponding to samples with particle size greater than 1 nm, were expected to fall below the correlation or "calibration" line to a degree which depends on particle size [eqn. (1)]. The appearance of a point above the calibration line, *i.e.*, showing excess Pt intensity in the *X*-ray photoelectron spectrum, is discussed below. Though differences between the ion-exchanged and the impregnated samples can be seen by application of eqn. (1), they are more apparent in Fig. 3, a plot of normalized $I_{Pt(4f)}/I_{Si(2p)}$ against

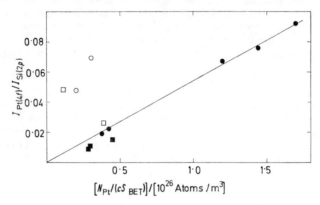

FIGURE 3   Relative XPS peak intensity ratio against $N_{Pt}/(cS_{BET})$ for Pt/SiO_2. Symbols as in Figure 2.

relative surface area $(N_{Pt}/(cS_{BET})]/[10^{26}$ atoms m^{-3}]. The importance of the variable $N_m/(cS_{BET})$ is apparent from eqn. (3), for the limit of low coverage and $c \gg \lambda$. It should be noted that since N_m/c is proportional to the metal surface area, Fig. 3 is a plot directly comparable to that presented by Sharpen.[2] The solid line in Fig. 3 is analogous to the linear correlation he found for the ion-exchanged catalysts. It is immediately clear from this Figure that incorporating S_{BET} into the analysis did not put the data for the impregnated samples on the correlation line. Examination of the data for the Northwestern catalysts shows fair agreement between data sets for the ion-exchanged samples. Comparison of the open squares with the filled squares, however, again indicates excess $Pt(4f)$ intensity for impregnated Pt/SiO_2.

DISCUSSION

The data for the alumina supported samples, as depicted in Fig. 1, represent a successful qualitative test of the proposed model. Quantitative consistency was checked by calculation of $\sigma_{Na(1s)}/\sigma_{Al(2p)} = 43$ from the slope of the upper line in Fig. 1 and eqn. (2). A value of 74 was estimated for this parameter from the equation $I_x^\infty/I_y^\infty = [\sigma_x \lambda_x(E_x)n_x]/[\sigma_y \lambda_y(E_y)n_y]$, the assumption that $\lambda_x(E_x) \propto E_x^{1/2}$, and the average $I^\infty_{Na(1s)}/I^\infty_{Al(2p)} = 40 \pm 2$ measured for NaX and NaY zeolites (Linde 13X and SK40) and Na mordenite (Norton zeolon). Uncertainty in surface stoicheiometry[16] as well as $\lambda(E)$ is symptomatic of the difficulty of finding suitable calibration procedures for quantitative XPS work. Nevertheless, the two estimated cross-section ratios are of the same order of magnitude. The lower value from the Na/Al$_2$O$_3$ data may suggest the loss of a fraction of the Na to the Al$_2$O$_3$ bulk.

The results for Pt/SiO_2 give further support for the model but also demonstrate the need for some refinement. The slope of the calibration line in Fig. 2 gives a value of $\sigma_{Pt}/\sigma_{Si} = 15$ as compared with a value of 8 estimated from Pt foil and pure SiO_2 data, again showing order of magnitude agreement between two independent estimates.

The linear correlation for the ion-exchanged Pt/SiO_2 samples from Stanford, represented by the solid line in Fig. 3, has been suggested by Sharpen[2] as a method for estimating dispersion. The correlation is better than expected on the basis of the present model, however, which predicts $I_m/I_s \propto N_m/(cS_{BET})$ only in the limit of $c/\lambda \gg 1$. While this discrepancy must be investigated further, it may be due in part to the varied state of oxidation of the Pt particles during the XPS measurement. Sensitivity of XPS data to the chemical state of the catalyst was also suggested by an increase of 16% in $I_{Pt(4f)}/I_{Si(2p)}$ when the 1·48 wt $\%$ Pt/SiO_2 was reduced in H_2 at 773 K for 1 h just prior to XPS evaluation. Thus, differences in pretreatment or preparation procedure could account for the filled squares falling slightly below the correlation line. In principle, binding energy shifts in the XPS spectra should reflect changes in the chemical state of the catalyst. Apparent changes in the degree of oxidation[3] and the size of final state relaxation effects[17] as a function of metal particle size complicate interpretation, however. *In situ* studies of catalysts with known dispersion and chemical state are needed to unravel this problem.

We turn now to the impregnated catalysts, which are characterized by unexpectedly high $I_{Pt(4f)}/I_{Si(2p)}$ ratios and dispersion less than 40%. These properties raise questions concerning the uniformity of the distribution of Pt metal particles on the SiO_2 surface and the efficiency with which XPS probes the surface in pores of various sizes. Clearly, as metal particle size increases, Pt will be excluded from pores of diameter smaller than c. In addition, the initial distribution of Pt on an impregnated catalyst will depend on whether Pt ions concentrate in the smallest pores, which are the last to be dehydrated, or deposit preferentially in the largest pores, which contain the most Pt per unit surface area when all the pores are just filled with an impregnating solution of uniform concentration. Since XPS sees only the outer surface of a catalyst particle, in cases where smaller pores open into the larger ones XPS would analyse the outer surface and larger pores preferentially. In this light, the best explanation for the high $I_{Pt(4f)}$ relative to $I_{Si(2p)}$ for the impregnated samples in both Fig. 2 and 3 appears to be a nonuniform distribution of Pt with enhanced Pt concentration on the outer surface or larger pores of the catalyst particles. The source of the nonuniformity cannot be established with certainty, but since excess Pt is observed even when the mean Pt particle diameter is smaller than the mean pore diameter in Davison No. 62 silica, the data suggest that the Pt distribution is associated with the impregnation/dehydration method itself.

Several interesting trends are established by the above results. For a series of similarly prepared supported catalysts with varied loading of an active component, a plot of I_m/I_s against N_m/S_{BET} can be expected to be linear when dispersion approaches 100%. I_m/I_s corresponding to catalysts with active component particles larger than 1·5 nm will normally fall below the 100% dispersion line, but uniformly distributed catalysts should show a linear correlation between I_m/I_s and $N_m/(cS_{BET})$ for all samples, particularly when $c \gg \lambda$. A nonuniform distribution of active component on the support surface complicates the analysis, but when data from such samples are compared with those for more uniform distributions, the discrepancy should be apparent in deviations from the correlation line in the I_m/I_s against $N_m/(cS_{BET})$ plot. As shown in Fig. 2, points above the 100% dispersion line in an I_m/I_s against N_m/S_{BET} plot indicate an excess of m in the outer surface of th catealyst. Refinements of the quantitative measurements will require further research, but the approach has particular promise for the study of supported alloys. Plots of I_m/I_s against N_m/S_{BET} and I_q/I_s against N_q/S_{BET} for a supported q, m bimetallic catalyst should indicate relative dispersion of q and m and, therefore, also give information on phase separation during sintering.

ACKNOWLEDGEMENTS. The authors gratefully acknowledge M. Boudart and R. L. Burwell, Jr. for the use of their Pt/SiO_2 samples and enlightening discussions of results. Special thanks are also due to N. Winograd and the Department of Chemistry at Purdue University for generosity in the extended use of the Purdue ESCA Facility, and to J. S. Brinen, L. H. Sharpen, G. L. Haller, H. A. Benesi, R. Madon, and J. B. Butt for helpful discussions. One of us (P.J.A.) would like to thank Texaco Incorporated for a fellowship at Yale and NSF for a graduate traineeship at Purdue. Acknowledgement is also made to the Donors of the Petroleum Research Fund, administered by the American Chemical Society, and to the National Science Foundation (grant No. ENG 74-23442) for support of this work.

REFERENCES
1. W. N. Delgass, T. R. Hughes, and C. S. Fadley, *Catalysis Rev.*, 1971, 4, 179.
2. L. H. Sharpen, *J. Electron Spectroscopy*, 1974, 5, 369.
3. J. S. Brinen, J. L. Schmitt, W. R. Doughman, P. J. Achorn, L. A. Siegel, and W. N. Delgass, *J. Catalysis*, 1975, 40, 295.
4. C. D. Wagner, *Analyt. Chem.*, 1972, 44, 1050.
5. R. Bowman and P. Biloen, *Surface Sci.*, 1974, 41, 348.
6. R. S. Swingle, II, *Analyt. Chem.*, 1975, 47, 21.
7. C. S. Fadley, R. Baird, W. Siekhaus, T. Novakov, and S. A. L. Bergström, *J. Electron Spectroscopy*, 1974, 4, 111.
8. P. J. Angevine and W. N. Delgass, unpublished results.
9. H. A. Benesi, R. M. Curtis, and H. P. Studer, *J. Catalysis*, 1968, 10, 328.
10. T. A. Dorling, B. W. J. Lynch, and R. L. Moss, *J. Catalysis*, 1971, 20, 190.
11. S. Brunauer, P. H. Emmett, and E. Teller, *J. Amer. Chem. Soc.*, 1938, 66, 309.
12. J. E. Benson and M. Boudart, *J. Catalysis*, 1965, 4, 704.
13. J. Freel, *J. Catalysis*, 1972, 25, 149.
14. P. H. Emmett and S. Brunauer, *J. Amer. Chem. Soc.*, 1937, 59, 310.
15. P. N. Ross, Jr. and W. N. Delgass, *J. Catalysis*, 1974, 33, 319.
16. J. Fr. Tempere, D. Delafosse, and J. P. Contour, *Chem. Phys. Letters*, 1975, 33, 95.
17. K. S. Kim and N. Winograd, *Chem. Phys. Letters*, 1975, 30, 91.

DISCUSSION

P. Canesson (*Catholic Univ. of Louvain*) said: There is an explicit assumption in equations (2) and (3) used to calculate dispersion. With XPS it is necessary to differentiate surface and bulk effects. It is possible to use BET surface areas in the calculations only after some verifications, because the surface area is a bulk measurement. If there is a monolayer of any dispersed phase or of adsorbate molecules at the surface, all these molcules are supposed to be sensitive to XPS analysis, but it is not the same for the support. It is important to be sure that the surface to volume ratio analysed by XPS is the same as the real surface to volume ratio. We have made calculations on the assumption that the porous material is constituted of identical nonporous particles; the total surface area is then a consequence of the particle sizes. Assuming three models for the particles (spheres, cubes, and sheets), it is possible to calculate the theoretical curve I_m/I_s as a function of particle sizes, that is, for various surface areas. The results show that, for low surface areas, the signal originating from the surface species is enhanced. The results obtained are valuable for a surface area of a carrier greater than 200 $m^2 g^{-1}$, depending upon the choice of the model. This is the case with your catalysts.

W. N. Delgass (*Pardue Univ.*) replied: In the region of low coverage of the support by metal, the model we present, using the BET surface area, is correct as long as the distribution of metal over the support surface is uniform. I agree, however, that when shading of the support by a metal overlayer becomes important, I_m/I_s can well be dependent on the model. For this case, the cube-on-a-flat-surface model that we have employed is not an accurate approximation and I look forward to seeing the results of your calculations. Another obstacle to interpretation arises when the metal particles are covered by a layer of adsorbed gas. Perhaps your model can be applied to this case as well.

J. Fraissard (*Univ. Pierre et Marie Curie, Paris*) said: You have analysed the influence of the particle size on XPS intensity, but the electronic properties of the surface atoms of the metal also change with the particle size. Have you detected any influence of the distributions of the particle size on the width of the XPS signal?

W. N. Delgass (*Purdue Univ.*) replied: The object of this study has been to examine the information contained in the XPS peak areas. We have noted significant changes in the shapes of the Pt $(4f)$ envelopes but there were no clear trends with particle size. Since the chemistry of the particle surface was not carefully controlled just before and during the measurements, we have not tried to draw chemical conclusions from the spectra.

F. Bozon-Verduraz (*Univ. de Paris VI*) said: First, in relation to Prof. Fraissard's question, I would point out that recent work by Escard *et al.*[1] has shown that metal particle size in Pt/Al_2O_3 greatly influences line intensities in XPS spectra.

Secondly, with the Pt/SiO_2 system, you observed a broadening of the Pt $(4f)$ lines which you attribute to a multiplicity of Pt chemical states. In work on palladium deposited on various carriers, we have noticed a broadening of the Pd $(3d)$ peaks which depends on the nature of the support and the preparation mode.[2] Do you observe any shift in the maximum of the Pt lines together with their broadening? Could you give any information about the chemical states involved?

W. N. Delgass (*Purdue Univ.*) replied: As I said to Dr. Fraissard, we have not tried to draw chemical conclusions from our spectra. In general, we have observed that supported metals give broader peaks than unsupported metals and this trend is present in the work reported here. Overlapping peaks from Pt in clearly different oxidation states causes additional broadening. Extraction of the chemical state of a metal from the electron binding energy requires an accurate account of final state relaxation effects which may depend on the particle-size and on the support. This is a difficult situation even when the surface condition of the sample is well controlled during measurement.

I appreciate your reference to the work of Escard and look forward to seeing if our model can be applied to his results.

A. Brenner (*Wayne State Univ.*) said: You conclude that, even when XPS experiments are carefully performed and interpreted, they cannot yield absolute dispersions. For example, lines passing through the origin may be drawn through two other sets of points in Figure 2, besides the points relating to 100% dispersion. As you note, standards are required in which catalysts have a known and constant dispersion over a range of loading. In this respect it has been claimed recently[3] that Mo/Al_2O_3 catalysts can be prepared in which the molybdenum is 100% dispersed over an approximately 30-fold change in loading, the maximum loading being 8×10^{17} Mo atoms min *vs* m^{-2}.

W. N. Delgass (*Purdue Univ.*) replied: I agree that if the distribution of metal on

[1] J. Escard, B. Pontvianne, M. T. Chenebaux, and J. Cosyns, *Bull. Soc. chim. France*, 1976, 349.
[2] A. Omar, Dr. es Sc. thesis, Univ. Paris (1976); F. Bozon-Verduraz, A. Omar, J. Escard, and B. Pontvianne, to be published.
[3] R. L. Burwell and A. Brenner, *J. Mol. Catalysis*, 1975, **1**, 77, and in a further paper submitted to *J. Amer. Chem. Soc.*

the catalyst is not known to be uniform, XPS data alone cannot yield absolute dispersions. We have found, however, that estimates of the slope of the 100% dispersion line can be useful for qualitative evaluation of limiting cases of dispersion and/or uniformity. I am particularly interested in your mention of the Mo/Al_2O_3 system. We have recently obtained some semi-quantitative results for Mo/SiO_2 and $Mo–Ru/SiO_2$ which show that Ru significantly improves the reducibility of the Mo but also reduces its dispersion by a factor of about one-third.

The Study of Zeolite Catalysts Containing Transition Elements by Means of X-Ray Photoelectron Spectroscopy

Kh. M. Minachev,* G. V. Antoshin, E. S. Shpiro, and Yu. A. Yusifov

Zelinskii Institute of Organic Chemistry, Academy of Sciences of the U.S.S.R.,
Moscow, U.S.S.R.

ABSTRACT By application of X-ray photoelectron spectroscopy it has been shown that the cations of transition elements in dehydrated zeolites form nearly pure ionic bonding with the zeolite framework.

Zeolite pretreatment under vacuum, H_2, or CO causes reduction of Fe, Co, Ni, Pd, Pt, Cu, Ag, and Zn to the metallic state. The reducibility is affected by the chemical nature of the transition element, the degree of cationic exchange, and zeolite composition. The reduction is accompanied by migration of the metal on to the external surface; the mobility of the metals in the zeolites studied decreases in the series: Ag > Pd > Zn > Cu > Ni > Pt $\geqslant$ Co. In the polycationic forms (Ag–Cu, Ni–Cu, Ni–Co) the migration of the metals is hindered by their interaction. The oxidation of reduced samples results in the formation of the corresponding oxides on the external surface and of cation associates with oxgyen anions which do not belong to the zeolite framework. The bonding (M–O) is more covalent in nature than that of isolated cations with the framework.

From catalytic studies and XPS data it is concluded that the state of the transition element is very important for zeolite activity in CO oxidation, NO reduction (Ni forms), low-temperature homomolecular oxygen exchange (Ni, Co, Pd, Pt forms) and catalyst selectivity (Ni zeolite) in the synthesis of s-butylbenzene from benzene and ethylene.

Introduction

It is generally accepted that the sites responsible for catalysis on zeolites are the cations compensating the excessive negative charge of Al–Si–O framework and the fragments of the structure in the vicinity of cations.

Hence the elucidation of the mechanism of catalytic action and the selection of most efficient zeolite catalysts require information concerning the valence and physical state of the exchangeable cations. For this reason the study of the effect of pretreatment and reaction media on the state of transition elements in the zeolite subsurface layer is of considerable interest. However, the relevant information available in the literature[1,2] is limited.

The aim of this work was to study the nature of bonding between transition elements and the zeolite framework as well as to investigate the physical and valence state of transition elements in mono- and poly-cationic forms of zeolites, subjected to redox pretreatments (vacuum, H_2, CO, NO, O_2, air) and treatments with reaction mixtures ($CO + O_2$, $CO + NO$). X-Ray photoelectron spectroscopy (XPS) was used for this purpose. No XPS studies of zeolite cationic forms, except for EuY,[3] have been reported. The XPS data for several samples studied have been compared with their catalytic properties with respect to homomolecular oxygen isotopic exchange, CO oxidation with oxygen, NO reduction with carbon monoxide, and benzene alkylation with ethylene.

Experimental

1 Catalysts

Zeolites containing transition elements were prepared by the conventional ion-exchange of sodium faujasite. The techniques used for zeolite preparation, analysis and control of their structure stability have been described earlier.[4,5]

2 *Procedure and apparatus*

The catalysts were dried after ionic exchange either at room temperature or at 120°C and then placed in a preparation chamber connected with a spectrometer. The samples were trained *in vacuo* or in gases under static conditions ($p = 5$—50 torr) and in flow conditions ($p = 1$ atm). After adsorption experiments performed at 250—550°C the samples were cooled in the gases to room temperature, evacuated up to $p = 10^{-6}$— 10^{-7} torr, and placed in a spectrometer chamber *in vacuo*.

CO oxidation and NO reduction were studied with a circulatory-static unit, the starting pressure of stoicheiometric mixtures being about 10—20 torr. The reaction kinetics were followed by monitoring the pressure drop. The reaction products were analysed by mass spectrometry and chromatography.

X-Ray photoelectron spectra were recorded by means of an A.E.I. ES-100 spectrometer with Al and Mg anodes.[6,7] The C 1S line of organic admixtures was used to calibrate energetic positions of the photoelectron lines. In some experiments the Si 2S line of zeolite framework Si was employed as an internal standard. The line intensity of transition elements was determined relative to that of Si 2S.

RESULTS AND DISCUSSION

1 *Effect of redox treatments on the valence and physical state of transition elements in zeolite mono-forms*

Here, the main results obtained will be considered, since the relevant data have been reported in detail earlier.[7-9]

1.1 *Nature of cation–zeolite framework bonding*

The nature of interaction between the transition elements and the framework has been studied by comparing the energetic position and structure of the photoelectron spectra from the cations in zeolites and the corresponding parameters in compounds with various chemical bonds. The comparison made for Cr, Mn, Fe, Co, Ni, Cu, and Pd has shown that the cations of transition elements form nearly pure ionic bonding with the zeolite framework. The ionic character of the bonding becomes more pronounced with an increase of the degree of Na$^+$ substitution by a transition-metal cation. The energetic non-equivalency of the cations located in various sites inside the zeolite structure can be seen from the broadening of the corresponding photoelectron lines.

The average effective charges on several cations in zeolite structures have been determined. To illustrate this, Fig. 1 shows the correlation of chemical shifts in Co and Cr spectra and the cation effective charges in some compounds calculated according to Pauling's electronegativity model.[10] The values of cation charges in zeolites are found to be close to those in ionic crystals, such as fluorides and chlorides.

Thus, in this work the concept of cation–framework ionic bonding which was earlier used *a priori* for calculations of the Madelung potentials in zeolite cavities[11] and for interpretation of catalytic data[12] has been substantiated experimentally.

1.2 *Redox properties of transition elements*

Upon treatments of catalysts *in vacuo* or under H$_2$ at high temperatures, the spectra of transition elements Ni, Co, Fe, Pd, Pt, Cu, Ag, and Zn show new lines or negative shifts characteristic of the reduction of cations to metallic state (Fig. 2). Under mild conditions of treatment Cu and Pd can be stabilized in the zeolite structure in intermediate monovalent states (Fig. 2, spectra 3 and 5). The reduction temperatures depend on a number of factors, such as degree of exchange, cation location, and thermostability of structural hydroxyls, arising during reduction. But the most important factor is the chemical nature of the transition element.

Minachev, Antoshin, Shpiro, Yusifov

In most cases the reduction is accompanied by migration (diffusion) of the metals on to the external surface of the zeolites, which is observed as an increase of relative intensity of the transition element lines in the spectra (Fig. 2). The ability of reduced metals to migrate decreases in the series: $Ag > Zn > Pd > Cu > Ni > Pt \geqslant Co$. From the XPS (migration kinetics) and X-ray phase analysis (mean metal crystallite sizes on the external surface) it has been found that the number of silver atoms (the most mobile element of elements studied), migrating on to the external surface of the

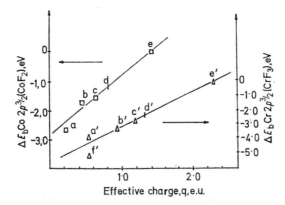

FIGURE 1 The dependences of chemical shifts in the spectra of Co $2p_{3/2}$ and Cr $2p_{3/2}$ on effective charges of Co and Cr, respectively.
a: CoI_2; b: $CoBr_2$; c: $CoCl_2$; d: CoNaY; e: CoF_2;
a′: CrI_3; b′: $CrBr_3$; c′: $CrCl_3$; d′: CrNaY; e′: CrF_3; f′: Cr_2S_3.

zeolite, corresponds to 25—50% of the total number of Ag atoms in the zeolite. From this one can suggest that in the cases of other zeolite forms, metals are mostly captured in the zeolite cavities.

Subsequent oxidation of zeolites containing reduced metals gives rise to positive shifts in the spectra (Fig. 2). But in this case the state of the transition elements is different from the starting state and is similar to that of the cations in the corresponding oxides. The increase of M–O bonding covalency after redox treatment of zeolites is due to formation of oxide phases on the external surface, and of amorphous oxides inside the zeolite structure (or associates of transition elements with oxygen which does not belong to the zeolite framework). Treatment of the starting Co^{2+}, Fe^{2+}, and Cr^{3+} forms at 450—550°C with O_2 or air gives rise to Fe^{3+}, Co^{3+}, and Cr^{5+}, Cr^{6+}.[5] Here, association of transition elements with oxygen which does not belong to the zeolite framework may also arise.

2 Reduction and migration of transition elements in polycationic zeolite forms

The mutual influence of transition elements on their ability for reduction and migration has been studied using Ag–Cu, Ni–Cu, Ni–Co, and Ni–Cr forms with the degree of Na^+ exchange for each cation being about 25—30%.

Treatment of Ag–Cu zeolite at 150—400°C *in vacuo* or under H_2 leads to Ag reduction and Ag^0 migration. At the same time no increase of Cu $2p_{3/2}$ line intensity, characteristic of Cu^0 migration, is observed. A small increase of Cu concentration in the zeolite subsurface layer takes place at 500°C. The study of Cu–Ni and Ni–Co forms has also shown that during reduction the more mobile metal of the pair migrates on to the external surface (although to a smaller extent than when in the corresponding mono-form) and the less mobile one distributes itself largely inside the zeolite cavities. In our opinion, the mutual influence of transition elements on their reduction and migration in poly-cationic forms is likely to be due to interaction of the metals occurring both inside the zeolite structure and on the external surface. The occurrence of the interaction is evident from the valence band spectra of Ag, Cu, and Ag–Cu forms (Fig. 3). The valence band

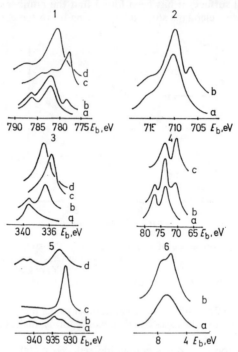

FIGURE 2 The changes in the XPS spectra of transition elements after redox treatments of zeolites.

1. Co $2p_{3/2}$ spectrum of 0·59 CoNaY zeolite a: 380°C, vacuum; b: 450°C, H_2; c: 550°C, H_2; d: 550°C, O_2.
2. Fe $2p_{3/2}$ spectrum of 0·41 FeNaY zeolite a: 20°C, vac.; b: 550°C, H_2.
3. Pd $3d_{5/2}$ spectrum of 0·15 PdNaX zeolite a: 20°C, vac., 20 min; b: 20°C, vac., 2 h; c: 400°C, vac.; d: 400°C, O_2. (note spectra "c" and "d" are drawn with reduced intensity scale, 1:2).
4. Pt $4f$ spectrum of 0·15 PtNaY zeolite a: 20°C, vac.; b: 300°C, vac + H_2; c: 400°C, vac + H_2 (note spectra 4 are obtained by subtraction of Al_{2p} line from Al_{2p}^+ Pt_{4f} spectrum).
5. Cu $2p_{3/2}$ spectrum of 0·26 CuNaY zeolite a: —100°C, vac.; b: 20°C, vac.; c: 500°C, vac. + H_2; d: 500°C, O_2.
6. Ag $4d$ spectrum of 0·14 AgNaY zeolite a: 20°C, vac.; b: 400°C, vac. (note spectrum "b" is drawn with reduced intensity scale, 1:5).

spectrum of the reduced Ag–Cu zeolite differs markedly as regards position and peak profile from the spectra of mono Ag and Cu forms, and also from the curve obtained by simple addition of spectra "a" and "b". The changes observed in valence band spectra of Ag–Cu zeolite seem to be due to the formation of copper–silver alloys or clusters. It should be mentioned that during formation of metastable Ag–Cu alloys, considerable changes in the valence band photoelectron spectra were also observed.[13]

It is difficult to obtain unambiguous evidence of Cu–Ni and Co–Ni alloy formation from XPS data because of the low intensity of the valence band spectra. But the validity of this assumption can be supported by the reference. The formation of copper–nickel alloys in Cu–Ni zeolite was found by the magnetic method.[14]

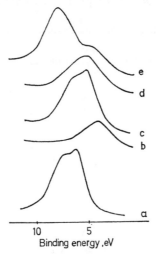

10   5
Binding energy ,eV

FIGURE 3   Valence band XPS spectra of Ag, Cu, and Ag–Cu zeolites treated in H_2 at 500°C

a: 0·14 AgNaY; b: 0·26 CuNaY; c: 0·31 Cu 0·25 AgNaY; d: 0·31 Cu 0·01 AgNaY; e: curve obtained by addition of spectra "a" and "b".

The results presented in Table 1 show that introduction of a second transition element into the zeolite structure causes significant changes in Ni reducibility. One of the reasons may be different location of Ni^{2+} ions in polycationic forms. The sites S_1 are the preferred centres for Ni and Co location in the monoforms,[15,16] whereas Cu and Cr tend to occupy

Table 1

Degree of nickel reduction (%) in polycationic zeolite forms
Treatment conditions: 400°C, air, 500°C, He + H_2

Zeolites	400°C	500°C	550°C
NiNay	50(90)[a]	85	>90
CoNiNaY	60(>90)	90	>90
CrNiNaY	45(60)	70	>90
CuNiNaY	40(40)	65	>90

[a] The figures in parentheses relate to samples which were not dehydrated before reduction.

sites S_1'.[17,18] It is possible that in presence of Co some nickel ions locate themselves in sodalite units and supercages where Ni reduction proceeds more easily. On the other hand, introduction of Cu or Cr may favour location of Ni^{2+} ions in hexagonal prisms where Ni reduction is hindered. It should be taken into account that the stability of structural hydroxyls, in which form the protons compensate the excessive negative charge of the framework during reduction in H_2, may be distinguished in the catalysts studied.

3 Relations between transition elements state and zeolite catalytic properties in some reactions

3.1 *Interaction of* Ni *containing zeolites with reaction mixtures* $CO + O_2$, $CO + NO$

Fig. 4 shows the changes in Ni $2p_{3/2}$ spectra observed after 0·33 NiNaY* was treated with carbon monoxide and $CO + O_2(1:1,M)$ mixture, at 250—550°C. CO adsorption

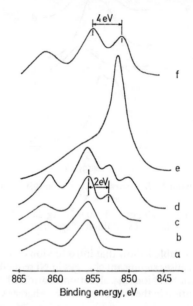

FIGURE 4   XPS Ni $2p_{3/2}$ spectrum of 0·33 NiNaY zeolite.
a: starting sample, 20°C vacuum; b: sample after dehydration in air at 400°C and in helium flow at 500°C; c: CO adsorption at 250°C; d: CO adsorption at 400°C; e: CO adsorption at 550°C; f: $CO + O_2$ mixture adsorption at 400°C.

at 250°C gives rise to a new line in the spectrum taking an intermediate position between the lines corresponding to the cation and the metal†). With an increase of temperature of zeolite treatment in CO up to 400°C, Ni^{2+} is reduced to Ni^+ as well as to Ni^0, and at 550°C practically all Ni becomes reduced to a metallic state. With polycationic forms similar treatment also leads to reduction of Ni^{2+} to Ni^+ and Ni^0 (Cu^{2+} in CuNiNaY is reduced to Cu^+ and Cu^0).

NiNaY treated with $CO + O_2$ mixture gives a line in the Ni $2p_{3/2}$ spectrum characteristic of Ni^0 (Fig. 4). But the spectrum is of lower resolution as compared with that from Ni zeolite treated in H_2.[7,8] This may be due to partial reduction of Ni, also, to an intermediate state during interaction with $CO + O_2$. It is also possible that a portion of the Ni^0 undergoes re-oxidation to an oxidized state,[7] Ni_{oxid}^{2+}. Upon adsorption of an

* The number before the zeolite symbol designates the degree of cationic exchange.
† The Ni intermediate state will be referred to as Ni^+.

NO + CO mixture on NiNaY cationic form, no lines characteristic of Ni^+ and Ni^0 formation were observed in the Ni $2p_{3/2}$ spectrum. In the case of interaction of pre-reduced Ni form with NO + CO, NO, or CO + O_2, a positive shift of the Ni $2p_{3/2}$ line by 1·2—1·4 ev takes place, which corresponds to the transition of Ni^0 into an oxi-dized and intermediate state.

The formation of the intermediate state of Ni in the course of zeolite treatment described above is likely to be due to interaction of the nickel with the adsorbed re-agents. CO adsorbtion was determined from the changes in the shape and width of the background C $1S$ peak. When zeolites were exposed to an NO + CO mixture, two strong lines with binding energies of 399 and 405·5 ev, respectively, appeared in the N $1S$ spectrum. The peak with a lower binding energy can be attributed, according to the published data,[19] to the nitrogen in the nitrosyl complex. These spectra represent strongly adsorbed NO forms, since the lines disappear from the spectra upon outgassing the samples at temperatures as low as 400°C.

Formation of stable surface nickel compounds with the adsorbate may be responsible for the lack of shifts in the Ni $2p_{3/2}$ spectrum of the NiNaY cationic form upon NO + CO adsorbation. Nickel may undergo partial reduction. But because of complex formation the effective positive charge on Ni increases, and this results in the reverse shift in the spectrum.

3.2 *Catalytic studies*

(a) *Alkylation of benzene with ethylene*

The comparison of the results obtained from the catalytic studies with the XPS data has demonstrated[20] that catalyst selectivity with respect to s-butylbenzene forma-tion depends significantly on the Ni state. On the cationic form, the reaction proceeds to give mainly s-butylbenzene. With a decrease of Ni^{2+} upon reduction of the zeolite in H_2 the yield of s-butylbenzene decreases, and the principal reaction products are ethyl- and diethyl-benzenes. These data and the results previously published[20] may suggest that Ni^{2+} ions and the Brønsted acidic centres conjugated with them are the sites for the synthesis of s-butylbenzene.

(b) *Low-temperature homomolecular oxygen exchange*

It has been shown[21] that low-temperature activity of Ni, Co, Pd, and Pt forms of zeolites in the $^{16}O + {}^{18}O \rightleftarrows 2 \ {}^{16}O^{18}O$ reaction is due to reduction of cations to a metallic state. When the samples were exposed to oxygen at a temperature above room temperature, the activity disappeared. The analysis of the kinetic dependence of the reaction on temperature shows thtt homomolecular exchange catalysis by reduced zeolites is characterized by low activation energies (3—5 kcal/mole) which are typical of this reaction over metallic catalysts.

(c) CO *oxidation and* NO *reduction.*

The cationic form of 0·33 NiNaY zeolite became active in CO oxidation only at 250°C. A complicated pattern of kinetic curves from the first experiments is observed. These results and XPS data (see 3.1 above) indicate that modification of surface com-position of the catalyst occurs as a result of the action of the reaction mixture. Then the catalyst shows stable and reproducible activity in CO oxidation. When Cu is intro-duced into Ni zeolite the activity becomes twice as high, The pre-reduced and pre-oxidized NiNaY forms are active at lower temperatures (80—120°C) than the cationic forms. In this case the reaction also proceeds at higher rates.

N_2 and CO_2 formation from NO + CO is observed over NiNaY which was pre-reduced in H_2 at 500°C. The cationic form is not active at 250—400°C.

From catalytic and XPS experiments one can see that the zeolites exhibit activity in $CO + O_2$ and $CO + NO$ reactions provided Ni (Cu in CuNiNaY) is converted into oxidized or intermediate states as a result of pretreatment or of the action of reaction mixtures. No activity is observed when the transition element remains as an isolated cation in the zeolite structure.

Finally one can draw some general conclusions about the properties of transition elements in zeolites and about the role of their state in the activation of zeolite catalysts. The cations of transition elements located inside the zeolite structure form nearly pure ionic bonding with the framework.

Reducing or redox treatments of zeolites cause considerable changes of the physical and valence state of the transition elements, such as the formation of metal and oxide phases, increase of metal–oxygen bond covalency, and stabilization of the intermediate forms. The changes also affect the properties of the zeolite matrix, such as the number and strength of acidic centres. The investigation of some reactions has demonstrated that such changes affect significantly the catalytic properties of zeolites. This suggests that the state of the transition element is an important factor responsible in many cases for the activity and the selectivity of zeolite catalysts.

The further study of polycationic forms shows great promise. By varying their cationic composition and by controlling the state of transition elements in the sub-surface layer it seems possible to prevent metal migration to the external surface, that is, to attain a highly dispersed state of the metal.

REFERENCES

[1] D. W. Breck, *Zeolite Molecular Sieves* (John Wiley and Sons, N.Y. 1974), Ch. 6, p. 99.
[2] G. V. Antoshin, *Ph. D. Thesis*, Zelinskii Institute of Organic Chemistry, Moscow, 1975.
[3] W. N. Delgass, T. R. Hughes, and C. S. Fadley, *Catalysis Rev.*, 1970, **4**, 179.
[4] Kh. M. Minachev and Ya. I. Isakov, *Preparation, activation and regeneration of zeolite catalysts* (TsNIITE Neftekhim, Moscow, 1971).
[5] E. S. Shpiro, *Thesis*, Zelinskii Institute of Organic Chemistry, Moscow, 1975.
[6] Kh. M. Minachev, G. V. Antoshin, and E. S. Shpiro, *Problemy kinetiki i kataliza*, 1975, **16**, 189.
[7] Kh. M. Minachev, G. V. Antoshin, and E. S. Shpiro, *Izvest. Akad. Nauk S.S.S.R., Ser. khim.*, 1974, 1015.
[8] Kh. M. Minachev, G. V. Antoshin, E. S. Shpiro, and Ya. I. Isakov, *Izvest. Akad. Nauk S.S.S.R., Ser. khim.*, 1973, 2131; Kh. M. Minachev, G. V. Antoshin, E. S. Shpiro, and T. A. Novrusov, *ibid.*, 1973, 2134.
[9] Kh. M. Minachev, G. V. Antoshin, and E. S. Shpiro, *Second Japan-Soviet Seminar on Catalysis*, Tokyo, 1973, prep. N3.
[10] K. Siegbahn *et al.*, *ESCA: Atomic, Molecular and Solid State Structure Studied by Means of Electron Spectroscopy*, Nova Acta Regiae Soc. Sci. Upsaliensis, Ser. IV, vol. 20, 1966.
[11] E. Dempsey, *J. Phys. Chem.*, 1969, **73**, 3660.
[12] G. K. Boreskov, *Proc. 5th Internat. Congr. Catalysis*, Miami Beach, 1972, vol. 2, p. 981.
[13] N. J. Shevchik and A. Goldman, *J. Electron. Spectr. Relat. Phenom.*, 1974, **5**, 631.
[14] W. G. Reman, A. H. Ali, and G. C. Shuit, *J. Catalysis*, 1971, **20**, 374.
[15] P. Gallezot, Y. Ben Taarit, and B. Imelik, *J. Catalysis*, 1973, **77**, 652.
[16] T. A. Egerton, A. Hagan, F. S. Stone, and J. C. Vickerman, *JCS Faraday I*, 1972, **68**, 723.
[17] P. Gallezot, Y. Ben Taarit, and B. Imelik, *J. Catalysis*, 1972, **26**, 481.
[18] Yu. S. Khodakov *et al.*, *Izvest. Akad. Nauk S.S.S.R., Ser. khim.*, 1969, 523.
[19] K. Kishi and Sh. Ikeda, *Bull. Chem. Soc. Japan*, 1974, **47**, 2532.
[20] Kh. M. Minachev, Ya. I. Isakov, G. V. Antoshin, V. P. Kalinin, and E. S. Shpiro, *Izvest. Akad. Nauk S.S.S.R., Ser. khim.*, 1973, 2522.
[21] Kh. M. Minachev, G. V. Antoshin, E. S. Shpiro, and E. N. Savostyanov, *Third Soviet-Japan Seminar on Catalysis*, Alma-Ata, 1975, prep. N 18.

DISCUSSION

H. Noller (*Tech. Univ. of Vienna*) said: This comment concerns the ionic state of cations in zeolites and other compounds. We have studied magnesium compounds. The XPS binding energies (in eV) of the Mg (2p) orbitals for the following compounds are as stated: MgO, 48·0; MgSO₄, 48·8; MgHPO₄, 49·4; MgX, 49·9; MgY, 49·7 (54·7 after treatment at 300 °C, removal of water). Magnesium in the zeolites is more ionic than in the other compounds, ionic character being most pronounced in the Y zeolite. This coincides well with catalytic properties of these compounds. MgO gives mainly dehydrogenation of alcohols with little dehydration. The electron pair acceptor (EPA) strength of magnesium is low and the electron pair donor strength of oxygen is high. The other catalysts give dehydration and the activity increases with the XPS binding energies, which reflect EPA strength. The higher activity of zeolites can clearly be seen from the fact that dealkylation does not proceed on MgO, MgSO₄, or MgHPO₄ but does proceed on MgX and MgY. Even the higher activity of MgY is clearly reflected in the XPS binding energies. A dramatic increase is observed when the water is recovered. This last finding is in good agreement with the decrease observed in the Ni ($2p_{3/2}$) energy in NiNaY when carbon monoxide, an electron-pair donor, is adsorbed (a line appears between that for the cation and that for the metal).

Kh. M. Minachev (*Zelinskii Inst. of Org. Chem., Moscow*) replied: Prof. Noller shows XPS results for magnesium-containing zeolites which confirm our conclusion concerning the ionic bonding between cations and the zeolite framework. We have also studied zeolites containing alkaline and alkaline earth cations including MgY (see ref. 5 in this paper) along with zeolites containing transition metals. It was shown for the former that the bonding between cations and the zeolite framework is again ionic in type. The large chemical shift in the Mg (2p) spectrum (5 eV) observed by Noller after dehydration of a zeolite is very surprising. The chemical shifts in cation spectra which we observed after similar pretreatment of zeolites did not exceed 1 eV.

As for the appearance of the new line in the Ni ($2p_{3/2}$) spectrum after adsorption of carbon monoxide on NiNaY, we also observed a similar effect (Fig. 4) and believe it is due to the transition of some nickel atoms into an intermediate univalent state.

B. Delmon (*Catholic Univ. of Louvain*) said: There is a diversity of views with respect to the places to which reduced atoms migrate. You assume migration to the surface whereas Gallezot *et al.*[1] assume a migration to supercages, where metal atoms accumulate and eventually cause a rupture of some supercages and a breaking of the crystallites.

I wonder whether the real situation is not somewhere in between these two extremes and that accumulation takes place at crystal defects (growth defects, or defects brought about by exchange, or dislocations, *etc.*). The usual morphology of commercial and many laboratory-prepared samples indeed suggests such a possibility. Migration would be more limited in distance, atoms would find energetically favourable sites, and sufficient space would be available for nucleation of crystallites of critical size. Have you any argument to prove that hypotheses other than migration to the surface must be ruled out?

Kh. M. Minachev (*Zelinskii Inst. of Org. Chem., Moscow*) replied: Prof. Delmon has directed our attention to the important problem of the distribution of reduced metal atoms in zeolites. Naturally, a consideration of the problem should not be limited to the migration of metal to the external surface. In fact the intermediate situation occurs. A proportion of the metal atoms migrates to the external surface with the formation of a separate crystal phase, and another proportion is distributed inside zeolite cavities.

[1] P. Gallezot, J. Datka, J. Massardier, M. Primet, and B. Imelik, this Congress, p. 696.

The increase in the relative intensity of metal:silicon signals which is observed characterizes the former process. But as we found earlier, on the basis of a kinetic study of migration (ref. 7), the intensity of the line due to metal rapidly reaches a maximum. One of the causes of this effect is probably the aggregation of some atoms inside the zeolite cavities which prevents the rest of the atoms leaving the bulk of zeolite. The ratio between the numbers of metal atoms on the external surface and the numbers inside the cavities depends on several factors: the nature of the cation, the type of zeolite structure, and the conditions of zeolite synthesis and treatment. On the basis of our results we conclude that reduced silver, copper, zinc, palladium, and nickel prefer to migrate to the external surface, whereas platinum (this form was studied by Gallezot *et al.*) and cobalt mainly aggregate inside the cavities. Polycationic forms are a special case; here the probability of the formation of dispersed metals inside the cavities may sometimes be higher due to the interactions between the reduced metals.

P. Canesson (*Catholic Univ. of Louvain*) said: I am somewhat surprised that you did not observe spontaneous reduction of Cu^{2+} to Cu^{+} or to Cu^{0} under vacuum conditions. Similar XPS experiments on the same systems in our laboratory[2] clearly show that copper is reduced spontaneously, as is nickel, but more slowly under vacuum. This phenomenon is probably a consequence of X-ray bombardment.

Kh. M. Minachev (*Zelinskii Inst. of Org. Chem., Moscow*) replied: I think a misunderstanding has arisen which may be due to the brief presentation of our results for copper zeolites. It follows from results presented here, and others published in more detail elsewhere (see ref. 7) that copper is partially reduced under the action of X-radiation during the recording of the spectra. This process can be retarded by decreasing the temperature of the measurement. Many of our results show without doubt that there is no appreciable reduction of nickel at room temperature under vacuum.

H. Tominaga (*Univ. of Tokyo*) said: In our laboratory, a new heterogeneous catalytic system has been developed which is remarkably active for oxidation of olefins to carbonyl compounds in the presence of oxygen and water. The catalyst is faujasite in which Na^{+} is ion-exchanged for Cu^{2+} and Pd^{2+}. Although neither of these two ions is catalytically active in the absence of the other, by combining the two in a zeolite a high activity is obtained. This is due to the prevention of Pd metal formation by reoxidation of Pd^{0}, formed by the carbonylation reaction, by Cu^{2+}. In view of this and other findings in our laboratory, I share the opinion, expressed in your last paragraph, as to the prospects for the development of transition-metal ion-exchanged zeolite as catalysts and for their use in organic synthesis.

Kh. M. Minachev (*Zelinskii Inst. of Org. Chem., Moscow*) replied: I am pleased that our conclusions, concerning the prospects for the application of polycationic zeolite forms containing transition elements in catalysis, are supported by your work. I would like to use this opportunity to present some results on the catalytic activity of bimetallic cobalt–nickel forms for nitric oxide reduction by carbon monoxide. Forms containing only nickel or cobalt did not show any activity in this reaction in the temperature range 473–723 K, whereas a zeolite containing both cobalt and nickel was the most active catalyst among those studied. This, and the example which Prof. Tominaga presented, demonstrate the far-reaching effect of the mutual influence of transition elements in zeolites on catalytic activity. We can certainly conclude that the applications of zeolites in heterogeneous catalysis are very extensive.

H. K. Beyer (*Centr. Res. Inst. Chem., Budapest*) said: You have found a reduction of transition-metal ions in zeolites during vacuum treatment. If these ions are reduced something must be oxidized; have you any idea what is oxidized in this case?

---

[2] C. Defossé, R. M. Friedman, and J. Fripiat, *Bull. Soc. chim. France*, 1975, 1513.

Kh. M. Minachev (*Zelinskii Inst. of Org. Chem., Moscow*) replied: The reduction in vacuum of transition-metal cations in zeolites may be due to several factors, (i) the presence of organic materials in the vacuum of 10^{-6}–10^{-7} Torr, (ii) the presence of adsorbed organic molecules, (iii) the decomposition of ammonium complex cations (in the case of palladium- and platinum-zeolites) and so on. Since the high degrees of reduction can be achieved, the reducing agents are apparently hydrogen molecules formed in the destruction of the residual hydrocarbon molecules. Also, during treatment with hydrogen, the reduction proceeds with the formation of water and the compensation of framework negative charge is provided by the addition of protons and the formation of structural hydroxyl groups.

P. A. Jacobs (*Catholic Univ. of Louvain*) said: We have investigated the same systems with a combination of volumetric and spectroscopic techniques, and generally there is agreement between the results of these two different approaches. However, in your paper you mention the reduction of zinc ions in zeolite Y, whereas from volumetric measurements we find no hydrogen uptake by this system. Do you think that this discrepancy is due to the different techniques used, or is it possibly related to differences in methods of sample preparation?

Kh. M. Minachev (*Zelinskii Inst. of Org. Chem., Moscow*) replied: It appears probable that your failure to detect the reduction of zinc in your zeolite is due to a difference in zeolite treatment and not to a difference in the method of investigation. Zinc reduction has also been observed by other methods.[3]

W. N. Delgass (*Purdue Univ.*) said: Extensive Mössbauer spectroscopic studies of FeY zeolites have shown that Fe^{2+} or Fe^{3+} in the bulk does not reduce to metallic iron when the zeolite structure is preserved. Your results suggest a marked difference in the chemistry of the external surface of the zeolite particles as compared to the internal surface. Such a difference might have important implications for catalysis.

Kh. M. Minachev (*Zelinskii Inst. of Org. Chem., Moscow*) replied: X-Ray photoelectron spectroscopy allows us to study the external surface of zeolites, and in some cases the chance of cation reduction in this surface layer may be different from that in the bulk. In many cases the study of the external surface of zeolites will be of special interest from the standpoint of catalysis. However, the literature suggests that the escape depth of photoelectrons from solids in such that, in the case of zeolites, the analysed layer is about 30 Å deep; that is, the interior of the zeolites is also analysed by XPS.

J. C. Vedrine (*CNRS, Villeurbanne*) (communicated): I would like, first, to comment on Prof. Delmon's remarks to Prof. Minachev, and on the work of Gallezot presented elsewhere in this Congress. I have performed experiments on Gallezot's samples, using a Vacuum Generators ESCA-3 spectrometer equipped with a side reaction chamber. Some samples were unreduced, some contained 20 Å crystallites, some contained 10 Å agglomerates, and some atomically dispersed platinum. I observed only small changes in the $Pt(4f)$: $Si(2p)$ and $Pt(4f)$: $O(1s)$ ratios, in contrast to Minachev's results.

Secondly, binding energies of platinum referred to inner elements of the zeolite [$Si(2p)$, $Al(2s)$, $O(1s)$] increase as the particle size of PtO decreases by as much as 1 eV for atomically dispersed platinum.[4] This may be wrongly interpreted as 'non-reduced' platinum and lead experimentalists to reduce samples at higher temperatures, and thus obtain sintering. Therefore differences in conditions of treatment may lead to discrepancies in results from different workers.

Kh. M. Minachev (*Zelinskii Inst. of Org. Chem., Moscow*) (communicated reply):

[3] D. J. C. Yates, *J. Phys. Chem.*, 1965, **69**, 1676.
[4] J. C. Vedrine, *J. Microsc. Electr. Spectrosc. Electron*, 1976, in press.

In answering Prof. Delmon I said that, in the case of platinum-zeolite, we did not observe any appreciable increase in the $Pt(4f):Si(2s)$ intensity ratio and we believe that this is due to the absence of metal migration to the external surface. When this is taken into account our results are in qualitative agreement with those of other workers[1,5] and with your ESCA results for finely dispersed platinum in zeolite.

[5] R. A. Dalla Betta and M. Boudart, *Proc. 5th Internat. Congr. Catalysis*, Miami Beach, 1972, p. 1329.

Application of N.M.R. to the Study of Heterogeneous Catalysis: I. Determination of the Brønsted Acid Strength of Oxides, II. Chemisorption of Hydrogen on Silica-supported Platinum

J. L. BONARDET, J. P. FRAISSARD, and L. C. DE MENORVAL

Laboratoire de Chimie des Surfaces (Chimie Générale), Université P. et M. Curie, 4 Place Jussieu, 75230 Paris, Cedex 05, France

ABSTRACT   Interactions between gases and solids can now be studied by n.m.r. Thus, as in homogeneous phases, it is possible to characterize Brønsted acidity of oxides by calculating the dissociation coefficient of the OH bonds in the presence of other molecules or by determining the chemical shift of the acidic hydrogen. For example, the dissociation coefficient of the OH groups in a silica gel is 0·5 in the presence of NH_3 and the proton chemical shift $\delta(OH)$ is 2 p.p.m. relative to gaseous Me_4Si.

Interactions between hydrogen and supported Pt have also been investigated by n.m.r. The irreversibly adsorbed hydrogen is shown to be both superficial and interstitial; the frequency of the rapid exchange between the latter form and reversibly adsorbed hydrogen is $>3 \times 10^3$ Hz at room temperature.

INTRODUCTION

Among the oldest problems concerning heterogeneous catalysis, the Brønsted acid strength of oxides and the nature of hydrogen adsorbed on metals are very important and are still not well resolved. N.m.r. is a suitable method for studying gas–solid interactions[1] and can be used to investigate the above problems.

The Brønsted acidity of homogeneous samples may be characterized either by the dissociation constant of the OH group or by the n.m.r. chemical shift of the ^{1}H nucleus which is determined by its electronic environment. The intrinsic Brønsted acidity of a solid catalyst has never been determined because it is impossible to measure the chemical shift of nuclei in a rigid lattice. We propose a new method for determining the dissociation constant of the OH groups on a solid surface and for calculating the chemical shift $\delta(OH)$ of the corresponding hydrogen atom.

The second application of n.m.r. concerns hydrogen adsorbed on supported platinum; this has been extensively studied with the aim of establishing the nature of the active form in the catalysis of exchange and hydrogenation reactions. Surface potential and conductivity measurements,[2-4] and theoretical considerations[5] suggest that hydrogen exists on platinum in two forms, a strongly adsorbed and a weakly adsorbed species. The heats of adsorption of these species have been quoted as 23 kcal mol^{-1} and 10 kcal mol^{-1} respectively.[6] Up to five forms of adsorbed hydrogen have been detected by desorption studies[7] in the temperature range 77—573 K, and by analysis of isotherms.[8] Two i.r. bands, at ca. 2120 and 2060 cm^{-1} were assigned to the stretching vibrations of Pt–H,[9-11] though some workers believe that the 2060 cm^{-1} band is due to chemisorption of CO impurities.[12,13] In addition, it now seems that the 2120 and 2060 cm^{-1} bands result from species which are reversibly bound at room temperature.[8,11,12] Hence the irreversible form cannot be detected by i.r. spectroscopy.

Following the first n.m.r. results[14] concerning the adsorption of hydrogen on platinum metal under high pressure, we have undertaken the study of H_2 chemisorbed irreversibly on Pt/SiO_2.

EXPERIMENTAL

The silica xerogel was prepared by Planck's method.[15] Samples initially treated for 15 h under 10^{-4} Torr at 150, 400, and 600°C are designated as Si 150, Si 400, and Si 600, respectively. The samples of NaY and NH_4Y used to check the proposed method were treated for 15 h under vacuum at 140°C in order to remove adsorbed water, but without any loss of ammonia from the NH_4Y. The zeolite HY was obtained by decomposition of NH_4Y at 400°C under 10^{-4} Torr.

The silica-supported platinum was prepared by impregnation of Davison silica with sufficient chloroplatinic acid solution to give 10 wt% platinum after reduction.[16] The particle diameter of platinum was about 200 Å. Before hydrogen adsorption, the samples were heated at 350°C under 10^{-4} Torr for 12 h.

The ^1H and ^{15}N experiments were performed with DP 60 and Fourier transform XL 100 Varian spectrometers at 60 and 10·1 MHz respectively. Chemical shifts were measured with an external reference. Bulk magnetic susceptibility corrections were made by the method of Bonardet *et al.*[17]

RESULTS AND DISCUSSION

I. *Determination of the Brønsted acid strength*

Theory In solids, atomic movements are very slow in comparison to liquids and gases, and so the strong dipolar interactions of spins are not reduced to zero by time-averaging. N.m.r. lines are very wide (even broader than the sweep field) and prevent any chemical shift measurements. The principle of our work is to make the surface proton take part in a heterogeneous equilibrium with an adsorbed proton acceptor molecule AH, eqn. (1):

$$S\text{–OH} + AH \rightleftharpoons S\text{–O}^- + AH_2^+ \tag{1}$$

When rapid exchange occurs between the proton of the surface S–OH and that of the adsorbed molecule AH, the acid proton must affect the chemical shift of the adsorbed phase. The high resolution spectrum should contain only one line at frequency ν_e, due to the coalescence of the lines at frequencies ν_{AH}, $\nu_{AH_2^+}$, and ν_{OH}. The observed chemical shift δ_{obs} is then

$$\delta_{obs} = P_{OH}\delta(OH) + P_{AH}\delta(AH) + P_{AH_2^+}\delta(AH_2^+) \tag{2}$$

where P_i is the concentration of hydrogen atoms in the group i. Knowing the chemical shift of the two nuclear types in the AH and AH_2^+ species (*e.g.*, ^1H and ^{15}N or ^{17}O), we can calculate the relative concentration of P_{AH} and $P_{AH_2^+}$, the dissociation coefficient of OH in equilibrium (1), and the chemical shift $\delta(OH)$ of the surface proton.

Chemical shifts of NH_3 and NH_4^+

The ^1H chemical shift of gaseous NH_3 at 50 Torr is 0·06 p.p.m. relative to gaseous Me_4Si at a pressure of 1 Torr. $\delta(NH_4^+)$ is measured using an aqueous solution of NH_4NO_3 containing dioxan, as the internal reference, and nitric acid, to prevent any proton exchange between NH_4^+ and H_2O. Relative to NH_3 gas, $\delta(NH_4^+) = 6·9$ p.p.m.

The ^{15}N signal of $^{15}NH_3$ has been detected at a pressure of 100 Torr. $\delta(^{15}NH_4^+)$ has been measured using $^{15}NH_4^{15}NO_3$ in aqueous solution. Relative to $^{15}NH_3$ gas, $\delta(^{15}NH_4^+) = 43·5$ p.p.m.

Experimental verifications of the proposed method

Initially our hypothesis was checked by means of ammonia adsorption on dehydrated NH_4Y zeolite. Ammonium cations were found inside the silicoaluminate cages. In a rigid lattice, the half-width of the NH_4^+ signal was about 32 kHz. At room temperature the hydrogen atoms in NH_4Y are able to exchange between different sites; the nuclear dipolar interactions were time-averaged and the line width was only 8 kHz. However, since the maximum sweep field of our apparatus is 4 kHz, the NH_4^+ signal could not be detected at high resolution. However, as soon as the zeolite had adsorbed a small amount of ammonia, a signal was detected whose position δ and width ΔH depended on the relative concentrations of the NH_4^+ and NH_3 species. This signal characterizes the exchange

$$N_iH_4^+ + N_jH_3 \rightleftharpoons N_iH_3 + N_jH_4^+ \tag{3}$$

between the protons of ammonium cations and those of the adsorbed molecules, the exchange frequency being higher than

$$\Delta\nu = \nu(NH_4^+) - \nu(NH_3) = 420 \text{ Hz.}$$

Thus, from the chemical shift point of view, the influence of nuclear movement upon the spectral shape is similar, regardless of any overlapping of the lines related to the different sites.

By extrapolating to zero concentration of adsorbed ammonia, it was found that the real chemical shift of the ammonium zeolite protons was 7.0 ± 0.1 p.p.m.

Using the same conditions, we have studied ammonia adsorption on NaY zeolite. Compared to the above results, the width and chemical shift of the signal were very low, indicating that only physisorption occurred in this case (e.g., at a pressure of 3 Torr, $\delta = 0.65$ p.p.m. and $\Delta H = 220$ Hz).

Brønsted acidity of a silica gel

[1]H n.m.r. results   On a more or less dehydrated silica gel surface, ammonia can be adsorbed either on OH groups (site I) or on other sites (site II). Possible reactions between NH_3 and the solid surface are:

transfer of a surface proton to a molecule $(NH_3)_I$ adsorbed on OH

$$SiO-H + (NH_3)_I \rightleftharpoons SiO^- + NH_4^+ \tag{4}$$

proton transfer inside the adsorbed phase

$$(NH_3-H)_I^+ + (NH_3)_{II} \rightleftharpoons (NH_3)_I + (NH_4^+)_{II} \tag{5}$$

exchange of adsorbed molecules on the different sites

$$(NH_3)_I + (NH_3)_{II} \leftrightharpoons (NH_3)_{II} + (NH_3)_I \tag{6}$$

In the case of a rapid exchange, the spectrum contains only one line whose position δ depends on the concentration of surface OH, on the SiO–H dissociation, and on the relative concentrations of NH_4^+ and NH_3 adsorbed either on OH or other sites.

For each sample, the chemical shift increased with decreasing concentration of adsorbed NH_3. In a previous paper we studied the influence of each equilibrium on δ and ΔH.[18] Here we consider only the δ value at low coverage. At 30°C, following ammonia desorption at 10^{-4} Torr, the chemical shifts all tended to a similar limit: 4·2, 3·9, and 3·7 p.p.m. for Si 150, Si 400, and Si 600 respectively. After such desorption the number of physically adsorbed molecules must be negligible and the δ limit must be typical of equilibrium (4) only.

$$\delta(^1H)_{NH_4^+} > \delta(^1H)_{limit} > \delta(^1H)_{NH_3}$$

$\delta(^1H)_{11mlt}$ decreased a little with decreasing OH concentration of the surface, which shows that the OH dissociation, and the average acidity of one OH, increase with the OH density.

^{15}N *n.m.r. results* In eqn. (4), the chemical shift $\delta(OH)$ and the constant K_4 are unknown. To calculate them, the detection of a spin other than ^1H is necessary. Because of the quadrupole moment of ^{14}N, we have chosen the ^{15}N nucleus. With the sample Si 400, $\delta(^{15}N)$ increases from 17·8 to 22·0 p.p.m. when the adsorbed NH_3 decreases from 50 to 5 mg g^{-1} (corresponding to 3 h desorption under 10^{-4} Torr).

Coefficient α of OH dissociation At the limit of signal detection, the shift $\delta(^{15}N)_{11mlt}$ = 22·0 p.p.m. characterizes eqn. (4) with sample Si 400.

$$\delta(^{15}N)_{11mlt} = P_{NH_3}\delta(^{15}NH_3) + P_{NH_4^+}\delta(^{15}NH_4^+) \tag{7}$$

where P_{NH_3} and $P_{NH_4^+}$ are the concentrations of the species NH_3 and NH_4^+:

$$P_{NH_3} + P_{NH_4^+} = 1.$$

With ammonia as the reference, $\delta(^{15}NH_3) = 0$,

then $\qquad P_{NH_4^+} = \dfrac{\delta(^{15}N)_{11mlt}}{\delta(^{15}NH_4^+)} = \dfrac{22\cdot0}{43\cdot5} \sim 0\cdot5, \qquad$ and $\qquad \dfrac{[NH_3]}{[NH_4^+]} = 1.$

Thus the coefficient of OH dissociation $\alpha = 0.5$. For comparison, the pH of an equimolar NH_3 and NH_4^+ buffer solution is *ca.* 4·7.

Chemical shift $\delta(OH)$ With Si 400, $\delta(^1H)_{11mlt} = 3\cdot9$ p.p.m. Using eqn. (2) and the coefficient α, the proton shift of the OH group can be calculated:

$$\delta(OH) = 2 \text{ p.p.m.}$$

It should be noted that this value is independent of the adsorbed phase. For comparison, $\delta(OH)$ of ethanol diluted in CCl_4 is 0·5 p.p.m.

Brønsted acidity of Ketjen silica–aluminia and zeolite

Ammonia was adsorbed at a pressure of 500 Torr on a silica–alumina sample previously heated at 250°C at 10^{-4} Torr. After ammonia desorption at room temperature the ^1H chemical shift of the chemisorbed phase was 5·15 p.p.m., which shows that this sample was more strongly acidic than silica gel. However, in this case δ(limit) was due to both Brønsted and Lewis acidities. In order to calculate $\delta(OH)$ precisely it is necessary to measure the chemical shift at AL ← NH_3 samples with Lewis sites only.

With zeolite, reaction (1) is

$$HY + NH_3 \underset{b}{\overset{a}{\rightleftharpoons}} NH_4^+ + Y^- \tag{8}$$

However, after ammonia adsorption at a pressure of a few Torr, a high proportion of ammonia could not be desorbed at room temperature; the shift of the signal corresponding to the larger degree of desorption (shift measured as previously) was that of the NH_4^+ species. In the presence of ammonia, the coefficient α of the OH groups of HY zeolite was 1, which shows that the acidity of HY is greater than that of silica–alumina. $\delta(OH)$ cannot be determined however since reaction (8a) is complete. Another base will have to be tried.

II. Adsorption of hydrogen on silica supported platinum

Results. For monolayer coverage (2×10^{19} adsorbed molecules g^{-1} under 5×10^{-2} Torr), the spectrum recorded at room temperature can be broken down into three lines. One (line 3) is due to the OH groups situated on the silica surface and will be used in what follows as the standard. The second (line 1) and the third (line 2) are shifted upfield, slightly (1 p.p.m.) and very much (46 p.p.m.), respectively.

After pumping at 10^{-4} Torr at room temperature for 12 h, the lines 1 and 2 are slightly weaker and the shift of line 2 is $\delta_1 = 49$ p.p.m. (Fig. 1C). This spectrum, named 1, corresponds to irreversibly chemisorbed hydrogen at 25°C. After heating the sample

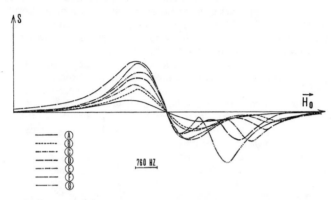

FIGURE 1   Change of the spectrum with the hydrogen pressure.
(A) Without hydrogen.
(B) Hydrogen desorbed at 48°C under vacuum for 2 h, after (G).
(C) Hydrogen desorbed at 25°C under vacuum for 12 h.
(D) Hydrogen desorbed at 25°C under vacuum for 1 h.
(E) H_2 adsorbed at 0·32 Torr.
(F) H_2 adsorbed at 2 Torr.
(G) H_2 adsorbed at 49 Torr.

under vacuum, the strength of lines 1 and 2 decreases and the shift of the latter increases. At the limit of signal detection, corresponding to a desorption temperature of 100°C, the shift of line 2 is $\delta_0 = 57$ p.p.m. (named spectrum 0). When the hydrogen pressure is increased, line 3 is unchanged. The intensity of line 1 increases with the pressure until it reaches about 1·5 Torr (corresponding to $1·06 \times 10^{20}$ adsorbed molecules g^{-1}), after which it remains constant. Line 2 increases continuously with the hydrogen pressure and is shifted downfield; its δ value varies approximately homographically with the number of adsorbed molecules.

After hydrogen adsorption at a few Torrs and 25°C, the n.m.r. spectrum was detected at 77 K. It can be represented approximately by spectrum 1 on which a relatively narrow non-shifted signal has been superimposed. However, line 2 is less shifted than in the spectrum 1.

Adsorption of various gases.   Let us consider the supported platinum with irreversibly chemisorbed hydrogen (spectrum 1). When this sample adsorbs successively small amounts of oxygen or ethylene, line 2 decreases in intensity and shifts upfield. At the detection limit its shift is close to δ_0. The water or hydrogenated species formed by the action of O_2 or C_2H_4 on chemisorbed hydrogen are adsorbed on the support and give a non-shifted signal which obscures a little the disappearance of line 1. Even at high pressure, the presence of deuteriated benzene never causes the elimination of more than 50% of lines 1 and 2. This result is only in partial agreement with those of Basset *et al.*[19] who found no catalytic hydrogenation of benzene on a surface covered only with irreversibly adsorbed hydrogen.

When the platinum surface is initially covered with strongly chemisorbed CO, neither line 1 nor 2 appears in the spectrum whatever the hydrogen pressure. However, the hydrogen is reversibly adsorbed and is indicated by a narrow signal which is shifted about 7 p.p.m. downfield.

Discussion. Let us consider firstly irreversible adsorption. The existence of two signals proves that it occurs in two different forms. The dominant contribution to the shift in the proton resonance arises from the spin polarization of the conduction electrons. Indeed, the effect upon δ of the bulk magnetic susceptibility X_V of the Pt/SiO_2 sample (determined following ref. 2) is a few Hz downfield. The diamagnetic or orbital effects of the conduction electrons produce shifts to higher field. However, they are of the same order as the ordinary chemical shift and must be negligible here.

The Knight shift in metals arises from the interactions of the conduction electrons near the Fermi surface of the metal with the metal nuclei. These s unpaired electrons, polarized in an external applied magnetic field, can have a finite density at the H adatoms on platinum which would give rise to a positive shift δ_K opposite to that observed. It seems likely that the 7 p.p.m. shift of the line associated with H_2 on a surface covered with CO corresponds to the order of magnitude of δ_K.

Direct overlap between the conduction electrons and electrons localized in s orbitals centred on the protons would pair the spins of the latter antiparallel to those of the conduction electrons, resulting in an upfield proton shift δ_K. The conduction electrons can also take part in exchange interactions with the bonding electrons in d orbitals centred on platinum, polarizing these latter parallel to their own direction. The d electrons transmit the spin polarization with a change of sign to the proton *via* the exchange interactions in the covalent part of the Pt–H bond. Then the proton shift is opposite to the Knight shift.

Therefore the sign and magnitude of δ depend on the relative contributions of the Knight shift and the mechanisms responsible for the negative shift. It could therefore be said that for the species corresponding to line 1, the Knight shift neutralizes almost completely the effects of polarization of the bonding electrons, whereas for the species revealed by line 2 the negative spin density is overwhelmingly dominant.

We know that the structure of the bands of a transition metal is characterized by broad s–p-bands with low state density, whereas the d-bands are narrow with particularly high state density. Moreover in platinum, the $5d$-band is close to the Fermi level. On the other hand the shift of line 1 is small and is not changed by increase in the H_2 concentration; then this line must correspond to the hydrogen atoms $H_{(1)}$ located above the metal surface and bonded with the superficial Pt_s atoms through s electrons. On the contrary, the δ value of line 2 is very large and decreases to a limit value δ_1 when the concentration of irreversible hydrogen increases. Line 2 therefore represents the hydrogen atoms $H_{(2)}$, preferentially adsorbed by means of the d-electrons of the metal, in the interstices between the Pt_s atoms of the surface. For example, on the (100) face the $H_{(1)}$ species will be bonded above the surface to each Pt_s atom, the species $H_{(2)}$ to one Pt atom below the surface and to four Pt_s atoms. As the interstices of the different planes are being filled, the density of the negative spin at each $H_{(2)}$ atom decreases up to the complete occupation of sites. These results show not only the possibility of a localized metal–adsorbate bond, as Bond has suggested,[20] but also the collective electronic properties of the metal.

At high pressures, the reversible form is characterized by line 2, which shifts markedly downfield as the hydrogen concentration increases. The evolution of this line is typical of a chemical exchange (at a frequency higher than 3×10^3 Hz) between the "irreversible" interstitial species $H_{(2)}$ resonating at high field (shift δ_1) and another species $H_{(3)}$, much less shifted. This is confirmed by the separation of line 2 into two components (of which one is very slightly shifted) when the spectrum is recorded at 77 K. However we have not enough evidence to state that the $H_{(3)}$ form is that which is adsorbed at room temperature on a CO-covered surface and situated at 7 p.p.m.

The above results show that the irreversibly adsorbed atoms $H_{(2)}$ interact with several Pt atoms. Thus the i.r. band for the Pt–H bond is then so broad that it cannot be detected. (It could also be undetectable for symmetry reasons.) But this irreversible form is absolutely necessary for detection, after further reversible hydrogen adsorption, of an absorption at 2120 cm^{-1}, which is usually attributed to this latter form. A slightly different interpretation of this band is possible. We have just seen that the $H_{(3)}$ atoms of the reversible form exchange rapidly only with the "irreversible" interstitial $H_{(2)}$ atoms. Because of the formation of an intermediate complex containing several H atoms, the atoms $H_{(2)}$ can be displaced from their initial position during a short time just long enough for i.r. detection. Therefore they are interacting with a reduced number of Pt atoms, in a low symmetry site. The line corresponding to the Pt–H bond can then be detected. After desorption of the reversible form, $H_{(2)}$ returns to its interstitial position and the 2120 cm^{-1} band can no longer be detected.

CONCLUSION

It has been shown in two major problems that n.m.r. spectroscopy is now suitable for the study of gas–solid interactions. Thus the Brønsted acidity of oxides can be characterized, as in the homogeneous phase, either by calculation of the dissociation constant of the O–H bond in the presence of other molecules, or by determining the chemical shift of the acidic hydrogen, *i.e.* by "evaluting" its electronic environment. It becomes therefore possible to go beyond comparative measurements such as the study of the temperature dependence of desorption of previously adsorbed bases.

N.m.r. can also be applied successfully to problems of gas–metal interactions. Thus, we have been able to detect spectroscopically the irreversibly adsorbed hydrogen on supported platinum and to describe it accurately: hydrogen atoms located on the surface and interstices. The hydrogen adsorbed subsequently is adsorbed reversibly and exchanges at frequencies above 3×10^3 Hz with the interstitial H atoms only. These results clarify those of Taro Ito *et al.*[14] for hydrogen adsorption on Pt metal at relatively high pressure. Finally, they lead to the suggestion of an alternative interpretation of the 2120 cm^{-1} i.r. band associated with Pt–H bonds.

REFERENCES

[1] E. G. Derouane, J. Fraissard, J. J. Fripiat, and W. E. Stone, *Calalysis Rev.*, 1972, **7**, 121.
[2] J. C. P. Mignolet, *J. Chim. phys.*, 1957, **54**, 19.
[3] W. M. H. Sachtler and G. J. H. Dorgelo, *Z. phys. Chem.* (*Frankfurt*), 1960, **25**, 69.
[4] R. Suhrmann, G. Wedler, and H. Gentsch, *Z. phys. Chem.* (*Frankfurt*), 1958, **17**, 350.
[5] I. Toya, *Prog. Theor. Phys.* (*Kyoto*), *Suppl.*, 1962, **23**, 250.
[6] Y. Kubokawa, S. Takashima, and D. Toyama, *J. Phys. Chem.*, 1964, **68**, 1244.
[7] S. Tsuchiya, Y. Amenomiya, and R. J. Cvetanovic, *J. Catalysis*, 1970, **19**, 245; 1971, **20**, 1.
[8] L. T. Dixon, R. Barth, R. J. Kokes, and J. W. Gryder, *J. Catalysis*, 1975, **37**, 376.
[9] W. A. Pliskin and R. P. Eischens, *Z. phys. Chem.* (*Frankfurt*), 1960, **24**, 11.
[10] D. D. Eley, D. M. Moran, and C. H. Rochester, *Trans. Faraday Soc.*, 1968, **64**, 2158.
[11] L. T. Dixon, R. Barth, and J. W. Gryder, *J. Catalysis*, 1975, **37**, 368.
[12] M. Primet, J. M. Basset, .M V. Mathieu, and M. Prettre, *J. Catalysis*, 1973, **28**, 368.
[13] D. J. Darensgourg and R. P. Eischen, *Proc. 5th Internat. Congr. Catalysis*, Miami Beach, 1972.
[14] T. Ito, T. Kadowaki, and T. Toya, *Jap. J. Appl. Phys.*, 1974, *Suppl.* 2, Pt. 2, 257.
[15] C. J. Planck, *Catalysis*, 1955, 1.
[16] T. A. Dorling, B. W. J. Lynch, and R. L. Moss, *J. Catalysis*, 1971, **20**, 190.
[17] J. L. Bonardet, A. Snobbert, and J. P. Fraissard, *Compt. rend.*, 1971, **272**, *C*, 1336.
[18] J. L. Bonardet and J. P. Fraissard, *Jap. J. Appl. Phys.*, 1974, *suppl.* 2, Pt. 2, 319.
[19] J. M. Basset, G. Dalmai-Imelik, M. Primet, and R. Mutin, *J. Catalysis*, 1975, **37**, 22.
[20] G. C. Bond, *Discuss. Faraday Soc.*, 1966, 200, and refs therein.

Bonardet, Fraissard, de Menorval

DISCUSSION

V. B. Kazansky (*Zelinskii Inst. of Org. Chem., Moscow*) said: I agree with Prof. Fraissard that studies of rapid exchange between chemisorbed and physically adsorbed molecules by n.m.r. is an interesting approach to the study of adsorption. In our paper with Borovkov and Zhidomirov[1] n.m.r. was used to study adsorption of hydrogen, olefins, and saturated hydrocarbons on surfaces containing co-ordinatively unsaturated Co^{2+} and Ni^{2+} ions. Equilibrium constants and the heats of specific adsorption were calculated. In my view the model used by Prof. Fraissard for the description of ammonia adsorption on silica-gel is inadequate. He postulates only two forms of ammonia adsorption on a silica surface: physically adsorbed ammonia molecules and ammonium ions formed by the reaction $OH + NH_3 \rightarrow O_{surf} + NH_4^+$. However, ammonia molecules could also exist co-ordinatively bonded to surface hydroxyls: $OH + NH_3 \Leftrightarrow OH \cdots NH_3$. This is very probable for such a weak acid as silica-gel. In this case the equilibrium constant calculated by Prof. Fraissard describes specific adsorptions rather than acidic properties of surface hydroxyls. Also, the proton transfer to ammonia molecules, and proton abstraction from ammonium ions, needs some activation energy and therefore could not occur in a millisecond, which is the time available for the fulfilment of rapid exchange.

J. Fraissard (*Univ. Pierre et Marie Curie, Paris*) replied: We have reported elsewhere (ref. 18 of my paper) the dependence of the spectrum of adsorbed ammonia on n.m.r. frequency, on experimental temperature, and in particular on the surface coverage. We have analysed all the possible reactions which can occur at the silica gel surface, (i) exchange of hydrogen atoms between OH and NH_3, (ii) dissociative chemisorption of ammonia with formation of $Si-NH_2$ groups, and (iii) the equilibria (4), (5), and (6) of the present paper.

We proved that equilibrium (4) is absolutely necessary to explain such a high chemical shift. For example the chemical shift of ammonia molecules co-ordinately bonded by the surface hydroxyls as $OH \cdots NH_3$ should be lower than 1 p.p.m. On the other hand, the same δ_{11mlt} value obtained at very low coverage, whatever the surface density of hydroxyl groups, cannot be explained if we do not accept that this value is typical of hydroxyl dissociation only [equilibrium (4)]. It would also be possible to explain it if we assume that the ratio of the concentrations of the different forms adsorbed remains constant whatever the hydroxyl concentration, but this would be a miracle!

R. C. Pink (*Queen's Univ., Belfast*) said: Like Professor Kazansky, I think that the chemical model may be oversimplified. For example, with amorphous silica–aluminas and with zeolites there is reason to believe that there may be a range of sites with rather widely varying acid strength. If this is so, equation (2) of the paper should contain a large number of terms and the measured acidity is essentially an average acidity in which the value for the catalytically important sites is masked by the contributions from large numbers of sites of moderate or weak acidity.

J. Fraissard (*Univ. Pierre et Marie Curie, Paris*) replied: You are right when you say that our results correspond to the average acidity of the silica gel surface. But the small decrease of δ_{11mlt} with the decrease of the surface coverage of hydroxyl groups shows that the average acidity increases with the OH density. Consequently it should be possible to determine each type of acidity in a stepwise manner. Indeed, in the case of amorphous silica–alumina we report in the present paper that δ_{11mlt} is due to both Brönsted and Lewis acidities. To determine δ_{OH} it is necessary to measure the chemical shift of ammonia molecules co-ordinately bonded by the surface Lewis sites.

I do not agree with the last part of your comment. At very low coverage, only the

[1] V. B. Kazansky, V. Yu. Borovkov, and G. M. Zhidomirov, *J. Catalysis*, 1975, **39**, 205.

most acidic sites are detected. For example, with HY zeolite, reaction (8a) is complete; δ_{11mit} is 7 p.p.m. That value is exactly the chemical shift of the ammonium ion.

R. Haul (*Technischen Univ., Hannover*) said: The determination of the Brönsted acidity of surface OH groups by means of measurements of the chemical shift requires sufficiently rapid proton exchange with the adsorbate. A drastic narrowing of the OH proton resonance line due to a much higher jump frequency caused by hydrogen bonding was also found by Staudte[2] and Freude *et al.*[3] for the systems benzene–silica gel and pyridine–Y zeolite.

On the other hand we have found in a recent study[4] of n.m.r. relaxation behaviour of cyclohexane, benzene, and pyridine adsorbed in microporous thoria that in all cases OH proton relaxation times T_2 were not enlarged by the adsorbed molecules. In such a case it is not possible to measure a chemical shift. This implies that the dissociation coefficient of the O—H bond is negligibly small for thoria which is considered as a typical non-acid catalyst. Nevertheless, the mobility of the protons between adjacent OH groups as determined from the onset of motional narrowing is just as large for thoria as for Y zeolites. Typical jump frequencies at 473 K are: for thoria 5×10^3 and 3×10^5 s^{-1} at OH group concentrations of 3·6 and 14·5 nm^{-2} respectively and 3×10^4 s^{-1} for Y zeolite pretreated at 673 K.

Although in the case of thoria the lack of appreciable proton exchange between OH groups and adsorbates prevents chemical shift measurements, just this behaviour made it possible to determine the mobility of the adsorbed molecules. For this purpose the T_1 relaxation times for OH and adsorbate protons were evaluated separately. A considerable shortening of T_1 values for the OH protons was found on adsorption, *e.g* from 10 s to 1 s for 0·2 monolayers cyclohexane at room temperature and 60 MHz.

Irrespective of the fact that the mobile adsorbate has no influence on the proton mobility the additional magnetic interaction now determines the T_1 relaxation of the OH protons. From the observed distinct minima at 140 and 90 K for the above example, translational and rotational correlation times of the adsorbed cyclohexane molecules were deduced. These are smaller by many orders of magnitude than those for OH protons.

Information on translational mobility of adsorbates is commonly obtained by means of the isotope dilution method. However, this procedure requires that the contributions of inter- and intra-molecular interaction are comparable. The method is not applicable, if the intramolecular second moment is large and/or the surface coverage is low, as was the case in the system cyclohexane/thoria studied here. The method outlined above offers an alternative approach to the study of microdynamics in adsorption layers provided that the OH protons remain in practically fixed surface positions.

J. Fraissard (*Univ. Pierre et Marie Curie, Paris*) replied: The narrowing of the OH proton resonance line with the adsorption of different molecules has been found by many authors (ref. 1 of my paper). It is true that the determination of the mobility of the adsorbed molecules from measurements of relaxation time is easier when the OH protons do not exchange with those of the adsorbed phase. It is certain that the chemical shift of OH groups located on the surface of non-acidic catalysts cannot be determined by the proposed method. But this method is specifically designed to give a precise value (the chemical shift) to the intrinsic acidity of a solid, and also to get rid of the purely comparative methods which have been used to date.

H. Charcosset (*CNRS, Villeurbanne*) said: You have observed two different forms of irreversibly adsorbed hydrogen on platinum. Would you expect one or two peaks

[2] B. Staudte, *Z. phys. Chem.* (*Leipzig*), 1974, **255**, 158.
[3] D. Freude, W. Oehme, H. Schmiedel, and B. Staudte, *J. Catalysis*, 1974, **32**, 137.
[4] E. Riensche, Dissertation, Hannover, 1976.

during temperature programmed desorption under vacuum? We have observed a single hydrogen TPD peak; however, most of our catalysts have a mean platinum particle size lower than 100 Å.

J. Fraissard (*Univ. Pierre et Marie Curie, Paris*) replied: We have studied the decrease of the intensity of the two n.m.r. lines as a function of hydrogen desorption and have shown that these two lines disappear at about the same desorption temperature under vacuum. Consequently, I think that it is normal to observe a single hydrogen TPD peak. We are now studying the influence of the platinum particle size on the n.m.r. spectrum. The results obtained so far show that line 1, the less shifted one, does not exist when the particle diameter is lower than 50 Å.

C. J. Wright (*AERE, Harwell*) said: Inelastic neutron scattering is another technique which has recently been used successfully in the investigation of hydrogen adsorbed upon a polycrystalline platinum surface.

Measurements with my co-workers Mr. J. Howard and Prof. T. C. Waddington of the University of Durham of the vibration spectrum of the irreversibly adsorbed hydrogen adsorbed upon a previously cleaned surface show an excitation at 412 cm^{-1} together with its first, second and third harmonics. It is interesting to make two speculations concerning this result. Firstly the vibration frequency is much lower than any previously recorded for a covalently bonded molecular transition metal hydride, and secondly the frequency is very close to the fundamental excitation frequency of hydrogen in the β-phase of palladium hydride at 460 cm^{-1}. Both these arguments support the growing body of evidence concerning the interstitial nature of the irreversibly adsorbed state of hydrogen upon platinum.

J. Fraissard (*Univ. Pierre et Marie Curie, Paris*) replied: I am glad to hear that by use of inelastic neutron scattering you arrive at the same conclusion concerning the interstitial nature of hydrogen irreversibly adsorbed on platinum. However, as the platinum particle size of your sample is probably larger than 200 Å, I am surprised that you detected only one kind of Pt–H bond.

M. Gigot, A. Gourgue, J. B. Nagy, and E. G. Derouane (*Univ. de Namur*) said: Different species formed by but-1-ene and *trans*-but-2-ene adsorption have been observed on NaY and HY zeolites by ^{13}C-n.m.r. spectroscopy. The chemical shifts are reported in the Table. For NaY zeolite a positive shift is found ($\sim$8 p.p.m.) for the —CH= group, while negative shifts characterize the other groups, the values for CH$_2$=, —CH$_2$— and CH$_3$— being -3.5, -3, and -4 p.p.m. respectively. The line widths are larger for species in the adsorbed form than for species in the liquid state ($\Delta H_{ads} = 50$ Hz; $\Delta H_{liq} = 20$ Hz). Adsorption can be explained by the formation of a π-complex between the molecule and the surface.[5] Theoretical studies in which the active site was supposed to be Na$^+$ ion, gave satisfactory results.[5] Nevertheless, similar chemical shifts have been found experimentally for adsorption on silica and on sodium-enriched silica.[6] It is thus not possible at present to determine unambiguously the nature of the active sites. The results for *trans*-but-2-ene adsorption on NaY zeolite are similar, but smaller chemical shift differences are observed.[7]

For HY zeolite evacuated at 400 °C (HY-400) there is a great variation in the band widths and the relative intensities of the lines. Clearly, an adsorbed species is present which is different from the π-complex on the NaY zeolite or on the SiO$_2$ surface. Surprisingly, very small chemical shifts are observed in the case of but-1-ene, which denotes an insignificant charge-transfer between the adsorbed molecule and the

[5] D. Deininger, D. Geschke, W.-D. Hoffmann, *Z. Phys. Chem. Leipzig*, 1974, **255**, 273.
[6] I. D. Gay, J. F. Kriz, *J. Phys. Chem.*, 1975, **79**, 2145.
[7] D. Michel, *Surface Sci.*, 1974, **42**, 453.

Table

Chemical shifts (in p.p.m. vs TMS) and line widths[a] of butenes adsorbed on NaY and HY zeolites[d]

Compound	Zeolite[c]	T (K)	δ (p.p.m.)				ΔH (Hz)			
			C-1	C-2	C-3	C-4	C-1	C-2	C-3	C-4
but-1-ene (l)	—	293	113·1	140·4	26·9	13·1	23	23	23	23
but-1-ene	NaY	300	109·5	147·9	23·8	9·2	70	70	60	50
	HY-400	203	112·6	139·7	24·8	12·3	100	100	200	120
	HY-750	203	114·0	140·1	28·5	14·9	120	130	130	120
trans-but-2-ene (l)	—	293	17·0	125·6			33	33		
trans-but-2-ene	HY-400	183	19·8	127·7			280	170		
	HY-750	193	18·4	126·1			100	90		
	HY-400	273	—	128·8[b]			—	160[b]		

[a] Except for NaY, the band widths were not corrected for instrumental line broadening (20 Hz).
[b] trans-but-2-ene obtained from but-1-ene.
[c] Numbers indicate evacuation temperature (°C).
[d] Surface coverage: $\simeq$ 0·5—2.

643

surface. To explain the absence of charge transfer and the restricted motion of the molecules, we propose a cyclic adsorbed species:

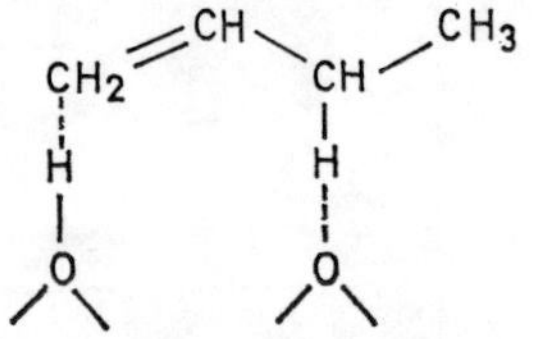

We have also observed the double bond shift starting from but-1-ene: the product formed is a cyclic adsorbed species close to *trans*-but-2-ene.

J. Fraissard (*Univ. Pierre et Marie Curie, Paris*) replied: We have proposed a mechanism for butene isomerization on acidic surfaces,[8] based on the intermediate phases detected by proton n.m.r.; your interpretation of butene chemisorption on HY zeolite does not agree exactly with ours for butene chemisorption on silica–alumina.

[8] J. Fraissard, S. Bielikoff, and B. Imelik, 4th International Congress on Catalysis, First Symposium Moscow 1968; J. Fraissard, S. Bielikoff, and B. Imelik *Compt. rend.* 1970, **271** *C*, 897.

Zeolite Chemistry I. The Role of Aluminium in the Thermal Treatment of Ammonium Exchanged Zeolite Y

D. W. BRECK and G. W. SKEELS

Union Carbide Corporation, Tarrytown Technical Center, Tarrytown, New York 10591, U.S.A.

ABSTRACT   Ammonium exchanged zeolite Y was calcined in shallow beds in air at 473 to 973 K and samples slurried in NaCl and KF solutions. Potentiometric titrations of the acidic (NaCl) solution and basic (KF) solution together with chemical analyses, thermal analyses, and infrared spectroscopic studies of the hydroxyl regions were carried out. The results show that about 16 Al atoms (per unit cell containing 56 Al atoms) are removed by the calcination and subsequent ion exchange and hydrolysis to form $AlOH^{2+}$, $Al(OH)_2^+$, and $[Al-O-Al]^{4+}$ ions. No silicon was found in the salt solutions indicating no dissolution of the zeolite, and no fluoride was found in the zeolite phase, thus indicating that no appreciable framework hydroxyl was formed. It is concluded that the acidity of hydrogen Y arises from hydroxoaluminium cations rather than from framework hydroxyl. The hydroxyl bands in the infrared spectra are attributed to the hydroxoaluminium cations.

INTRODUCTION

The decomposition of highly ammonium exchanged zeolite Y (NH_4Y), the reaction stoicheiometry, and the stable and unstable products that are formed as the result of various chemical and thermal treatments have been extensively studied.[1,2] Some of these products are highly active catalysts in hydrocarbon reactions. It was initially suggested that the thermal decomposition of NH_4Y results in a hydrogen Y (HY) by the loss of NH_3 at about 600 K. At higher temperatures hydrogen atoms and framework oxygen atoms are removed by dehydroxylation to form "decationized" zeolite Y. The acidic nature of calcined NH_4Y is well known and has been investigated by many techniques including infrared spectroscopy and catalytic property measurements. Despite these extensive studies, the source of the acidity has not been conclusively established. It has recently been suggested that calcination of NH_4Y in air causes the formation of hydroxoaluminium cations, leaving tetrahedral site vacancies.[1] It is the purpose of this study to confirm this by utilizing potentiometric titrations and other methods to verify the formation of hydroxoaluminium cations, and to determine the reaction stoicheiometry and suggest possible mechanisms.

Ion exchange of the hydrogen ions in HY by treatment with an NaCl solution should produce an acidic solution and NaY.

$$\underset{\text{zeolite}}{HY} + Na_{aq}^+ \rightleftharpoons \underset{\text{zeolite}}{NaY} + H_3O^+ \tag{1}$$

If acidity develops due to formation of hydroxoaluminium cations, ion exchange with NaCl will also yield an acidic solution as well as cationic aluminium.

$$\underset{\text{zeolite}}{Al(OH)_{3-x}^{z+}} + xNa_{aq}^+ = \underset{\text{zeolite}}{Al(OH)_{3-x}^{z+}} + Na^+ \tag{2}$$

$$Al(OH)_{3-x}^{z+} + xH_2O = Al(OH)_{3_{am}} + xH^+ \tag{3}$$

For $x = 1$, $\log K = -0.65$; $x = 2$, $\log K = -4.9$ at 298 K.

If the hydrogen Y, (HY) is treated with an aqueous KF solution, different results can be obtained. Hydronium ion can also be formed as the result of K^+ exchange, as in equation (1). However, if fluoride substitutes for OH in the structure, then a basic solution will result.

$$OH + F^-_{aq} = F + OH^-_{aq} \tag{4}$$
$$\text{zeolite} \qquad\qquad \text{zeolite}$$

With hydroxoaluminium cations, free OH should also result.

$$Al(OH)^{z+}_{3-z} + 3KF_{aq} = AlF_{3aq} + xK^+\text{zeolite} + (3-x)KOH_{aq} \tag{5}$$
$$\text{zeolite}$$

For $x = 2$, $\log K = 7.7$.

Reactions (2) and (3) are written in terms of mononuclear species which are present at low metal concentrations. However, polynuclear species such as $Al_2(OH)^{4+}_2$ are present at higher concentrations.[3]

Earlier potentiometric titrations were carried out on NH_4Y calcined at various temperatures with samples of the zeolites placed in NaCl solution and titrated with dilute NaOH.[4] Maximum acidity developed at 773 K. It was concluded in this work that Al cations such as $Al(OH)^{2+}$ were formed during the stabilization treatment.

We have carried out potentiometric titrations with HCl on the the fluoride-treated samples in addition to NaOH titrations of the samples treated with NaCl solution.

EXPERIMENTAL

The composition of the NH_4Y zeolite was $Na_8(NH_4)_{48}[(AlO_2)_{56}(SiO_2)_{136}] \cdot nH_2O$. Samples (1 g) in shallow dishes (30 mm × 40 mm × 4 mm) were placed in a preheated oven (volume of 3300 cm³) under a thermocouple. Three samples were calcined at each temperature; $T/K = 473, 573, 623, 673, 723, 773, 873,$ and 973. Air dried by a column of activated molecular sieve (4A) was swept over the surface of the samples at 80 cm³ s⁻¹. The samples were removed from the oven after 2 h and cooled in a desiccator.

One of the calcined samples was slurried as rapidly as possible in 50 cm³ of 3·4 mol dm⁻³ NaCl solution, stirred at room temperature for 2 h, filtered, and washed with 25 cm³ of the NaCl solution, and finally washed with 25 cm³ distilled water. The combined filtrates were titrated with 0·1 mol dm⁻³ NaOH in 1 cm³ increments to a pH of about 11. A second sample was slurried in 50 cm³ of 3·4 mol dm⁻³ KF solution, similarly equilibrated for 2 h and the slurry titrated with 0·1 mol dm⁻³ HCl in 1 cm³ increments. The third sample was used to determine the extent of NH_4^+ removal at each calcination temperature by conventional analysis.

The samples treated with NaCl were further analysed by X-ray powder diffraction and infrared spectroscopy; the chemical composition (SiO_2, Al_2O_3, M_2O) was determined by standard wet chemical methods. The filtrate was also analysed for SiO_2 and Al_2O_3 following the titration.

Thermal analyses were obtained with a Mettler Thermoanalyzer simultaneous DTA–TG microbalance. About 25 mg of loose powder was placed in the platinum crucible with a typical sample depth of *ca.* 2 mm. Preheated dry air was passed over the surface of the sample at a rate of 200 cm³ min⁻¹. The sample chamber volume was 300 cm³. The samples were heated rapidly to the desired temperature at 36 K min⁻¹, held for 2 h, and the temperature then raised to $T/K = 1273$ at 8 K min⁻¹. The temperatures employed were $T/K = 398, 473, 573, 673, 773,$ and 873.

RESULTS

The titration curves for the NaCl-treated samples are shown in Figure 1. The unit cell compositions calculated from the chemical analyses, together with the titration endpoints calculated as H⁺/unit cell in the NaCl treatment, or OH⁻/unit cell in the KF treatment, are shown in Table 1. Chemical analyses of the titrated filtrates showed that a negligible amount of silica was removed from the zeolites during NaCl treatment. Analysis of the solid and liquid showed that aluminium was removed from the zeolite. The maximum number of aluminium atoms removed was 9·5 Al/u.c. from the NH_4Y sample calcined at 673 K, the temperature at which maximum acidity was produced. Above this temperature the number of aluminium species removed decreases to near zero after calcining at 873 K. In addition, a significant amount of cation deficiency

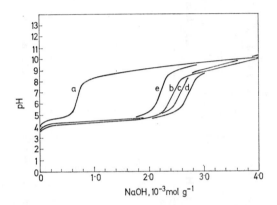

FIGURE 1 Potentiometric titration curves for calcined NH_4Y, titrated after treatment in NaCl, 3·425 mol dm^{-3} with NaOH, 0·1 mol dm^{-3}. $T/K = 473(a)$, 573 (b), 623 (c), 673 (d), 773 (e).

develops following NaCl treatment (in addition to the amount of cation deficiency accounted for by aluminium removal). This value rises gradually as the calcination temperature is increased to 773 K. An abrupt increase in cation deficiency occurs at 873 K. This is the same temperature at which the acidity of the NaCl solution and the amount of aluminium removed from the zeolite decrease abruptly to zero. However, the base formed in the KF slurry does not decrease after the 873 K calcination. Figure 2 shows a typical potentiometric titration curve for NH_4Y fired in dry air and titrated in KF solution.

The hydroxyl region of the infrared spectra of the NaCl-treated samples are shown in Figure 3. The infrared sample wafers were vacuum-activated at 473 K to remove only physically adsorbed H_2O. All samples show a very broad OH band, similar to that found in samples of aluminium-depleted NaY.[5] The treated samples of NH_4Y calcined at temperatures below 873 K show a very small, sharp 3646 cm^{-1} band. Activation of the infrared sample wafers at 773 K removes most of the broad OH band as well as a fraction of the 3646 cm^{-1} band.

Thermogravimetry (TG) data on samples of NH_4Y calcined in dry air for 2 h at $T/K = 398$, 473, 573, 673, 773 and 873, are shown in Figure 4. The TG curves indicate that the major portion of the dehydroxylation occurs above 873 K, except for the sample of NH_4Y calcined at 873 K, which shows no weight loss due to dehydroxylation. Examination of the expanded TG scale showed that the initial dehydroxylation begins between 833—848 K. Since chemical analysis indicated that, at 673 K, a shallow bed, dry air-calcined NH_4Y contains insignificant residual NH_4^+, it was assumed that the weight loss was due only to constitutive water, which is shown in the curves by the abrupt loss of weight near 873 K. It is apparent that calcining the NH_4Y at different temperatures

Table 1

Unit cell compositions, titration data, and X-ray data for calculated NH_4Y treated in NaCl, 3·425 mol dm^{-3} o titrated in KF, 3·425 mol dm^{-3}

| | Unit cell composition, NaCl treated samples[a] | | | | | | Calculated from titration endpoints | | | | X-ray |
| | | | | | | | NaCl treatment | | KF treatment | | |
T/K	Si/u.c.	Al/u.c.	M$^+$/u.c.	Removed Al/u.c.	Additional cation deficiency/u.c.	Residual NH$_4{}^+$/u.c.	H$^+$/u.c.	H$^+$/Removed Al	OH$^-$/u.c.	H$^+$/OH$^-$	I/I_s[b]
None	135·8	56·2	55·6	None	None	48·0	—	—	—	—	100
473	135·8	55·0	49·9	1·2	5·1	34·2	8·1	7·0	—	—	99
573	135·8	48·4	42·9	7·8	5·5	10·7	27·2	3·5	11·9	2·3	71
623	135·8	47·6	40·1	8·6	7·5	3·3	29·1	3·4	—	—	73
673	135·8	46·6	37·6	9·5	9·0	0·5	30·7	3·2	16·3	1·9	74
723	135·8	47·4	35·1	8·7	12·3	—	28·0	3·2	—	—	76
773	135·8	49·5	31·6	6·7	17·9	—	25·0	3·8	15·5	1·6	70
873	135·8	55·8	9·2	0·3	46·7	0·4	2·6	7·7	21·0	0·1	<10
973	135·8	56·2	6·7	0·0	49·5	—	0·0	0	—	—	<10

[a] Based on Si/u.c. = 135·8

[b] I/I_s = Crystallinity of NaCl-treated sample based on the intensities of 5 reflections relative to the untreated sample of NH$_4$Y. Oxygen adsorption isotherms show no loss of structure up to T/K = 873.

648

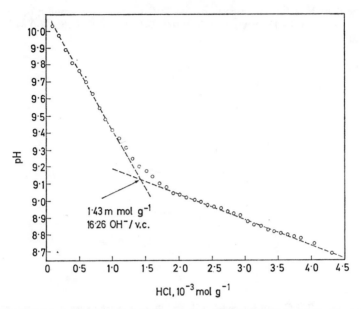

FIGURE 2 Potentiometric titration curve for calcined NH_4Y at 673 K, titrated in KF solution, $3\cdot425$ mol dm^{-3} with HCl, $0\cdot1$ mol dm^{-3}.

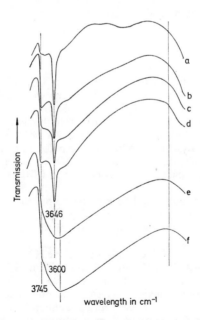

FIGURE 3 Hydroxyl infrared spectra of calcined NH_4Y following treatment with $3\cdot425$ mol dm^{-3} NaCl; wafers activated *in vacuo* at 473 K. Dry air calcination temperature $T/K = 473(a)$, 573 (b), 673 (c), 773 (d), 873 (e), 973 (f).

below the dehydroxylation temperature has little effect on the amount of constitutive water loss and that the H_2O/Al_2O_3 ratio is only 0·4—0·5, rather than 1·0 as previously suggested by others.[6-8]

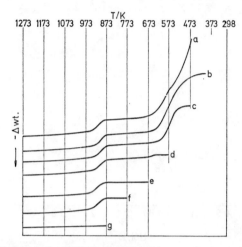

FIGURE 4 TG Curves for NH_4Y calcined in air at various temperatures (heating rate = 8 K min^{-1}). No preheat, 2·32% wt. loss = 14·7 H_2O/u.c. (a); Calcined 398 K, 2·17% wt. loss = 13·7 H_2O/u.c. (b); Calcined 473 K, 2·20% wt. loss = 13·9 H_2O/u.c. Calcined 573 K, 1·92% wt. loss = 12·1 H_2O/u.c. (d); Calcined 673 K, 2·12% wt. loss = 13·3 H_2O/u.c. (e); Calcined 773 K, 2·28% wt. loss = 14·4 H_2O/u.c. (f); Calcined 873 K, wt. loss none (g).

DISCUSSION

Hydrogen ions responsible for the acidity produced by calcined NH_4Y in aqueous NaCl solution are removed from the zeolite together with a soluble aluminium cation species. This suggests that the acidity is produced by the exchange and hydrolysis of hydroxoaluminium cations. In KF solution, the pH of the slurry increases by as much as 3 pH units. After removal of the zeolite from the KF solution, the free OH^- can be titrated. By chemical analysis the zeolite phase was found to contain no fluoride, indicating that F did not substitute for framework OH. These results eliminate the possibility that the species produced during shallow bed (SB) calcination is $Al(OH)_3$ or Al^{3+}.

Hydroxoaluminium cations would be expected to produce acidity in NaCl solution and free OH^- in KF solution as shown by Equations (2), (3), and (5). For $x = 1$, the H^+/OH^- ratio in the two titrations should be equal to 1/2. For $x = 2$, the H^+/OH^- ratio is 2.

The results in column 11 of Table 1 indicate that the H^+/OH^- ratio is 2 and that the major cation species is $AlOH^{2+}$. However, data in column 9 indicate the ratio of $H^+/Al_{removed}$ is about 3. Also, the cation deficiency increases (col. 6) with calcination temperature and is equivalent to a major fraction of the aluminium after dehydroxylation. It is concluded that during the NaCl treatment some aluminium is hydrolysed to $Al(OH)_3$ or AlOOH within the sodalite cages and, therefore, trapped. The summation of the cation deficiency (col.6) and Al removed (col. 5), compared with the acidity (H^+/u.c.) in col. 8, indicates a cation charge of +2 as previously deduced from col. 11.

Above the dehydroxylation temperature, 873 K, little acidity is produced but a considerable amount of free OH^- is formed in KF solution. This may be caused by condensation of monohydroxoaluminium cations to an oxygen-bridged species.

$$2\ AlOH^{2+} \rightarrow [Al\text{-}O\text{-}Al]^{4+} + H_2O \tag{6}$$

$$[Al\text{-}O\text{-}Al]^{4+} + 6\ KF + H_2O \rightarrow 4K^+ + 2\ \underset{\text{zeolite}}{AlF_{3aq}} + 2KOH_{aq} \tag{7}$$

The formation of bridged species in stabilized zeolite Y has also been suggested by Jacobs and Uytterhoeven.[9] The data in Table 1 can be used to calculate average unit cell compositions. As an example, at $T/K = 673$, 9·5 Al atoms were removed from the unit cell during NaCl treatment leaving 46·6 Al remaining, with 37·6 charge-balancing univalent ions and a deficiency of 9·0. The NaCl filtrate contained the equivalent of 30·7 H^+/u.c. Assuming a charge of $+2$ (for $AlOH^{2+}$), this is equivalent to 15·4 $AlOH^{2+}$/u.c. Since 9·5 Al/u.c. were removed, then 5·9 Al are presumed to be trapped in sodalite cages as $Al(OH)_3$. The remaining 3·1 Al/u.c. may be condensed to $[Al\text{-}O\text{-}Al]^{4+}$.

From the KF titration, the measured free base was 16·3 OH^- per unit cell as compared to an expected 15·4 from the 15·4 $AlOH^{2+}$/u.c. The difference is attributed to OH^- formed by equation (7). After calcination at 673 K, the sample lost 2·1 wt. % water (Figure 4), equal to 13·3 H_2O molecules/u.c. (based on the deammoniated and dehydrated zeolite). The weight loss of 12—14 H_2O/u.c. occurs at the same dehydroxylation temperature of *ca.* 823 K for all samples except those which apparently lost crystallinity (873 and 973 K). Sixteen H_2O/u.c. is equivalent to 2·46 wt %. This is much less than the quantity previously reported and corresponds to a ratio of $H_2/OAl \approx 13/48 \approx 0·3$ rather than the value of 0·5 required for the earlier model of decationization.[10]

Suggested mechanism

It has been previously concluded that hydroxoaluminium cations are formed during the "stabilization" process in NH_4Y.[11] The present results demonstrate that spontaneous framework dealumination occurs during the calcination of NH_4Y in dry air in a shallow-bed configuration. The driving force for the removal of aluminium from tetrahedral sites in the framework and formation of cations involves the reaction of the cationic hydroxoaluminium group and mobile acidic hydrogens. As dehydration of the zeolite proceeds to a point where the H_2O/NH_4^+ ratio is about unity, hydroxoaluminium cations are formed. They are probably located in site within the β cages, as indicated by changes in specific X-ray reflections which are characteristic of cation movement and positioning in β cage sites.[12] Vacant tetrahedral sites are probably occupied by H_4 groups as evidenced by the broad OH band in the infrared spectra (Figure 3). The initial cations, probably $Al(OH)_2^+$ at low temperature, react with framework protons to form $AlOH^{2+}$. Since there are eight β cages per unit cell, this mechanism will allow for the removal of 16 Al as $Al(OH)_2^+$, two in each β cage, which balance the charge for 16 framework negative sites. Further reaction with 16 framework protons results in 16 $AlOH^{2+}$ cations, which balance the charge for 32 framework negative sites. This is summarized in the following schemes:

$$Na_8(NH_4)_{48}[Al_{56}Si_{136}O_{384}] \cdot 234\ H_2O$$
$$\downarrow$$
$$Na_8(NH_4)_{32}[Al(OH)_3]_{16}[Al_{40}(H_4)_{16}Si_{136}O_{384}] + 16\ NH_3\uparrow\ + 186\ H_2O\uparrow$$
$$\downarrow$$
$$Na_8(NH_4)_{16}[Al(OH)_2{}^+]_{16}[Al_{40}(H_4)_{16}Si_{136}O_{384}] + 16\ NH_3\uparrow\ + 16\ H_2O\uparrow$$
$$\downarrow$$
$$Na_8[Al(OH)^{2+}]_{16}[Al_{40}(H_4)_{16}Si_{136}O_{384}] + 16\ NH_3\uparrow\ + 16\ H_2O\uparrow$$

By further dehydroxylation, a bridged cation, $[Al-O-Al]^{4+}$ can result and for structural reasons is probably located within the sodalite cage.

$$Na_8[Al(OH)^{2+}]_{16}[Al_{40}(H_4)_{16}Si_{136}O_{384}]$$
$$\downarrow$$
$$Na_8[Al-O-Al]_8{}^{4+}[Al_{40}(H_4)_{16}Si_{136}O_{384}] + 8\ H_2O \uparrow$$

The maximum number of $AlOH^{2+}$ ions per unit cell is sixteen—eight are probably located in site I′ with the OH in site U (centre of sodalite cage). Since the sodalite cage can accommodate only one $Al(OH)^{2+}$ species in tetrahedral co-ordination, the remaining eight must be located externally in the supercage, probably in site II with the OH in site II* (projecting away from site II into the supercage), (Figure 5a). All aluminium

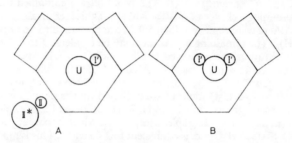

FIGURE 5   Projection of the species $Al(OH)^{2+}$ and $[Al-O-Al]^{4+}$, showing the possible positions in the zeolite framework. Small circles, Al; large circles, OH, or oxygen. (A) Projection of $Al(OH)^{2+}$ in either site I′ or site II. Al is in tetrahedral co-ordination. (B) Projection of $[Al-O-Al]^{4+}$ in the sodalite unit following final dehydroxylation. Al is in site I′ with the oxygen in site U. Al is in tetrahedral co-ordination.

atoms are in tetrahedral co-ordination, in agreement with X-ray fluorescence results.[13] Dehydroxylation at 823 K requires movement of cations from site II into the sodalite cage to form the bridged $[Al-O-Al]^{4+}$ (Figure 5b).

Since the hydroxoaluminium cations produce the acidity in deammoniated NH_4Y, they also may account for the OH bands in the infrared spectra. The 3640 cm^{-1} band in hydrogen Y can be assigned to $AlOH^{2+}$ cations in the site II positions in the supercage and the 3550 cm^{-1} band to $AlOH^{2+}$ cations in the sodalite cage. The small 3646 cm^{-1} band observed in the NaCl-treated samples can be assigned to some framework OH. The increase in the 3745 cm^{-1} band for calcined NH_4Y[14,15] is explained by the formation of new SiOH groups in the tetrahedral vacancies. The broad (3745—3000 cm^{-1}) band shown in Figure 4 is due to $(OH)_4$ in tetrahedral vacancies. Dehydroxylation of these groups at 773 K results in increased intensity of the 3745 cm^{-1} band.[5] These conclusions consequently have considerable implications for catalysis.

REFERENCES

[1] D. W. Breck, *Zeolite Molecular Sieves: Structure, Chemistry and Use*, (Wiley-Interscience, New York, 1974), pp. 474–483, 507–518.
[2] G. T. Kerr, *Adv. Chem. Ser.*, Amer. Chem. Soc., 1973, **121**, Mol. Sieves, 3rd Internat. Conf., 219.
[3] R. E. Mesmer and C. F. Baes, Jr., *Inorg. Chem.*, 1971, **10**, 2290.
[4] J. B. Peri, *Proc. 5th Internat. Congr. Catalysis*, Miami Beach, 1972, vol. 1, pp. 329–341.
[5] J. M. Bennett, D. W. Breck, and G. W. Skeels, to be published.
[6] D. A. Hickson and S. M. Csicsery, *J. Catalysis*, 1968, **10**, 27.
[7] A. P. Bolton and M. A. Lanewala, *J. Catalysis*, 1970, **18**, 154.
[8] G. T. Kerr, *J. Catalysis*, 1969, **15**, 200.

[9] P. A. Jacobs and J. B. Uytterhoeven, *JCS Faraday I*, 1973, **69**, 373.
[10] J. Cattanach, E. L. Wu, and P. B. Venuto, *J. Catalysis*, 1968, **11**, 342.
[11] G. T. Kerr, *J. Catalysis*, 1969, **15**, 200.
[12] J. V. Smith, J. M. Bennett, and E. M. Flanigen, *Nature*, 1967, **215**, 241.
[13] G. H. Kühl, *Proc. 3rd Internat. Conf. on Mol. Sieves*, (Leuven University Press, 1973), *Recent Progress Reports No.* 127, p. 227.
[14] J. W. Ward, *J. Catalysis*, 1967, **9**, 225; 1970, **19**, 348.
[15] P. A. Jacobs and J. B. Uytterhoeven, *JCS Faraday I*, 1973, **69**, 359.

DISCUSSION

L. D. Rollman (*Mobil, Princeton*) said: This comment has been prepared with the assistance of Dr. G. T. Kerr. The authors conclude that the acidity of hydrogen Y arises from hydroxoaluminium cations rather than from framework hydroxyl groups; this view is a significant departure from that published in the literature. There are two well-known facts which argue against that conclusion but which are not mentioned in the paper. First, ammonia can be desorbed from an NH_4Y on heating to about 673 K in an inert purge and *reversibly* sorbed on cooling; secondly, ion-exchange capacity of hydrogen Y zeolite is lost on simple slurrying in water, presumably by hydrolysis. I suggest that the exchangeable aluminium found in these samples results from slurrying with acqueous solution and not from calcination.

The only evidence presented to counter this suggestion is the low H_2O/Al_2O_3 ratio observed on dehydroxylation above 600 °C, and this evidence differs from that cited in references 7, 8, and 10. Kerr has measured the H_2O/Al_2O_3 ratio in an NH_4Y calcined under conditions duplicating those of Figure 4(a). The sample lost 3·89 g chemical water/100 g ash at 670 °C. That corresponds to 0·51 (mole H_2O)/(mole NH_3) or 25 H_2O/unit cell. Note that the sample contained 26 Al_2O_3/unit cell. Can the authors explain the difference between their results and those presented here and in the publications of Bolton, Kerr, Cattanach *et al.*, and of Benesi and Uytterhoeven *et al.*?

W. O. Haag (*Mobil, Princeton*) said: The authors suggest that the acidity of deammoniated NH_4Y is due to hydroxoaluminium cations formed by spontaneous framework dealuminization during calcination. While the chemistry of HY is complex, there is considerable evidence that a 'true' HY with unaltered framework exists, contrary to the authors' conclusion. We have determined the hydroxyl content of a calcined NH_4Y (450 °C, <0·1 wt% Na) by exchange with molecular deuterium as well as by gravimetry and have obtained the following values:

H by deuterium exchange	4·20 mmol g^{-1} HY
H by water loss	4·15 mmol g^{-1} HY
H calculated from NH_3 titration	4·18 mmol g^{-1} HY
H calculated from Al content	4·28 mmol g^{-1} HY

There can be little doubt that the sample contained one hydroxyl group per framework aluminium. That the framework is unaltered has been shown previously by Venuto who was able to reconstitute the original NH_4Y by treatment with dry ammonia. The freshly calcined HY zeolite (without exposure to moisture) had excellent catalytic activity for a variety of acid catalysed reactions, including hexane cracking, olefin polymerization, xylene isomerization and aromatic alkylation. It appears, therefore, that an HY zeolite with an unaltered framework and with the expected hydroxyl content can indeed be prepared and used as a catalyst.

G. W. Skeels (*Union Carbide, Tarrytown, N.Y.*) replied to Rollman and Haag thus: In reply to the first point raised by Rollman and Kerr, it is possible that ammonia can sorb on the hydroxyls of hydroxoaluminium cations, or they may sorb or react in a dehydroxylated tetrahedral vacancy, with the resulting ratio of $NH_3/Al = 1$.

(A)   (B)

In B, each edge of the tetrahedron would contain one proton resulting in hydrogen-bonded NH and OH. In recent work, when pure dry ammonia gas was admitted to a dehydroxylated aluminium-deficient NaY, a broad i.r. adsorption band appeared in the $3,750-3,000$ cm^{-1} region; this can be considered as evidence for species B. The reconstitution experiment of Venuto *et al.*[1] to which Haag refers (see next comment) can also be explained by the above model.

The second point raised by Rollman and Kerr is that the ion-exchange capacity of HY zeolite is lost on simple slurrying in water. What they do not point out is that the *X*-ray crystallinity of HY exposed to water vapour is almost completely lost, a fact already reported in the literature.[1] Our samples of NH$_4$Y calcined below 873 K and treated in sodium chloride solution are highly crystalline, maintaining 80–90% of *X*-ray crystallinity and full retention of oxygen adsorption.

With regard to our thermal analysis data I quote from Uytterhoeven, Christner, and Hall[2] '. . . OH content was a maximum at 290 °C and heating to higher temperatures resulted in secondary dehydroxylation'. The data of their Table 1 with respect to thermal analysis is plotted here in Figure 1. Following calcination at 440 °C the ratio of hydroxyls per available aluminium was $25\cdot7/21\cdot1$ (a value greater than unity). If a maximum concentration of hydroxyls was found following calcination at 290 °C, then this ratio must have been significantly greater than unity. On the other hand, if only the dehydroxylation step is considered, then the OH/Al ratio is reduced to 0·69, similar to the present result.

Using the data of Benesi[3] about 2·54 wt% H$_2$O was lost due to dehydroxylation where chemical analyses predicted a weight loss of 3·79 wt%, a ratio of 0·67. Bolton and Lanewala[4] reported that the dehydroxylation weight loss from 650 to 700 °C '. . . is insufficient to account for the calculated value', and further stated, 'that observed weight loss derived from thermogravimetric analysis only matches the calculated value if the former is taken over the temperature range of 500–900 °C'. Kerr[5] reports a ratio of water to available alumina of 0·97 when the weight loss is taken over the range 500–1000 °C. Flank[6] also comments on the lack of stoicheiometric correspondence of hydroxyl loss with the alumina content of the zeolite.

In the present work we have taken only the weight lost during the actual dehydroxylation step. There was additional weight loss in the range 773–1273 K due, we believe,

[1] J. Cattanach, E. W. Wu, and P. B. Venuto, *J. Catalysis*, 1968, **11**, 342.
[2] J. B. Uytterhoeven, L. G. Christner, and W. K. Hall, *J. Phys. Chem.*, 1965, **69**, 2117.
[3] H. A. Benesi, *J. Catalysis*, 1967, **8**, 368.
[4] A. P. Bolton and M. A. Lanewala, *J. Catalysis*, 1970, **18**, 154.
[5] G. T. Kerr, *J. Catalysis*, 1969, **15**, 200.
[6] W. H. Flank, *Analytical Calorimetry*, 1974, **3**, 649.

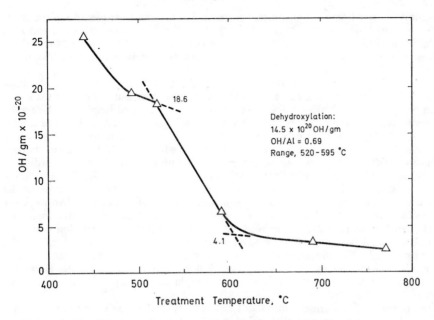

FIGURE 1 Thermogravimetric data from Table 1 in ref. 2.

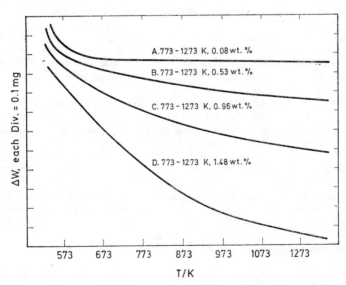

FIGURE 2 Thermogravimetric curves for NaY and aluminium-deficient NaY.
A: NaY; B: 25% Al-deficient; C: 50% Al-deficient; D: 75% Al-deficient.

to the dehydroxylation of tetrahedral vacancies. The i.r. spectra in Figure 3 of the paper clearly show $(OH)_4$ groups. Figure 2 (here) shows the TG curves for 25, 50, and 75% aluminium-deficient NaY compared to a non-deficient NaY. While most of the $(OH)_4$ sites are dehydroxylated below 573 K, there is still substantial water lost between 773 and 1273 K.

From the new results provided by Rollman and Kerr we agree that about 25 H_2O/unit cell were lost, but not at 670 °C as they suggest, but rather over the whole temperature range of 500–1000 °C. Our estimation of the actual weight lost during the dehydroxylation step is 2·92 wt% H_2O, corresponding to 18·1/H_2O unit cell or a ratio of $H_2O/Al_2O_3 = 0·69$, in excellent agreement with our data. It has been generally accepted that dehydroxylated Y zeolite is formed when the HY form loses its constitutive water leaving one out of every two aluminium atoms in a tri-co-ordinated state, and a like number of silicon atoms in a tri-co-ordinated state. The remaining aluminium and silicon atoms are tetra-co-ordinated. The X-ray fluorescence results of Kühl,[7] who studied the AlK_B line shift and the SiK_B line shift in dehydroxylated Y zeolite, gave no indication of tri-co-ordinated silicon or of tri-co-ordinated aluminium.

D. Barthomeuf (*ESCIL, Villeurbanne*) said: I agree with your conclusion that close to about 15 Al per unit cell are easily removed from a Y zeolite. This confirms the results described in the literature for ultrastable zeolites obtained by heating, and for Al-deficient zeolites obtained by chemical extraction. According to Kerr[8] the aluminium atoms that are easily removable should be the same whatever the treatment.

As to the acidity, I am not sure that I have clearly understood your conclusion that the acidity of hydrogen Y zeolite is due to the 16 hydroxoaluminium cations. Several spectra have been published in the literature which show the 3640 cm^{-1} and 3550 cm^{-1} i.r. bands in zeolites from which the 16 aluminium atoms have been extracted. Our own results[9] show (see Figure 3) that the extraction of the early removable Al atoms reduces neither the strong acidity nor the catalytic activity.

G. W. Skeels (*Union Carbide, Tarrytown, N.Y.*) replied: The zeolite to which Dr. Barthomeuf refers was an ammonium-exchanged aluminium-depleted Y zeolite. We would expect that calcination of this sample would produce additional hydroxoaluminium cations, thus maintaining an acidic character. The number and positions of these cations within the zeolite structure would of course be determined by both framework aluminium content and the degree of ammonium exchange. It is reasonable then, that strong acidity and catalytic activity are maintained even when 30–40% of the framework aluminium atoms have been removed.

H. Bremer (*Tech. Hochschule, Merseburg*) said: What has been shown by the authors of this paper is probably right in the case of wet zeolites HY or with zeolites HY (or what we call so) *suspended in aqueous solutions*. It has not been demonstrated that the transport of Al-species from the anionic lattice to cationic sites occurs to any great extent during deammoniation. Is it not possible to explain the results by an attack of *hydrated* protons on the lattice by which Al-species are extracted? Furthermore I do not agree with the assignment of the i.r. band at 3640 cm^{-1} to the hydroxyls of hydroxoaluminium-ions. On the adsorption of pyridine the intensity of this band is strongly decreased. Simultaneously the band of the pyridinium ion at 1550 cm^{-1} occurs. This shows that the corresponding OH-groups must have acidic properties. With OH-groups of $Al(OH)^{2+}$ cations this is surely not the case.

J. B. Uytterhoeven (*Catholic Univ. of Louvain*) said: Stated in extreme terms, this

[7] G. H. Kühl, *Proc. 3rd Internat. Conf. on Molecular Sieves* (Louven University Press, 1973), Recent Progress Reports No. 127, p. 227.
[8] G. T. Kerr, *Adv. Chem. Ser.*, 1973, **121**, 219.
[9] R. Beaumont and D. Barthomeuf, *Compt. rend.*, 1971, **272**, C, 363.

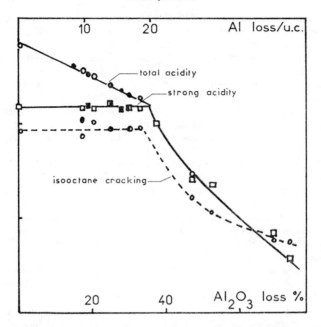

FIGURE 3 Changes in the catalytic activity and acidity (arbitrary units) with the framework Al content of Y zeolites. (Reproduced by permission from *Compt. rend.* 1971, **272** *C*, 363.)

paper leads to the conclusion that a real HY zeolite does not exist. In my view the experiments reported here are not adequate to prove this. HY has been described as a very delicate material. Immersion of an HY zeolite into aqueous solution is certainly provoking reactions (such as extraction of Al) but can one draw conclusions from this concerning the structure of the HY zeolite?

Strong arguments for the existence of HY zeolite formed by stoicheiometric deammoniation has been given by us on the basis of quantitative i.r. spectroscopy.[10] The reversibility of deammoniation is another strong proof for the existence of HY zeolite.

If aluminium–hydroxy cations were formed during deammoniation, I would expect them to go inside the cubo-octahedra and be unexchangeable with sodium ions. In my opinion solubilization of aluminium suggests a serious attack during contact with the solution rather than during deammoniation.

G. W. Skeels (*Union Carbide, Tarrytown, N.Y.*) in reply to the two previous comments said: There is a great deal of concern that hydroxoaluminium cations are formed when we put calcined NH_4Y into the salt solution, and are not formed prior to this treatment. We must consider several additional important observations. If true HY was formed then, at some calcination temperature, about 48 zeolite H^+/unit cell should have been obtained from the sodium chloride titration. The maximum experimental value obtained was 30 H^+/unit cell substantially below the expected value. With potassium fluoride solution, two alternatives exist for a true HY zeolite. In the first case, treatment with potassium fluoride solution would produce acid as was observed

[10] J. B. Uytterhoeven, P. Jacobs, K. MaKay and R. Schoonheydt, *J. Phys. Chem.*, 1968, **72**, 1768.

with the sodium chloride treatment. This did not occur. In the second case the fluoride ion would substitute for lattice hydroxyl, free base would be produced and fluoride would be found in the solid zeolite product. While free base was produced in our experiments, fluoride was not found in the zeolite. These observations, including the observation that there is insufficient weight loss during dehydroxylation (TG), are strong indications that HY zeolite as described by Kerr and others does not exist. Our suggested mechanism may not be completely correct but we feel that it is consistent with the present results and also with the data in the literature. Unfortunately, space limitation has prevented a more detailed discussion in the present paper.

Divalent and trivalent cation exchanged forms of Y zeolite do exhibit distinct i.r. OH absorption frequencies. With pyridine there is an interaction of co-ordinately bound pyridine with the respective cation as well as an absorption band at 1545 cm^{-1} due to Brönsted acidity. We see no reason why $Al(OH)_2^+$ or $Al(OH)^{2+}$ should not interact in a similar manner with pyridine, giving rise to similar infrared absorption bands.[11]

J. Turkevich (*Princeton Univ.*) said: Stoicheiometric relationships established between the amount of ammonia removed and the amount of water removed by thermal treatment of NH_4Y zeolite permit a simple interpretation of Brönsted acid sites being associated with the zeolite lattice. Recent work by Mastiklun and Turkevich on the proton resonance of water in NaY zeolite and HY zeolite obtained by thermal treatment of NH_4Y shows a chemical shift in the HY which can be attributed to the presence of H_3O^+. This again supports the idea of protons being associated with the zeolite lattice.

The nuclear resonances of aluminium in hydrated zeolite and that of sodium are sharp. There is evidence for symmetric octahedral ligand positions around each nucleus, since an unsymmetric environment would lead to a broad line because of the existence of a quadruple moment in the Al and Na nucleus. The presence and the magnitude ($> 50\%$ of all Al) of the sharp line of the aluminium resonance in hydrated zeolite must indicate that the aluminium leaves the zeolite carcass and goes into the cavity in a dynamic fashion. Thus aluminium is mobile even in ordinary hydrated zeolite.

Could the loss of aluminium be associated with loss of crystallinity: for example (Table 1), 8·6 Al removed from 55·6 in unit cell with 27% loss in crystallinity by X-ray?

J. C. McAteer (*ICI Corporate Lab., Runcorn*) (communicated): HY Zeolite and amorphous silica–alumina react with methylsilanes to form products which have no protonic acidity but still possess Lewis acidity.[12] How do you explain the similarity in acidity characteristics of crystalline and amorphous aluminosilicates in terms of your conclusion that the acidity of HY zeolite is due to the presence of hydroxoaluminium ions?

G. W. Skeels (*Union Carbide, Tarrytown, N.Y.*) replied: The similarity in acidity characteristics of crystalline zeolites and amorphous aluminosilicates may be explained by Peri's report[13] that species such as:

[11] J. W. Ward, *J. Colloid Interface Sci.*, 1968, **28**, 269.
[12] J. J. Rooney and J. C. McAteer, *Int. Congr. Mol. Sieves, Zurich*, 1973.
[13] J. B. Peri, *J. Catalysis*, 1976, **41**, 227.

are present on the surface of silica–alumina. He suggests that, at high calcination temperatures these species may condense to

Y. Ben Taarit (*CNRS, Villeurbanne*) said: As you may know, Pd^{II}- and Pt^{II}-exchanged Y zeolites can be reduced in hydrogen at 25 °C (following proper pretreatment) to Pd^0 HY zeolites and Pt^0 HY zeolites. Apart from effects attributable to the presence of the metal, the resulting material behaves rigorously as an HY zeolite originating from an NH_4Y form. Do you expect that under these mild conditions, aluminium ions are likely to be extracted to form the hydroxoaluminium species that you have put forward in order to account for the acidic properties of deammoniated Y zeolite?

G. W. Skeels (*Union Carbide, Tarrytown, N.Y.*) replied: I do not know whether aluminium cations are produced during this treatment since we have not performed experiments of this type.

J. C. Vedrine (*CNRS, Villeurbanne*) (communicated): The authors propose a mechanism involving species such as $Al(OH)^{2+}$, $(Al\text{—}O\text{—}Al)^{4+}$. In a previous work[14] we have shown that the latter species can be characterized unambiguously by e.p.r. after gamma-irradiation, since an unpaired electron is equally distributed on both Al nuclei giving eleven hyperfine lines. Introduction of oxygen at 77 K also showed that the species involved was located in the supercages. I would suggest the authors try such a method which is unambiguous if they succeed.

[14] J. C. Vedrine, A. Abou Kais, J. Massardier, and G. Dalmai-Imelik, *J. Catalysis*, 1973, **29**, 120.

Synthesis and Catalytic Properties of a 1,4-Diaza-bicyclo[2,2,2]octane–Montmorillonite System—a Novel Type of Molecular Sieve

Joseph Shabtai,* Norbert Frydman, and Rachel Lazar

Department of Organic Chemistry, The Weizmann Institute of Science, Rehovot, Israel

ABSTRACT   A new type of molecular sieve has been synthesized by subjecting montmorillonite to ion-exchange with a rigid dication derived from the cage-like diamine 1,4-diazabicyclo-[2,2,2]octane (DABCO). The resulting cross-linked DABCO–montmorillonite derivative (DABCO–M) shows permanent porosity, good thermal stability, and resistance to swelling in organic solvents. The basal spacing of DABCO–M is 14·8 Å, corresponding to an interlayer spacing of 5·3 Å, while the calculated lateral free distance between the immobilized cationic species is *ca.* 6 Å. DABCO–M shows markedly higher catalytic activity for esterification of carboxylic acids compared to ordinary alkylammonium- or α,ω-alkylenediammonium-exchanged montmorillonites. Comparison of relative rates of esterification of acetic acid with isomeric pentanols and hexanols, in the presence of DABCO–M, H-montmorillonite (H–M), 1-dodecylammonium montmorillonite (DA–M), and 1,12-dodecylenediammonium mont-morillonite (DDA–M) as catalysts, shows that DABCO–M has pronounced molecular sieve properties, as expressed in a sharp drop in esterification rate with increase in the critical dimension of the alcoholic reactant. No such selectivity is observed with the other montmorillonite systems.

Introduction

It has been reported[1,2] previously that highly porous forms of montmorillonite, hectorite, and synthetic fluorhectorites can be prepared by ion-exchange with small globular cations, *e.g.* tetramethylammonium and ethylenediammonium ions. Smectite derivatives of this type, however, are not suitable as fixed solid frameworks for catalyst systems since (a) the interlayer slit-shaped channels are too narrow, *e.g.* 2·8 Å in ethylene-diammonium hectorite, to allow for reasonably fast penetration and diffusion of organic substrates, and (b) the structures are not resistant to swelling, *viz.* slow uptake of guest molecules with a critical cross-sectional dimension of > 3 Å is accompanied by gradual distortion or destruction of the fixed interlayer network.

The properties of α,ω-alkylenediammonium-exchanged montmorillonites, hectorites, and beidelites have been also examined.[3,4] It was found that in sorbate-free derivatives of this type the α,ω-alkylenediammonium ions lay flatwise, *i.e.* parallel to the layers, since the charge density of the smectities is relatively low, while, on the other hand, there is no steric element in the conformation of the α,ω-dications which would force them to acquire a vertical orientation relative to the layer surface. Such orientation is attained only in layer silicates of very high charge density, *e.g.* batavite, although in this case the lateral distance between the vertically stretched α,ω-alkylenediammonium ions is too small to allow for use of the compound as a fixed porous medium for organic reactions.

A novel approach for construction of swelling-resistant molecular sieve systems, outlined in the present paper, involves cross-linking of layer silicates with di- or poly-cations derived from rigid, preferably cage-like amines, which acquire a single stable orientation in the interlayer space due to steric requirements imposed by their particular configuration. A uniform "pillar" network is obtained by two-point or multi-point attachment of such cations to opposite layer surfaces. The interlayer spacing in such structures is determined by the molecular dimensions of the cross-linking cation, while the lateral ("interpillar") distance can be regulated by the use of smectites possessing different charge densities. Ammonium ions derived from di- or higher amines, *e.g.* 1,4-diazabicyclo[2,2,2]octane, 1,4- or 2,6-diaminoadamantanes, tetrakis-(*p*-amino-phenyl)methane, and some 2,2′,6,6′-tetrasubstituted benzidines which are sterically hindered from a flatwise orientation, are found to serve as suitable cationic species for preparation of cross-linked smectities. The first synthesis of such a system was achieved by cross-linking of montmorillonite unit layers with the diammonium ion of 1,4-diazabicyclo[2,2,2]octane.* The present paper deals with the structural characteristics, as well as the catalytic and molecular sieve properties of the resulting DABCO-montmorillonite compound.

EXPERIMENTAL

Preparation of diazabicyclo[2,2,2]octane–montmorillonite (DABCO–M).

Commercially available montmorillonite (Volclay Bentonite SPV, American Colloid Co.) was subjected to fractionation by gravity sedimentation, and the particle size fraction of $< 2 \mu$m was converted into the homoionic sodium form by treatment with sodium chloride, followed by repeated washing with deionized water and centrifuging.[5] The chloride-free sodium montmorillonite sample obtained (C.E.C. = 88 ± 2 meq/ 100 g clay) was dispersed in distilled water to give a starting colloidal solution (A). Diazabicyclo[2,2,2]octane dihydrochloride (m.p. 310°C; lit.[6] 310°C) was prepared by passing dry gaseous hydrogen chloride through an ethanolic solution of the free diamine (supplied by Schuchardt München).

The cross-linking step was carried out by adding 1·67 g (18·0 meq) of DABCO · 2HCl to 600 ml of the colloidal dispersion A (containing 18·0 meq of sodium montmorillonite) and stirring the mixture for 4 h at room temperature. The precipitate formed was first repeatedly washed with distilled water and centrifuged until chloride-free, then freeze-dried and, finally, dried at 0·5 torr at 60°C for 12 h to give a nearly white flaky sample.

Analysis. Calc. for a 1:1 diazabicyclo[2,2,2]octane–montmorillonite complex: C, 3·55; N, 1·40. Found: C, 3·60; N, 1·43.

Oriented films of DABCO–M were examined by *X*-ray diffractometry and the d(001) spacing found was 14·8 Å. Allowing 9·5 Å for the thickness of a unit layer, this value corresponds to an interlayer distance of 5·3 Å. The sharpness of the *X*-ray diffraction pattern was consistent with a highly uniform basal spacing. No change in the latter was observed in measurements with DABCO–M films saturated with water, or with organic solvents, *e.g.* acetonitrile, toluene, *p*-xylene, C_2—C_{12} alcohols, and heptan-2-one.

Oriented translucent films of DABCO–M, prepared by evaporation of 1—3% aqueous suspensions of the compound on the surface of a thin sheet of polyethylene, were used for measurement of the infrared spectrum, which showed the following absorption maxima (s-strong; m-medium, w-weak): 340 (m), 440—470 (s), 505—520 (s), 775 (w), 794 (m), 840 (m), 880 (m), 915 (m), 1000—1060 (s), 1318 (w), 1355 (w), 1400 (w), 1464 (m) 1623 (m), 1700 (w), 1840 (w), 2640 (m), 2830 (w), 2900 (w), 2950 (w), 3010 (m), 3210 (w), 3450 (m), 3630 (s) cm^{-1}. The i.r. band at 3010 cm^{-1} is assigned to the C–H stretching, while the band at 1464 cm^{-1} to the C–H deformation of the CH$_2$ groups in the ring system. The band at 2640 cm^{-1} can be tentatively assigned to the $^+$N–H stretching.

D.T.A. measurements, performed on undiluted samples of DABCO–M showed that the compound is stable up to 345°C, a temperature at which fast decomposition was observed.

---

* This commercially available amine, known also as triethylenediamine, is denoted as DABCO in the following text.

In a separate preparation the cross-linking step was performed by repeated treatment of the montmorillonite dispersion A (*vide supra*) with a large excess of DABCO dihydrochloride (40 meq of salt/meq of sodium montmorillonite), applying otherwise exactly the procedure given above. The thoroughly washed product was identical with the 1:1 DABCO–M complex, obtained by using equivalent amounts of the two components.

Preparation of reference montmorillonite derivatives

Homoionic H–montmorillonite (H–M)[7] was prepared by the same procedure as described for homoionic sodium montmorillonite, except that 0·1 N-aqueous hydrochloric acid was used in the ion-exchange step. The work-up time was shortened in order to avoid changes in the crystalline structure of the smectite.

1-Dodecylammonium montmorillonite (DA–M) and 1,12-dodecylenediammonium montmorillonite (DDA–M) were prepared by subjecting sodium montmorillonite (colloidal dispersion A) to ion-exchange with the corresponding amine hydrochlorides, using the same procedure as described for the preparation of DABCO–M. The $d(001)$ spacings of DA–M and DDA–M, in the sorbate-free state, were 17·6 Å and 13·3 Å, respectively.

Catalytic reactions

Esterification experiments were performed in a flow system consisting of a veritcally mounted 50 cm × 10 mm (i.d.) Pyrex glass reactor, a calibrated Sage syringe pump, and a series of coolers for quantitative collection of products. Heating was provided by means of a cylindrical glass furnace, equipped with an Eurotherm temperature controller and permitting easy establishment of an isothermal ($\pm$ 2°) zone.

A fresh catalyst bed, consisting of 0·2 g of montmorillonite derivative in admixture with 1·0 g of an inert catalyst carrier (Carborundum Co., type AMC), was used in each experiment. The catalyst was preactivated *in situ* at 200°C for 2 h, and then cooled down to the selected reaction temperature (190°C). The liquid feed, consisting of a carboxylic acid–alcohol mixture (1:2 molar ratio) was introduced in the reactor at a constant rate (0·13 ml/min) and the products were collected and analysed by gas chromatography on a phenylmethylsilicone (OV–17) column. Pure grade (> 99%) C_2—C_6 alcohols and C_2—C_5 carboxylic acids were used as starting materials, and products were identified by comparison with purified samples (> 98%) of the corresponding esters.

RESULTS AND DISCUSSION

Structure of the 1:1 DABCO–Montmorillonite compound

The product obtained by ion-exchange of sodium montmorillonite with the diammonium ion derived from diazabicyclo[2,2,2]octane possesses a uniform basal spacing of 14·8 Å. This value corresponds to an interlayer distance of *ca.* 5·3 Å, which is close to the molecular dimension of the cage-like DABCO diammonium ion in a vertical orientation relative to the montmorillonite unit layers. Due to the rigid configuration of the dication, any alternative orientation in the sorbate-free state, *e.g.* parallel to the unit layers, would be energetically unfavourable since it would require separation of the positively charged ammonium nitrogens from anchoring sites on the negatively charged internal surface. In the fully exchanged 1:1 compound the distribution of the cross-linking dications should be uniform (Fig. 1). Measurements by means of Cahn-Riley-Robertson orbit molecular models show that the dimension of the dication along a straight line connecting the two nitrogens, and including the two ammonium hydrogens, is 6·05 Å. This value is somewhat higher than the observed interlayer distance of 5·3 Å, indicating some penetration, *i.e.* embedding of the $^+$N–H group's vertically oriented

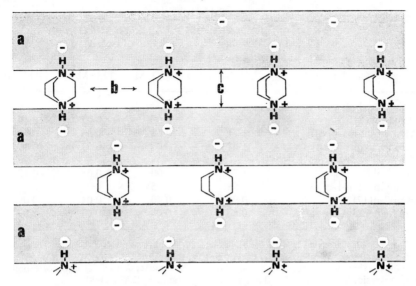

FIGURE 1   Schematic view of a cross-linked DABCO–M system: (a) montmorillonite unit layer; (b) lateral (interpillar) distance; (c) interlayer distance.

hydrogen inside oxygen 6-rings of the internal surface. Such embedding of ammonium groups in the interlamellar surface has been observed previously in structural studies of conventional alkylammonium-exchanged montmorillonites.[4] In the present case, simultaneous embedding of the two ammonium hydrogens into oxygen 6-rings of opposite internal surfaces apparently results in complete immobilization, *viz.* "locking-in" of the rigid DABCO dication, with consequent stabilization of the entire framework. From the equivalent area of the montmorillonite sample (*ca.* 90 $Å^2$/electronic charge) and the effective width of the cyclic system in the vertically oriented DABCO diammonium ion (*ca.* 4·0 Å), it is calculated that the free lateral distance between the ring edges of two adjacent cross-linking cations is *ca.* 6 Å. On this basis the channel system of the 1:1 DABCO–M compound is characterized by an essentially uniform pore size of approximately 5·3 × 6·0 Å. It should be noted, however, that lateral distances larger than 6 Å can be obtained by using lithium-modified montmorillonites, possessing lower charge densities.[8]

Catalytic activity and molecular shape selectivity of DABCO–M

The catalytic activity and molecular sieve properties of DABCO–M were examined using esterification of alkanoic acids as a model reaction. Comparative experiments with DABCO–M, H–montmorillonite (H–M), 1-dodecylammonium montmorillonite (DA–M), and 1,12-dodecylenediammonium montmorillonite (DDA–M) as catalysts were performed under selected conditions (see Experimental) using C_2—C_5 normal and branched carboxylic acids, and C_2—C_6 normal and branched alcohols, as reactants. Results obtained by esterification of acetic acid (1) with isomeric pentanols (2—4) and hexanols (5—8) are summarized in Table 1. Conversion levels were kept below 50% and relative conversions obtained under identical experimental conditions were taken as an approximate measure of relative esterification rate (V_{rel}), permitting evaluation of differences in catalyst activity.

663

Table 1

Relative Rates of Esterification $(V_{rel})^{a,b}$ of Acetic Acid (1) with
Isomeric C_5 and C_6 Alcohols^{c,d} as a Function of Catalyst Type

Catalyst	H–M	DABCO–M	DA–M	DDA–M
Starting alcohol:				
n-Pentanol (2)	1·80	1·00	0·07	0·19
3-Methylbutan-1-ol (3)	1·71	0·68	0·08	0·19
2,2-Dimethylpropan-1-ol (4)	1·67	<0·02	0·07	0·17
n-Hexanol (5)	1·72	1·00	0·05	0·12
2-Methylpentan-1-ol (6)	1·58	0·65	0·05	0·11
Hexan-2-ol (7)	0·96	0·40	<0·04	0·09
2-Methylpentan-2-ol (8)	0·85	0·08	<0·04	0·07

[a] For esterification of (1) with C_5 alcohols: V_{rel} = conversion of (1) into corresponding ester/conversion of (1) into n-pentyl acetate in the presence of DABCO–M.

[b] For esterification of (1) with C_6 alcohols: V_{rel} = conversion of (1) into corresponding ester/conversion of (1) into n-hexyl acetate in the presence of DABCO–M.

[c] Molar ratio of alcohol/(1) in liquid feed = 2:1.

[d] LHSV = $6·0\,h^{-1}$; reaction temperature: 190°.

As seen from Table 1, the H–M catalyst shows higher esterification activity compared to DABCO–M, but it lacks molecular shape selectivity, *i.e.* it is essentially insensitive to the extent of branching in the alcoholic reactant, as illustrated by the negligible differences ($< 8\%$) in V_{rel} values observed with the isomeric pentanols (2—4). Similar minor differences in V_{rel} values are noticed in comparing the two primary C_6 alcohols (5) and (6), or the pair of less reactive secondary C_6 alcohols (7) and (8). In contrast, the DABCO–M catalyst shows pronounced molecular-shape selectivity as expressed in a sharp drop in V_{rel} values with increase in branching, *viz.* in critical cross-sectional dimension, of the alcoholic reactant, For example, a drastic ($> 5,000\%$) drop in V_{rel} is observed on going from pentan-1-ol (2) with a kinetic diameter = 4·4 Å, to neopentyl alcohol (4), with a kinetic diameter = 6·2 Å. Further, a considerable drop in V_{rel} is found in passing from the straight-chain primary hexanol (5) to the branched primary alcohol (6), while an even more pronounced drop (400%) in V_{rel} is observed in passing from the straight-chain secondary hexanol (7) to the branched isomer (8).

As seen from the data in Table 1 the DA–M and DDA–M reference systems show in general much lower esterification activity compared to the H–M and DABCO–M catalysts. Furthermore, these systems do not show any significant molecular shape selectivity toward C_5 or C_6 isomeric alcoholic reactants under the experimental conditions.

The molecular sieving effect of the DABCO–M system is also operating when the extent of branching is changed, even to a minimal extent, in the carboxylic acid reactant, as found for instance in comparative esterifications of isomeric C_4 and C_5 carboxylic acids with n-butanol (Table 2). As seen, there is only a minor drop (*ca.* 10%) in V_{rel} in going from n-butyric acid (kinetic diameter = 4·4 Å) to isobutyric acid (kinetic diameter = 5·0 Å) in the H–M catalysed reaction, while in the presence of DABCO–M as catalyst the corresponding drop in the V_{rel} value is $> 100\%$. A similar difference is found in comparing the esterification of n-valeric acid (11) and isovaleric acid (12) in the presence of the two catalysts.

Table 2

Relative Rates of Esterification (V_{rel})[a] of C_4 and C_5 Carboxylic
Acids with n-Butanol, as a Function of Catalyst Type[b,c]

Catalyst	H–M	DABCO–M
Starting acid:		
$CH_3(CH_2)_2CO_2H$ (9)	1·65	1·00
$(CH_3)_2CH·CO_2H$ (10)	1·49	0·48
$CH_3(CH_2)_3CO_2H$ (11)	2·01	1·24
$(CH_3)_2CH·CH_2CO_2H$ (12)	1·95	0·67

[a] V_{rel} = conversion of carboxylic acid into corresponding ester/conversion of (9) into n-butyl butanoate in the presence of DABCO–M.
[b] Molar ratio of n-butanol/acid in liquid feed = 2:1.
[c] LHSV = $6·0\,h^{-1}$; reaction temperature: 190°.

The results obtained can be rationalized in terms of differences in the structure and mode of catalytic action of the montmorillonite systems examined. The very high esterification sctivity of H–M is consistent with the presence of protonic acidity in this catalyst.[9] Since H–M undergoes easy swelling by sorption of polar organic substrates (*e.g.* sorption of heptan-2-one results in a $d(001)$ spacing of 18·0 Å), it is probable that under reaction conditions there is sterically unrestricted diffusion of reactant molecules in the interlayer space of the catalyst, wherein interlamellar protons are available. An alternative possibility is that esterification takes place on the outside surface of H–M particles, since acidity is available at the particle edges or could be supplied by diffusion of interlamellar protons.[9] In either case, the catalyst system does not impose any significant steric constraint upon reactant or product molecules and, therefore, should not be expected to show molecular shape selectivity.

The very low catalytic activity of DA–M is probably due not only to the absence of protonic acidity, but also to effective shielding of Lewis acid sites on the internal surface by the C_{12}-alkyl chains of the intercalated ions. In the swollen state of the system, caused by sorption of reactant molecules in the interlamellar space, these chains have no fixed conformation and could effectively interfere with the diffusion of reactant molecules, as well as compete with the latter for adsorption sites. This assumption is in line with the finding that montmorillonites exchanged with very bulky alkylammonium ions, *e.g.* dimethyldi-n-octadecylammonium montmorillonite (Bentone 34), lack any catalytic activity.[10] The slightly higher activity of DDA–M as compared to DA–M may be due to vertical orientation of a part of the 1,12-dodecylenediammonium ions during the swelling process,[3,4] resulting in some increase of available internal surface.

The considerably higher catalytic activity of the swelling-resistant DABCO–M compound as compared to DA–M and DDA–M can be ascribed to the fact that in the former system there is essentially complete availability of the internal surface, except for the surface fraction occupied by the immobilized dications. On the other hand, the pronounced molecular shape selectivity of DABCO–M is clearly related to its fixed, zeolite-like channel system.

It should be noted that synthesis of cross-linked analogues of DABCO–M could provide a series of molecular sieve frameworks with a wide range of pore sizes and channel geometries. Furthermore, introduction of specific functionality in such systems, by attachment of catalytically active components to the cross-linking agent, could produce catalysts with less limitations for organic synthesis in comparison with zeolites.[11]

REFERENCES

1. R, M. Barrer, in *Molecular Sieves*, Advances in Chemistry Series No. 121 (Amer. Chem. Soc., Washington, D.C., 1973), Ch. 1, p. 1.
2. R. M. Barrer and D. L. Jones, *J. Chem. Soc.(A)*, 1971, 2594.
3. A. Weiss, *Chem. Ber.*, 1958, **91**, 487; A. Weiss, *Angew. Chem. Internat. Edit.*, 1963, **2**, 134.
4. R. M. Barrer and A. D. Millington, *J. Colloid Interface Sci.*, 1967, **25**, 359.
5. S. Yariv, L. Heller, Z. Sofer, and W. Bodenheimer, *Israel J. Chem.*, 1968, **6**, 741; A. M. Posner and J. P. Quirk, *Proc. Roy. Soc.*, 1964, **A278**, 35.
6. *Dictionary of Organic Compounds* (Oxford University Press, New York, 1965), Vol. 2, p. 886.
7. D. J. Greenland, R. H. Laby, and J. P. Quirk, *Trans. Faraday Soc.*, 1962, **58**, 829; N. T. Coleman and D. Craig, *Soil Sci.*, 1961, **91**, 14.
8. J. Shabtai, U. Shani, and R. Lazar, unpublished results.
9. A. Banin and S. Ravikovitch, *Clays Clay Miner.*, 1966, **14**, 193.
10. L. H. Klemm, J. Shabtai, and F. H. W. Lee, *J. Chromatog.*, 1970, **51**, 433; J. Shabtai, L. H. Klemm, and K. C. Bodily, *Israel J. Chem.*, 1970, **8**, 74p.
11. J. Shabtai, R. Lazar and G. M. J. Schmidt, *Proc. 3rd Internat. Conf. Molecular Sieves*, Zürich (Leuven University Press, Leuven, 1973), p. 383.

DISCUSSION

J. B. Uytterhoeven (*Catholic Univ. of Louvain*) said: The dimensions of the inter-lamellar pores given in the paper must be taken as average values. Clay is indeed heterogeneous with respect to charge distribution. The *d*-spacings are not affected by this, but the 'interpillar distances' are variable between broad limits. Is there any proof that reaction takes place between the sheets, and do the H-montmorillonite and the DABCO-montmorillonite catalysts swell in the reaction media?

J. Shabtai (*Univ. Munich*) replied: Since the charge density in montmorillonite changes to some extent with particle size, it is obvious that preparation of a perfectly uniform DABCO-montmorillonite system would require the use of a very narrow particle size fraction of the starting smectite. However, for most applications of DABCO-montmorillonite as a molecular sieve, complete uniformity in the *b*-dimension may not be critical.

The pronounced molecular shape selectivity shown by DABCO-montmorillonite in the esterification experiments proves that reaction takes place in the interlayer channel system. Reaction on the outside surface could not lead to such selectivity.

As indicated in the paper, H-M swells by sorption of heptan-2-one. Therefore, it is probable, although not certain, that it swells during reaction. DABCO-Montmorillonite, on the other hand, does not swell, or undergo very small changes in basal spacing by sorption of organic solvents including alcohols. Also, the observed molecular sieve effect of this system indicates that no significant swelling occurs during reaction at 190 °C.

H. Heinemann (*Mobil, Princeton*) (communicated): How can you distinguish between shape-selective catalysis in your catalytic experiments and pore blockage? Could this not be checked by experiments with normal compounds at long reaction times in comparison with H-montmorillonite?

J. Shabtai (*Univ. Munich*) replied: In the determination of relative esterification rates conversion levels were kept below 50%. However, some experiments at prolonged reaction time indicate that esterification of acetic acid with the C_5-alcohols (2) and (3) can be brought essentially to completion, whereas no reaction occurs with neopentyl alcohol (4), which is apparently excluded from the interlayer space. Since the critical cross-sectional dimensions of the ester products are similar to those of the reactants no blockage as a result of esterification can be expected. However, in the case of substrates having kinetic diameters very close to the indicated pore size of DABCO-montmorillonite gradual blockage is not excluded, and your suggestion to examine this possibility in more detail is appreciated.

Investigation into the Nature of the Interaction between the Metal Component and Support in Metal–Zeolite Catalysts

G. D. Chukin,* M. V. Landau, V. Ya. Kruglikov, D. A. Agievskii, B. V. Smirnov, A. L. Belozerov, V. D. Asrieva, N. V. Goncharova, E. D. Radchenko, O. D. Konovalchikov, and A. V. Agafonov

All-Union Research Institute of Oil Refining—VNII NP, Aviamotornaya 6, E-250 Moscow, U.S.S.R. and Elektrogorsk Affiliated Branch of the All-Union Research Institute of Oil Refining, Elektrogorsk, U.S.S.R.

ABSTRACT   The chemical properties of Pd–zeolite catalysts based upon an untreated and a dealuminized zeolite Y have been studied by i.r.-spectroscopic and X-ray diffraction methods. The interaction of palladium with strongly acidic sites of the dealuminized zeolite has been observed. The metal does not interact with the weakly acidic protonic sites of the untreated zeolite. It has been shown that this interaction occurs by a donor–acceptor mechanism causing electron deficiency on the metal atoms. The increased sulphur resistance of the catalyst derived from a dealuminized zeolite in the hydrogenation of aromatics is attributed to the electron deficiency of palladium.

Introduction

In the paper by Rabo and co-workers[1] an increased resistance towards sulphur has been demonstrated for the Pt–CaY catalyst prepared by introduction of the metal into the zeolite crystalline lattice by ion-exchange followed by cation reduction. The authors assumed the increased sulphur resistance of the catalyst to be attributable to an atomic dispersion of the metal in the zeolite. However, according to refs 2—5 the metal forms aggregates in the bulk and on the surface of the zeolite crystals. Thus, the question as to why the metal–zeolite catalysts exhibit increased sulphur resistance in hydrogenation processes still remains open.[5]

Della Betta and Boudart[5] have suggested that the increased sulphur resistance of metal–zeolite catalysts is associated with a deficiency of electrons in the metal due to transfer of electrons from the metal to the support. The electron-deficient nature of metals in acidic zeolites has been established.[6]

It has been shown[7] that the introduction of platinum into the CaY zeolite brings about a considerable reduction in the surface acidity. At the same time it has been found that removal of alumina from the Y-zeolite along with exchange of Na^+ cations for poly-valent cations causes a sharp increase in sulphur resistance of Pd–zeolite catalysts for the hydrogenation of aromatic hydrocarbons.[8] It is known that in the hydrogen form of the dealuminized Y-zeolite the number of strongly acidic sites is higher as compared with the hydrogen form of the untreated zeolite.[9-11] It may be suggested that there is a correlation between the acidity of the zeolite and the sulphur resistance of the metal–zeolite catalyst predetermined by the interaction of the metal and the support. However, it is not clear so far with what element of the zeolite crystalline lattice the metal atoms are interacting. We have therefore carried out a study of the chemical properties of Pd–zeolite catalysts.

EXPERIMENTAL

Catalyst preparation. Pd–zeolite catalysts based upon untreated ($NaY_{4.5}$) and de-aluminized (Deal-$NaY_{6.4}$) Y-type zeolites, each containing 4·46 wt. % of metal have been studied. Palladium was introduced by ion-exchange from an aqueous solution of $Pd(NH_3)_4Cl_2$, containing a certain amount of NH_4^+ ions; in this case simultaneously with the introduction of the metal, a proportion of the Na^+ cations were exchanged for NH_4^+. For comparison the untreated and dealuminized zeolites were studied in Na^+ and NH_4^+ forms. The composition of compensating cations in the samples under investigation is shown in Table 1. The zeolite was dealuminized with EDTA.

Table 1

Compositions of the samples

No.	Catalyst	Relative amount of Na^+ taken for treatment[a]	Composition of compensating cations[b]				
			H^+	Na^+	NH_4^+	Pd^{2+}	$Pd(NH_3)^{2+}$
1	$NaY_{4.5}$	—	9	91	—	—	—
2	$NH_4NaY_{4.5}$	—	9	47	38	—	—
3	$Pd-NH_4NaY_{4.5}$	—	9	42	29	—	20
4	$Deal-NaY_{6.4}$	—	2	98	—	—	—
5	$Deal-NH_4NaY_{6.4}$	—	2	36	62	—	—
6	$Pd-Deal-NH_4NaY_{6.4}$	—	2	34	38	—	26
7	$Pd-Deal-NH_4NaY_{6.4}$	0·15	34	40	—	26	—
8	$Pd-Deal-NH_4NaY_{6.4}$	0·30	28	46	—	26	—
9	$Pd-Deal-NH_4NaY_{6.4}$	0·75	10	64	—	26	—
10	$Pd-Deal-NH_4NaY_{6.4}$	1·50	2	72	—	26	—

[a] mg-equiv. Na^+/mg-equiv. H^+; the ratio of the amount of Na^+ in the NaOH solution to the amount of H^+ compensating cations.
[b] (mg-equiv. cation/g) $\times$ 100/(mg-equiv. Al/g).

Neutralization of the protonic sites of the zeolite with Na^+ cations was performed by treating the Pd–Deal-$NH_4NaY_{6.4}$ catalyst with 0·1 N-NaOH solution after calcination in air at 500°C to decompose the ammonium ions and the complex $Pd(NH_3)_4^{2+}$ cations. Table 1 presents data on the reagent ratio when the catalysts are treated with alkali, and on the composition of the samples (samples 7—10).

Experimental techniques. For spectroscopic analysis the samples were pressed into pellets without any binding agent, the thickness of the pellets amounting to 4·5 mg/cm². The weight of the pellets fluctuated within only $\pm$ 2%, thus permitting comparison of the absorption band intensities. The spectra were recorded on a Perkin-Elmer 225 spectrophotometer.

X-Ray diffraction patterns of the samples were obtained on a DRON-1·5 diffracto-meter using Cu-K_α radiation with a nickel filter. The sample was pressed into a cuvette made of plexiglass.

The activity and sulphur resistance of the catalysts were determined on a downstream unit at a hydrogen pressure of 50 atm. in a hydrogenation of *o*-xylene containing 0·5% sulphur in the form of thiophen. The value of stable *o*-xylene conversion after 10 h of catalyst operation was taken as a measure of sulphur resistance. Before experiments, all the samples were calcined for 3 h at 500°C under air, and then subjected to reduction with hydrogen for 10 h at the same temperature.

The spectral analysis of the surface OH groups was performed after thermal vacuum treatment of the catalysts. However, the results of this analysis may be compared sufficiently well with the catalytic properties. It has been shown[12] that on thermal vacuum treatment palladium in the zeolite assumes the metallic state, as is the case after reduction with hydrogen. The X-ray diffraction study we have carried out has revealed the metallic palladium phase both in vacuum-treated and reduced samples.

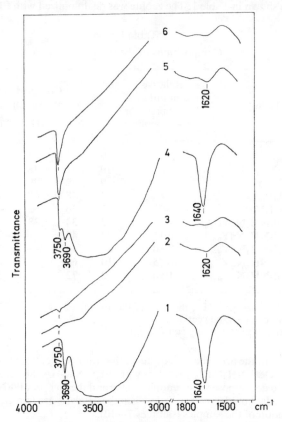

FIGURE 1   Infrared spectra of $NaY_{4.5}$ and $Deal-NaY_{6.4}$ zeolites in the process of water desorption:
$NaY_{4.5}$; $t°C$. 1: 20; 2: 300; 3: 450.
$Deal-NaY_{6.4}$; $t°C$: 4: 20; 5: 300; 6: 450.

PROTONIC ACIDIC SITES AND STATE OF METAL IN THE ZEOLITE

Infrared spectra of the untreated and dealuminized Y-zeolite in the Na^+ form subjected to thermal vacuum treatment are shown in Fig. 1. It is seen from comparison of the spectra that in the region associated with valence and deformation vibrations of water molecules (3700—3400, 1700—1600 cm^{-1}) they are very similar. The difference consists of the presence in the spectrum of $Deal-NaY_{6.4}$ zeolite of a very intense band at 3750 cm^{-1} which we attribute to vibrations of free Si–OH groups formed in the process of dealuminization.[13] In the spectra of both the untreated and dealuminized zeolites there are no bands at 3550—3650 cm^{-1} associated with protonic acidity.[14] Thus, dealuminization of the zeolite with EDTA does not result in the formation of protonic acidic sites.

Bands at 3640 and 3650 cm^{-1} appear in the process of thermal vacuum treatment of ammonium forms of the untreated and dealuminized zeolites (Fig. 2, curves 1,3). It is evident that the band of 3650 cm^{-1} in the spectrum of the untreated zeolite (ammonium

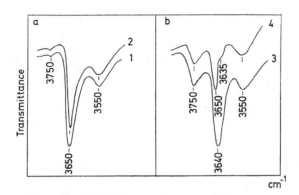

FIGURE 2   Absorption bands of OH groups of the ammonium and palladium forms of the untreated and dealuminized zeolite (400°C):
(a) Untreated zeolite. 1: NH$_4$NaY$_{4.5}$; 2: Pd–NH$_4$NaY$_{4.5}$;
(b) Dealuminized zeolite. 3: Deal–NH$_4$NaY$_{6.4}$;4: Pd–Deal–NH$_4$NaY$_{6.4}$.

form) is shifted towards the low-frequency region by 10 cm^{-1} in the spectrum of the Deal-NH$_4$NaY$_{6.4}$ sample. The frequency decrease corresponds to the decrease in the force constant of the O–H bond, indicating the availability of stronger protonic sites in the zeolite.[11]

Similar bands are observed in the spectra of the dealuminized zeolite treated with Pd(NH$_3$)$_4$Cl$_2$ solution (Fig. 3a, b). Bands at 3260, 3180, and 1440 cm^{-1} refer to the compensating NH$_4^+$ cation.[15] In the spectrum there is a band at 3340 cm^{-1} (Fig. 3a, curves 1,3) which is not present in the spectrum of the ammonium form and disappears on vacuum treatment at 350—400°C. On the basis of data in the literature[16,17] we suggest that the band at 3340 cm^{-1} is caused by NH$_3$ molecular vibrations in the Pd(NH$_3$)$_4^{2+}$ complex ion.

As the temperature of vacuum-treatment is increased, the presence of two components is observed in OH group absorption band at 3650—3630 cm^{-1}; these components are associated with the protonic acidity of the zeolite. At temperatures ranging from 300 to 500°C (Fig. 3b, curves 4—6) a decrease in the intensity of the band at 3635 cm^{-1} is observed, and a band at 3650 cm^{-1} shows up clearly. At the same time the complex cations decompose within the same temperature range, and metallic palladium is formed.

The spectra of the Pd–NH$_4$NaY$_{4.5}$ sample are virtually identical with those of Pd–Deal-NH$_4$NaY$_{6.4}$, but for the band due to acidic OH groups. In this case a much more intense component is observed at 3650 cm^{-1} (Fig. 2, Curves 2,4).

Comparison of the spectra from ammonium- and palladium-containing samples of the untreated and dealuminized zeolites at approximately equal content of Na$^+$(Fig. 2, a, b) indicates that introduction of palladium into the untreated zeolite is accompanied by virtually no change in the intensity and shape of the band at 3650 cm^{-1}. In the case of a dealuminized zeolite the intensity of this band is sharply decreased when palladium is introduced, and its shape is changed by a decrease in the intensity of the low-frequency component at 3635 cm^{-1}.

The disappearance of the low-frequency component in the spectrum of acidic OH groups when metal is introduced is indicative of an interaction of the palladium atoms with strongly acidic protonic sites, while in the untreated zeolite no interaction of this

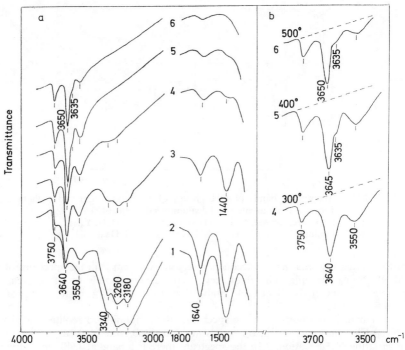

FIGURE 3   Infrared spectra of Pd–Deal-NH$_4$NaY$_{6.4}$ catalyst:
(a) t°C. 1: 20; 2: 100; 3: 200; 4: 300; 5: 400; 6: 500.
(b) Bands of zeolite OH groups t°C. 4: 300; 5: 400; 6: 500.

kind is observed. The hydroxyl groups in the dealuminized zeolite prove to be stronger proton donors. The same sites may be considered as electron acceptors. In the presence of a strong acceptor palladium can function as an electron donor.[18] This gives rise to the suggestion that the interaction between palladium and the strongly acidic sites of the zeolite occurs by a donor–acceptor mechanism.

Experiments involving CO adsorption were conducted to check this suggestion. From the shift in the maximum of the high-frequency absorption band in the 2000—2160 cm^{-1} region corresponding to the linear structure M—C≡O, it is possible to draw conclusions about the state of the metal.[6,19] A shift to higher frequency in the CO vibration is indicative of decreasing electron density on the metal atoms.

CO adsorption was performed on zeolite samples containing Pd^{2+} cations (calcined sample of Pd–Deal-NH$_4$NaY$_{6.4}$); on reduced metal; and on pure palladium without support. It is seen from Table 2 that the highest frequency (2153 cm^{-1}) is observed when

CO interacts with palladium cations, while the lowest frequency (2085 cm^{-1}) is observed on the pure metal without support. Metal reduction is accompanied by a decrease in CO vibration frequency, which corresponds to a decrease in electron deficiency on Pd atoms. In this case a considerable difference in the position of CO absorption bands on Pd–NH$_4$NaY$_{4.5}$ and Pd–Deal NH$_4$NaY$_{6.4}$ is observed ($\Delta\nu = 13$ cm^{-1}). The higher-frequency value of the absorption band due to CO adsorbed on Pd–Deal-NH$_4$NaY$_{6.4}$ as compared with that on Pd–NH$_4$NaY$_{4.5}$ indicates that palladium atoms in a dealuminized zeolite are characterized by lower electron density.

Table 2

Position of absorption bands of CO adsorbed on Pd–zeolite catalysts
($t = 25°$C $P_{CO} = 12$ mm Hg)

Catalyst	Relative amount of Na$^+$ taken for treatment[a]	Absorption bands, cm^{-1}
Pd–Deal-NH$_4$NaY$_{6.4}$ Calcinated in air flow, 500°C	—	2153 2135 2115
Pd film	—	2085 1910
Pd–Deal-NH$_4$NaY$_{6.4}$ reduced with H$_2$, 350°C	—	2108 1970 1930
Pd–NH$_4$NaY$_{4.5}$ reduced with H$_2$ 350°C	—	2095 1995 1910
Pd–Deal-NH$_4$NaY$_{6.4}$ reduced with H$_2$ 350°C	0·75	2090 2042 1880
Pd–Deal-NH$_4$NaY$_{6.4}$ reduced with H$_2$ 350°C	1·50	2090 2000 1880

[a] mg-equiv Na$^+$/mg-equiv. H$^+$.

The increase in electron deficiency on palladium atoms in the dealuminized Y-zeolite we ascribe to the interaction of the metal with strongly acidic OH groups by a donor–acceptor mechanism. This conclusion can be verified independently. It is, for instance, necessary to neutralize by means of Na$^+$ cations the strongly acidic sites (OH groups) able to interact with Pd atoms. In this case there should be little difference between the electronic state of the palladium atoms and that of the pure metal without support.

As the amount of Na$^+$ introduced into the zeolite is increased, the band intensity of the acidic OH groups in the i.r. spectra gradually diminishes and eventually disappears completely. As can be seen from Table 2, the absorption bands due to CO on Pd–Deal-$NH_4NaY_{6.4}$ treated with NaOH are similar to those on pure metal and on the Pd–$NH_4NaY_{4.5}$ catalyst, as far as their frequency is concerned.

The study of the i.r. spectra in the region of crystalline lattice vibrations (Table 1) has shown that the zeolite structure is not damaged by the processes of synthesis, thermal vacuum treatment, calcination, and reduction. These results are confirmed by *X*-ray diffraction data.

SULPHUR RESISTANCE AND STATE OF THE METAL IN THE ZEOLITE

Palladium–zeolite catalysts operating on sulphur-containing feedstocks lose their activity over a period of time, as shown in Fig. 4 (curves 1,2) but after 3—5 h of operation

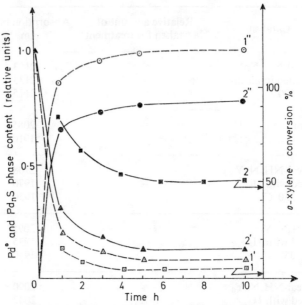

FIGURE 4 Change in activity of Pd–zeolite catalysts and state of the metal in the process of sulphurized *o*-xylene hydrogenation.
($t = 300°C$; $P_{H_2} = 50$ atm., Space velocity $= 1$, $V_{H_2} = 1250$).
Pd–$NH_4NaY_{4.5}$: 1: *o*-xylene conversion; 1′: $Pd°$ content; 1″: Pd_nS total content.
Pd–Deal-$NH_4NaY_{6.4}$: 2: *o*-xylene conversion; 2′: $Pd°$ content; 2″: Pd_nS total content.

they become stabilized. A portion of the palladium is transformed into the sulphides, $Pd_{2.8}S$ and Pd_4S. Hence, the decrease in the $Pd°$ content and the increase in the total content of Pd_nS are correlated with the change in activity with time (Fig. 4, curves 1, 1′, 1″ and 2, 2′, 2″).

On the basis of these data it can be concluded that the drop in activity during the initial period of catalyst operation is associated with sulphidation of the palladium. After stabilization, the hydrogenation occurs on the remaining metal.

The sulphur resistance of Pd–Deal-$NH_4NaY_{6.4}$ catalyst is higher by a factor of 10 than that of Pd–$NH_4NaY_{4.5}$ (Fig. 4, curves 1,2). Hence more metal and less sulphide are present after stabilization in the sample produced from the dealuminized zeolite.

One may suggest that the increased sulphur resistance of Pd–Deal-$NH_4NaY_{6.4}$ is connected with the availability of electron-deficient palladium in the sample. The deficiency of electrons on Pd atoms prevents them from interacting with electron-acceptor atoms of sulphur. In the presence of sulphur, the protonic sites and the sulphur atoms are likely to compete for the electrons from the metal. To protect the metal against sulphidation, the support has to contain acidic sites, which are superior to sulphur atoms in electron-acceptor capacity. This accounts for the low sulphur resistance of the catalyst derived from the untreated $NaY_{4.5}$ zeolite, in which no strongly acidic sites are available.

The concept of the necessity of strongly acidic protonic sites on the surface of the support to protect the metal against sulphur has been tested by poisoning these sites

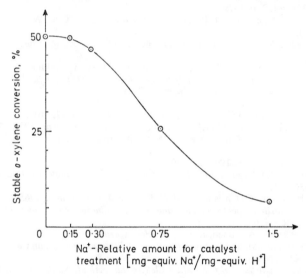

FIGURE 5 Dependence of Pd–Deal-$NH_4NaY_{6.4}$ catalyst sulphur resistance on the degree of neutralization of protonic sites of the zeolite in the process of sulphurized *o*-xylene hydrogenation ($t = 300°C$; $P_{H_2} = 50$ atm, space velocity $= 1$; $V_{H_2} = 1250$).

with Na^+ cations. A decrease in the number of protonic sites in the Pd–Deal-$NH_4NaY_{6.4}$ catalyst upon treatment with NaOH brings about a drop in its sulphur resistance (Fig. 5).

CONCLUSIONS

As a result of these studies, the interaction of metal atoms with strongly acidic protonic sites in the zeolite has been demonstrated. It has been shown that the interaction of the metal with strongly acidic sites (OH groups) occurs by a donor–acceptor mechanism.

Partial migration of free electrons from the metal atom to the proton of the strongly acidic OH group results in the formation of a charge-transfer complex:

$$\overset{\delta+}{Pd}----\overset{\delta-}{H}—O<zeolite$$

The electron deficiency on the palladium atom results from this kind of interaction, thus protecting the hydrogenating metal against sulphur poisoning.

The above considerations are valid for individual metal atoms or for atomic clusters formed in the bulk of zeolite crystals, as shown in refs 4 and 5. That portion of the metal which is aggregated into large crystallites on the surface of the zeolite crystals apparently does not interact with the protonic sites, and is converted into the sulphide form when the catalyst is used on sulphur-containing feedstocks.

REFERENCES

[1] J. A. Rabo, V. Shomaker, and P. E. Pickert, *Proc. 3rd Internat. Congr. on Catalysis*, (Amsterdam, 1965), vol. 2, p. 1264.

[2] P. H. Lewis, *J. Catalysis*, 1969, **11**, 162.

[3] C. R. Wilson and W. K. Hall, *J. Catalysis*, 1970, **17**, 190.

[4] H. P. Studer, *Microscopie Elekton.* v.**1**, 309, (1970).

[5] R. A. Della Betta and M. Boudart, *Proc. 5th Internat. Congr. on Catalysis*, (Palm Beach, Florida, 1972), vol. 2, p. 1329.

[6] F. Figueras, R. Gomez ,and M. Primet, *Adv. Chem. Series*, A.C.S., Washington, No. **121**, p. 480 (1973).

[7] K. V. Topchieva, V. A. Dorogochinskaya, and Kho Shi Tkhoang, *Zhur. fiz. Khim.*, 1974, **48**, 182.

[8] M. V. Landau, V. Ya. Kruglikov, F. A. Pogrebnoi, D. A. Agievskii *et al.*, *"Materials of the second scientific and technical conference of young scientists in the field of oil refining and petrochemistry"*, "Esh gvardiya", Tashkent, 1974, p. 44.

[9] R. Beamont and D. Barthomeuf, *J. Catalysis*, 1972, **27**, 45.

[10] K. V. Topchieva and Kho Shi Tkhoang, *Zhur. fiz. Khim.*, 1973, **47**, 2103.

[11] T. Tsutsumi, H. Kajiwara, and H. Takahashi, *Bull. Chem. Soc. Japan*, 1974, **47**, 801.

[12] Kh. M. Minachev, G. V. Antoshin, and E. S. Shpiro, *Izvest. Akad. Nauk SSSR, Ser. Khim.*, 1974, 1012.

[13] G. T. Kerr, *J. Phys. Chem.*, 1968, **72**, 2594.

[14] A. V. Kiselev and V. I. Lygin, *Infrared spectra of surface active compounds* (Nauka, Moscow, 1972), pp. 336–381.

[15] J. B. Uytterhoeven, L. G. Christner, and W. K. Hall, *J. Phys. Chem.*, 1965, **69**, 2117.

[16] K. Nakamoto *Infrared spectra of inorganic and coordination compounds*, (Mir, Moscow, 1966), p. 197. (Russian translation).

[17] W. Wendlandt and L. A. Funell, *J. Inorg. Nuclear Chem.*, 1969, **26**, 1879.

[18] J. Turkevich, F. Nozaki, and D. Stamires, *Proc. 3rd Internat. Congr. on Catalysis*, (Amsterdam, 1965) vol. 1, p. 586.

[19] N. M. Popova, *Trudy Inst. org. Kataliz i Elektrokhim. Akad. Nauk Kazakh. S.S.R.*, 1974, **9**, pp. 32–51.

Production of Supported Copper and Nickel Catalysts by Deposition–Precipitation

Jos A. van Dillen, John W. Geus,* Leo A. M. Hermans, and
Jan van der Meijden

Division of Inorganic Chemistry, University of Utrecht, The Netherlands

ABSTRACT After a discussion of the drawbacks of the usual methods of applying a catalytically active material on to a support, it is argued that precipitation from a homogeneous solution can lead to the desired uniform distribution of active particles over the support. A method is elaborated in which precipitation is effected by a homogeneous increase of the hydroxyl ion concentration of a suspension of the support in a solution of the active material.

It is demonstrated experimentally that the interaction of partially hydrolysed copper(II) ions with silica surfaces depends strongly on the pH-value at which the precipitation proceeds. Depending on the temperature, precipitating nickel ions can react more or less extensively with a silica support. At high temperatures the reaction leads to finely divided nickel silicates. The results presented show that the method of deposition–precipitation can lead to excellent supported catalysts.

INTRODUCTION

Generally a high activity and a long life is a prerequisite for catalysts to be used in industrial processes. These requirements can only be met by catalysts having a large and thermostable active surface area. Many catalytically active solids, however, sinter at so great a rate at the elevated temperatures of many catalytic reactions that they cannot be used without admixtures that prevent sintering to big particles. Application of small active particles on to the surface of a highly porous, thermostable support can suppress sintering almost completely. The carrier, being itself mostly catalytically inactive, dilutes the catalyst. Nevertheless, its effect on the rate of sintering causes supported catalysts to display a stable active surface area generally surpassing that of the corresponding unsupported solid.

When the active material is not expensive, the activity of the catalyst per unit volume is generally most important. On increasing the degree of loading of the support, the activity per unit volume of supported catalysts goes through a maximum. The maximum is usually observed at a content of active material above about 20 wt %. Apart from the effects of the characteristics of the support, the magnitude of the maximum is determined largely by the size and the distribution of the active particles over the surface of the support. A homogeneous distribution of small active particles over the surface of the support leads to the most efficient use of the carrier. Since contact between the particles mostly results in coalescence and, hence, in a decrease of the active surface area, a high thermostability also calls for a homogeneous distribution of active particles.

Two methods, viz. impregnation and drying, and precipitation, are generally used to apply an active component on to a support. Either the active material itself or a compound that can yield the active component by a thermal treatment (catalytic precursor) are deposited on to the support. When the carrier must be loaded to only a small extent, impregnation can give rise to a uniform distribution of the active material, provided the support adsorbs the dissolved active ions or atoms. The metal content of precious-metal catalysts is usually small. These catalysts are therefore often produced by impregnation. Since the adsorption remains restricted to a monolayer or even less,

677

only a small amount of active material can be deposited adequately by this method. A higher degree of loading requires evaporation of the solvent or precipitation to deposit the active material. These methods, however, do not lead to a uniform distribution.

When an impregnated carrier is dried, the solid active material crystallizes where the solvent evaporates. Since it is impossible to accomplish evaporation of the solvent homogeneously throughout the support, impregnation and drying do not result in the desired distribution of the active material. Normal precipitation also does not lead to a homogeneous distribution. The precipitating active material generally nucleates rapidly. It consequently precipitates where the precipitant enters into the suspension of the support. The precipitant contacts the suspension at the liquid–gas interface, where large shear stresses cannot be applied. A rapid distribution of the precipitant into the suspension will therefore not be obtained. This leads to a high local supersaturation. As a result the usual precipitation method gives rise to an inhomogeneous distribution over the support.

Better results are obtained when a catalytically active precursor is precipitated together with the carrier (coprecipitated catalysts[1,2]). This procedure can, for example, be utilized with nickel-on-silica and copper-on-zinc oxide catalysts. The porous structure of coprecipitated supported catalysts is, however, difficult to control and coprecipitated metal catalysts are often difficult to reduce completely.

Precipitation of an active component or its precursor only on to the surface of a suspended support makes it possible to obtain a distribution analogous to that obtained with impregnation of ions adsorbing on to the support even at higher loadings of the carrier. To avoid generation of crystallites of the precipitate in the solution itself, the precipitation must be carried out at low degrees of supersaturation throughout the suspension. As dealt with, e.g., by Blaedel and Meloche,[3] precipitation from a homogeneous solution can be used to prevent local high degrees of supersaturation. Analytical chemistry has developed this method to produce precipitates that can be filtered easily. Well known is the use of the hydrolysis of urea[4] to precipitate compounds by a homogeneous increase in the hydroxyl ion concentration.[5,6] Since at room temperature a urea solution does not hydrolyse markedly, the hydroxyl ion concentration does not increase during dissolution of urea into a solution of the ions to be precipitated. After homogenizing at room temperature, maintaining the solution above about 70°C brings about rapid hydrolysis of urea. A homogeneous increase in the hydroxyl ion concentration results.

Without a solid with a large surface area that markedly interacts with the precipitating compound, precipitation from a homogeneous solution mostly leads to a limited nucleation. Growth of the small number of nuclei gives large particles, which are easy to filter. Precipitation from homogeneous solution can lead, on the other hand, to a uniform, continuous layer or a homogeneous distribution of small particles over a support suspended in the solution. To get the uniform layer or the small particles one of the following two conditions must be met:

(a) nucleation of the precipitating species uniformly over the surface of the support at a lower concentration than in the bulk of the solution;
(b) strong adherence of nuclei that do not grow or grow slowly, to the surface of the support.

If either of these conditions is fulfilled, precipitation on to a suspended support (deposition–precipitation) allows the preparation of thermostable supported catalyst containing a high fraction of active material.[7] These catalysts consequently have a high activity per unit volume. An additional advantage of the use of a homogeneous solution is that the scale of the preparation easily may be enlarged. The procedure with a 2,000 l vessel led to a catalyst that did not differ from that prepared in a 3 l vessel.

678

As mentioned above, the interaction of the support with the precipitating material is of paramount importance. To illustrate the effect of this interaction, we shall present results with copper ions precipitating in a suspension of silica particles. Experiments with nickel ions precipitating on to silica particles will also be described. The results show the effect of nucleation of an insoluble compound at the surface of a support.

EXPERIMENTAL

Apparatus.[8]

The precipitations have been performed in two double-walled Pyrex vessels having volumes of 1·2 and 4·8 l. The temperature of the vessels is controlled by circulating thermostatted water between the inner and outer walls. To assure the homogeneity of the suspensions, the vessels are equipped with three vertical glass baffles. The suspension is agitated by two stirrer blades attached to a shaft mounted centrally in the vessel.

Besides changing the hydroxyl ion concentration by the hydrolysis of urea, a method is utilized in which an alkaline or acid solution is injected into the suspension through a capillary tube having its end below the level of the suspension. Around the tip of the tube, the large shear stresses required to obtain a rapid mixing can be generated easily. Diffusion of ions from the suspension into the capillary is prevented by maintaining an uninterrupted flow through the narrow tip (*ca.* 0·1 mm). To prevent nucleation of precipitate at the tip the capillary is constructed of PVC.

With the small vessel, injection is carried out by means of a Radiometer Autoburet ABU12. To handle the larger volumes to be used in the 4·8 l vessel a Perspex tube (diameter 12 cm) was manufactured that could be emptied by a piston driven at an accurately adjustable speed. The pH of the suspension is measured by a Pye Unicam pH meter model 292 Mk 2 using Ingold electrodes (Type 465-35).

Materials

As a silica support Aerosil (Degussa, West Germany) grades 200 V and 380 V were used. The surface areas were determined as 200 and 380 m²/g, respectively.

Procedure

After dissolving the metal ions to be precipitated, the pH-value of the solution is adjusted so as to prevent premature hydrolysis. In experiments where the precipitation must be carried out in the presence of a suspended carrier, the support is added. Next the suspension is heated to the desired temperature, after which the precipitation can commence. With urea the reaction is started by addition of the required amount of solid urea.

RESULTS AND DISCUSSION

Precipitation of copper(ii) ions

The precipitation of copper(ii) by a homogeneous increase of the hydroxyl ion concentration by the hydrolysis of urea at 90°C is represented in Figure 1. The upper part of this Figure shows results with copper(ii) nitrate. As explained in the introduction, the pH-curve displays a distinct maximum. When the concentration reaches the value of the supersolubility curve,[3] nucleation and growth of the solid phase commences rapidly. The hydrolysis of urea being too slow to keep up with the rate of the hydroxyl ion consumption leads to a temporary decrease in the pH-value. According to the X-ray pattern the precipitate consists of big particles of $Cu_2(OH)_3NO_3$ (gerhardtite). Investigation in the scanning electron microscope showed the presence of hexagonal platelets of a thickness of 1–2 μm and a diameter of 15–20 μm. From Figure 1 it is apparent that suspended silica does not affect the precipitation of basic copper(ii) nitrate. As is to be expected, the X-ray pattern of the precipitate does not show any difference.

When the pH-value of a copper(ɪɪ) perchlorate solution is homogeneously raised, a maximum at a higher value of the pH is traversed. The pH-level at which the copper ions precipitate after the nucleation is also higher than that observed with copper(ɪɪ) nitrate. The higher pH-value points to a solubility of basic copper(ɪɪ) perchlorate or copper(ɪɪ) hydroxide that is higher than that of basic copper(ɪɪ) nitrate. The precipitate

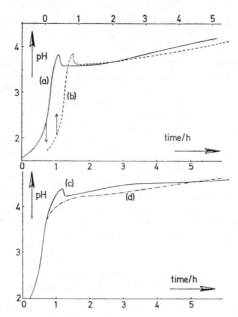

FIGURE 1   Precipitation of copper(ɪɪ) at 90°C from a homogeneous solution by hydrolysis of urea.

Curve (a) copper nitrate 0·167 M, urea 0·502 M.
Curve (b) copper nitrate 0·167 M, urea 0·502 M. 29·4 g Aerosil 200 V.
Curve (c) copper perchlorate 0·19 M, urea 0·56 M.
Curve (d) copper perchlorate 0·19 M, urea 0·56 M. 29·4 g Aerosil 200 V.

is blue at first, but it decomposes rather suddenly to copper(II) oxide at higher pH-values, as can be concluded also from the X-ray pattern. Suspended silica has a pronounced effect on the precipitation from a copper(ɪɪ) perchlorate solution. The pH-curve shown in the lower part of Figure 1 does not exhibit a maximum, and the precipitation of the copper ions proceeds at an appreciably lower pH-level. The copper(ɪɪ) species deposited on to the silica does not dehydrate to copper oxide at 90°C. The loaded carrier remains blue. The fact that the precipitate does not exhibit an X-ray pattern shows the particles to be very small.

The maxima in the pH-curves with copper(ɪɪ) nitrate and with copper(ɪɪ) perchlorate without silica show that to get nucleation in the bulk solution a barrier must be crossed. The presence of the barrier implies that the supersolubility curve[3] runs markedly below the solubility curve. At the surface of suspended silica the nucleation proceeds much more easily but only with the pH-value of the solution above about 4·0. As will be explained elsewhere,[9] condensation of hydroxyl groups of the silica surface and of partially hydrolysed metal ions is likely to be accelerated by hydroxyl ions. The concentration of the supersolubility curve of basic copper(II) nitrate is already reached before the hydroxyl ion concentration is sufficiently high to bring about a marked rate of condensation. The interaction of copper ions with the silica surface consequently remains low. As a result basic copper(II) nitrate precipitates in the bulk of the solution. With copper(II) perchlorate, on the other hand, the hydroxyl ion concentration can attain a value where the condensation proceeds rapidly before the concentration of the supersolubility curve of basic copper perchlorate or copper hydroxide is reached.

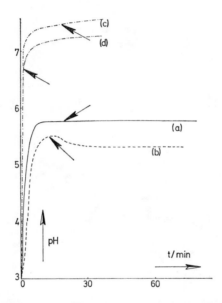

FIGURE 2   Precipitation of nickel from a homogeneous solution by hydrolysis of urea at 90°C (curves a and b) and by injection of ammonia at 25°C (curves c and d).

Nickel nitrate 0·14 M
Curve (a) Urea 0·42 M
Curve (b) Urea 0·42 M; 7·6 g Aerosil 200 V.
Curve (c) Injection of 0·5 N-NH$_3$; rate 2·8 ml/min.
Curve (d) 7·6 g Aerosil 200 V; injection of 0·5 N-NH$_3$; rate 2·8 ml/min.

Precipitation of nickel

Deposition–precipitation has been carried out at 90° and 25°C. Figure 2 shows the pH *versus* time curves. The onset of precipitation, which is indicated by arrows in Figure 2, has been established by recording the light-scattering of the suspensions. We shall discuss first the precipitation at 90°C, which was done by hydrolysis of urea.

As observed with the precipitations of copper(II) from a perchlorate solution, the curve observed with silica suspended in the solution is situated appreciably below that found without silica. Whereas precipitation of copper(II) not interacting with silica leads to a sharp maximum, the pH-curve of Figure 2 (without silica) is smooth. The unsupported nickel hydroxide precipitated from homogeneous solution is not as well crystallized as basic copper(II) nitrate prepared in the same way. The X-ray pattern displays broadened bands, and the scanning electron microscope shows small thin platelets that are highly clustered. The nickel hydroxide lattice consists of a stacking of layered units, each unit consisting of two layers of hydroxyl ions enclosing a layer of nickel ions. Whereas the interaction within a unit is strong, that between the composite layers is small. Growth parallel to the direction of a layer will consequently proceed much more rapidly than growth perpendicular to the layer. As the edges of a layer expose a rather limited surface area, crystallization will be slow. As a result the generation of hydroxyl ions can keep up with the crystallization.

The curve obtained at 90°C with suspended silica differs also from the corresponding curve with copper in exhibiting a maximum. Since with suspended silica the pH-value

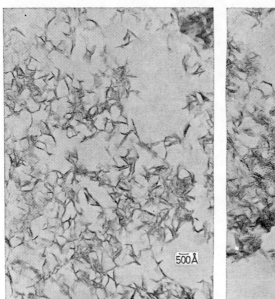
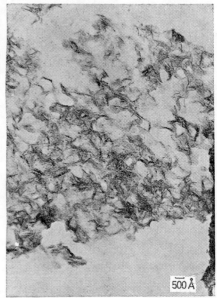

FIGURE 3 Electron micrographs of a nickel-on-silica catalyst prepared by hydrolysis of urea at 90°C.

Left: Dried catalyst, nickel content 26·1 wt. %
Right: Reduced catalyst, nickel content 34·2 wt %.
Original magnification 140,000 × ; ultrathin section after impregnation with methyl methacrylate.

remains also at the maximum considerably below the curve observed without silica, precipitation in the bulk of the solution must be rejected. Moreover, the X-ray diffraction pattern and investigation in the (transmission) electron microscope (see Figure 3) show an extremely uniform distribution of nickel ions over the support. Reduction of the loaded carrier leads to a highly active catalyst, which remains stable at 550°C in hydrogen for prolonged periods.

Adsorption of nickel ions on to silica just before the precipitation, as well as the solubility and structure of the precipitated nickel have been investigated extensively. It appears that the maximum in the pH-curve is connected with nucleation of nickel hydrosilicate, $Ni_3(OH)_4Si_2O_5$.[10,11] At 90°C nickel hydrosilicate only dissolves at pH-values as low as 3·8. Nucleation of nickel hydrosilicate therefore leads to a fast growth during which hydroxyl ions are consumed so rapidly that the pH displays a transient decrease. The rate of nucleation of nickel hydrosilicate depends strongly on the temperature. At room temperature nickel hydrosilicate develops too slowly to affect the precipitation process.

This can be inferred from the curves of Figure 2 measured at room temperature. In these experiments the hydroxyl ion concentration was increased by injection of ammonia. Again precipitation of nickel ions without silica proceeds at higher pH-values than that with silica. Accordingly the nickel ions are distributed homogeneously over the silica surface. At room temperature the pH-curve with suspended silica displays, however, no maximum. Separate experiments show that the adsorption of nickel ions (presumably from a $NiOH^+$ species) increases gradually with the pH-value. Redissolution experiments demonstrate a slow (taking some days) conversion of nickel ions contacting the silica into nickel hydrosilicate.

Formation of nickel hydrosilicate is evident from electron micrographs. Figure 3 represents micrographs of dried (left) and a reduced (right) nickel-on-silica catalyst. The micrograph of the dried catalyst shows thin continuous streaks, which are due to nickel hydrosilicate layers having their planes parallel to the electron beam. Reduction of the catalyst converts the continuous (hydr)oxide layer into many small nickel particles.

Nickel hydrosilicate has a layer structure.[10] After nucleation the layers grow laterally by capturing nickel ions from the solution that reacts with the silica. It can be shown that at 90°C the formation of hydrosilicate does not remain restricted to the layer of nickel ions deposited directly on to the silica surface. Penetration of nickel ions into the silica or migration of silica leads to thicker nickel hydrosilicate layers.[12] As can be seen from Figure 3 the layers cannot be bent, but are rather flat. At higher nickel-to-silica ratios a substantial fraction of the support reacts to nickel hydrosilicate. Conversion of an appreciable fraction of the support into a compound having a layer structure profoundly affects the texture of the support.

When the deposition of nickel on to silica is performed at room temperature and the loaded carrier is not aged in the aqueous solution, formation of thicker nickel hydrosilicate layers does not occur. The structure of the support is consequently much less affected.

PROPERTIES OF THE CATALYSTS

Copper catalysts prepared as described above show large free copper surface areas. Some results for a catalyst containing, before reduction, about 25 wt % copper will be given. The copper surface area was roughly estimated from the sorption of molecular oxygen at room temperature, which was measured gravimetrically and volumetrically. Assuming a number of 1.65×10^{15} copper atoms cm^{-2} and one oxygen atom per metal surface atom, copper surfaces of 148 and 138 m^2/g copper were obtained for catalysts reduced at 220° and, for a short time, at 530°C, respectively. After reduction at 220°C and reoxidation at 450°C, the catalyst was used in the oxidation of carbon monoxide. The activity per g of copper agreed with that calculated from the value measured with unsupported copper oxide (3 $m^2\ g^{-1}$) within experimental error. For a number of industrial copper catalysts, Scholten and Konvalinka[13] mentioned copper surface areas of 12 to 71 $m^2\ g^{-1}$.

Van Hardeveld and van Montfoort[14] investigated adsorption of molecular nitrogen as an infrared-active species by nickel-on-silica catalysts. They demonstrated that adsorption as an infrared-active species was characteristic of very small nickel particles. Catalysts prepared by precipitation from homogeneous solution displayed an amount of adsorbed infrared-active nitrogen that largely exceeds that found with impregnated catalysts. Van Hardeveld and Hartog[15] mentioned nickel surface areas of 130 to 150 m^2/g nickel for catalysts, containing about 25 wt % nickel, prepared by deposition precipitation. The catalysts were reduced at 425 to 450°C. Coenen and Linsen[11] described impregnated nickel-on-silica catalysts containing 33 wt % nickel which have nickel surface areas of 38 to 70 m²/g nickel after reduction only at 300°C.

Besides copper and nickel catalysts the method of deposition–precipitation has been used to procedure a large number of supported catalysts. Excellent catalysts have been obtained the properties of which can be effectively controlled.

ACKNOWLEDGEMENTS. The authors are indebted to Messrs. J. van Hofwegen and H. de Rooij for experimental assistance. Mr. J. J. M. G. Eurlings is gratefully thanked for his invaluable aid in the development of the described method.

REFERENCES

[1] J. J. B. van Eijk van Voorthuijsen and P. Franzen, *Rec. Trav. chim.*, 1951, **70**, 793.
[2] K. Morikawa, T. Shirasaki, and M. Okada, *Adv. Catalysis*, 1969, **20**, 98.
[3] W. J. Blaedel and V. W. Meloche, *Elementary Quantitative Analysis*, 2nd edn. (Harper and Row, New York, 1963), p. 177, p. 719.
[4] W. H. R. Shaw and J. J. Bordeaux, *J. Amer. Chem. Soc.*, 1955, **77**, 4729.
[5] H. H. Willard and N. K. Tang, *Ind. Eng. Chem. Analyt. Edn.*, 1937, **9**, 357.
[6] P. F. S. Cartwright, E. J. Neuman, and D. W. Wilson, *Analyst*, 1967, **92**, 663.
[7] J. W. Geus, *Dutch Patent Application*, 1967, 67 05, 259; *Chem. Abs.*, 1970, **72**, 36325k.
[8] A. C. Vermeulen, J. W. Geus, R. J. Stol, and P. L. de Bruyn, *J. Colloid Interface Sci.*, 1975, **51**, 449.
[9] J. W. Geus and L. A. M. Hermans, to be published.
[10] W. Feitknecht and A. Berger, *Helv. Chim. Acta*, 1942, **25**, 1543.
[11] J. W. E. Coenen and B. G. Linsen, *Physical and Chemical Aspects of Adsorbents and Catalysts.* ed. B. G. Linsen, (Academic Press, London and New York, 1970) p. 471.
[12] D. Stigter, J. Bosman, and R. Ditmarsch, *Rec. Trav. chim.*, 1958, **77**, 430.
[13] J. J. F. Scholten and J. A. Konvalinka, *Trans. Faraday Soc.*, 1969, **65**, 2465.
[14] R. van Hardeveld and A. van Montfoort, *Surface Sci.*, 1966, **4**, 396.
[15] R. van Hardeveld and F. Hartog, *Proc. 4th Internat. Congr. Catalysis*, Moscow, 1968 (Akademiai Kiado, Budapest, 1971) vol. 2, p. 295.

DISCUSSION

G. A. Martin (*CNRS, Villeurbanne*) said: In this paper, it has been assumed that basic nickel silicate formed was $Ni_3(OH)_4Si_2O_5$ (nickel antigorite). In fact, another basic nickel silicate may be obtained from reaction of silica with nickel compounds, *i.e.* the nickel talc, $Ni_3(OHSi_2O_5)_2$. Its formation is favoured when the silica concentration is large enough. Have the authors evidence that the silicate is antigorite rather than talc?

J. van der Meijden (*Univ. of Utrecht*) replied: The extent of the reaction of the silica support to nickel hydrosilicate depends on the nickel:silica ratio, the temperature, and the time that the support is exposed to precipitated nickel and water. When nickel hydrosilicate is formed it is badly crystallized and the ordering of the layers is small. X-Ray diffraction therefore is not very helpful in the characterization of our preparations. However, we measured the dehydration of the catalysts in a thermobalance in an inert or oxidizing atmosphere. With preparations for which one or more of the above conditions for the formation of nickel hydrosilicate were very favourable, we observed a nickel:water ratio of about 1·5. This ratio points to nickel antigorite; talc should have given a ratio of 3. We therefore ascribed the reaction with silica to formation of antigorite.

N. D. Parkyns (*British Gas Corporation, London*) said: A few years ago, Dr. Drake prepared catalysts containing 25% nickel on silica by the urea hydrolysis method described in this paper. We found that, after evacuation at 450 °C, the i.r. spectrum of the product had absorption bands not only at the expected value of 3750 cm^{-1} for the silica carrier but also at 3614 cm^{-1}. The new band was as sharp as that at 3750 cm^{-1}. On reduction with hydrogen, the band at 3614 cm^{-1} disappeared whereas that at 3750 cm^{-1} remained unchanged. Examination of the unreduced product by X-ray diffraction failed to show any pattern characteristic of nickel oxide or hydroxide but the reduced catalyst showed broadened lines due to nickel crystallites about 40 Å in size, a value which agreed well with that derived from measurements of hydrogen chemisorption. These anomalies are very conveniently explained by the finding in the present paper that nickel hydrosilicate is the product of homogeneous hydrolysis of nickel nitrate on silica. We have recently examined the temperature-programmed reduction of the hydrolysis product and find that it is almost as hard to reduce as a co-precipitated nickel on alumina catalyst. This would support the view that a bulk nickel hydrosilicate is involved in the reduction process.

J. Turkevich (*Princeton Univ.*) asked: What is the size of the copper and nickel particles as determined by electron microscopy? Is the distribution in size large or small?

J. W. Geus *and* **J. van der Meijden** (*Univ. of Utrecht*) replied: The particle size distribution of nickel catalysts prepared according to this method has been studied carefully. Van Hardeveld and van Montfoort[1] published a size distribution determined from electron micrographs; it was very narrow. Wösten, Osinga, and Linsen[2] calculated the particle sizes of similarly prepared catalysts from magnetic, X-ray line-broadening, and hydrogen adsorption data. The values, displaying differently-averaged particle sizes, show small differences which also indicate a narrow particle size distribution. For copper catalysts, X-ray diffraction patterns and oxygen adsorption results show the particles to be of the order of 30–40 Å in size. No particle size distribution has been measured as yet.

[1] R. van Hardeveld and A. van Montfoort, *Surface Sci.*, 1966, **4**, 396.
[2] Wösten, Osinga, and Linsen, in *The Structure and Chemistry of Solid Surfaces*, ed. G. A. Somorjai (Wiley, New York, 1969) p. 54.

The Effect of the Gas Atmosphere on the Growth of Platinum Particles Dispersed on Alumina

Z. A. Furhman and G. Parravano*

SNAM Progetti S.p.A., 20097 San Donato Milanese, Italy, and Department of Chemical Engineering, University of Michigan, Ann Arbor, Michigan 48104, U.S.A.

ABSTRACT   The rate of growth of platinum particles embedded in a porous alumina matrix was followed in the temperature range 700—1,000°C. The experimental variables investigated were the gas atmosphere (O_2, H_2, N_2, CO, H_2O) and platinum loading (0·72 and 1·44 wt %). Particle growth was followed by differential frontal sorption of H_2 and by electron microscopy. Under gas phase oxidizing conditions there was a higher growth rate than under reducing conditions. From the type of kinetic expression which best described the experimental results, and from the influence of platinum loading on the growth rate, it is argued that migration of atoms was the prevailing transport process, and that it was kinetically limited by interphase transfer (reducing conditions) and by surface diffusion (oxidizing conditions).

Introduction

The influence of thermal treatments on the growth of platinum particles supported on refractory oxides has been extensively studied in recent years.[1] Two kinds of mass transport have been invoked to rationalize the results. In one, metal atoms migrate from small to larger particles in a process akin to precipitate ageing, while in the second small metal crystallites provide the material carriers. Due to the statistical nature of some of these analyses, the relatively large experimental error, and the many-variable problem that supported particle growth represents, the conclusions reached have not been definitive and conflicting views have developed.[1] To provide additional insight into this problem and develop further correlations with the experimental variables, we have measured the growth of platinum particles supported on alumina under various gas atmospheres and temperatures. The results of the study together with an analysis of their significance are reported.

Experimental

A commercial preparation containing 0·72 wt % Pt on γ-Al_2O_3 (surface area 175 $m^2 \, g^{-1}$, average pore diameter 210 Å, width at half height 60 Å) was employed for one series of runs: a second series was conducted on samples containing 1·44 wt % Pt. These were prepared by impregnation of the 0·72 wt % Pt samples with chloroplatinic acid (reagent grade) followed by drying in air, heating at 450°C for 1 h, and reducing in H_2 at 450°C for 1 h. Measurements on the loss of platinum surface area were conducted on aliquots placed in a ceramic tube positioned in an electrically heated oven (temperature profile < 1°C) with different gases flowing (60 $cm^3 \, min^{-1}$) around the sample. Heating and cooling were carried out in N_2 for periods of $\sim$ 15 min each. The total surface area of the samples was determined by standard BET measurements. Some decrease in the alumina surface area took place during the experiments. The loss of area amounted to 5, 12 and 18% at 700, 800 and 900°C, respectively, measurements of the accessible platinum area being carried out by means of hydrogen adsorption using differential frontal sorption (DFS). The method employed two separate glass ampoules, one filled with the platinum on alumina preparation and the other with the alumina support only, arranged in parallel in a flow train that permitted rapid change from N_2 to $N_2 + H_2$. The gas streams issuing from the ampoules were diverted into different branches of a thermal conductivity cell, the two branches being connected electrically to two arms of a Wheatstone bridge. While He flowed through the filled ampoules the bridge was zeroed electrically. After establishing steady conditions, the gas stream was switched from N_2 to $N_2 + H_2$ and the signal resulting from the bridge inbalance was recorded on a potentiometer. With a suitable calibration, the area under the signal was used to calculated the amount of H_2 selectively adsorbed on the platinum and hence its surface area (assuming a $1:1$ stoicheiometry). DFS has distinct advantages over pulse[2] and frontal[3] methods, which are generally employed in flow measurements of adsorbed gases.[4] In DFS, support adsorption is eliminated, adsorption takes place at constant H_2 pressure, and calibration is independent of desorption flow rate. The experimental procedure employed in DFS was as follows. The sample temperature was raised to $400°C$ in N_2 (60 cm^3 min^{-1}), followed by H_2 for 30 min. After cooling to $20°C$, adsorption measurements were carried out at 40 mbar H_2 partial pressure in N_2 (1013 mbar total pressure) with a H_2 flow rate of $2·5$ cm^3 min^{-1}. The DFS method agreed closely with conventional static volumetric measurements of H_2 adsorption (Table 1). Platinum particle size and size distribution were also measured directly by electron microscopy (Table 1).

Table 1

Average metal particle diameters (Å),[a] in Pt($0·72$ wt %)-Al$_2$O$_3$ by electron microscopy (EM), differential frontal sorption (DFS), and static volumetry (SV) of H_2. Thermal treatment $900°C$

Treatment gas time (h)		EM	DFS	SV
fresh		34	32	28
N_2	6	93	104	98
O_2	6	295	264	256

[a] Computed assuming spherical shape.

RESULTS

The growth of the average platinum particle diameter, d, in N_2 at 800 and $900°C$ for two platinum loadings is reported in Fig. 1, together with the results obtained in an atmosphere of $H_2 + 10$ vol % H_2O at $800°C$. In this instance the growth rate has increased by a factor of *ca.* 3. Particle growth in H_2 at 800, 900 and $1,000°C$ is presented in Fig. 2. Addition of 740 p.p.m. of O_2 to H_2 at $1,000°C$ did not appreciably alter the rate. Fig. 3 presents the results obtained in O_2; the accellerating influence of O_2 over N_2 and H_2 is apparent. Fig. 3 also demonstrates the effect of small additions of CO and H_2 to O_2. By adding H_2 at $700°C$ the growth rate was decreased by a factor of $2·5$, while there was no effect on addition of CO. The results of the influence of platinum loading in an O_2 atmosphere at $800°C$ are also shown in Fig. 3. Particle size histograms from electron microscope pictures for a fresh preparation and for two preparations treated in O_2 and N_2 at $900°C$ for 6 h each are reproduced in Fig. 4. As sintering proceeded, there was a gradual shift of the distribution to higher average diameters and an increase in polydispersity. Growth in oxidizing conditions tended to even out the distribution. This effect strongly suggests a different type of material transport

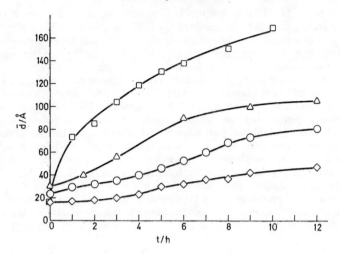

FIGURE 1   Platinum particle average diameter, *d*, against time, in N_2 atmosphere for Pt–Al_2O_3 preparations.

◇, 0·72 wt % Pt, 800°C; ○, 1·44 wt % Pt, 800°C; △, 0·72 wt % Pt, 900°C; □, 0·72 wt % Pt, 800°C, N_2 + 10 vol. % H_2O.

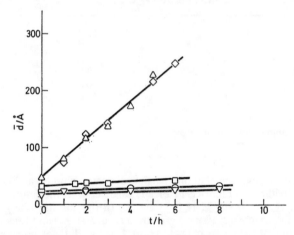

FIGURE 2   Platinum particle average diameter, *d*, against time in H_2 atmosphere for Pt (0·72 wt %)–Al_2O_3 preparations.

○, 800°C; □, 900°C; △, 1,000°C; ◇, 1,000°C, H_2 + 0·074 vol. % O_2; ▽, 800°C (1·44 wt % Pt).

under O_2 in contrast to that under N_2. It was found that the particle diameter distribution could be accurately fitted by a log normal curve.[5] Average diameters of platinum particles were calculated according to this distribution (Table 1).

Adsorption experiments were also carried out with 31·2 mbar of CO pressure at 20°C. Sample pretreatment was similar to that employed prior to H_2 adsorption. The volumes of CO adsorbed under various thermal treatments are collected in Table 2

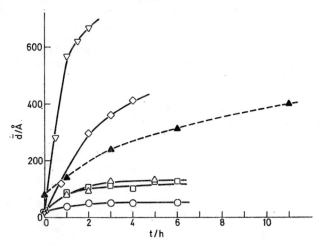

FIGURE 3 Platinum particle average diameter, $\bar{d}$, against time in O_2 atmosphere for Pt–Al_2O_3 preparations.

$\triangle$, 0·74 wt % Pt, 700°C; $\square$, 0·74 wt % Pt, 700°C, O_2 + 0·088 vol. % CO; $\bigcirc$, 0·74 wt % Pt, 700°C, O_2 + 0·074 vol. % H_2; $\Diamond$, 0·72 wt % Pt, 800°C; $\triangledown$, 1·44 wt % Pt, 800°C; — — —, replotted from ref. (1b) Fig. 5, p_{O_2} = 6·4 mbar.

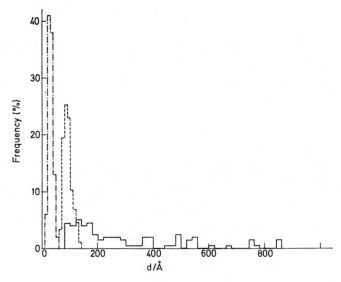

FIGURE 4 Average platinum particle diameter distribution for Pt(0·7 wt %)–Al_2O_3 preparations.

— · —, fresh; — — —, 900°C, 6 h, N_2 atmosphere; ——, 900°C, 6 h, O_2 atmosphere.

together with data on H_2 adsorption on similar samples. For H_2 and O_2 treatment-the CO/H_2 ratio while nearly unity, tended to decrease as the platinum particle diameter increased. It is likely that this is the consequence of a gradual change of adsorbed CO from a linear to a bridged configuration. This difficulty in using results from CO adsorption to derive metal particle area has already been noted.[6] Previous observations on CO adsorption on unsupported platinum also indicated this problem since a ratio $CO/Pt = 0.80$ was found.[7]

Table 2

H_2 and CO Adsorption on Pt(1·44 wt %)-Al_2O_3 subjected to thermal treatment at 800°C

Treatment time (h)	gas	Adsorption cm³ (g cat)⁻¹		Ratio CO/H_2
		H_2	CO	
fresh sample		0·76	0·83	1·09
0·5	O_2	0·02	0·02	1·12
1·0	O_2	0·01	0·01	1·00
2·0	O_2	0·01	0·00	0·92
1·0	H_2	0·31	0·37	1·16
2·0	H_2	0·31	0·31	0·99
4·0	H_2	0·28	0·30	1·08
6·0	H_2	0·26	0·28	1·06

Samples (0·7 wt % Pt), treated in N_2 at 800°C for 3·5 h, were subsequently subjected to different cooling rates. One sample was slow quenched (to room temperature in 15 min), while a second was subjected to a fast quench in liquid N_2 (< 30 s). The volumes of H_2 adsorbed were 13·19 and 13·05 cm³, respectively. Thus, contrary to previous findings,[8] no effect of the cooling rate on particle size was detected.

DISCUSSION

In agreement with earlier findings,[1,9] the results reported in the previous section indicate a large effect of the gas atmosphere on the growth of supported platinum particles. At constant temperature, growth rate increased in the sequence: $O_2 > N_2 \sim H_2$, the difference between oxidizing (O_2) and reducing (N_2, H_2) conditions being about 2 to 3 orders of magnitude. This fact leads one to believe that different transport phenomena provided the underlying mechanism for growth. It is worth mentioning that this behaviour is strikingly different from that of contacting, non-supported platinum spheres subjected to sintering. In this instance, the sintering rate in H_2 was found to be three to four orders of magnitude faster than in He.[10]

Empirically, the equation expressing the decrease of surface area, S, from the initial value, S_0, as a function of time t may be expressed as[1a]

$$\frac{1}{S^n} = \frac{1}{S_0^n} + kt \tag{1}$$

with n and k constant. The best fit of the results under reducing conditions is obtained with $n = 2$. In some instances the data cannot be fitted through a single straight line, but are best represented by means of two straight lines of different slope, the change in slope taking place at a critical value of the metal particle diameter. For particles with average diameter > 25—30 Å (area $ca.$ 60—70 m^2 g^{-1}) the growth rate was slower than that of particles whose diameter was < 25–30 Å. This effect is irrespective of temperature, gas atmosphere and platinum content, suggesting a growth rate which is dependent upon particle diameter, or which is structure-sensitive. The relation between the constant k in eqn. (1) and platinum content in the Pt-Al$_2$O$_3$ samples is also an important mechanistic criterion.[1a,11] The ratio of k to platinum content, z, was computed and is reported in Table 3. Inspection of Table 3 shows that the ratio k/z remained about constant for particle growth under N$_2$ and H$_2$.

Table 3

Platinum content, z, and constant, k, from eqn. (1) with $n = 2$, for the decrease of platinum surface area in supported platinum preparations at 800°C and 1013 mbar gas pressure

z $\left[\dfrac{\text{cm}^3 \text{ (Pt)}}{\text{cm}^2 \text{ (total surface)}}\right] \times 10^{10}$	Gas atmosphere	k $\left[\left(\dfrac{\text{m}^2}{\text{g(Pt)}}\right)^{-2} \text{h}^{-1}\right] \times 10^5$	k/z (g^2 h^{-1}) $\times 10^{-5}$
1·86	H$_2$	0·82	4·4
3·72	H$_2$	1·46	3·9
1·86	N$_2$	1·37	7·4
1·86	N$_2$	4·25	22·9
3·72	N$_2$	3·53	6·4
3·72	N$_2$	12·03	32·4

To sum up, the conclusions for particle growth under reducing atmosphere are: (a) the best fit of eqn. (1) is with $n = 2$; (b) the constant, k in eqn. (1) is dependent upon particle size and independent of platinum concentration; and (c) growth rate in N$_2$ is about equal to that in H$_2$ while that in O$_2$ is a few orders of magnitude faster.

These conclusions are consistent with a transport process involving diffusion of atomic species along the support surface, and a rate limiting stage akin to a transfer across an interphase.[1,12] Several analytical treatments of these kinetic conditions are described in the literature.[1] They are derived from an earlier analysis of precipitate ageing.[13] Predominant structural and morphological changes of small metal particles are concentrated in a narrow interval of their size, 10–30 Å.[14] In the present case, the most likely interphase processes include evaporation of single atoms (for surface diffusion) and formation of nuclei (for condensation): the rate of these processes is related to morphological factors of the metal surface (steps, kinks, crystallographic planes). This offers a possible explanation for the structure sensitivity of particle growth rate which is suggested by the exprimental results.

The data on particle growth under oxidizing conditions could be fitted by means of eqn. (1) with $n = 4$. The calculated values of the constant k are reported in Table 4.

In an effort to rationalize the results obtained under oxidizing conditions, we note that the observed growth rate was too slow for gas phase transport. This conclusion is arrived at by a comparison between the growth rate observed in this work and that recorded in a recent study under similar conditions (except for $p_{O_2} = 6\cdot4$ mbar),[1b] (see Fig. 3). Since the observed rate is slower than that previously reported by Wynblatt et al.,[1b] and since they show that their experimental rate is slower than that theoretically predicted for gas phase transport, we conclude that this transport mode cannot be identified with the rate controlling stage under the experimental conditions of the present investigation. In addition, the previous analysis of the results under oxidizing conditions has shown that: (a) the data may be fitted by means of eqn. (1) using the value of $n = 4$, and (b) the ratio k/z is approximately constant (Table 4). These conclusions indicate that, by analogy with the conclusion reached from the experiments

Table 4

Platinum content, z, constant, k, from eqn. (1) with $n = 4$, for the decrease of platinum surface area in supported platinum preparation at 800°C and 1013 mbar O_2

z $\left[\dfrac{\text{cm}^3 \text{ (Pt)}}{\text{cm}^2 \text{ (total surface)}}\right] \times 10^{10}$	k $\left[\left(\dfrac{\text{m}^2}{\text{g(Pt)}}\right)^{-4} \text{h}^{-1}\right] \times 10^5$	k/z $[\text{g}^2 \text{ h}^{-1}] \times 10^{-5}$
1·86	19·0	10·0
3·72	36·1	9·9

under reducing conditions, the predominant transport process under oxidizing conditions is a migration of platinum atoms along the support surface, kinetically limited by diffusion.[1] At the temperatures employed for the thermal treatments in the present work, the gas atmosphere may influence the surface oxygen vacancies of the alumina. At $T > 800$°C these vacancies are known to originate by loss of OH^- groups.[15] Thus a support effect on platinum particle growth, dependent upon the reducing or oxidizing conditions in the gas atmosphere, cannot be ruled out. Small additions of CO and H_2 to O_2 have a different influence on the growth rate (Fig. 3). This is probably related to subtle redox effects which contribute to the energy barrier of the interphase transport. This effect was also evident in the case of additions of H_2O to N_2 (Fig. 1).

To sum up, the experimental results of the present work on the growth rate of small platinum particles (20 to 300 Å) embedded into high area alumina are consistent with a surface transport process of metal atoms (or small atomic clusters), kinetically limited by surface diffusion under oxidizing conditions and by interphase transfer under reducing conditions. The physicochemical redox mechanism which interacts with the energetics of the kinetic process is very sensitive to the conditions of the gas phase and it may conceivably operate either by influencing the structure of the surface of the support and/or of the metal particles and their morphology.

ACKNOWLEDGEMENTS. We thank Dr. B. Notari for his interest in, and support for, this work and for many helpful discussions.

REFERENCES

[1] See among others (a) J. A. Betts, K. Kinoshita, and P. Stonehart, *J. Catalysis*, 1974, **35**, 307; (b) P. Wynblatt, R. A. Dalla Betta, and N. A. Gjostein, Paper Presented at the Battelle Colloquium on *The Physical Basis for Heterogeneous Catalysis* (Gstaad, 1967).
[2] T. E. Eberly, *J. Phys. Chem.*, 1961, **65**, 68; H. L. Gruber, *Analyt. Chem.*, 1962, **34**, 1828; N. Giordano and E. Moretti, *J. Catalysis*, 1970, **18**, 228; F. F. Roca, L. de Mourgues, and Y. Trambouze, *J. Gas Chromatog.*, 1968, **6**, 161; J. Freel, *J. Catalysis*, 1972, **25**, 139, 149.

3 O. Piringer and E. Tataru, *J. Gas Chromatog.*, 1964, **2**, 323; N. E. Buyanova, A. P, Karnankhov, L. M. Kefeli, I. D. Ratner, and O. N. Chernyavskaya, *Kinetika i Kataliz.* 1967, **8**, 868.

4 A. Renouprez, C. Hoang-Van, and P. A. Compagnon, *J. Catalysis*, 1974, **34**, 411; Kh. Shpindler, I. D. Ratner, and L. M. Kefeli, *Kinetika i Kataliz*, 1973, **14**, 1588; T. E. Whyte, Jr., P. W. Kirklin, R. W. Gould, and H. Heinemann, *J. Catalysis*, 1972, **25**, 407; G. R. Wilson and W. K. Hall, *J. Catalysis*, 1970, **17**, 190; N. M. Zaidman, V. A. Dzis'ko, A. P. Karnankkov, L. M. Kefeli, N. P. Krasilenko, N. G. Koroleva, and I. D. Ratner, *Kinetika i Kataliz*, 1969, **10**, 386.

5 R. D. Cadle, *Particle Size Theory and Industrial Applications* (Reinhold Publishing Corporation, 1965), p. 33.

6 T. R. Hughes, R. J. Houston, and R. P. Sieg, *Ind. Eng. Chem., Process Res. Develop.*, 1962, **1**, 95; T. A. Dorling and R. L. Moss, *J. Catalysis*, 1967, **7**, 328; D. Cormach and R. L. Moss, *J. Catalysis*, 1969, **13**, 1.

7 Y. Nishiyama and H. Wise, *J. Catalysis*, 1974, **32**, 50.

8 G. I. Emelianova and S. A. Hassan, *Preprints Proc. Fourth Inernat. Congress on Catalysis* (Moscow, 1968), p. 1326.

9 G. A. Mills, S. Weller, and E. B. Carnelius, *Proceedings 2nd Internat. Congress on Catalysis* (Editions Technip, Paris, 1961), p. 2221; G. R. Wilson and W. K. Hall, *J. Catalysis*, 1970, **17**, 190; P. Wynblatt and N. A. Gjostein, *Scripta Met.*, 1973, **9**, 969.

10 L. F. Norris and G. Parravano, *Reactivity of Solids*, ed. J. W. Mitchell, R. C. Devries, R. W. Roberts, and P. Cannon (J. Wiley and Sons, 1969), p. 149.

11 W. J. Dunning, *Particle Growth in Suspension*, ed A. L. Smith (Academic Press, London, 1973), p. 1.

12 B. K. Chakraverty, *J. Phys. and Chem. Solids*, 1967, **28**, 2401.

13 C. Wagner, *Z. Electrochem.*, 1961, **65**, 581.

14 O. M. Poltorak, V. S. Boronin, and A. N. Mitrofanova, *Preprints Fourth Internat. Congress on Catalysis* (Moscow, 1968), p. 1238.

15 B. C. Lippens and J. J. Steggerda, *Physical and Chemical Aspects of Adsorbents and Catalysts*, ed. B. G. Linsen (Academic Press, 1970).

DISCUSSION

J. J. Carberry (*Univ. of Notre Dame, Indiana*) said: I wish to draw attention to two points regarding Prof. Parravano's contribution to this much neglected yet highly important area. Firstly, as revealed in our work[1], the rate of sintering in nitrogen of a supported 0·035% platinum on α-alumina catalyst as a function of sintering temperature passes through a maximum; this maximum occurs at about 700 °C. Secondly, we utilized an α-alumina support to eliminate support sintering. Such sintering of γ-alumina was observed by Parravano. Such a process would give rise to the disappearance of platinum *via* pore collapse, thus distorting the inferred phenomonological kinetics of platinum sintering, *per se*. We find the sintering of platinum in air in the absence of support to be about second order in platinum. This contrasts with a higher order noted by Parravano, whose γ-alumina support sintered to the extent of about 15%. Of course, the surface characteristics of α- and γ-alumina differ, which may well account for the different kinetics should surface migration be rate determining.

G. Parravano (*Univ. of Michigan*) replied: Sintering measurements obtained by the conventional method used by Prof. Carberry and ourselves are semiquantitative. Only trends and order-of-magnitude conclusions are meaningful. Within these limitations we did not observe a maximum in sintering rate as a function of temperature. Obviously the support plays an important role in influencing the rate of sintering of metal particles. This role has still to be defined, and hence, in my opinion, meaningful comparisons of α- and γ-alumina as supports cannot yet be made.

1 J. Zahradnik, E. F. McCarthy, G. C. Kuczynski, and J. J. Carberry, *Sintering and Catalysis. Materials Science Research*, Vol. 10, ed. G. C. Kuczynski (Plenum Press, New York, 1975), p. 199.

H. Bremer (*Techn. Hochschule, Merseburg*) said: Prof. Parravano's results agree well with those of investigations on the sintering behaviour of alumina-supported platinum catalysts in various gas atmospheres which have been carried out by Dr. Spindler and his colleagues. They have found an inversion point in the sintering of supported platinum in an oxidizing atmosphere: above about 550 °C crystal growth takes place, below this temperature redispersion of platinum occurs. They have determined the distribution of platinum particle size by X-ray methods[2,3] and by electron microscopy[4] and have found that both the median and variance of the distribution function increase with increasing platinum concentration, provided the samples have the same history. The platinum size distribution, determined by means of electron microscopy, can be described by a log-normal function. This is indeed only the result of statistical tests for goodness of fit. So far as we know there is not a theoretical description for the type of distributions observed in the growth of supported metal particles.

G. Parravano (*Univ. of Michigan*) replied: Thank you for bringing to my attention the latest results of Dr. Spindler's group. I am happy to know that our results agree with theirs.

S. Wanke (*Univ. of Alberta*) said: The authors concluded that the sintering of platinum on alumina catalyst occurs *via* an atomic transport mechanism. Although I agree that atomic migration is the mechanism of supported metal catalyst sintering, I do not believe that the results presented can be used to discriminate between atomic and crystallite transport mechanisms. Both mechanisms result in equation (1) of the paper with various values of n.[5,6] Bett *et al.*[7] argue that an increase in sintering rates with an increase in metal loading supports the crystallite migration model,[3] but atomic migration can also show a loading dependence.[6] Furthermore, it was concluded recently[8] that log-normal crystallite size distributions result from a crystallite migration-coalescence mechanism. Thus sintering rate data are insufficient for discrimination between the two mechanisms. Other experiments, such as the *in situ* microscope sintering studies of Baker *et al.*[9] are needed to elucidate the sintering mechanism.

G. Parravano (*Univ. of Michigan*) said: Truly, sintering rates alone cannot conceivably yield definitive conclusions, particularly about the migration of metal atoms or crystallites (which cannot be easily defined) on a surface which is even less definable. However, within the validity of our conception of these phenomena and species, and considering the broad range of our experimental conditions and the internal consistency of the conclusions, it appears that the conclusions are qualitatively definitive.

J. Turkevich (*Princeton Univ.*) said: In all studies of platinum sintering, growth and loss, we should keep in mind the possibility of platinum transport as a volatile compound. Dr. Nambe and I carried out the following experiment. Shiny platinum foil was exposed to chlorine gas at room temperature. The foil turned grey. It was then heated to 150 °C in a stream of carbon monoxide. The foil lost the grey colour and became shiny. A white deposit of $Pt(CO)_2Cl_2$ was found in the cool end. $PtCl_2$ is stable to 500 °C and the action of carbon monoxide is confined to a temperature region of 120–250 °C. Below 120 °C no reaction with carbon monoxide takes place; above 250 °C the volatile platinum carbonyl chloride decomposes.

[2] H. Spindler, H. Weissenborn, and M. Kraft, *Tagungsbericht III, Internat. Katalysekonferenz der DDR*, Reinhardsbrunn 1974, p. 198.

[3] G. E. W. Schulze, R. Radosta, and H. Spindler, *Z. anorg. Chem.*, 1976, **421**, 183.

[4] H. Spindler, Th. Krajewski, and P.-E. Nau, *Tagungsbericht III, Internat. Katalysekonferenz der DDR*, Reinhardsbrunn 1974, p. 203.

[5] E. Ruckenstein and B. Pulvermacher, *Amer. Inst. Chem. Engineers J.*, 1973, **19**, 356.

[6] P. C. Flynn and S. E. Wanke, *J. Catalysis*, 1974, **34**, 400.

[7] J. A. Bett, K. Kinoshita, and P. Stonehart, *J. Catalysis*, 1974, **35**, 307.

[8] C. G. Granqvist and R. A. Buhrman, *J. Catalysis*, 1976, **42**, 477.

[9] R. T. K. Baker, C. Thomas, and R. B. Thomas, *J. Catalysis*, 1975, **38**, 510.

G. Parravano (*Univ. of Michigan*) replied: The possibility of gas phase transport as a mechanism for sintering supported platinum was considered. As indicated in the paper, calculated sintering rates for a gas phase transport model are higher than those observed in our work.

Unusual Catalytic Behaviour of Very Small Platinum Particles Encaged in Y Zeolites

P. Gallezot,* J. Datka†, J. Massardier, M. Primet, and B. Imelik

Institut de Recherches sur la Catalyse, C.N.R.S., 79, bd. du 11 Novembre 1918, 69626 Villeurbanne, France

ABSTRACT The catalytic activities for benzene hydrogenation and the poison sensitivity of platinum Y zeolites have been investigated. Two different Pt dispersions are considered: Pt agglomerates of 10 Å diameter fitting into the supercages and 15–20 Å crystallites occluded in the zeolite crystals. Pt agglomerates are only accessible in the vicinity of the supercage apertures; their activities are, however, comparable to those of platinum on conventional supports. Their sensitivity to NH_3 poisoning and low sensitivity to sulphur poisoning are attributed to an electron deficient character. Charge transfer is occurring between reagents or poisons and the metal as shown by i.r. spectroscopy using CO and NO as probe molecules. The formation of 15–20 Å Pt crystallites involves a limited breakdown of aluminosilicate framework; the properties of this system are different in respect to the agglomerates.

Introduction

Adsorption and catalytic properties of very small metal particles are a subject of growing interest. In addition to the effects on surface atom geometry[1] there is evidence[2–6] that their intrinsic electronic properties are different from those of larger particles. Moreover, the extent of charge transfer from the metal atom to the support becomes appreciable.[7,8] These factors are therefore expected to modify the catalytic activity and poison sensitivity of the metal. Metal loaded zeolites provided suitable materials for investigation in this field. Rabo et al.[9] claimed that highly dispersed platinum in Pt Y zeolite has a remarkable resistance to sulphur poisoning because of a weaker Pt–S bonding than for other metal dispersions. Dalla Betta and Boudart[10] reported that Pt clusters formed inside the zeolite supercages exhibit enhanced catalytic activities. This was attributed to support effects leading to electron deficient particles. Similar cause and effect were reported by Figueras et al.[11] for a Pd Y zeolite.

In our recent study on Pt Y zeolite,[12] different states of Pt dispersion have been characterized: isolated atoms in sodalite cages, 10 Å diameter agglomerates fitting into the supercages, and 15–20 Å crystallites occluded in the zeolite crystals. In this work the catalytic activities for benzene hydrogenation and the poison sensitivity of these well defined catalysts have been measured in order to detect any zeolite matrix effect as well as intrinsic properties of highly divided metal. The influence of poison and reagent adsorption on metal electronic levels was monitored by i.r. spectroscopy using CO and NO as probe molecules.

Experimental

Materials

Platinum zeolites were prepared from a Linde Na Y zeolite by ion-exchange as previously described.[12] The zeolite compositions determined by chemical analysis are: $Pt_{10} Na_{17} (NH_4)_{19} Y$ (Sample I) ($Y = Al_{56} Si_{136} O_{384}$), $Pt_{7.5} Na_{21} (NH_4)_{20} Y$ (Sample II),

---

† Present address: Institute of Chemistry, Jagiellonian University 30-060 Kraków, Poland.

$Pt_4 Na_{48}$ Y (Sample III), and $Pt_{4·5} Na_7 (NH_4)_{40}$ Y (Sample IV). After activation and reduction the compositions are $Pt^0_{10} Na_{17} H_{39}$ Y, *etc*. . . Platinum blacks from commercial source and platinum supported on alumina prepared by H_2PtCl_6 impregnation were also used, their metal dispersion measured by H_2 chemisorption are given in Table 1.

Table 1

Poisoning and regeneration of catalytic activities. Ratio: activity/activity before poisoning for different heating temperatures.

Poisons	Samples	Disper-sion (H/Pt)	Regeneration[a] temperature (°C)						
			25	150	200	300	400	500	600
NH_3	Pt-black	0·04	1						
NH_3	10% $Pt–Al_2O_3$	0·05	1						
NH_3	5% $Pt–Al_2O_3$	0·6	0·5						
NH_3	3% $Pt–Al_2O_3$	0·9	0·25						
NH_3	II 300 R 400	1 (ag.)[b]	0	0·09	1				
NH_3	III 300 R 700	1 (ag.)	0		1				
NH_3	III 600 R 650	0·6 (cr.)[b]	0			0·3		0·7	0·3
H_2S	5% $Pt–Al_2O_3$	0·6	0			0·02	0·04	0·15	0·3
H_2S	3% $Pt–Al_2O_3$	0·9				0·02	0·08	0·30	0·40
H_2S	II 300 R 400	1 (ag.)				0·05	0·1	0·5	0·75
Thiophen	5% $Pt–Al_2O_3$	0·6	0		0·02	0·1	0·1	0·16	
Thiophen	II 300 R 400	1 (ag.)	0		0·02	0·1	0·2	0·5	

[a] = Regeneration under vacuum for NH_3 and under flowing H_2 for H_2S and thiophen.
[b] = (ag.) = 10 Å agglomerates, (cr.) = 15–20 Å crystallites.

Samples treatments and nomenclature

Treatments giving reproducible and homogeneous platinum dispersion have been described in a previous paper.[12] The treatment conditions and device used in the present investigation were similar except that the adsorption cell was replaced by a microreactor equipped with grease-free stopcocks so that it can be connected to a vacuum or gas line and to the reaction system. After activation (in flowing O_2 from 25 to 300°C and *in vacuo* at 300 or 600°C overnight) and reduction (at 300°C under 300 Torr hydrogen pressure for 3 h) the zeolites were transferred into the reactor and then evacuated at 300°C or higher temperature before catalytic activity measurements or poison adsorption.

The samples were designated according to the following examples: I 600 R 800. I means sample I, 600 is the activation temperature, R means reduction at 300°C and 800 is the final evacuation temperature. Pt Y 300 R means any sample activated and reduced at 300°C. For poisoning experiments the zeolites were desorbed and contacted during 30 min with increments of poison vapours measured by volumetry. The various poisons (NH_3, H_2S, C_4H_4S) obtained from commercial sources have been purified by repeated distillations.

Metal dispersion measurements

Platinum dispersion thoroughly studied previously[12] was checked after catalytic runs. Particle sizes were also obtained from electron microscopy photographs of zeolite thin cuttings, a detailed account on this work will be published elsewhere.[13]

Catalytic Activity Measurements

Benzene hydrogenation was performed in a conventional flow reactor. Pure benzene (Merck) was further purified by distillation and dried on Na wire. Hydrogen was purified by passage over a deoxo catalyst and molecular sieve dryer. The activities were measured at 25°C for a hydrogen pressure 690 Torr and a benzene pressure of 70 Torr. In our experimental conditions the order in respect to benzene was zero. With 10–30 mg of catalyst the conversion is sufficiently low ($< 3\%$) to avoid mass and heat transfer limitations. Catalytic activities are expressed in turnover number (N) *i.e.*, number of benzene molecules converted per metal surface atom per hour.

Infrared spectroscopy

The zeolite powder is compressed (1000 kg cm^{-2}) to obtain wafers (weight 20–40 mg, diameter 18 mm) which are placed in a quartz sample holder fitting into a grease-free i.r. cell (windows sealed by Viton joints). Treatment conditions for samples were similar to those described above. I.r. spectra were recorded with a Perkin-Elmer grating spectrophotometer model 125 (resolution 3 cm^{-1} in the spectral range investigated: 4000–1300 cm^{-1}).

RESULTS

Platinum dispersion

The differents states of Pt dispersion occurring in Y zeolites have been previously described in detail.[12] The final dispersion of Pt depends upon the activation temperature. In samples activated and reduced at 300°C the platinum forms 10 Å agglomerates fitting into the zeolite supercages. The dispersion measured by H$_2$ chemisorption is 100% up to 800°C. In samples activated at 600°C and reduced at 300°C, about 75% of the platinum is dispersed as isolated atoms in sodalite cages and the other part forms 10 Å agglomerates and 20 Å crystallites. On heating at higher temperature the isolated atoms migrate and form 15–20 Å crystallites. Plate 1(a) corresponding to the electron

a   b

PLATE 1   Electron microscopy photographs of Pt Y zeolite thin cuttings.[13] Magnification 10^6 (1 mm = 10 Å). (a) I 300 R 800 (10 Å Pt agglomerates) (b) I 600 R 750 (15–20 Å Pt crystallites).

microscopic examination of sample I 300 R 800 confirms the homogeneous distribution of Pt-agglomerates, of which the diameters range from 8 to 12 Å. Plate 1(b) shows that in sample I 600 R 750, 95% of the platinum is organized in 15–20 Å crystallites occluded in the bulk of the zeolite crystals. This means a partial destruction of the zeolite framework around the crystallites, since they are too large to fit in the supercages (12·5 Å diameter).

Catalytic activity measurements

The rates of benzene hydrogenation at 25°C have been measured on samples Pt Y 300 R (agglomerates) and Pt Y 600 R (crystallites) evacuated at increasing temperatures. The turnover numbers N are given in Figure 1. The apparent activation energy deduced

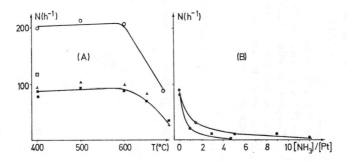

FIGURE 1   Catalytic activities at 25°C for benzene hydrogenation (turnover number N in h^{-1}). (A) N(h^{-1}) against temperature of evacuation T(°C). (B) N(h^{-1}) against the ratio (NH$_3$)/(Pt). ○: II 600 R (15–20 Å Pt crystallites), ▲: II 300 R, ■: III 300 R, ●: IV 300 R (10 Å Pt agglomerates), □: 3% Pt–γAl$_2$O$_3$ (H/Pt = 0·9).

from Arrhenius plots is 11 ± 1 kcal mol^{-1}; N measured on a 3% Pt–Al$_2$O$_3$ is also given. Its value (125 h^{-1} at 25°C) is comparable with those obtained by Dorling and Moss[14] on Pt/SiO$_2$ catalysts (87–145 h^{-1}). Results of poisoning and regeneration experiments are given in Figure 1 and in Table 1. In the first set of experiments the samples were contacted with increasing amounts of NH$_3$, the corresponding values of N have been plotted in Figure 1 as a function of the ratio: (NH$_3$ molecules)/(Pt surface atoms). In other experiments the catalysts were contacted with a large excess of poison (10 times the amount required to equilibrate all the metal atoms and acid sites of the support). Then the samples were heated at different temperatures for 1 h *in vacuo* or in flowing H$_2$ for regeneration. The ratios, activity/activity before poisoning, are given in Table 1.

Infrared spectroscopy

The stretching frequencies of CO and NO adsorbed on platinum are given in Table 2 for various samples. The interactions of C$_6$H$_6$, NH$_3$, H$_2$S, and thiophen with the metal have been investigated using preliminarily adsorbed CO at low coverage. The shift of ν_{CO} gives information on the overall flow of charge resulting from chemisorption on platinum because the ν_{CO} vibration is very sensitive to the extent of electron back-donation from the metal to the antibonding orbital of CO.[15] Results are given in Table 2. The spectrum corresponding to NH$_3$ adsorption on sample III 300 R 300 (agglomerates) partially covered by CO ($\theta = 0.1$) is given in Figure 2. On progressive addition of NH$_3$ the main features are: the shift of ν_{CO} from 2062 to 2025 cm^{-1}, the increase of the broad band at 1400–1480 cm^{-1} attributed to NH$_4^+$ ions, the correlative increase of the bands in the region 3100–3400 cm^{-1} corresponding to ν_{NH} frequencies,[16] and the decrease of the ν_{OH} bands at 3640 and 3540 cm^{-1}. On evacuation from 25 to 250°C these features are reversible and the spectrum becomes progressively similar to that observed before adsorption. The behaviour of samples containing Pt crystallites is similar, except that on evacuation the platinum is already free of ammonia at 150°C as shown by the spectrum of CO subsequently adsorbed at full coverage on this sample (Table 2).

Table 2

Infrared spectroscopy data

Samples and dispersion	Adsorbate and treatments	Probe molecule (coverage)	Stretching frequencies (cm^{-1})	Remarks and additional features
II 300 R 300 (ag.)[a]	CO[c]		ν_{CO} 2078	+ 1855 and 1910 cm^{-1} (bridged CO)
II 600 R 650 (cr.)[b]	CO[c]		ν_{CO} 2078	+ 1855 cm^{-1} (very weak)
II 300 R 300 (ag.)	NO[c]		ν_{NO} 1828	Sample with Brønsted acidity
II 300 R 650 (ag.)	NO[c]		ν_{NO} 1828	Sample with Lewis acidity
II 600 R 300 (cr.)	NO[c]		ν_{NO} 1815	Very weak (few crystallites)
II 600 R 650 (cr.)	NO[c]		ν_{NO} 1815	Strong
II 300 R 300 (ag.)	{ C₆H₆[c] / 100 Torr H_2 added }	CO (0·1)	ν_{CO} 2069 / 2040 / 2059 / 2071 / 2045	2930–2850 cm^{-1} (cyclohexane)
II 600 R 650 (cr.)	{ C₆H₆[c] / 100 Torr H_2 added }	CO (0·1)	ν_{CO} 2058 / 2065	2930–2850 cm^{-1} (cyclohexane)
II 300 R 300 (ag.)	NH₃[c] + 150°C desorption	CO[d]	ν_{CO} 2078	
II 300 R 300 (ag.)	NH₃[c] + 250°C desorption	CO[d]	ν_{CO} 2078	
II 600 R 650 (cr.)	NH₃[c] + 150°C desorption	CO[d]	ν_{CO} 2078	
II 300 R 300 (ag.)	{ H₂S[c] / 600°C H_2 and vacuum }	CO (0·1) / CO[d]	ν_{CO} 2100 / ν_{CO} 2078	
II 300 R 300 (ag.)	{ Thiophene[c] + flowing H_2 200°C / + flowing H_2 800°C }	CO[d] / CO[d]	ν_{CO} 2095 / ν_{CO} 2078	Weak

[a] 10 Å agglomerates.

[b] 15–20 Å crystallites.

[c] Molecule irreversibly adsorbed at room temperature.

[d] Subsequent addition of CO followed by desorption at room temperature.

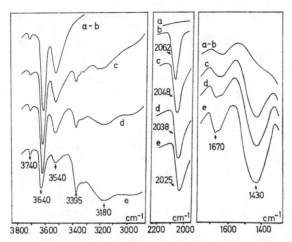

FIGURE 2   Infrared spectra of sample III 300 R 400 (10 Å Pt agglomerates) after adsorption of various amount of ammonia $\rightarrow R = (NH_3)/(Pt)$: (a) $R = 0$, (b) $R = 0.2$, (c) $R = 1$, (d) $R = 2$, (e) $R = 3.5$.

Discussion

Properties of platinum agglomerates

In Pt Y 300 R samples platinum forms agglomerates of about 10 Å diameter fitting into supercages homogeneously scattered throughout the zeolite crystals. Their size and their thermal stability up to 800°C suggest that the particle structure could be the stable icosahedral packing of 13 atoms.[17] All the Pt atoms are accessible to hydrogen (H/Pt = 1) however the agglomerate size is only slightly smaller than the supercage diameter (12·5 Å) so that a benzene molecule may only interact with encaged metal in the vicinity of the four apertures (7·5 Å) of the supercage (at most one-half of the agglomerate surface would be accessible to a benzene molecule). Benzene adsorption on platinum partially covered by CO produces a shift of ν_{CO} from 2069 to 2040 cm^{-1} (Table 2). This is due to a charge transfer from the π-electrons of benzene to the platinum which increases the electron density of the metal and subsequently the back-donation to the π^*-antibonding orbital of CO. After the treatment under hydrogen, ν_{CO} recovers almost its initial frequency (Table 2) while bands at 2930 and 2850 cm^{-1} corresponding to cyclohexane are observed. These experiments prove that benzene is adsorbed on the accessible surface of the agglomerate (probably as a π-complex) and that the site of adsorption is also the site of hydrogenation.

In addition to the reduced accessibility, diffusion limitation must be taken into consideration when comparing the activity of zeolite encaged platinum with that of platinum on a conventional support. The effective diffusivity of C_6H_6 in Y zeolite is very low (10^{-11} cm^2 s^{-1})[18] so that the platinum in inner shells has probably a smaller effectiveness factor than platinum encaged near the external surface. Although the turnover number (N) at 25°C on agglomerates (Pt Y 300 R samples) is slightly smaller (80–90 h^{-1}; Figure 1) than N for 3% Pt–Al$_2$O$_3$ (125 h^{-1}), accessibility and mass transfer considerations would indicate that the real activity of Pt atoms is higher for agglomerates than for Pt supported on alumina. The intrinsic activity of Pt Y 300 R samples does not depend upon the concentration in platinum or protons (Figure 1). It does not change from 400 to 600°C while the protons are eliminated by dehydroxylation. The creation of Lewis electron-acceptor sites has therefore no appreciable effect on the agglomerate activity. The subsequent fall of N at higher temperatures is probably due to a progressive increase of lattice defects (X-ray patterns show a broadening and decrease of line intensities) which restricts the diffusion of molecules by slight blockage of passageways.

Poisoning and regeneration experiments provide valuable information on the properties of Pt agglomerates. Figure 1 indicates that the deactivation of the zeolite is not linear and depends upon the concentration of Brønsted sites. The shape and the relative position of the poisoning curves obtained for sample III 300 R 400 and IV 300 R 400 (only differing by their proton concentration) suggest that there is a competitive adsorption between platinum and Brønsted sites for ammonia. The partition between them is complex; however, the extrapolation of the initial slope of the curves indicates that the true poisoning effect corresponds to $NH_3/Pt = 1$. Desorption of ammonia between 150 and 200°C completely regenerates the activity (Table 1). The i.r. investigation corroborates these findings. Figure 2 shows that NH_3 adsorption on Pt agglomerate (shift of ν_{CO} to lower frequencies) occurs simultaneously with NH_3 adsorption on Brønsted sites (increase of the 1430 cm^{-1} band and decrease of 3540 and 3640 cm^{-1} hydroxyl bands). The features are reversible; upon evacuation, NH_3 molecules are simultaneously desorbed from the metal and from the acid sites. The energy of Pt–NH_3 bonding is therefore comparable to the energy of formation of NH_4^+ ions, which accounts for the high sensitivity of Pt agglomerate to NH_3 poisoning. The behaviour of larger Pt particles is different. Table 1 shows that poorly dispersed Pt blacks and 10% Pt–Al_2O_3 are almost insensitive to NH_3 poisoning (at most, the activity exhibits a small decrease but regenerates under reaction course). On the other hand, there is a two-fold decrease for 5% Pt/Al_2O_3 (H/Pt = 0·6) and a four-fold decrease 3% Pt–Al_2O_3 (H/Pt = 0·9). The whole set of results can be interpreted by assuming that the smallest Pt particles and especially the Pt agglomerates are electron deficient and can therefore interact strongly with Lewis bases such as NH_3 which deactivate the metal.

The behaviour of Pt agglomerates towards sulphur-containing poisons supports these conclusions. When the electronegative sulphur atoms are fixed on Pt after H_2S dissociation or thiophen thermal decomposition, there is a charge transfer from Pt to S which results in a shift of ν_{CO} toward higher wave-numbers (2100 cm^{-1}, Table 2) since there are less electrons available for back donation to the π^*-antibonding orbital of CO. As the strength of the Pt–S bonding depends on the extent of charge transfer from Pt to S, electron-deficient particles must be more loosely bonded to sulphur. This actually happens for Pt agglomerates: although they are completely poisoned by H_2S or thiophen, their activities are more readily regenerated on heating in flowing H_2 than larger Pt particles on alumina (Table 1).

Additional evidence for the electron-deficient character of Pt agglomerates are given by NO adsorption experiments. The stretching frequency ν_{NO} of NO adsorbed on metal is shifted towards high frequencies on decreasing the number of electrons available for back-donation to the π^*-antibonding orbital of NO acting as NO$^+$ ligand.[6] Indeed, Table 2 indicates that ν_{NO} on Pt agglomerates is higher than for any other metal dispersion previously studied.[6] Moreover the fact that agglomerates encaged in a dehydroxylated zeolite (II 300 R 650) or in a hydroxylated zeolite (II 300 R 300) have similar ν_{NO} frequencies (Table 2) proves that the presence of Lewis electron-acceptor sites has no effect on the electron properties of the metal. This is in accordance with catalytic data showing no change in activity on heating from 400 to 600°C. Therefore the electron-deficient character of Pt agglomerates appears more as an intrinsic property of very small particles rather than a support effect. It must be stressed that the concept of electron-deficient character used so far, although very convenient to account for the interactions between molecules and metal clusters, has an imprecise physical meaning; however, further speculation would be hazardous since the physics of these metal clusters is hardly known.

The present investigation clearly demonstrates the implications of the electron-deficient character of Pt agglomerates on their sensitivity towards poisons. On the other hand, the effects on catalytic activities for benzene hydrogenation are not so marked. However, a study under way in the laboratory shows important effects for hydrogenolysis of ethane and propane.[19] Moreover Dalla Betta and Boudart,[10] who probably used similar samples, reported greatly enhanced catalytic activities for demanding reactions such as isomerization or hydrogenolysis of neopentane.

Properties of platinum crystallites

Previous studies[12] have shown that in samples Pt Y 600 R 300 about 75% of the platinum is dispersed as isolated atoms in sodalite cages. The remainder forms 15–20 Å diameter crystallites and there are few agglomerates in supercages. On heating at higher temperatures the isolated atoms migrate out of their cage and form 15–20 Å crystallites occluded in the bulk of the zeolites; thus at 750°C, 95% of the platinum is organized in crystallites. Since the isolated atoms do not chemisorb hydrogen[12] and are inaccessible to benzene the activity of the catalyst is merely that of the crystallites. As the number of crystallites increases at the expense of isolated atoms the chemisorption and the catalytic activity are enhanced in the same proportion up to 600°C. Therefore, the turnover number is approximately constant up to 600°C (Figure 1). The decrease observed after heating at higher temperatures (similar to that observed for Pt agglomerates) is probably due to a partial loss of zeolite crystallinity leading to pore blockage.

The growing of crystallites inside the zeolite crystals gives rise to a partial destruction of the cage walls since these crystallites are too large to fit in the supercage. The resulting arrangement of the destroyed aluminosilicate around the crystallites and its implication for the overall properties of the system are so far not clear. Although no definite conclusions on the properties of occluded Pt crystallites can be drawn, some speculations can be made. The main problem is to explain why the intrinsic activity of Pt crystallites (230 h^{-1} at 25°C) is higher than those of Pt agglomerates or of Pt/Al$_2$O$_3$. One explanation could be that the accessibility is improved with respect to that of Pt agglomerates. However, volumetric measurements indicate that Pt crystallites adsorb less CO than H$_2$, while i.r. spectroscopy shows that few bridged species are present (Table 2). Therefore the metal accessibility for CO and *a fortiori* for C$_6$H$_6$ should be poor, and cannot account for the enhanced activity. Another explanation could be that crystallites have a strong electron-deficient character, leading to enhanced activity. This seems not to be true, since NH$_3$ molecules are less strongly bonded on Pt crystallites than on agglomerates (Table 2 indicates that the initial ν_{CO} frequency is already restored at 150°C). Moreover, ν_{NO} is lower than for Pt agglomerates (Table 2). This also suggests a less pronounced electron-deficient character (see discussion about NO adsorbed on agglomerates).

One may assume than that the destroyed aluminosilicate material surrounding the crystallites could play a role in the catalytic process. In fact, the wall of a supercage is formed by 48 SiO$_4$ and AlO$_4^-$ tetrahedra, the growing of a 20 Å crystallite fed by the Pt atoms migrating from sodalite cages involves the breakage of many Al–O–Si bonds, leading probably to numerous trico-ordinated aluminium atoms. These unshielded aluminiums in the immediate vicinity of the metal surface could act in two ways in the catalytic reaction. They may provide additional sites for benzene adsorption, since they have electron-acceptor properties, or their electric field may strongly polarize the benzene molecules which become more reactive (this effect could be similar to the effect of bivalent cations leading to enhanced activity for ethylene hydrogenation).[10] This hypothesis is supported by the results of regeneration experiments after NH$_3$ poisoning. Although NH$_3$ is readily desorbed from the metal surface at 150°C, only 70% of the activity is regenerated at 500°C (whereas the activity of Pt agglomerates was completely restored at 200°C). Therefore it seems that the strong poisoning effect of NH$_3$ merely involves the electron-acceptor sites (probably trico-ordinated aluminium) surrounding the crystallites; the corrolary is that these sites could play a role in enhancing the activity of Pt crystallites. Additional experiments are, however, needed to support this hypothesis.

ACKNOWLEDGEMENTS. The authors gratefully thank Dr. Dalmai-Imelik for taking electron microscopy photographs.

REFERENCES

[1] R. Van Hardeveld and F. Hartog, *Surface Sci.*, 1969, **15**, 189.
[2] P. N. Ross, K. Kinoshita, and P. Stouhart, *J. Catalysis*, 1974, **32**, 163.
[3] P. H. Lewis, *J. Catalysis*, 1968, **11**, 162.
[4] R. C. Baetzold, *J. Chem. Phys.*, 1971, **55**, 4363.
[5] M. Cini, *J. Catalysis*, 1975, **37**, 186.
[6] M. Primet, J. M. Basset, E. Garbowski, and M. V. Mathieu, *J. Amer. Chem. Soc.*, 1975, **97**, 3655.
[7] R. C. Baetzold, *Surface Sci.*, 1972, **36**, 123.
[8] R. C. Baetzold, *J. Catalysis*, 1973, **29**, 129.
[9] J. A. Rabo, V. Schomaker, and P. E. Pickert, *Proc. 3rd Internat. Congr. Catalysis*, (North Holland, Amsterdam, 1965), Vol. 2, p. 1264.
[10] R. A. Dalla Betta and M. Boudart, *Proc. 5th Internat. Congr. Catalysis*, Palm Beach, 1972, Vol. 2, p. 1329.
[11] F. Figueras, R. Gomez, and M. Primet, *Adv. Chem. Ser.*, Amer. Chem. Soc., 1973, **121**, 480.
[12] P. Gallezot, A. Alarcon Diaz, J. A. Dalmon, A. J. Renouprez, and B. Imelik, *J. Catalysis*, 1975, **39**, 334.
[13] P. Gallezot, G. Dalmai-Imelik, and B. Imelik, *J. Microscopy, Electron Spectroscopy*, 1976, **1**, 1.
[14] T. A. Dorling and R. L. Moss, *J. Catalysis*, 1966, **6**, 111.
[15] M. Primet, J. M. Basset, M. V. Mathieu, and M. Prettre, *J. Catalysis*, 1973, **29**, 213.
[16] J. B. Uytterhoeven, L. E. Christner, and W. K. Hall, *J. Phys. Chem.*, 1965, **69**, 2117.
[17] J. J. Burton, *Catalysis Rev.*, 1974, **9**, 209.
[18] C. N. Satterfield, *Mass Transfer in Heterogeneous Catalysis*, (M.I.T. Press, Cambridge, Mass., 1970), p. 54.
[19] N. Kaufherr, J. Bandiera, M. Dufaux, and C. Naccache, to be published.

DISCUSSION

R. L. Moss (*Warren Spring Lab.*, *Stevenage*) said: The authors referred to our specific rate measurements of benzene hydrogenation over platinum on silica catalysts.[1] Expressed as a turnover number, our constant value was about 140 h^{-1} and their standard value is 125 h^{-1} at 298 K. They also found, as we did, that severe heat treatment greatly decreased the specific activity. In our work, using silica-supported catalysts, this occurred at 773–873 K and above. The onset of changes in support porosity also seemed to accompany this decline in activity and the authors advance a similar explanation, based on partial loss of zeolite crystallinity and pore blockage. Subsequently we extended our crystallite size range downwards using catalysts prepared by ammine-adsorption. These platinum on silica catalysts[2] gave no X-ray diffraction pattern, crystallites were just detectable by electron microscopy, and chemisorption indicated a mean size of about 13 Å. In contrast to the results reported by the authors, our specific activity remained at the 'standard' value,[3] whereas they found a more active

[1] T. A. Dorling and R. L. Moss, *J. Catalysis*, 1966, **5**, 111.
[2] T. A. Dorling, M. J. Eastlake, and R. L. Moss, *J. Catalysis*, 1969, **14**, 23.
[3] T. A. Dorling and R. L. Moss, unpublished work.

series of catalysts, containing 15–20 Å crystallites. Therefore, the authors are probably correct to seek an explanation based on some interaction between platinum crystallites and the zeolite support (or some alternative explanation), rather than any effect *per se* of platinum crystallite size.

P. Gallezot (*CNRS, Villeurbanne*) replied: We would draw your attention to the fact that on the same catalysts, an effect of crystallite size has been observed for a structure-sensitive reaction. At the Fifth Ibero-American Symposium on Catalysis in Lisbon later this month we shall report that the hydrogenolysis of isobutane is five times faster on the 10 Å agglomerates than on the 15–20 Å crystallites. This is attributed to stronger adsorption of hydrocarbon on the agglomerates than on the crystallites. This interpretation is consistent with the present results, which show that the smaller particles are strongly bonded to electron-donor species.

P. A. Jacobs (*Catholic Univ. of Louvain*) said: You clearly state in your paper that the platinum agglomerates are electron-deficient and that this is a property of small particles in zeolites rather than a support effect. Recently we advanced a mechanism for the reduction of transition ions in Y-zeolite.[4] Based on kinetic arguments, it seems that the rate-limiting step in the reduction is the collision of metal ions with metal agglomerates which generates positively loaded metal agglomerates. Such a reduction mechanism also applies to the case of platinum, thus giving rise to electron-deficient platinum agglomerates.

P. Gallezot (*CNRS, Villeurbanne*) replied: We do not believe that the formation of platinum agglomerates primarily requires the migration of Pt^{2+} cations towards a metal seed where the reduction would proceed. More likely, the Pt^{2+} are readily reduced because of the high reduction potential of platinum; then the reduced atoms, no longer electrostatically bonded to the framework oxygens, readily migrate from cage to cage and meet other atoms to form an agglomerate. Anyway, as stated in the paper, the concept of electron deficiency that we have used has an imprecise physical meaning. It is even misleading in so far as it suggests that the metal particles have lost electrons. In order to describe the intrinsic electronic properties of these very small particles, it is better to use a concept of electron availability which is directly related to the ionization potential of metals. The ionization potential of a given metal is higher for the element than for the bulk, and recent theoretical work demonstrates that it gradually increases from the bulk to the element value as the diameter of a very small particle decreases. Thus, because the ionization potential is higher there are fewer electrons available in the 10 Å agglomerate than in the 15–20 Å crystallite or larger particles for back-donation to the π^*-orbitals of nitric oxide. Also the electron transfer from platinum to the electronegative sulphur atom must be lower, resulting in a weaker Pt—S bonding, whereas an electron donor such as an ammonia molecule can be more strongly bonded.

R. C. Hansford (*Union Oil Co., Brea*) said: Speculations in this and other recent papers, ascribing unusual catalytic behaviour of metals supported on or in zeolites to electron deficiency of the metal particles, do not consider a probable alternative explanation for the observed enhanced activity and sulphur resistance of such catalysts. It is possible that the electron-accepting (acid) surface is not interacting at all with the metal agglomerate to produce a sort of 'transmutation' of the metal to the electronic structure of the next lower element (*e.g.* platinum is said to become more like iridium in catalytic properties). Perhaps the hydrocarbon reactant is interacting with the acid centres of the surface, producing a highly active carbonium ion. This activated species

[4] P. A. Jacobs, W. De Wilde, R. A. Schoonheydt, J. B. Uytterhoeven, and H. Beyer, *JCS Faraday I*, 1976, 1221.

then can react very rapidly with hydrogen, which may dissociate on the metal surface to form hydride ions and protons. Hydride ions add rapidly to the carbonium ion and protons simultaneously propagate the activation process.

We reported work at the First North American Chemical Congress in Mexico City in 1975, showing a forty-fold increase in turnover number for the hydrogenation of aromatic hydrocarbons over platinum supported on silica–alumina in an alumina matrix, compared to the specific rate for platinum supported on the bare alumina matrix. The density of electron-accepting sites in the surface of the silica–alumina component cannot be sufficient to interact with more than about 1% of the platinum atoms in the 2 nm particles in both catalysts as calculated from dispersion measurements with hydrogen. Therefore it was concluded that, in this case, it is very unlikely that electron-deficient platinum agglomerates can account for the high specific activity of the acid-supported metal. The alternative explanation of activation of the reactant molecules seems more reasonable. Similar conclusions were reached in studies with various cationic forms of Y-zeolite loaded with palladium and used as catalysts for the hydrogenation of tetralin. The more acidic the cation, the higher is the specific hydrogenation rate.

P. Gallezot (*CNRS, Villeurbanne*) replied: Our paper has been misunderstood. We have not concluded that there is an electron transfer from the metal to the support (although this may conceivably occur in other systems). On the contrary, we have emphasized that the concept of electron-deficient character, although bearing little physical meaning, is very convenient to account for the interactions between poisons or probe molecules with the 10 Å agglomerates, and that it is an intrinsic electronic property of the very small particles rather than a support effect. The concept of electron deficiency is not appropriate here and must be replaced by other concepts connected with the particular ionization potential of very small particles (see my reply to Dr. Jacobs). In the case where a higher catalytic activity is observed (for the 15–20 Å crystallites), this behaviour is attributed to the presence in the vicinity of the metal surface of acid sites which could play a role *per se* in the catalytic process. To a certain extent this corresponds to your own interpretation.

M. Boudart (*Stanford Univ.*) said: Our observations on PtY zeolites, which we reported at Palm Beach are very similar to yours. This prompts me to ask whether you have encountered the same difficulty in preparing these systems as we did. Attempts to prepare more than 500 mg of catalyst at a time failed. We attributed this to the need to maintain very low partial pressures of water vapour during the reduction of the catalyst precursor. How much catalyst have you succeeded in preparing in a single batch?

P. Gallezot (*CNRS, Villeurbanne*) replied: Thermal activation before reduction is indeed a very critical step in preparing these systems because reduction to give large crystallites is liable to occur when the $[Pt(NH_3)_4]^{2+}$ complex cations decompose within the 473–573 K temperature range. Together with the space velocity of oxygen and the rate of heating, the amount of treated powder is a very important factor in this untimely reduction process. We never prepared more than 200 mg of catalyst in a single batch. We believe that it is the residual pressure of ammonia rather than the pressure of water vapour which plays a critical role in this reduction and sintering process. In a paper to be presented at the Fifth Ibero-American Symposium on Catalysis, we show that the 10 Å agglomerates can be converted into large crystallites by heating at 573 K in presence of a low pressure of ammonia; this process involves the complexation and the transport of platinum atoms from the disintegrating agglomerates to the growing crystallites.

F. Figueras (*CNRS, Villeurbanne*) (communicated): The results presented here support the view that no support effect exists with platinum on zeolites. By contrast we proposed the existence of a support effect for zeolite-supported palladium,[5] as was also proposed for supported iridium;[6] in that case a shift of the binding energy of iridium was detected by ESCA when silica–alumina was used as a support. Palladium and iridium have ionization potentials of 8·3 eV compared with a value for platinum of 9·2 eV, and it is therefore likely that the effects of the support are greater in the cases of the first two metals. As the effect on catalytic activity was found to be rather small for palladium, no real disagreement exists between these results and our previous findings.

[5] F. Figueras, R. Gomez, and M. Primet, *Molecular Sieves* (Advances in Chemistry Series No. 121, The American Chemical Society, Washington, 1973), p. 480.
[6] J. Escard, C. Leclère, and J. P. Contour, *J. Catalysis*, 1973, **29**, 31.

Nitrogen Chemisorption on Reduced Cobalt Oxide with and without K_2O

HISAO SUZUKI and ISAMU TOYOSHIMA*

Research Institute for Catalysis, Hokkaido University, Sapporo Japan, 060

ABSTRACT It has been shown that both reduced cobalt oxide catalysts and those promoted with potassium oxide will chemisorb nitrogen at about 200—350°C. In the case of the promoted catalyst, nitrogen is adsorbed even at room temperature; at least two different types of adsorption occur at temperatures ranging from room temperature to 220°C, and the rate of adsorption is proportional to (nitrogen pressures)$^{\frac{1}{2}}$, whereas it is proportional to nitrogen pressure on the unpromoted catalyst at around 200°C.

The co-adsorption of hydrogen and nitrogen has also been investigated. Above 200°C, no effect of hydrogen on adsorption of nitrogen is observed on the promoted catalyst. Below 200°C, adsorption of nitrogen is inhibited by preadsorbed hydrogen. In addition, a partial blocking effect due to nitrogen adsorption on the adsorption of hydrogen is also observed.

When nitrogen adsorbed on the promoted catalyst is brought into contact with hydrogen, ammonia is readily formed even at 50°C.

INTRODUCTION

The effect of the promoter on iron catalysts for ammonia synthesis has been studied in connection with the elucidation of the mechanism of ammonia decomposition[1–4] and synthesis,[5] and with investigations of adsorbed states of nitrogen.[6–8] In particular, the important role of the potassium oxide in the catalyst in the mechanism of ammonia decomposition has been pointed out.[1,2,4] However, work along the same lines has not previously been carried out in detail for nickel and cobalt catalyst, since these are currently accepted as being inactive in respect of the chemisorption of nitrogen, although they belong to the same group of the Periodic Table as iron.[9–11] Among these transition metals, it was found that chemisorption of nitrogen occurs on nickel at —195°C as well as at higher temperatures.[11–14] Hence it is difficult to think of a reason why chemisorption of nitrogen could not be detected on a cobalt catalyst. We have therefore examined nitrogen adsorption on such a catalyst, and recently found evidence for nitrogen chemisorption on well reduced cobalt oxide at about 200°C, in the presence of potassium oxide as a promoter and in its absence,[15] and have also shown that nitrogen chemisorbed on a cobalt catalyst promoted with potassium oxide is very readily hydrogenated to ammonia at temperatures as low as 70°C.[16]

We have therefore subjected to detailed study the adsorbed states of nitrogen as well as the reactivity of adsorbed nitrogen both on promoted and on unpromoted cobalt catalysts.

EXPERIMENTAL

Catalyst Preparation. Basic cobalt carbonate was precipitated by adding a solution of ammonium carbonate (110 g) in distilled water (2 l) to cobaltous chloride (248 g) distilled water (500 ml) over 1h with stirring at room temperature. The precipitate was filtered, washed until the filtrate was free from chloride ion, and then dried overnight at 140–170°C. The carbonate was heated at 320°C for 5 h and then at 200°C overnight in air to give Co_3O_4. The promoted catalyst was prepared by soaking 15 g of this oxide in 250 ml of 1 N-KOH solution for 5 h at room temperature, then filtering and drying at 100°C for 29 h. The promoted catalyst thus prepared contained 3·54% of potassium oxide. In the preparation of the catalyst, guaranteed grade reagents were used throughout.

These two forms of Co_3O_4 (*i.e.* the unpromoted and promoted forms) were degassed *in vacuo* at 200°C and then reduced at 300°C for 72 h under purified hydrogen at a flow rate of 200 ml/min. The unpromoted catalyst was further reduced at 400°C for 24 h, because it did not show appreciable nitrogen adsorption at 200°C.

Purification of gas. Gases used were carefully purified in the same manner as in the previous work.[3]

Adsorption of nitrogen. The adsorption of nitrogen was performed volumetrically at pressures of 4·0–20·0 × 10^3 Pa and temperatures ranging from 25 to 350°C. Before each adsorption run, the catalyst was reduced overnight as described above and then evacuated to 133 μPa.

The surface areas of the catalysts in reduced form were determined by the BET method with argon, since nitrogen chemisorption occurs on cobalt surface at —195°C.[11] The surface areas of unpromoted and promoted catalysts were 1·20 and 1·16 m^2 g^{-1}, respectively.

Hydrogenation of adsorbed nitrogen. Adsorbed nitrogen was removed as ammonia by passing a stream of hydrogen (atmospheric pressure) over the catalyst at the temperature at which nitrogen had been adsorbed at the required pressure. After ammonia formation ceased at this temperature, the temperature of the catalyst was raised to 300°C. The amount of ammonia thus formed was determined by absorption in sulphuric acid solution with a pH meter. In a few experiments, the nitrogen in the gaseous phase and/or of weakly adsorbed states was pumped out at the adsorption temperature before hydrogen was passed over the catalyst. This procedure gives information about the type of adsorbed nitrogen that takes part in the reaction.

RESULTS AND DISCUSSIONS

Adsorption of Nitrogen. Fig. 1 shows typical results for nitrogen adsorption as a function of time. The unpromoted catalyst showed low adsorption at 250°C and 300°C, at 13·6 × 10^3 and 11·3 × 10^3 Pa, respectively, while the promoted catalyst adsorbed nitrogen appreciably even at 200°C, at 13·3 × 10^3 Pa. For both catalysts, the amount of nitrogen adsorbed increases at higher temperature, indicating chemisorption as shown in Fig. 1. The initial rate of adsorption on the promoted catalyst was ten times higher than that for the unpromoted catalyst at 250°C. It should be noted that the promoted catalyst has the lower reduction temperature and the higher adsorption ability, whereas the unpromoted one has the higher reduction temperature and the lower nitrogen adsorption ability. In the latter catalyst, therefore, adsorption at 200°C could not be detected. The unpromoted catalyst prepared from cobalt nitrate also showed similar adsorption behaviour.

On the other hand, the promoted actalyst showed excellent activity for nitrogen adsorption above room temperature. It should be noted that the catalyst containing 15·6% potassium oxide showed similar adsorption behaviour to this catalyst. Fig. 2 shows typical results for nitrogen adsorption at 70°C as a function of time. It can be seen that the initial uptake of nitrogen is very rapid, and then this is followed by slow adsorption. The rate curves shown in this Figure are different from those obtained at higher temperatures, as shown in Fig. 1, in which no marked initial uptake of nitrogen can be observed.

At 70°C, the promoted catalyst was exposed to nitrogen for 5 h, then the nitrogen was pumped off at the same temperature and adsorption of nitrogen again followed on the surface covered with irreversibly bound nitrogen. The results are shown in the curve marked by the filled circles in Fig. 2. The rate curve is quite similar to that over the virgin catalyst. These results show that at least two different states of adsorbed nitrogen exist on promoted catalysts, one that is adsorbed irreversibly and another that is adsorbed reversibly on the surface. The amount of the former relative to the latter increases with increase in temperature, as shown in Fig. 3.

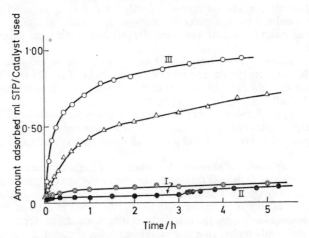

FIGURE 1   Nitrogen adsorption as a function of time. I: Adsorption on unpromoted catalyst at 250°C and 13·6 × 10³ Pa. II. Change of adsorption temperature: from 250 to 300°C. Catalyst used: 4 g (as oxide). III: Adsorption on promoted catalyst at 200°C and 13·3 × 10³ Pa; catalyst used: 8 g (as oxide).

△: promoted catalyst at 140°C and 11·3 × 10³ Pa
⊙: unpromoted catalyst at 350°C and 14·0 × 10³ Pa.

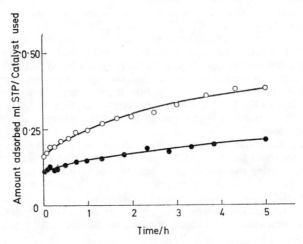

FIGURE 2   Nitrogen adsorption on promoted catalyst as a function of time at 70°C and 18·6 × 10³ Pa.

○: Total adsorption of nitrogen
●: Adsorption of nitrogen after evacuation for 30 min.

710

The irreversible adsorption of nitrogen is chemisorption, while the reversible adsorption may be partly physical adsorption or a weak chemisorption. Above 200°C, only irreversibly adsorbed nitrogen is present, and below 200°C both types of adsorbed nitrogen can be found. In addition, further results distinguish the behaviour of the two types; the irreversibly adsorbed nitrogen is easily removed by hydrogen, while the rate of irreversible nitrogen adsorption is proportional to (nitrogen pressure)$^{\frac{1}{2}}$, as will be shown in the next section.

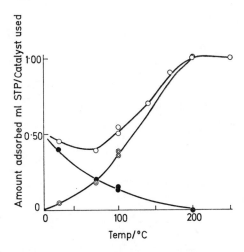

FIGURE 3 Adsorbed amount of nitrogen as a function of temperature at 18·6 × 10³ Pa.

○: Total adsorption
⊙: irreversibly adsorbed nitrogen
●: reversibly adsorbed nitrogen

Order of adsorption rate and activation energy for adsorption. Nitrogen adsorption over the promoted catalyst was performed at different pressures in order to study the dependence of the rate upon nitrogen pressure, as well as the adsorbed states of nitrogen. The pressure dependency was estimated from the log of the initial rates against the log of pressures of nitrogen over the promoted catalyst at 200°C, at which temperature only chemisorption of nitrogen occurred. On the other hand, the unpromoted catalyst could not chemisorb nitrogen at this temperature; therefore the rate was measured at 250°C. The pressure dependence of nitrogen upon the rate was close to 0·5 over the promoted catalyst, whereas it was close to unity over the unpromoted catalyst. With the former catalyst, it was also confirmed by plots of the amount of adsorption against $\sqrt{}$(pressure × time of adsorption); *i.e.* all points measured at the same temperature and three different pressures lay on the same line.

The one-half-order rate law observed is of particular interest with the promoted catalyst, because a simple first-order adsorption occurs on the unpromoted catalyst. This fact suggests that potassium oxide helps to dissociate adsorbed nitrogen molecules.

The activation energy for adsorption was obtained from Arrhenius plots on the two catalysts by using the initial rates. The values obtained were 58·6—71·2 and 42·7 kJ mol^{-1} over the unpromoted and the promoted catalysts, respectively. The order of reaction and the activation energy for nitrogen adsorption also support the view that the adsorption is chemisorption, with the adsorbed species probably molecular on the unpromoted catalyst, but atomic on the promoted catalyst. With the latter catalyst, the rate may be controlled by surface diffusion of nitrogen atoms, as already observed in the case of nitrogen adsorption on iron films.[17] However, the possibility of lattice penetration of atoms cannot be ruled out.

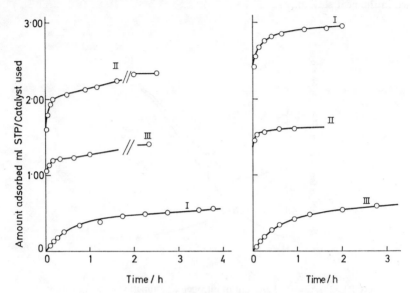

FIGURE 4   Co-adsorption of nitrogen and hydrogen at 200°C.

Left: Hydrogen adsorption on the surface covered with preadsorbed nitrogen.

I: N_2 at $14 \cdot 4 \times 10^3$ Pa
II: H_2 adsorption after evacuation at $9 \cdot 18 \times 10^3$ Pa
III: H_2 adsorption after evacuation at $10 \cdot 9 \times 10^3$ Pa

Right: Nitrogen adsorption on the surface covered with preadsorbed hydrogen.

I: H_2 adsorption on at $9 \cdot 18 \times 10^3$ Pa
II: H_2 adsorption after evacuation at $10 \cdot 9 \times 10^3$ Pa
III: N_2 adsorption at $14 \cdot 8 \times 10^3$ Pa

Simultaneous adsorption of nitrogen and hydrogen. In order to study the interaction of hydrogen and nitrogen, each component was adsorbed on the promoted catalyst covered with the other component irreversibly adsorbed, *i.e.* after adsorption of hydrogen (or nitrogen), the system was evacuated to $13 \cdot 3$ μPa for 30 min and then nitrogen (or hydrogen) adsorption was followed at the same pressure and temperature at which hydrogen (or nitrogen) adsorption had been carried out.

Fig. 4 shows typical data for nitrogen adsorption on the bare surface and on the surface covered with irreversibly bound hydrogen, as well as the effect of nitrogen on subsequent hydrogen adsorption at 200°C. Similar measurements were also carried out at 100—200°C. It was found that hydrogen either exerts no effect, or else completely inhibits nitrogen adsorption, depending on the temperature. On the other hand, preadsorbed nitrogen partly inhibits hydrogen adsorption on the surface at any temperature. In either case, however, adsorption of hydrogen is initially very rapid, and is followed by slow adsorption; the amount of reversibly adsorbed hydrogen is not much influenced by whether preadsorbed nitrogen is present or not. But the amount of irreversibly bound hydrogen is decreased appreciably in the presence of preadsorbed nitrogen, and the amount almost corresponds to the amount of nitrogen adsorbed. On the basis of these results, it may be concluded that nitrogen could be adsorbed on the same site as hydrogen. However, above 200°C, more sites on which nitrogen could be adsorbed are still free of hydrogen adsorption, whereas below 200°C, all the sites available for nitrogen adsorption could be occupied by preadsorbed hydrogen.

It is interesting to compare the present results with those obtained on iron catalysts used in ammonia synthesis. Brunauer and Emmett, and Zwietering *et al.*, observed that preadsorbed nitrogen enhanced Type B chemisorption of hydrogen over a singly promoted catalyst,[18,19] whereas it decreased hydrogen chemisorption over a doubly promoted ion catalyst.[19] Conversely, as to the effect of chemisorbed hydrogen, the nitrogen exchange reaction over an iron catalyst is accerelated by adsorbed hydrogen.[20-24] Tamaru also noted the accerelating effect of hydrogen on nitrogen adsorption.[25] However, Takezawa and Emmett recently showed definitely that hydrogen exerted no influence on the adsorption of nitrogen on the iron catalysts used in ammonia syntheses.[26] More recently, Kazusaka and one of the present authors showed that the effect of hydrogen on the nitrogen exchange reaction was observed over a singly promoted iron catalyst, while no effect was found on doubly promoted catalysts, and pointed out that the relative amounts of alumina and potassium oxide in the catalyst must be the starting point in consideration of promoter action.[4,27] So hydrogen either exerts no effect or else accelerates nitrogen adsorption, depending on the temperature and on the nature of the catalyst.

Turning now to the present experimental results with the promoted cobalt catalyst, preadsorbed nitrogen inhibits hydrogen adsorption, whereas preadsorbed hydrogen exerts no effect on nitrogen adsorption above 200°C. However, below 200°C, preadsorbed hydrogen prevents nitrogen adsorption completely. Thus the formation of NH or NH_2 on the surface of the present catalyst would not be expected, and the promoted cobalt catalyst resembles the doubly promoted iron catalyst very closely in adsorption behaviour.

Hydrogenation of adsorbed nitrogen. From the point of view of ammonia synthesis, it appeared to be of interest to investigate the reactivity of nitrogen adsorbed on the surface of promoted cobalt, on which two different types of adsorbed state exist even at room temperature.

When a stream of hydrogen was passed over the catalyst covered with preadsorbed nitrogen, the nitrogen was readily hydrogenated to ammonia even at 50°C. At this temperature, ammonia was detected in the outlet hydrogen stream by Nessler's reagent. Above 70°C, the amount of ammonia formed was followed by the method described in the experimental section. At 70°C, the total amount of adsorbed nitrogen was 0·314 ml at $19·2 \times 10^3$ Pa. The amount of nitrogen removed as ammonia was equal to 0·150 ml in 30 min, which was close to the amount of irreversively adsorbed nitrogen. However, if the temperature of the catalyst was gradually increased from 70 to 300°C, the amount removed reached 0·258 ml, which was less than the total amount of adsorbed nitrogen. It appears that reversibly bound nitrogen may be transformed into the reactive form with increase of temperature during the hydrogenation.

Table 1 shows the results obtained under the various experimental conditions. As seen from the Table, the amount of nitrogen removed as ammonia at 300°C was less than that of adsorbed nitrogen at every temperature. Also, removal of nitrogen adsorbed on the surface by hydrogenation at the adsorption temperature appears to be not such an easy process, as seen from the results of the experiment at 200°C, where the amount of nitrogen removed as ammonia amounted to only 0·75 ml for 2h. The balance of the nitrogen probably remains adsorbed on the surface. When the promoted cobalt catalyst on which nitrogen and hydrogen were coadsorbed at 200°C was brought in contact with stream of He, no NH_3 could be detected in the gas phase. Then, on replacement of He with hydrogen, NH_3 formation was observed. Thus it seems unlikely that NH or NH_2 are formed during the coadsorption. Attempts have been made to

Table 1

Hydrogenation of adsorbed Nitrogen.

Temperature °C	Pressure Pa	N_2 adsorbed (Total) ml(STP)/Cat. used	N_2 removed as ammonia at adsorption temperature ml(STP)	at 300°C ml(STP)
100	$(13·8 \times 10^3)$	0·428	0·109(20)[a]	0·289
	$(5·46 \times 10^3)$	0·365	0·185(45)	0·287
170	$(5·22 \times 10^3)$	0·773	0·752(120)	—
200	$(4·92 \times 10^3)$	0·957	0·756(120)	0·967
250	$(11·2 \times 10^3)$	1·012	0·795(45)	0·910
	$(11·2 \times 10^3)$	1·017	0·825(25)	0·844

[a] Figure in parentheses shows the time(min) at the adsorption temperature during hydrogenation of adsorbed nitrogen.

produce ammonia by using ammonia synthesis gas with the stoicheiometric composition at above 70°C, so far without success. This may be due to the inhibiting effect of hydrogen adsorption on activated nitrogen adsorption.

The present work suggests that K_2O helps to dissociate nitrogen molecules, and nitrogen chemisorbed on K_2O-promoted catalyst is readily hydrogenated to ammonia at low temperatures.

ACKNOWLEDGEMENT. We thank Dr. N. Takezawa for valuable discussions during the course of this work.

REFERENCES

[1] K. S. Love and P. H. Emmett, *J. Amer. Chem. Soc.*, 1941, **63**, 3297.
[2] N. Takezawa and I. Toyoshima, *J. Catalysis*, 1966, **6**, 145.
[3] N. Takezawa and I. Toyoshima, *J. Catalysis*, 1970, **19**, 271.
[4] N. Takezawa, I. Toyoshima, and A. Kazusaka, *J. Catalysis*, 1972, **25**, 118.
[5] A. Ozaki, K. Aika, and Y. Morikawa, *Proc. 5th Internat. Congr. Catalysis*, Miami Beach, 1972, (North-Holland, Amsterdam, 1973) Vol. 2, p. 1251.
[6] N. Takezawa and P. H. Emmett, *J. Catalysis*, 1968, **11**, 131.
[7] N. Takezawa, *J. Phys. Chem.*, 1966, **70**, 597.
[8] N. Takezawa, *J. Catalysis*, 1972 **24**, 417.
[9] G. C. Bond, *Catalysis by Metals*, (Academic Press, London and New York, 1962).
[10] D. O. Hayward and B. M. W. Trapnell, *Chemisorption*, (Butterworth, London, 1964).
[11] R. J. Kokes, *J. Amer. Chem. Soc.*, 1960, **82**, 3018.
[12] R. J. Kokes and P. H. Emmett, *J. Amer. Chem. Soc.*, 1958, **80**, 2082.
[13] R. J. Kokes and P. H. Emmett, *J. Amer. Chem. Soc.*, 1960, **82**, 1037.
[14] N. Takezawa, *Nippon Kagaku Kaishi*, 1970, **91**, 43.
[15] I. Toyoshima, N. Takezawa and H. Suzuki, *J. C. S. Chem. Comm.*, 1973, 270.

[16] N. Takezawa, I. Toyoshima, and H. Suzuki, *Z. phys. Chem. (Frankfurt)*, 1974, **89**, 323.
[17] E. Greenhalgh, N. Slack, and B. M. W. Trapnell, *Trans. Faraday Soc.*, 1956, **52**, 865.
[18] S. Brunauer and P. H. Emmett, *J. Amer. Chem. Soc.*, 1940, **62**, 1732.
[19] C. Bokhoven, C. van Heerden, R. Westrik, and P. Zwietering, *Catalysis*, (Reihold, New York, 1955), ed. P. H. Emmett, Vol. 3, Ch. 7, p. 265.
[20] G. G. Joris and H. S. Taylor, *J. Chem. Phys.*, 1939, **7**, 893.
[21] J. T. Kummer and P. H. Emmett, *J. Chem. Phys.*, 1951, **19**, 289.
[22] J. P. McGeer and J. S. Taylor, *J. Amer. Chem. Soc.*, 1951, **73**, 2743.
[23] G. K. Boreskov, A. I. Gorbunov, and O. L. Masanov, *Doklady Akad. Nauk S.S.S.R.*, 1958, **123**, 90.
[24] A. I. Gorbunov and G. K. Boreskov, *Problemy Kinet. Katal.*, 1960, **10**, 192.
[25] K. Tamaru, *Proc. 3rd Internat. Congr. Catalysis* (North-Holland, Amsterdam, 1965), Vol. 1, p. 664.
[26] N. Takezawa and P. H. Emmett, personal communication.
[27] A. Kazusaka, *Doctoral Thesis*, Hokkaido Univ., 1972.

DISCUSSION

J. J. Fripiat (*CNRS, Orleans*) said: We reported on the grafting of methylvinylsilane on the surface of partially hydrolysed magnesium silicates in 1966, 1969, and 1975 and have evidence (unpublished) that the vinyl group is able to fix osmium tetroxide at a molecularly dispersed level. After reduction, a structure can be formed, which is able to chemisorb dinitrogen between 0 and 200 °C. Adsorption isotherms are linear; equilibrium is obtained rapidly, the rate being limited by diffusion from the gas phase. It is proposed that dinitrogen forms a bridge between two osmium atoms.

I. Toyoshima (*Hokkaido Univ.*) replied: The rate of adsorption at about 200 °C is proportional to the square root of the nitrogen pressure for promoted catalysts whereas it is proportional to the first power of the nitrogen pressure for unpromoted catalysts. Therefore, the adsorbed nitrogen may be atomic on the former catalyst, and molecular on the latter. We are not sure whether the molecular form is a bridged structure.

A. B. Stiles (*Univ. of Delaware*) said: You report a marked increase in the extent of nitrogen chemisorption by cobalt oxide when it is promoted with potassium oxide. Did you investigate a range of potassium oxide content and, if so, what was the optimum content?

M. Schiavello (*Univ. Rome*) asked: Was molecular nitrogen adsorbed on potassium oxide or on metallic potassium in these specimens? Was potassium oxide reduced during the reduction procedure? Have you studied the adsorption of molecular nitrogen on (i) potassium oxide as such, and (ii) potassium oxide treated with molecular hydrogen in the manner used in the preparation of the promoted catalysts?

I. Toyoshima (*Hokkaido Univ.*) replied: In reply to Prof. Stiles, I can say that we observed that a catalyst containing 15·6% potassium oxide adsorbed almost the same amount of nitrogen as one containing 3·54% K_2O. This observation serves also to show Prof. Schiavello why we consider that molecular nitrogen is not chemisorbed on potassium oxide or on metallic potassium in our system. Potassium oxide may not be reduced at our low reduction temperatures. We have not investigated the adsorption of nitrogen on potassium oxide previously exposed to hydrogen.

J. G. van Ommen (*Tech. Univ. Twente, Enschede*) said: You state that potassium oxide effects the dissociation of nitrogen molecules which chemisorb on the promoted cobalt catalyst. How sure are you that potassium is present as the oxide? From the way you prepare your catalyst I suggest that potassium is not present as K_2O but as KOH. Thermodynamic calculations show that potassium hydroxide is more stable than potassium oxide or potassium metal, even if there is less than 0·01 p.p.m. water present in the gases used for reduction.[1] (**Dr. Z. M. George** (*Alberta Research Council, Edmonton*) also enquired about the possible uniform distribution of potassium hydroxide over the catalyst.)

I. Toyoshima (*Hokkaido Univ.*) replied: We do not know the form of potassium in the catalyst because structural studies were not performed. Your opinions about potassium hydroxide may be right.

[1] J. G. van Ommen, W. J. Bolink, J. Prasad, and P. Mars, *J. Catalysis*, 1975, **38**, 120.

Forms of Interaction of Oxygen, Nitrogen, and Hydrogen with the Ammonia Synthesis Catalyst

A. V. KRYLOVA,* S. S. LACHINOV, N. S. TOROCHESHNIKOV, and
N. N. VOKHMYANIN

D. I. Mendeleyev Institute of Chemical Engineering, Moscow, U.S.S.R.

ABSTRACT   Forms of interaction of oxygen, nitrogen, and hydrogen with industrial ammonia catalyst have been studied to investigate the mechanisms of pyrophoric catalyst stabilisation, and the sequence of the stages in ammonia synthesis. A mechanism of stabilisation is suggested which involves the interaction of the oxygen with hydrogen dissolved in the catalyst. The stabilised catalyst may be activated not only in hydrogen, but also *in vacuo* and in currents of inert gases or nitrogen at 720—820 K. The interaction of adsorbed species of nitrogen and hydrogen at catalysis temperatures indicates that chemisorption of nitrogen is the first stage of the ammonia synthesis, whereas the second stage lies in nitrogen-induced chemisorption of hydrogen.

INTRODUCTION

The active solid surface is built up on exposure to the reaction mixture.[1,2] Adsorption processes may also modify the surface while preparing the catalyst or poisoning it.

An intestigation of the interaction of oxygen with the ammonia synthesis catalyst seemed of interest in order to explain the mechanism of the stabilisation of pre-reduced pyrophoric catalysts and the activation of stabilised catalysts in the converter. In an effort to remove the pyrophoric nature of the reduced catalysts, Temkin and Pyzhev stabilised catalysts at a low temperature in a stream of a nitrogen–hydrogen mixture containing a small quantity of oxygen.[3] The stabilisation of commercial catalysts is carried out nowadays in a stream of nitrogen containing admixed oxygen; activation is performed in a nitrogen–hydrogen mixture. Stabilised catalysts require special storage conditions.[4]

The same decrease in the work function of stabilised ammonia catalysts during re-reduction in hydrogen and during their high temperature treatment *in vacuo*[5] is an indication that the stabilised layer can be removed in the absence of hydrogen. In this connection the present paper deals with the nature of the stabilising layer and the possibility of activating stabilised catalysts in the absence of hydrogen.

Although there has already been extensive investigation of the interaction of nitrogen and hydrogen with the ammonia catalyst, it is only recently that the various forms of adsorption have been discussed.[6-8] In the present work the interaction of various forms of chemisorption of nitrogen and hydrogen was studied to determine the role of reactants in the formation of active centres on the catalyst surface and to help elucidate the sequence of stages in the mechanism.

EXPERIMENTAL

The industrial fused iron catalysts containing 0·8 mass % K_2O, 2·6% CaO, and 2·9% Al_2O_3 were generally used. The pre-reduction was carried out with a current of a nitrogen–hydrogen mixture under industrial or laboratory conditions at a pressure of 5—10 MPa. The pre-reduced catalysts were stabilised in a stream of nitrogen containing 0·1—0·2 volume % of oxygen. The specific surface was determined by the BET method from the low temperature adsorption of nitrogen or argon. Mercury porometry was used to examine the samples' porous structures.

In order to study the adsorption properties of catalysts, the stabilised samples were re-reduced in a current of hydrogen on stepwise heating to 800 K. The activation of stabilised catalysts was also carried out in the absence of hydrogen by heating *in vacuo* (*ca*. 0·1 mPa) to 870 K in a stream of inert gases or nitrogen. The interaction of oxygen, nitrogen, and hydrogen with the catalyst was studied by a volumetric method in a static system and by pulse chromatography. In the chromatographic experiments the rate of the gas-carrier flow varied over a range of $0·2—1·2 \ \mu m^3/s$; the pulse of adsorbate was introduced by a control cock or by syringe. Strong chemisorption was determined by the amount of adsorbed gas, and weak chemisorption by comparing the time neces-sary to retain equal portions of the adsorbate and non-adsorbing gas. Thermal desorp-tion was investigated in a linear heating schedule to 870—970 K and at a rate (β) of 0·05—3·30 K/s by various methods. The gases from cylinders (He, Ar, H_2, N_2) were purified from oxygen and dried. In static systems, nitrogen was made by decomposing sodium azide; hydrogen was obtained in the Varian Aerograph model 9652; oxygen was produced by decomposition of potassium permanganate.

Measurement of the electron work function of catalysts was performed by the contact potential difference method with a glazed molybdenum vibrating electrode as the stan-dard. Exoelectronic emission was registered *in vacuo* with a secondary electron multiplier. Derivatographic measurements were conducted on the MOM type deriva-tograph. Depending on the experiment, the weighed portion of the catalyst varied from 10 mg to 10 g, and the grain size from 1—2 mm to 0·01 mm.

RESULTS

The interaction of the catalyst with oxygen

Deuterium adsorption at 293 K on the reduced and evacuated catalyst, with sub-sequent heating of the samples to 770 K, gives rise to HD and HDO peaks in the mass spectra of the products, which is indicative of the probability of hydrogen being in the atomic state in the catalyst. Oxygen, therefore, during stabilisation comes into contact with a catalyst which contains active atomic hydrogen.

On heating the stabilised catalysts to 870 K in a pumping apparatus with a McBain balance, the decrease in weight was most intensive at 300—470 K (*ca*. 1%) and at 670—820 K (*ca* 0·5%). Mass spectrometry revealed among the products of heat–vacuum treatment of stabilised catalysts with different promoters H_2O, CO_2, CO, N_2, and H_2. Small amounts of oxygen were found only at $\beta = 3·3$ K/s.

Fig. 1 shows thermal desorption spectra of individual products resulting from treat-ment of the catalyst in the ion-source reactor[9] of an MI-1309 mass spectrometer. The relatively low temperatures corresponding to the initial and maximum rates of CO_2 evolution suggest that CO_2 was mainly adsorbed on the catalyst during contact with the air. The evolution of CO (partly of CO_2), and of N_2 and H_2 was caused, respectively, by the interaction between oxygen of the stabilised layer with the admixed carbon in the catalysts, and by the decomposition of traces of ammonia. The separation of water from the catalyst is characterised by three various adsorption forms.

Experiments with isotopes may serve as an indication of the immediate formation of water in the process of stabilisation. After deuterium had been adsorbed on the catalyst (reduced and evacuated at 870 K) stabilisation was effected by exposure to oxygen at low pressure under static conditions. In the mass spectrum of the products of heating the catalyst, the peak of D_2O is more intense than that of HDO, which is in contrast to the exchange spectrum of D_2 with H_2O in the reduced catalyst. In other experiments, the catalyst was stabilised with oxygen enriched with ^{18}O. The products' mass spectrum

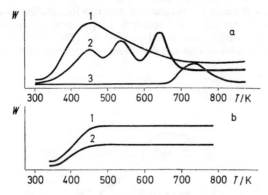

FIGURE 1 Dependence of the desorption rate W (arbitrary units) of the product
from the stabilised catalyst on temperature T/K.
(a) 1, CO_2; 2, H_2O; 3, CO, N_2.
(b) 1, H_2; 2, O_2.

is characterised by the presence of $H_2^{18}O$ and ^{18}OH. The occurrence of D_2O and $H_2^{18}O$
may be indicative of the interaction between hydrogen and oxygen.

On heating the stabilised catalysts in a current of helium, argon, nitrogen, or hydrogen,
it was found by thermal volumetric and chromatographic analysis that the same products
are produced as in the case of heat–vacuum treatment (Table 1).

Table 1

Quantitative analysis of catalyst activation products

Experimental method	Activation conditions	Amount of products per g of catalyst			
		$H_2O + CO_2$ mg	H_2	CO μm^3	N_2
McBain balance	Vacuum	8–10	$\Sigma = 4\cdot4$		
Thermal desorption	Vacuum	10	$\Sigma = 2\cdot4$–$4\cdot2$		
Thermal volumetry	Helium	—	$\Sigma = 5\cdot1$–$5\cdot3$		
	Nitrogen		$2\cdot8$	—	
	Hydrogen	—	—	$\Sigma = 2\cdot4$	
Chromatographic analysis	Helium				
	Argon	—	$2\cdot7$	$1\cdot2$	$0\cdot5$

Fig. 2a presents the changes in work function (curve 1) on heating stabilised catalysts
in vacuo. In accordance with ref. 5, an appreciable decrease in the work function may
be explained by desorption of CO_2, and, to a lesser degree, of CO. A second heating of
the catalyst to 600 K (curve 2) has practically no effect on the work function, whereas
at higher temperatures, a substantial decrease is observed without the evolution of
products to the gas phase. This shows that, in addition to desorption, heat–vacuum
treatment is accompanied by a change in the surface composition above 600 K.

It is known that the energy spectrum of electron localization levels characterised by exoemission is determined by the adsorption layer composition on the surface of a solid.[10] Fig. 2b gives thermal stimulated exoemission (TSE) curves. Together with the TSE peaks (curve 1) velocity maxima for thermal desorption of water were observed,

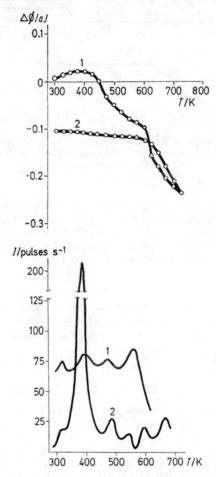

FIGURE 2   (a) The change in electron work function for the stabilised catalyst
$\Delta\phi/aJ$ against temperature T/K.
1, initial curve; 2, curve obtained on repeated heating following vacuum treatment
at 720 K.
(b) Emission intensity I/pulses s^{-1} against temperature T/K.
1, without excitation; 2, after electron bombardment excitation.

in agreement with thermal desorption data in Fig. 1a. After excitation, electron emission generally takes place. The intense emission peak at 390 K is especially stable in repeated "heating–excitation" cycles and explained by the presence of a surface oxygen layer. The TSE peaks and the breaks on the curve of the work function are obviously caused by the same processes on the surface. The TSE data show that before the initial heating of the stabilised catalysts, water is present, and that on heating the oxygen takes part in desorption processes.

Fig. 3 (curve 1) shows that raising the vacuum-treatment temperature to 800 K brings about an increase in the specific surface of the catalyst; this is due to the appearance of fine pores (5—6 nm radius) and may be explained by the removal of water from the pores, and by solid-phase restructuring. The specific surface for singly promoted samples (Al_2O_3 or K_2O) is independent of the vacuum-treatment temperature.

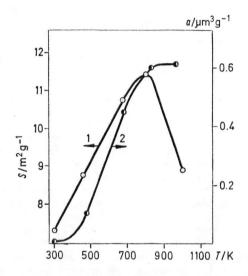

FIGURE 3   The relationship between the specific surface $S/m^2\ g^{-1}$ (1) and the quantity of rapidly chemisorbed oxygen $a/\mu m^3\ g^{-1}$ (2) and temperature (T/K) of vacuum treatment of the catalyst.

According to our derivatography data, the stabilised layer is stable in air to 400 K. The exoeffect initiated at 400 K is accompanied by an increase in the weight of the catalyst which is characteristic of catalyst oxidation.

It is of interest that an investigation of the removal of the stabilising layer in hydrogen by derivatography revealed only one endothermal peak, with its maximum at 350 K, which can be explained by catalyst "drying".

Heat treatment of the catalysts in the absence of hydrogen leads to surface activation with regard to chemisorption of CO, CO_2, O_2, N_2, and H_2. On the activated catalysts a fast and a slow oxygen chemisorption were observed. Fast chemisorption may be considered to be adsorption, while slow chemisorption characterises absorption and oxidation.[5,11] Fig. 3 (curve 2) shows that the amount of fast chemisorbed oxygen at 293 K increases as the temperature of the vacuum-treatment is increased up to 800 K, and this temperature was accepted as optimal for catalyst activation. At 720 K the degree of ammonia decomposition on catalysts activated in helium and hydrogen is the same and totals 70%. The effect of the mode of activation of the catalyst on ammonia synthesis was studied in a laboratory converter. Table 2 shows that no significant difference in catalyst activity is found.

Table 2

Catalytic activity in ammonia synthesis

Activation conditions	Synthesis temperature K	% NH$_3$ in exit gas	
		$P = 6\cdot0$ MPa	$P = 10\cdot0$ MPa
Nitrogen and hydrogen mixture; 5·0 MPa	773	5·3	10·0
	748	4·5	—
	723	2·8	12·5
Helium, nitrogen 0·1 MPa	773	5·4	10·3
	748	4·7	—
	723	3·8	11·7

The interaction of nitrogen and hydrogen with the catalyst

To characterise the behaviour of nitrogen and hydrogen which had been adsorbed on an activated catalyst at 5·33 kPa and, respectively, at 570 and 293 K, the catalyst was evacuated at 293 K and heated at $\beta = 0\cdot33$ K/s in a closed volume of a static arrangement.

From the data obtained it can be said that two forms of nitrogen and hydrogen chemisorption are found. Various species are possible on the ammonia catalyst.[6,7] By pulse and thermal desorption methods, only one form of strongly bound nitrogen was detected at ammonia synthesis temperature (720 K). The pulse method has shown that the amount of adsorbed hydrogen at 720 K exceeds that of a monolayer, which is indicative of an absorption process. The thermal desorption of hydrogen adsorbed at 720 K proceeds very slowly and is characteristic of evolution of absorbed hydrogen. The different retention times of helium and hydrogen in argon confirm the existence of weak hydrogen adsorption (Fig. 4). One may assume that absorption takes place through this weak adsorption.

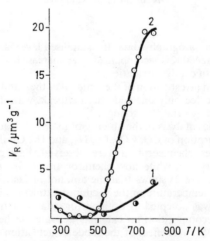

FIGURE 4 Dependence of hydrogen retention volume $V_R/\mu m^3\ g^{-1}$ on adsorption temperature T/K in argon (1) and nitrogen (2).

To identify the forms involved in the process of catalysis, the interaction of nitrogen with the catalyst was studied by a pulse method in hydrogen, as well as the interaction of hydrogen in nitrogen. Nitrogen adsorption was the same in hydrogen and in helium.

As can be seen from Fig. 4, the retention volumes of hydrogen in nitrogen above 400 K are considerably greater than in argon and, therefore, nitrogen presumably induces that form of adsorption. The weak hydrogen adsorption in the presence of a nitrogen flow is activated. The approximate value of the chemisorption activation energy of this form is of the order of 33·5 kJ/mol.

DISCUSSION

Catalyst stabilisation and activation

Stabilisation,[5] as well as fast oxygen chemisorption,[11] is accompanied by a considerable increase in the work function of the ammonia catalyst ($\Delta\varphi = 0.16$—0.24 aJ). As shown in ref. 5, and confirmed by the data reported above (Fig. 2a), the removal of the stabilising layer is, on the contrary, characterised by a decrease in φ of the same order. It can be concluded that a layer of negatively charged oxygen dipoles is involved in stabilisation. Activation in the absence of hydrogen is confirmation of the prevailing adsorptive nature of the stabilising layer. In the formation of the stabilising layer, however, it is not the adsorbed oxygen alone that is involved. Slow chemisorption is related to partial oxygen diffusion into the catalyst volume, or to oxidising bulk iron, since it is accompanied by a decrease in the work function.[11] This suggests that the stabilising layer contains some dissolved oxygen or oxide. Oxygen interaction in the process of stabilisation differs from the effect of oxygen at synthesis temperatures, when the chemisorption of the negatively charged oxygen is just a fast initial stage leading to oxide formation. That is why surface poisoning and regeneration at a sufficiently high concentration of oxygen-containing compounds follow the oxidation–reduction mechanism.[12]

Our results show that stabilisation may be accompanied by the formation of OH groups and adsorbed water. The probability of the formation of these products during stabilisation is confirmed by thermal desorption spectra and by exoemission data. The protective action of the stabilising layer is evidently associated with both the presence of a layer of adsorbed water and of negatively charged oxygen dipoles which prevent further oxygen adsorption on exposure to air. It is known that ammonia catalysts can be sealed with water to suppress their pyrophoric nature. However, the pyrophoric nature of the catalysts shows up again on drying.[4] The temperature of the initial destruction of the stabilising layer on exposure to air and *in vacuo* is *ca.* 400 K, which confirms that the protective action of the stabilising layer is partly due to the layer of adsorbed water.

The structure of the stabilizing layer of industrial catalysts and of those promoted with aluminium oxide alone may be different, there being only about half as much change in the work function during stabilisation (and activation) in the latter case[5] and no change observable in the specific surface during activation. The catalyst composition, therefore, affects the structure of the stabilising layer.

Thermal activation of stabilised catalysts *in vacuo* and in various gases is generally no accompanied by evolution of oxygen to the gas phase. Since the diffusion processes speed up with increase in temperature, the interaction of impurities and oxgyen of the stabilising layer is possible, with the formation of products which are evolved on activation. The optimal activation temperature coincides with the termination of gas evolution, thus leading to the conclusion that the evolution of products is associated with the formation of a bare metallic surface. Activation affects not only the surface, but also inner catalyst subsurface layers, since it is accompanied by texture rearrangement.

The results obtained show that activation of stabilised catalysts in hydrogen can be substituted by heat activation *in vacuo* or in an inert gas. On the other hand, activation of catalysts in hydrogen is accompanied by the separation of water from the stabilising layer and other gases and does not necessarily proceed by hydrogenating the layer, but may follow the activation *in vacuo* or with a variety of gases.

Initial stage of ammonia synthesis

Under optimal synthesis temperature (about 730 K), the separate adsorption of gases on commercial catalysts is characterised by two types of hydrogen interaction, namely weak adsorption and absorption, and one type of nitrogen interaction—strong chemisorption. Our studies have revealed one type of mutual influence of nitrogen and hydrogen under the conditions reported: weak hydrogen chemisorption induced by the strongly chemisorbed nitrogen. In the absence of nitrogen this kind of chemisorption was virtually absent. According to preliminary data the hydrogen liberated during ammonia decomposition in nitrogen is loosely bound and is characterised by an acitvation energy of 33·5 kJ/mol. Thus, the participation of loosely bound hydrogen in the catalytic process is confirmed. On the basis of the data reported it can be concluded that the first stage in ammonia synthesis under the conditions studied is strong nitrogen chemisorption, whereas the second stage consists of nitrogen-induced chemisorption of hydrogen. The sequence of stages is similar to that of the synthesis of alcohols from CO and H_2 on fused iron catalysts.[13]

ACKNOWLEDGEMENTS. The authors are grateful to V. V. Grigoriev for the mass-spectrometric and isotope study and to I. V. Krylova for the exoemission study.

REFERENCES

[1] G. K. Boreskov, *Zhur. fiz. Khim.*, 1958, **32**, 2739.
[2] S. S. Lachinov *et al.*, *Processes with the Participation of Molecular Hydrogen* (Novosibirsk, 1973), p. 53.
[3] M. I. Temkin and V. M. Pyzhev, *Zhur. fiz. Khim.*, 1946, **20**, 151.
[4] *Catalyst Handbook* (Wolfe Scientific Books, London, 1970), pp. 126 and 161.
[5] E. Kh. Enikeev and A. V. Krylova, *Kinetika i Kataliz*, 1962, **3**, 139.
[6] R. V. Chesnokova, A. I. Gorbunov, S. S. Lachinov, and G. K. Muravskaya, *Kinetika i Kataliz*, 1970, **11**, 1486.
[7] N. Takezawa, *J. Catalysis*, 1972, **24**, 417.
[8] Y. Amenomyia and G. Pelizier, *J. Catalysis*, 1973, **28**, 442.
[9] A. K. Baibakov and V. V. Grigoriev, *Zhur. fiz. Khim.*, 1972, **46**, 771.
[10] I. V. Krylova, *Phys. Stat. Sol.*, 1971, 7, 359.
[11] G. I. Kovalev, Yu. B. Kagan, and A. V. Krylova, *Kinetika i Kataliz*, 1970, **11**, 1505.
[12] S. S. Lachinov, *Zhur. fiz. Khim.*, 1940, **14**, 1260.
[13] G. I. Kovalev, Yu. B. Kagan, and A. V. Krylova, *Bulgarian Academy of Sciences, depart. chem.*, 1973, **6**, 235.

DISCUSSION

J. W. E. Coenen (*Unilever, Vlaardingen*) said: You explain the passivation process as oxidation of adsorbed hydrogen plus adsorption of oxygen. We have experience only with nickel, which is more noble than iron. Still, even in careful room temperature passivation with nitrogen containing some oxygen, after oxidation of adsorbed hydrogen, the process always goes further than simple oxygen chemisorption. At least two atomic layers of *oxide* are formed. With vacuum reactivation we made the same observations as you did: after desorption of water, the catalyst again becomes pyrophoric and catalytically active.

Passivation-reactivation cycles could be repeated up to four times. During these cycles the nickel metal is progressively oxidized. This undoubtedly will also happen with the less noble iron. So I doubt whether the passivation of iron is limited in the way you describe, and suggest that you measure the percentage of metallic iron remaining (*e.g.* by dissolving the catalyst in acid and measuring the hydrogen evolved). You will find that bulk oxidation occurs.

A. V. Krylova (*Mendeleyev Inst. of Chem. Eng., Moscow*) replied: We do not exclude the possibility of the formation of oxide during the passivation, but we consider that only a part of the active iron surface is covered with an oxide layer. The rest of the active surface is probably covered by adsorbed water and oxygen and may be easily reactivated. The quantity of oxide formed seems to depend on the conditions of passivation. Repeating the passivation-reactivation cycles many times may be accompanied by the progressive oxidation of metal due to the dissolution of oxygen in the bulk of the catalyst during heating in vacuum or in inert gas.

R. B. Anderson (*McMaster Univ., Hamilton*) said: Regarding reactivation of stabilized catalysts by inert gases or evacuation, I wish to ask if oxygen remains on the catalyst and is reduced almost instantaneously at the start of the synthesis. Catalysts partly oxidized by steam are rapidly reduced[1] and catalysts poisoned by water vapour during synthesis are rapidly restored to their original activity when water is removed from the feed gas.

A. V. Krylova (*Mendeleyev Inst. of Chem. Eng., Moscow*) replied: We think that most of the stabilized layer (adsorbed oxygen and water) may be removed during activation in the absence of hydrogen. The quantity of CO_2 and NH_3 which could be adsorbed increased after such treatment. However, we do not exclude the possibility that the sites covered with oxide are activated instantaneously during the synthesis.

I. Toyoshima (*Hokkaido Univ.*) said: Recently we have examined the thermal desorption of nitrogen from singly promoted catalysts (containing 2·69% alumina) and from doubly promoted catalysts (containing 2·55% alumina and 0·33% potassium oxide), by use of linear temperature sweep and recording the weight loss of the catalyst due to the desorption. The desorption was made after adsorbing nitrogen on an iron catalyst at room temperature. Both catalysts showed the existence of at least two forms of adsorbed nitrogen; one was desorbed below 100 °C and the other above 300 °C. The temperatures at which the desorption rate was a maximum were different for the two catalysts and were lower on the singly promoted than on the doubly promoted catalyst.

The desorption curves were analysed by a modified Wigner–Polanyi desorption equation

$$dn/dt = v_k n^k \exp -(E_d{}^0 - \alpha n)/RT \qquad (1)$$

where v_k is the pre-exponential factor in the rate constant, $E_d{}^0$ is the activation energy for desorption at zero coverage, k is the desorption order and α is a constant.

The rate parameters, $E_d{}^0$ and v_k were obtained by fitting the experimental evolution curves to computer-generated curves obtained by numerically integrating equation (1). Thus, $E_d{}^0 = 52$ kcal mol^{-1}, $10^{12}\alpha = 3\cdot7-5$ cal/mol/molecule/cm^2, $v_1 = 2\cdot5 \times 10^{12}$ s^{-1} and $k = 1$ for singly promoted catalyst, whereas $E_d{}^0 = 60-63$ kcal mol^{-1}, $10^{12}\alpha = 0\cdot4-1\cdot3$ cal/mol/molecules/cm^2, $v_2 = 6\cdot65 \times 10^{-2}$ s^{-1}/molecules/cm^2 and $k = 2$ for doubly promoted catalyst.

You have also found the two forms of nitrogen chemisorption. Did you determine the kinetic parameters and did you find any difference in stabilization due to the amount or the kind of the promoters used?

A. V. Krylova (*Mendeleyev Inst. of Chem. Eng., Moscow*) replied: The thermal desorp-

---

[1] W. K. Hall, W. H. Tarn, and R. B. Anderson, *J. Phys. Chem.*, 1952, **56**, 688.

tion curves for nitrogen on the industrial catalyst have been published.[2] Desorption of the first form of nitrogen begins at 400 K and that of the second at 600 K. We did not determine the activation energy of desorption.

We suppose the distinction in stabilization due to the kind of promoters used. The influence of potassium oxide was especially is significant. The exo-emission data and the quantitative analysis of the thermal desorption products during catalyst activation revealed a higher amount of water and carbon dioxide in the cases of K_2O-promoted catalysts, doubly-promoted catalysts, and industrial catalysts, than in the case of singly (Al_2O_3) promoted catalysts. These results indicate a distinction in the composition of the adsorbed layer for the various stabilized catalysts. We oxidized several stabilized catalysts by heating in air and have evidence that K_2O-promoted catalysts undergo deeper oxidation during stabilization and storage than either singly (Al_2O_3) promoted catalysts, or doubly-promoted catalysts, or industrial ammonia catalysts.

A. B. Stiles (*Univ. of Delaware, Wilmington*) said: You report the deactivation of an ammonia synthesis catalyst when it has adsorbed oxygen and water vapour. Did you investigate catalyst additives (additional components) that might have had the effect of minimizing or eliminating the effect of these well-known but reversible poisons?

A. V. Krylova (*Mendeleyev Inst. of Chem. Eng., Moscow*) replied: It would be very interesting to find catalyst additives that would minimize the effects of poisons, but we have not made this a subject for investigation.

[2] A. V. Krylova, N. N. Vokhmyanin, T. L. Koroleva, and V. V. Morozov. *Conference on Mechanism of Heterogeneous Catalytic Reaction*, Moscow, 9—13 September, 1974, Preprint of paper 9 (in Russian).

Heterogeneous–Homogeneous Reactions in Hollow Catalyst Pellets

ROBERT D. HOLTON and DAVID L. TRIMM*

Department of Chemical Engineering and Chemical Technology, Imperial College,
Prince Consort Road, London SW7 2BY

ABSTRACT Studies have been made of the mathematical modelling of hollow catalyst pellets operating in the absence and presence of a homogeneous reaction. A full analytical solution of the interaction of diffusion and reaction is very time consuming and so approximate solutions have been developed and compared. Agreement between the analytical and the approximate solutions depends strongly on the Thiele modulus. At intermediate values, an equivalent slab model gives a maximum discrepancy from the analytical solution of 15%, as compared to 10% for an equivalent sphere model.

For a first-order irreversible isothermal catalytic reaction affected by diffusion, the modelling shows that the catalyst material can be used more efficiently if hollow pellets are employed. For large Thiele moduli, the reaction rate is largely determined by pellet superficial area: hollowing increases this area for a given external radius. The effect is more pronounced for large pellets. However, this is only true provided the thickness of the annular shell is larger than the diffusional penetration distance for reactants. For concurrent heterogeneous–homogeneous reaction networks, selectivity to the heterogeneous product may also be improved by pellet hollowing. For consecutive reaction networks, values of pellet hollowing may be found to maximise the yield either of the heterogeneous or the homogeneous product.

INTRODUCTION

The use of hollow cylindrical catalyst pellets in an industrial reactor can offer significant advantages as a result of the elimination of "dead spaces" at the pellet centre and by reducing pressure drop across a packed bed. This paper focuses attention on the modelling of such pellets when they are used to catalyse a reaction in the absence and presence of a second concurrent or consecutive gas phase reaction. The study is concerned with isothermal pellets with negligible mass and heat transfer resistances at the surface: extension to other situations is comparatively easy, using the same arguments.

An analytical solution to the conservation equation and boundary conditions for simultaneous reaction and diffusion in hollow catalyst pellets has been presented by Gunn.[1] Although the approach gives accurate results, it is computationally long and an approximate method, the equivalent slab model, has been developed.[2-6] In this model, the relative importance of diffusion and reaction in an arbitrarily shaped pellet is characterised by a Thiele modulus in which the length parameter is the ratio of volume to external surface area of the pellet.

Comparisons of the predictions of the analytical solution and the equivalent slab model revealed some major discrepancies, and it was decided to develop an alternative approximate model. This is based on the assumption that a hollow cylinder can be modelled by an equivalent hollow sphere, with gas having equal access to the inner and outer surfaces of the sphere. The assumption allows the reduction of the problem to one dimension, with the balance equations for spherical geometry being more easily handled than those for cylindrical geometry.

The equivalence of the hollow sphere and the hollow cylinder is specified by an equal surface area of sphere and cylinder and by an equal volume of catalyst material in both forms: this results in the following equivalence equations:

$$r^2 (1 + a^2) = \frac{b^2}{2} [(1 + \alpha) \beta + (1 - \alpha^2)] \tag{1}$$

$$r^3 (1 - a^3) = \frac{3b^3}{4} [\beta(1 - \alpha^2)] \tag{2}$$

These may be solved simultaneously to give r and a.

It is widely reported (*e.g.*, ref. 1) that the interaction of diffusion and first-order isothermal chemical reaction in a spherical catalyst pellet is represented for dimensionless concentration, χ, and radius, y, by:

$$\frac{1}{y^2} \frac{d}{dy} \left(y^2 \frac{d\chi}{dy} \right) - \lambda^2 \chi = 0 \tag{3}$$

with boundary conditions for the hollow sphere

$$y = 1; \chi = 1$$
$$y = a; \chi = 1 \tag{4}$$

Solution of equations (3) and (4) gives the concentration profile in the equivalent hollow sphere as:

$$\chi = \frac{1}{y} \left[\frac{\exp[(1 + a)\lambda] - a \exp(2\lambda)}{\exp(a\lambda) - \exp[(2 - a)\lambda]} \exp(- \lambda y) + \frac{a - \exp[(1 - a)\lambda]}{\exp(a\lambda) - \exp[(2 - a)\lambda]} \exp(\lambda y) \right] \tag{5}$$

Defining the effectiveness factor, ε, as the ratio between the actual rate of reaction in the catalyst pellet and the rate of reaction in the absence of any diffusional resistance, then:

$$\varepsilon = \frac{3}{(1 - a^3)\lambda^2} [(\lambda - 1)B' \exp(\lambda) + (a\lambda + 1)A'\exp(- a\lambda) - (\lambda + 1)$$

$$A'\exp(- \lambda) - (a\lambda - 1)B'\exp(a\lambda)] \tag{6}$$

where $A' = \dfrac{\exp[(1 + a)\lambda] - a \exp(2\lambda)}{\exp(a\lambda) - \exp[(2 - a)\lambda]}$; $B' = \dfrac{a - \exp[(1 - a)\lambda]}{\exp(a\lambda) - \exp[(2 - a)\lambda]}$

Thus for given values of α, β, and the kinetics, a, b, and hence λ may be calculated for equations (1) and (2) and used in equation (6) to calculate the isothermal effectiveness factor.

COMPARISON OF THE MODEL PREDICTIONS

A utility factor is defined, after Gunn,[6] as the ratio of the rate of reaction per unit volume of a bed of hollow cylinders to the rate of reaction per unit volume of a bed of similar solid cylinders without diffusional resistance. Thus:

$$\eta = \varepsilon (1 - \alpha^2) \tag{7}$$

The utility factor for hollow cylindrical catalyst pellets has been calculated by all three approaches for various values of α, β and θ, where θ is defined as $b\sqrt{k/D_{eff}}$. An example of the results given by the equivalent hollow sphere model is shown in Fig. 1. This model can be shown to have the same asymptotic behaviour, with respect to θ, as the equivalent slab model.[7]

The results of the equivalent hollow-sphere model have been compared with those of the equivalent slab model and with those of Gunn's analysis. There appears to be no published experimental data for absolute comparison. The discrepancies between the three models are shown in Fig. 2 for typical values of α and β. The discrepancy for the slab and hollow sphere models is calculated as a percentage deviation from the

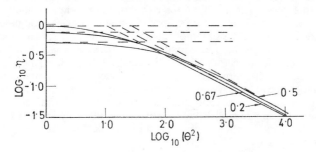

FIGURE 1 The utility factor calculated by the equivalent hollow sphere model $\beta = 2\cdot0$: α shown as a parameter.

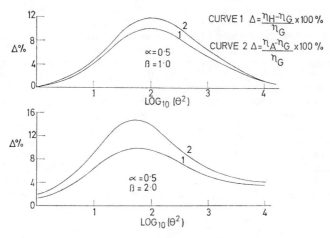

FIGURE 2 Comparisons between the full analytical solution and approximate models.

prediction of Gunn's analysis. For low values of Thiele modulus ($\theta^2 < \sim 1$) the diffusional influence is small and the effectiveness factor tends to unity. The discrepancy between all three models is $\sim 1\%$ in this, kinetic, region.

The discrepancies at high values of the Thiele modulus are less well defined. For $\theta^2 > \sim 3000$ both approximate models agree within $0\cdot5\%$ but differ from Gunn's results by up to 5%. This relatively good agreement between the models at high Thiele modulus is to be expected since this situation corresponds to reaction bands near to the inner and outer surfaces: these bands became thin as the active region approaches slab geometry.

For $1 < \theta^2 < 3000$ the phenomena of diffusion and reaction are at their most interactive and there is less agreement between the three models. Nevertheless the maximum discrepancy between the sphere model and the analytical solution is reduced to 10% from a 15% discrepancy between the slab model and the analytical solution. Maximum discrepancies are found in the region $100 < \theta^2 < 300$, depending on α and β.

The advantage, if any, to be gained from use of the equivalent hollow sphere model may be assessed by consideration of the accuracy of its predictions and the comparative computational effort required. The computational requirements of the analytical approach are considerably greater than for either of the two approximate models. The calculations for these models were performed on a CDC 6400 computer, and were found to take 0·15 ms for the slab model and 0·75 ms for the sphere model. It would thus appear that, whereas the sphere model gives a better estimate than the slab model for the utility factor for a bed of hollow cylindrical catalyst pellets, the increase in computational effort is rather large, However, in a chemical reactor, small errors in any of the rate parameters can lead to much larger cumulative errors in conversion, selectivity and temperature rise. Consequently the slab model could be useful for preliminary calculations but the sphere model, and eventually the full analytical approach, would be desirable for final optimisation in the design of chemical reactors using hollow cylindrical catalyst pellets.

Optimization of a single irreversible reaction

Optimum pellet performance may be defined in a number of ways. In particular one could attempt to optimise the use of catalyst material or of reactor volume. The choice of which policy to follow would probably be an economic decision.

In other to follow the former policy it is required to maximise the pellet effectiveness factor. It has been found that the pellet effectiveness factor increases with increasing pellet hollowing, and optimisation requires the use of the greatest degree of pellet hollowing consistent with the required strength characteristics of the catalyst material. The latter policy presents a more interesting problem. Fig. 3 shows the utility factor as a function of pellet hollowing. This reveals that for larger values of Thiele modulus, η shows a maximum with respect to α, and the greater the value of θ, then the greater is the degree of hollowing required to reach this maximum. Consideration of Gunn's results[6] shows that for $\beta < 2$, the degree of enhancement, if any, to be achieved by pellet hollowing is small and generally the enhancement to be gained by pellet hollowing increases with θ and β. At $\beta \sim 4$ and $\theta^2 \sim 10,000$, η may be doubled by pellet hollowing.

Using the equivalent slab model to define the pellet effectiveness factor as[3]

$$\varepsilon = \frac{\tanh |\Lambda|}{\Lambda} \tag{8}$$

where
$$\Lambda = b(1 - \alpha)\beta \sqrt{\frac{k}{D_{\text{eff}}}}.$$

Equation (8), with equation (1), may be differentiated to give

$$\frac{d\eta}{d\alpha} = \frac{(\beta - 2\alpha)(1 - \alpha + \beta)\sinh(2\Lambda)}{\beta^2 \theta} - (1 + \alpha) \tag{9}$$

Thus for a maximum in η with respect to α, $\dfrac{d\eta}{d\alpha} = 0$ and:

$$\alpha_{\max} = \frac{(\beta - 2\alpha_{\max})(1 - \alpha_{\max} + \beta)\sinh(2\Lambda_m)}{\beta^2 \theta} - 1$$

where
$$\Lambda_m = \frac{(1 - \alpha_{max})\beta\theta}{2(1 - \alpha_{max} + \beta)} \tag{10}$$

α_{max} was found for various values of β and θ from equation (10) using the memory method of Shackham and Kehat,[8] the equation being ill-conditioned to other iterative solution methods. The results, summarised in Fig. 4, show that for $\beta < 2$, α_{max} tends to $\beta/2$ as θ tends to infinity.

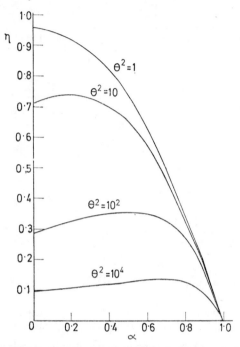

FIGURE 3 The utility factor as a function of the hollowness factor calculated by equivalent hollow sphere model: $\beta = 2\cdot0$.

The pellet geometry was also optimised using the hollow sphere model. The average discrepancy between the predictions of the two models was of the order of 10%, with a maximum of *ca.* 20%. Both approximate models predict low values for α_{max} relative to the analytical solution. This may be regarded as a 'safe' estimate, in that in the region of optimum hollowing, further increases in α rapidly cause the utility factor to fall towards zero.

OPTIMISATION OF HETEROGENEOUS–HOMOGENEOUS REACTIONS

The first type of heterogeneous–homogeneous reaction that will be considered involves a concurrent heterogeneous reaction and a homogeneous reaction.

Consider the reaction scheme:

$$A \begin{array}{c} \xrightarrow{\text{heterogeneous}} \\ \xrightarrow[\text{homogeneous}]{} \end{array} \begin{array}{l} B \text{ rate} = k_1 C_A{}^c \\ C \text{ rate} = k_2 C_A{}^g \end{array}$$

where the decomposition of nitrous oxide over a catalyst or in the gas phase[7] would be a case in point.

731

The relative rates of production of B and C are given by:

$$S = \frac{k_1 \varepsilon (1 - \phi_H) C_A^c}{k_2 \phi_H C_A^g} \tag{11}$$

ϕ_H, the voidage of the bed of hollow cylinders, is related to ϕ_S, the voidage of a bed of similar solid cylinders by:

$$\phi_H = \phi_S + \alpha^2 (1 - \phi_S) \tag{12}$$

The case where it is desired to enhance the production of the homogeneous product, C, is trivial, as the pellet may be hollowed as much as desired. If the heterogeneous reaction

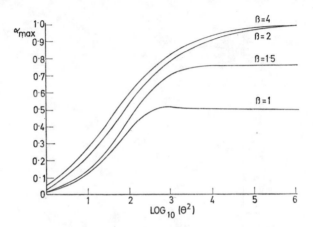

FIGURE 4 Optimum degree of hollowing as a function of θ^2.

is to be enhanced, it is necessary to enhance the catalytic reaction by hollowing to an extent greater than the enhancement afforded to the gas phase reaction by the increased gas space. Thus if $\left.\dfrac{dS}{d\alpha}\right|_\alpha > 0$, further hollowing will enhance selectivity to the hetero-

geneous product for a particular value of α.

It has been shown that the enhancement of heterogeneous reactions by pellet hollowing is most significant for high values of θ. Substituting equations (8) and (12) in equation (11), and taking high values of the Thiele modulus gives:

$$S = \frac{2k_1 (1 - \phi_S) C_A^c}{k_2 C_A^g \beta \theta'} \frac{(1 + \alpha)(1 - \alpha + \beta)}{[\phi_S + \alpha^2 (1 - \phi_S)]} \tag{13}$$

where $\theta' = \theta \sqrt{\dfrac{C_A^{c-1} (c + 1)}{2}}$ and is a generalised Thiele modulus for a reaction of order c.

Differentiating equation (13) gives:

$$\frac{dS}{d\alpha} = \frac{2k_1 (1 - \phi_S) C_A^c}{k_2 C_A^g \beta \theta'} \frac{[\phi_S + \alpha^2 (1 - \phi_S)](\beta - 2\alpha) - 2\alpha(1 - \phi_S)(1 + \alpha)(1 - \alpha + \beta)}{[\phi_S + \alpha^2 (1 - \phi_S)]^2} \tag{14}$$

whence it can be seen that $\dfrac{dS}{d\alpha}$ is always positive for $\alpha = 0$. Thus the selectivity to the heterogeneous product can always be improved upon by hollowing solid catalyst pellets. The optimum hollowing occurs when $\dfrac{dS}{d\alpha} = 0$ and:

$$\alpha_{max} = \frac{\frac{1}{2}[\phi_S + \alpha^2_{max}(1 - \phi_S)][\beta - 2\alpha]}{[1 - \phi_S][1 + \alpha_{max}][1 - \alpha_{max} + \beta]} \tag{15}$$

At high Thiele modulus, the optimum hollowing depends only on β and ϕ_S. However, the amount by which the selectivity is adjusted will also depend on the Thiele modulus and, as a result, on the kinetics. Values of α_{max} for various values of β and ϕ_S are given

Table 1

Value of α_{max} for optimal heterogeneous selectivity.

$$\frac{2k_1 C_A{}^c}{k_2 C^g{}_A \theta} = 0.2$$

ϕ_S \ β	1·0	2·0	3·0
0·35	0·104 (2·6)	0·146 (4·9)	0·169 (6·3)
0·45	0·142 (3·6)	0·205 (6·8)	0·237 (8·9)
0·6	0·208 (5·2)	0·312 (10·4)	0·375 (13·9)

in Table 1. The percentage increase in selectivity to the heterogeneous product is shown in brackets for

$$\frac{2k^1 C_A{}^c}{k_2 C_A{}^g \theta'}$$

If pellet hollowing is to be employed to reduce pressure drop in a reactor, similar calculations would reveal whether such a policy would be detrimental to the desired selectivity.

Attention may now be switched to consecutive reaction schemes of the general form:

$$A \xrightarrow{\text{heterogeneous}} B \xrightarrow{\text{homogeneous}} C$$

Catalytic oxidation, followed by gas-phase oxidation of the products would be a case in point.

The case where B is the desired product can be approached from an extension of the arguments advanced for concurrent reactions. When the Thiele modulus is high, the value for optimum pellet hollowing is identical to that for the concurrent system, as the value has been shown to be independent of gas- and solid-phase kinetics. The degree of improvement, which depends on kinetics, will differ between the two reaction schemes.

The selectivity to B may be written:

$$S' = \frac{2k_1(1 - \phi_S)C_A{}^c}{k_2 C_B{}^g \beta \theta'} \frac{(1 + \alpha)(1 - \alpha + \beta)}{[\phi_S + \alpha^2(1 - \phi_S)]} \tag{16}$$

The case where C is the desired product can be tackled by considering a number of optimal policies. In order to demonstrate how an optimum can be achieved by pellet

733

hollowing, let us assume that we may ensure a high yield of C by operating at the point where $\dfrac{\mathrm{d}B}{\mathrm{d}t} = 0$. That is to say

$$k_1 C_A{}^c \varepsilon (1 - \phi_H) = k_2 C_B{}^g \phi_H \tag{17}$$

The behaviour of the system may be controlled through the parameters ε and ϕ_H which are linked by α.

Using equation (12) in equation (17) yields:

$$\left[\frac{k_1 C_A{}^c (1 - \phi_S)(1 - \alpha^2)}{k_2 C_B{}^g \phi_S + \alpha^2 (1 - \phi_S)} \, \varepsilon \right] = 1 \tag{18}$$

Solution of equation (18) for α_{max} will give the optimum pellet hollowing for given values of C_A, C_B and k_1, k_2.

For the particular case at high Thiele modulus, equation (18) becomes:

$$\frac{2k_1 C_A{}^c (1 - \phi_S)(1 + \alpha)(1 - \alpha + \beta)}{k_2 C_B{}^g \beta \theta' \; [\phi_S + \alpha^2 (1 - \phi_S)]} \, 1 = 0 \tag{19}$$

Variation of the left-hand side of this equation is shown for various values of the pertinent parameters in Fig. 5.

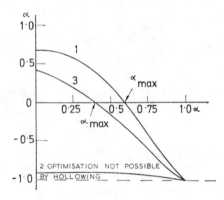

FIGURE 5 Left-hand side of equation.

Parameter values

	1	2	3
ϕ_S	0·4	0·4	0·4
θ	10·0	50·0	1·0
β	2·0	2·0	2·0
$\dfrac{k_1' C_A}{k_2' C_B}$	3·0	1·0	1·0

From equation (19), as $C_A \to 0$, $\alpha_{max} \to 1$, which confirms that the catalytic reactor bed should be terminated when A becomes insignificant. As $C_B \to 0$ (*i.e.* at the bed entrance) the gas phase reaction does not occur and $\alpha_{max} \to 0$.

NOMENCLATURE

a	internal to external radius ratio of hollow sphere
b	external radius of hollow sphere
C_A, C_B	concentration of species A, B respectively (kmol m^{-3})
c	order of catalytic reaction
D_{eff}	effective diffusivity of catalyst material (m^2 s^{-1})
g	order of gas-phase reaction
k	reaction rate constant
k_1, k_2	reaction rate constants for heterogeneous and homogeneous reactions, respectively
r	radial co-ordinate (m)
S, S'	selectivities as defined in text
y	dimensionless radial co-ordinate
α	ratio of internal to external radius of hollow cylinder
α_{max}	optimum α
β	length to external radius of cylinder
ε	effectiveness factor of hollow cylinder/sphere
λ	$r\sqrt{k/D_{eff}}$
θ	$b\sqrt{k/D_{eff}}$
θ'	$\theta \cdot \sqrt{\dfrac{C_A{}^{c-1}(C+1)}{2}}$
η	utility factor $= \varepsilon(1-\alpha^2)$
Λ	$b(1-\alpha)\beta\sqrt{k/D_{eff}}$
ϕ_H	voidage of a bed of hollow cylinders
ϕ_S	voidage of a bed of solid cylinders
χ	dimensionless concentration

REFERENCES

[1] E. E. Petersen, *Chemical Reaction Analysis*, Prentice-Hall, New Jersey, 1965.
[2] R. Aris, *Chem. Eng. Sci.*, 1957, **6**, 252.
[3] D. Basmadjian, *J. Catalysis*, 1963, **2**, 440.
[4] R. Rester and R. Aris, *Chem. Eng. Sci.*, 1969, **24**, 793.
[5] R. Rester, J. Jouven, and R. Aris, *Chem. Eng. Sci.*, 1969, **24**, 1019.
[6] R. England and D. J. Gunn, *Trans. Inst. Chem. Engrs.*, 1970, **48**, 265.
[7] R. D. Holton, *Ph.D. Thesis*, London University, 1975.
[8] M. Shackham and E. Kehat, *Chem. Eng. Sci.*, 1972, **27**, 2099.

DISCUSSION

J. W. Hightower (*Rice Univ., Houston*) said: I find your results surprising, particularly the increase in selectivity for the heterogeneous reaction in hollow pellets under conditions of severe mass transfer limitation. Of course the bed density is lowered by hollowing, which means there is more void space in which the homogeneous reaction can occur. Thus, the expanded external surface area together with the additional void volume is such that *both* the homogeneous and the heterogeneous reactions will be increased. Would you comment on this?

Secondly, I wonder if it is valid to compare the results at the same Thiele modulus when the length parameter is defined in different ways for the solid and hollowed pellets. Is it possible that this difference in definition may have accounted for some of your results?

Finally, since the catalyst strength is decreased by hollowing, the rather small increase in selectivity predicted may not make it worthwhile to use hollowed pellets in all internal mass transfer limited industrial applications.

D. L. Trimm (*Imperial College, London*) replied: Changes in selectivity result from competition between the heterogeneous reaction (which depends on the utility factor) and the homogeneous reaction (which proceeds in the voidage). Selectivity increases result from the fact that the gain in the utility factor outweighs the increase due to larger voidage. A modified Thiele modulus is used in the calculations, and comparisons are drawn for the same value of this quantity. Adjustments are made for the characteristic dimension in calculating the effectiveness factor [Eq. (7)].

The paper is based on kinetic information for specific reactions which was available in the literature. I agree that the increase in selectivity is small, but I would expect it to be larger where there is a larger difference in rate constants and activation energies for reactions in the two phases. Thus, for example, an oxidation should offer larger differences in selectivity. Unfortunately, appropriate kinetic data are not available, although we are now obtaining results for the oxidation of butene to butadiene and to carbon dioxide.

J. Happel (*Columbia Univ., New York*) said: Since the ratio of homogeneous to heterogeneous reaction rates is the important parameter in these studies, it would be of interest to know the specific reactions considered in the models developed in this paper.

D. L. Trimm (*Imperial College, London*) replied: The model for the consecutive reaction was the catalytic oxidation of methanol to formaldehyde, followed by the gas-phase oxidation of the aldehyde to carbon dioxide. The concurrent reaction was the decomposition of nitrous oxide. Full details are given elsewhere.[1]

J. B. Butt (*Northwestern Univ., Evanston*) said: The results presented here for isothermal systems are interesting, but a logical extension and perhaps most important from a realistic point of view, is the corresponding non-isothermal system. I would be interested to hear your comments on this, and in particular any thoughts you may have concerning heat transfer coefficients in the external boundary layer for hollow shell catalysts.

D. L. Trimm (*Imperial College, London*) replied: I agree that extension to the non-isothermal system would be of interest. In view of the small temperature differences recorded inside pellets, the heterogeneous reaction can probably be considered as isothermal. However, it would be necessary to consider film heat transfer coefficients in modelling a two-phase system, in view of the differences in temperature between the solid and the gas.

The choice of the correct transfer coefficient for use in a hollow sphere model is a problem. Since the hollow sphere models a hollow cylinder, it would seem that a heat transfer coefficient pertinent to the normal turbulent transfer in a packed bed would be appropriate, so long as the degree of hollowing was not so small as to reduce the central void to a pore.

S. P. S. Andrew (*ICI, Billingham*) said: Another example of the importance of the ratio between voidage and catalyst surface area, when the heterogeneous catalytic reaction course is affected by the presence of a homogeneous reaction, is seen in the steam reforming of naphtha.

Precursors of carbon formation are produced by homogeneous reaction in the gas phase, they migrate to the catalyst surface, and can then polymerize to form carbon and hence deactivate the catalyst. In industrial practise it is important to ensure that the ratio of homogeneous reaction volume to catalyst surface area is not excessive.

D. L. Trimm (*Imperial College, London*) replied: We have studied carbon formation in steam reforming, and find that processes leading to the deposition of carbonaceous matter can occur in the gas phase or on the catalyst surface. The relative importance of the two processes depends on the reaction conditions and the feedstock. Insufficient

[1] R. D. Holton and D. L. Trimm, *Chem. Eng. Sci.*, 1976, **31**, 549.

data are available to test quantitatively the importance of heterogeneous–homogeneous effects in the system.

J. R. Rostrup-Nielsen (*Haldor Topsøe, Lyngby*) (communicated): I wish to remark on the comment of S. P. S. Andrew. First we have shown that carbon formation in tubular steam reforming of naphtha need not take place *via* precursors in the gas phase if the catalyst is active at low temperature. We agree that carbon may result from homogeneous steam cracking (pyrolysis) of the naphtha, when it passes to a hot part of the tube, and we further agree on the role of the void fraction for that situation.

Your conclusion is different from what is expected from the paper. I think this may be related to the fact that the model does not consider the residence time distribution of the hydrocarbons. The conversion of the feed into carbon is much smaller than that of the catalytic reaction, and the variations of the residence time in the randomly packed bed of rings may easily include local areas where the residence time is high enough for carbon production affecting the performance of the reformer.

Assessment of Step Velocities and Concentrations of Surface Species by Transient Isotope Tracing

John Happel,* Miguel A. Hnatow, San Kiang, and Shoichi Oki†

Columbia University, New York, N.Y. 10027, U.S.A.

ABSTRACT   The overall objective of this work is to develop a new method of using transient tracing to determine rate controlling steps and concentrations of surface intermediates in heterogeneous catalysis systems. The method may be applicable to homogeneous catalysis to enzyme catalysed reactions as well as to non-catalysed solid-liquid reaction systems.

In the present study experimental data are presented and interpreted for the isotopic tracing by ^{13}C of the oxidation of carbon monoxide over commercial Hopcalite catalyst. It is shown that surface concentrations of adsorbed CO_2, and step velocities for the rates of desorption and adsorption of CO_2 formed, can be readily computed from the zeroth and first moments of the response curves derived from transient tracing. These moments are related to the Laplace transforms of the differential equations involved.

INTRODUCTION

We have studied a number of fairly complex industrial reactions using commercial catalysts under steady state reaction conditions, in which superposed tracer transfer was also at steady state.[1] Use of this technique often makes it possible to obtain complete resolution of individual step velocities in a postulated mechanism but not the concentrations of surface intermediates.

The present study is concerned with a new method for estimating the concentrations of surface species as well as the step velocities by extending the tracer technique to transient tracing. This approach has advantages over transient experiments involving changes in concentration of reactants. Concentration jumps in reactant rather than tracer must be kept small in order to linearise the correlating equations; even then such jumps could result in changes in catalyst behaviour.

If a gradientless recirculating reactor is employed, the material balances representing isotopic transfer, which result in a system of simultaneous linear algebraic equations for steady state isotopic transfer, become a system of first order linear differential equations with constant coefficients in the case of transient tracing. A variety of methods is available for solving such equations but the method of moments appears to be especially attractive.[2] Impulse or step changes result in response curves which can be conveniently interpreted by the moments developed from Laplace transforms of the differential equations.

In order to demonstrate the usefulness of this procedure it was necessary to choose an actual system with an appropriate postulated reation mechanism. The oxidation of carbon monoxide over Hopcalite catalyst was chosen for this purpose. This catalyst, which consists of a coprecipitate of oxides of manganese and copper has been studied by numerous investigators.[3] It has the advantage of being quite active at room temperature so that problems associated with heating, insulation and temperature control are minimal.

Generally the mechanism is assumed to proceed by an oxidation-reduction scheme as originally proposed by Rogers et al.[4] though a number of refinements have been added.[5]

† Visiting scholar from Utsunomiya University, Japan.

The following simplified mechanistic scheme was chosen to interpret the results of the present study:

$$2\ CO + O_2 = 2\ CO_2 \tag{1}$$

$$\left.\begin{array}{l}
O_2 + 2l \xrightarrow{\ v_{+1}\ } 2\ Ol \\[2mm]
CO + Ol \xrightarrow{\ v_{+2}\ } CO_2l \\[2mm]
CO_2l \underset{v_{-3}}{\overset{v_{+3}}{\rightleftharpoons}} CO_2 + l
\end{array}\right\} \tag{2}$$

The three steps in eqn. (2) will add up to the overall reaction eqn. (1), if the first step is taken once and the remaining two steps are each taken twice. $v_{\pm i}$, $(i = 1,2,3)$ represent step velocities; Ol and CO_2l represent intermediates on the active sites, l. These sites need not all be the same, as is required in methods of correlation based on kinetics, such as the Langmuir-Hinshelwood hypothesis, since in the treatment which follows only material balances are employed. The idea of participation of molecular oxygen can be accommodated by the assumption that Ol involves two types of adsorbed oxygen, namely $O_2(a)$ and $O(a)$, and the formation of intermediate carbonate ion by the assumption that CO_2l involves two types of adsorbed CO_2, namely $CO_2(a)$ and $CO_3(a)$. The irreversibility of steps 1 and 2 in eqn. (2) was confirmed by separate preliminary experiments involving the concentration jump technique.

The result of the transient tracer experiments, which form the main part of this investigation, was the simultaneous determination of $v_{\pm 3}$ and the concentration of CO_2l on the catalyst surface. Estimates of the concentrations of CO_2l and Ol were also obtained independently by concentration jump experiments. Determination of Ol by tracer experiments was inaccurate because of exchange of oxygen with the oxide lattice.

EXPERIMENTAL

For conducting the experimental program both a differential flow reactor and a gradientless recirculating reactor were employed, schematic diagrams of which are given in Fig. 1. The output gas was analysed by a Finnigan Quadrupole Mass Spectrometer, Type 1015C with the electron energy fixed at 70 eV. The Hopcalite catalyst (Mines Safety Appliance Co., Lot No. 21215) was crushed and screened to 30–60 mesh size. Before use it was pretreated with a mixture of $CO/O_2 = 2/1$ at room temperature for 24 h. Feed mixture was injected with a 100 cm³ injector with syringe pump (Sage Instruments, Model 351). Flow rate of the feed was 2 cm³ min⁻¹ and that of helium carrier gas was 100 cm³ min⁻¹.

He, CO, O_2, and CO_2 (99·99% pure, Matheson Gas Co. Ltd., New Jersey) were used without further purification. $^{18}O_2$ (Miles Yada R & M Co. Ltd., Kiyat, Rehovoth, Israel) and $^{13}CO_2$ and $C^{18}O_2$ (Bio-Rad Laboratories, Richmond, California) were used without purification.

RESULTS

Determination of adsorbed gases by desorption

Preliminary experiments were carried out using the differential flow reactor to determine the amount of CO_2 adsorbed on the catalyst during reaction. Three different gas feeds were employed: CO_2 alone, CO alone, and the mixture $CO/O_2 = 2/1$. After the steady state had been attained for each of these gases, the feed was shut off rapidly while still continuing the flow of helium. The effluent components were monitored for $M/e = 28, 32$, and 44 peaks by mass spectrometry. The amount of CO_2 desorbed at 25°C was essentially the same for all three mixtures, *ca.* 2·3 cm³ g⁻¹ corresponding to

6×10^{19} molecules g^{-1}. No CO or O_2 desorption could be detected, indicating that these materials are adsorbed or reacted irreversibly.

To determine oxygen adsorbed on the catalyst, a feed mixture containing CO and O_2 was passed over the catalyst until steady state reaction was obtained, and the feed mixture was shut off as before. After the helium flow had completely desorbed the CO_2, CO was introduced into the system as a titrating agent and the production of CO_2 was

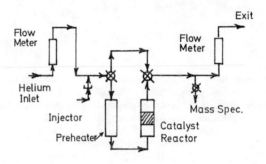

Differential Flow Reactor

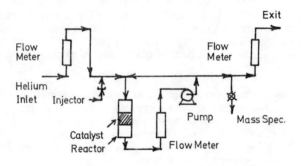

Gradientless Recirculating Reactor

FIGURE 1   Schematic diagram of reactors: differential flow reactor and gradientless recirculating reactor.

monitored by mass spectrometry. The oxygen adsorbed on the catalyst surface was then estimated from the CO_2 formation. In the case of a feed mixture consisting of $CO/O_2 = 2/1$ the amount of adsorbed oxygen was about 0.5 cm^3 g^{-1} of molecular oxygen corresponding to 2.5×10^{19} atoms g^{-1} or 1.25×10^{19} molecules g^{-1} of oxygen chemisorbed. When a feed mixture corresponding to $CO/O_2 = 8/1$ was used, the adsorbed oxygen was only one tenth of this amount, the CO_2 adsorption remaining unchanged. The surface of the Hopcalite catalyst was estimated to be 60 m^2 g^{-1} by means of the standard BET procedure. The area occupied by all adsorbed molecules was taken at 16Å^2 per molecule of adsorbate, equivalent to the N_2 molecule. With this assumption, the surface coverage by CO_2 is 9.6 m^2 g^{-1} and of oxygen (as molecular oxygen) 2.0 m^2 g^{-1}. Thus the fraction of surface covered is $11.6/60 \approx 0.19$.

Kinetic studies

Kinetic data were obtained at 25°C in the differential reactor and correlated by a simple power function model. The rate of oxidation of CO varied as the 0·8 power of CO partial pressure and was substantially zero order in oxygen and carbon dioxide, in qualitative agreement with the findings of Brittan et al.[3]

For determination of activation energy of the reaction the $CO/O_2 = 2/1$ mixture was used over a temperature range 15–150°C. The apparent activation energy was $\sim 5·55$ kcal mol^{-1}.

Isotopic exchange

In addition to transfer rates of isotopes corresponding to tracing of the reaction eqn. (1), the following isotopic exchange reactions were studied:

$$^{16}O_2 + {}^{18}O_2 = 2\ {}^{16}O\ {}^{18}O \tag{4}$$

$$C\ {}^{16}O_2 + C\ {}^{18}O_2 = 2\ C\ {}^{16}O\ {}^{18}O. \tag{5}$$

Mixtures of CO_2, CO, and O_2 were used at 25°C. No $^{16}O\ {}^{18}O$ formation was detected but $C^{16}O^{18}O$ formation was very fast and rapidly reached equilibrium.

After the adsorbed species were marked by ^{18}O in the exchange experiments, the feed mixture was changed to one containing non-tagged molecules and ^{18}O transfer from the catalyst surface to gaseous CO, O_2, and CO_2 was monitored. In these experiments marking was only observed in CO_2. Similar results were obtained by ^{13}C marked molecules. These results imply that oxygen is adsorbed on the catalyst surface irreversibly and that CO adsorbs on the surface to form an intermediate directly with adsorbed oxygen.

Correlation of transient tracing data using ^{13}C

Consider that a reaction involving the mechanism depicted by eqn. (2) is being conducted in an open gradientless recirculating reactor system and that temperatures are low enough to avoid intraparticle diffusional effects. The following system of equations may be written for atomic transfer of ^{13}C superposed on a steady state overall reaction in which CO_2 is marked by ^{13}C:

$$-\frac{F^{CO_2}\, z^{CO_2}}{W} = \frac{\beta}{W}\, C^{CO_2}\, \frac{dz^{CO_2}}{dt} - v_{+3}\, z^{CO_2 l} + v_{-3}\, z^{CO_2} \tag{6}$$

$$C^{CO_2 l}\, \frac{dz^{CO_2 l}}{dt} = -v_{+3}\, z^{CO_2 l} + v_{-3}\, z^{CO_2} \tag{7}$$

where C^{CO_2} = concentration of gas phase CO_2 in the reaction system (g mol cm^{-3}), $C^{CO_2 l}$ = concentration of adsorbed CO_2 on solid catalyst (g mol g^{-1}), F^{CO_2} = outlet flow rate of CO_2 (g mol min^{-1}), t = time (min), $c_{\pm 3}$ = unidirectional step velocities (g mol min^{-1} (g catalyst)$^{-1}$), W = weight of catalyst in system (g), z^i = fraction of component i ($i = CO_2, CO_{2l}$) containing tracer, in this case ^{13}C, β = volume of dead space, including that in catalyst pores, voids and apparatus (cm^3).

Note that the rate of inlet flow does not appear because the inlet stream is assumed to be unmarked beginning at time $t = 0$. Since the reactions are irreversible and ^{13}C is the marked species, balances corresponding to the reaction steps 1 and 2 do not appear.

The boundary conditions corresponding to a pulse of marked CO_2 are:

at $t = 0$	at $t = \infty$
$z^{CO_2} = z_0^{CO_2}$	$z^{CO_2} = 0$
$z^{CO\, l} = 0$	$z^{CO_2 l} = 0.$

The Laplace transforms of eqn. (6) and (7) are:

$$-\frac{F^{CO_2}\,\overline{z^{CO_2}}\,(s)}{W} = \frac{\beta}{W}\,\overline{C^{CO_2}}\,s\,z^{CO_2}\,(s) - \frac{\beta}{W}\,C^{CO_2}\,z_0^{CO_2} - v_{+3}\,\overline{z^{CO_2 l}}\,(s) + v_{-3}\,\overline{z^{CO_2}}\,(s)$$

(8)

and

$$C^{CO_2 l}\,s\,\overline{z^{CO_2 l}}\,(s) = -\,v_{+3}\,\overline{z^{CO_2 l}}\,(s) \times v_{-3}\,\overline{z^{CO_2}}\,(s)\,.$$

(9)

These equations may be solved for $\overline{z^{CO_2}}\,(s)$ to obtain

$$\overline{z^{CO_2}}\,(s) = \frac{\dfrac{\beta}{W}\,C^{CO_2}\,z_0^{CO_2}}{\dfrac{\beta}{W}\,C^{CO_2}\,s + v_{-3} + \dfrac{F^{CO_2}}{W} - \dfrac{v_{+3}\,v_{-3}}{sC^{CO_2 l} + v_{+3}}}\,.$$

(10)

The zeroth moment corresponds to the limit of the transform as s approaches zero:

$$M_0\,{}_{2CO_2} = \lim_{s\to 0}\,\overline{z^{CO_2}}\,(s) = \frac{\beta\,C^{CO_2}\,z_0^{CO_2}}{F^{CO_2}}\,.$$

(11)

The first moment is obtained by taking the derivative d $\overline{z^{CO_2}}\,(s)/ds$ of eqn. (10) and reducing it to the limiting form as s approaches zero:

$$M_1\,{}_{2CO_2} = -\lim_{s\to 0}\frac{d\overline{z^{CO_2}}\,(s)}{ds} = \left(\frac{W}{F^{CO_2}}\right)^2\frac{\beta}{W}\,C^{CO_2}\,z_0^{CO_2}\left(\frac{\beta}{W}\,C^{CO_2} + \frac{v_{-3}}{v_{+3}}\,C^{CO_2 l}\right)$$

(12)

Eqn. (12) can be solved for $v_{-3}/v_{+3}\,C^{CO_2 l}$. The moments are obtained from the response curves of tracer concentration in reactor effluent. Thus:

$$M_0 = \int_0^{\infty} z^{CO_2}\,dt$$

(13)

$$M_1 = \int_0^{\infty} z^{CO_2}\,t\,dt.$$

(14)

If a value of $C^{CO_2 l}$ can be determined from desorption experiments v_{-3}/v_{+3} can be calculated. Then, since $2V = v_{+3} - v_{-3} = v_{+2} = 2v_{+1}$ the step velocities can be determined.

Table 1 gives a series of runs conducted using a pulse of $^{13}CO_2$, assuming an approximate value of $2 \cdot 0$ cm^3 g^{-1} of CO_2 adsorbed corresponding to $C^{CO_2 l}$.

If it is convenient to use a larger amount of tracer so that tracer concentration reaches a steady state, a step function change corresponding to rapid shut-off of tracer in the feed is possible. This procedure furnishes enough data to make a simultaneous calculation of step velocities and the surface concentration $C^{CO_2 l}$, without the need for a separate determination of $C^{CO_2 l}$ by concentration jump (desorption).

For this procedure the boundary conditions corresponding to eqn. (6) and (7) are:

at $t = 0$	at $t = \infty$
$z^{CO_2} = z_0^{CO_2}$	$z^{CO_2} = 0$
$z^{CO_2} = z_0^{CO_2 l}$	$z^{CO_2 l} = 0.$

Table 1
Pulse experiments using $^{13}CO_2$ tracing

Run No.	022075	022475	022775	031175	031775
Catalyst wt W (g)	0·25	0·25	0·25	0·25	0·25
F^{CO_2} (cm³ min⁻¹) at 25°C	0·290	0·270	0·270	0·273	0·279
β (cm³)	90	90	90	90	90
$Z_0{}^{CO_2}$ (fraction)	0·0270	0·0270	0·0203	0·0325	0·0240
C^{CO_2} (vol. fraction)	$2·84 \times 10^{-3}$	$2·65 \times 10^{-3}$	$2·65 \times 10^{-3}$	$2·68 \times 10^{-3}$	$2·73 \times 10^{-3}$
$C^{CO_2 l}$ (assumed) (cm³ g⁻¹)	2	2	2	2	2
$M_0{}_2{}^{CO_2}$ (min)	0·0341	0·0265	0·0175	0·0398	0·0340
$M_{12}{}^{CO_2}$ (min²)	0·0449	0·0442	0·0289	0·0662	0·0523
Mixture ratio CO/O₂/CO₂	6/3/1	6/3/1	6/3/1	3/6/1/	3/6/1
P (atm)	1·0	1.0	1·0	1·0	1·0
T (°C)	25·0	25·0	25·0	25·0	25·0
V cm³ O₂ min⁻¹ g⁻¹	0·180	0·140	0·140	0·146	0·158
v_{+1} cm³ O₂ min⁻¹ g⁻¹	0·180	0·140	0·140	0·146	0,158
v_{+2} cm³ CO min⁻¹ g⁻¹	0·360	0·280	0·280	0·292	0·316
v_{+3} cm³ CO₂ min⁻¹ g⁻¹	0·860	0·537	0·462	1·313	2·856
v_{-3} cm³ CO₂ min⁻¹ g⁻¹	0·500	0·257	0·182	1·021	2·540
(v_{+3}/v_{-3})	1·72	2·09	2·54	1·286	1·13

Under these conditions, following the same procedure as was used to derive eqn. (11) and (12), we obtain:

$$M_0{}_2{}^{CO_2} = \frac{W}{F^{CO_2}} \left(\frac{\beta}{W} C^{CO_2} z_0{}^{CO_2} + C^{CO_2 l} z_0{}^{CO_2} \right) \tag{15}$$

and

$$M_{12}{}^{CO_2} = \frac{\frac{W}{F^{CO_2}} (C^{CO_2 l})^2 z_0{}^{CO_2 l}}{v_{+3}} + \frac{\left(\frac{\beta}{W} C^{CO_2} z_0{}^{CO_2} + C^{CO_2 l} z_0{}^{CO_2 l} \right) \left(\frac{\beta}{W} C^{CO_2} + \frac{v_{-3}}{v_{+3}} C^{CO_2 l} \right)}{\left(\frac{F^{CO_2}}{W} \right)^2}. \tag{16}$$

At steady state eqn. (7) also gives the relationship:

$$z_0{}^{CO_2} = \frac{v_{-3}}{v_{+3}} z_0{}^{CO_2}. \tag{17}$$

Simultaneous solution of eqn. (15)–(17) gives an explicit relationship for v_{-3} in terms of observable quantities:

$$v_{-3} = \frac{\left(\frac{W z_0{}^{CO_2}}{F^{CO_2}} \right) \left(\frac{M_0{}_2{}^{CO_2} F^{CO_2}}{z_0{}^{CO_2} W} - \frac{\beta}{W} C^{CO_2} \right)^2}{M_{12}{}^{CO_2} - \frac{(M_0{}_2{}^{CO_2})^2}{z_0{}^{CO_2}}}. \tag{18}$$

v_{+3} and $C^{CO_2 l}$ are then readily computed as well as the step velocities v_{+1} and v_{+2}.

Table 2 summarizes the results of step function tracing using this procedure. As the overall reaction progressed to steady state no ^{13}C was transferred from CO₂ to CO.

The rate of step 3 appears to be about ten times faster than step 2. The amount of adsorbed carbon dioxide species on the catalyst is about 2 cm^3 g^{-1} in agreement with the desorption experiments.

Table 2.

Step function change experiments using $^{13}CO_2$ tracing

Run No.	030675	031375
Catalyst wt. (g)	0·25	0·25
F^{CO_2} (cm^3 min^{-1}) at 24°C	0·275	0·274
β cm^3	90	90
Z^{CO_2} (fraction)	0·0991	0·0974
C^{CO_2} (vol. fraction)	2·69 × 10^{-3}	2·68 × 10^{-3}
$M_{0_2}{}^{CO_2}$ (min)	0·303	0·252
$M_{1_2}{}^{CO_2}$ (min^2)	1·120	0·791
Mixture ratio, CO/O$_2$/CO$_2$	6/3/1	6/3/1
P (atm)	1·0	1·0
T (°C)	24·0	24·0
V cm^3 O$_2$ min^{-1} g^{-1}	0·150	0·148
v_{+1} cm^3 O$_2$ min^{-1} g^{-1}	0·150	0·148
v_{+2} cm^3 CO min^{-1} g^{-1}	0·300	0·296
v_{+3} cm^3 CO$_2$ min^{-1} g^{-1}	2·98	2·61
v_{-3} cm^3 CO$_2$ min^{-1} g^{-1}	2·68	2·32
$C^{CO_2 l}$ (cm^3 g^{-1})	2·66	2·12

Fig. 2 illustrates the procedure employed in obtaining experimental values of the moments required. It corresponds to the step function concentration change, run No. 030675 given in Table 2. The zeroth moment is the area under the z^{CO_2} vs t curve while the first moment is the area under the z^{CO_2} t vs t curve. Excessive tailing, which sometimes presents a problem in evaluating higher moments, does not seem to occur.

Transient tracing with ^{18}O

Tracing with ^{18}O is more complicated than for ^{13}C marking because the oxygen may exchange with lattice oxygen contained in oxide catalysts. Also in the case of O$_2$ marking three simultaneous differential equations must be solved, although the same procedure involving the method of moments is applicable. Two sets of experiments were performed with such tracing; the results are only summarized here.

First, we made experiments with ^{18}O marked O$_2$ using the differential reactor to reduce residence time of marked species over the catalyst. The amount of adsorbed oxygen was estimated to be twice as much as that obtained by the titration method previously discussed.

Second, we used the gradientless recirculating flow reactor with ^{18}O marked CO$_2$. It was found that the rate of step 3 was about 10 times faster than that of step 2 in agreement with the ^{13}C tracer data. However, the amount of adsorbed carbon species estimated by ^{18}O tracer data was about 7–10 times larger than that obtained by ^{13}C tracer data. This large value is probably due to the fact that ^{18}O exchanges with unmarked oxygen in the catalyst structure. ^{13}C on the other hand cannot participate in exchange with the catalyst in this fashion. Results with ^{13}C agree with values obtained by desorption lending support to this explanation.

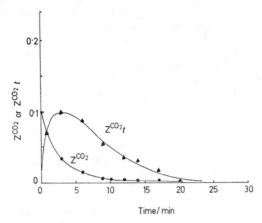

FIGURE 2   Response curves for $^{13}CO_2$ tracing:
run no. 030675;
catalyst used, 0·25 g;
temperature, 25°C;
mixture ratio, $CO/O_2/CO_2 = 6/3/1$.

DISCUSSION

As demonstrated above, a new method for simultaneous measurement of adsorption and kinetics by a transient tracing technique seems to provide access to fundamental information on catalyst structure. Step velocities can also be calculated from the data obtained by simple formulae. In the case of Hopcalite catalyst, results are in qualitative agreement with previous studies, especially as regards the reversibility of CO_2 desorption. Use of ^{13}C data provides a more straightforward procedure in the case of Hopcalite catalyst than ^{18}O, due no doubt to exchange of marked oxygen with the lattice oxygen of the catalyst. More detailed studies will be necessary to interpret fully the ^{18}O tracer results.

The procedure described should be a useful adjunct in catalyst evaluation studies because of its simplicity and applicability to industrial catalysts.

ACKNOWLEDGEMENT. Support of this project by the National Science Foundation is gratefully acknowledged.

REFERENCES

[1]   J. Happel and M. A. Hnatow, *Ann. New York Acad. Sci.*, 1973, **213**, 206.
[2]   J. Happel, *AIChE Journal*, 1975, **21**, 602.
[3]   M. J. Brittan, H. Bliss, and C. A. Walker, *AIChE Journal*, 1970, **16**, 305.
[4]   T. H. Rogers, C. S. Piggot, W. H. Bahlke, and J. M. Jennings, *J. Amer. Chem. Soc.*, 1921, **43**, 1973.
[5]   M. Kobayashi and H. Kobayashi, *J. Catalysis*, 1972, **27**, 100; 108; 114.

DISCUSSION

J. W. E. Coenen (*Unilever, Vlaardingen*) said: My problem may be semantic in origin. In the context of carbon monoxide oxidation on Hopcalite, what is meant by 'the surface concentration of adsorbed oxygen'? According to the accepted mechanism (Mars and van Krevelen) for such oxidations, oxygen from the lattice participates in the reaction. The catalyst is thus partially reduced and subsequently reoxidized. One can still speak of surface oxygen concentrations but a zero level has to be defined.

J. Happel (*Columbia Univ., New York*) replied: You are correct in considering the problem of oxygen adsorption to be one of semantics. The measurement which we made gives only the amount of oxygen adsorbed per gram of catalyst. Our reported surface concentration was calculated from this as described in the paper on the assumption that it occupies surface sites. However, if lattice sites are involved in the catalytic reaction, such oxygen would need to be very close to the surface because carbon monoxide and dioxide will not diffuse into the lattice.

R. W. Coughlin (*Lehigh Univ., Bethlehem, Pa.*) said: Has this tracer method been applied to reactions known to take place with self-oscillation (or flickering)? Such phenomena would ordinarily be too rapid for resolution by such a method but it is interesting to consider whether some information on self-oscillation might not be superimposed on the transient tracer results.

J. Happel (*Columbia Univ., New York*) replied: As far as we know the tracer method of the type discussed here has not been applied to oscillating reactions. Use of a Finnigan mass spectrometer enables up to 100 scans per second to be monitored so, in principle, such an application should be possible. We agree that interesting results might be obtained.

L. Guczi (*Inst. of Isotopes, Budapest*) said: The experimental conditions listed in Table 1 for Run 022075 and those in Table 2 for Run 030675 are essentially the same although the experimental method is different. Is there any reason why the values of v_{+3} and v_{-3} for the two experiments should differ?

J. Happel (*Columbia Univ., New York*) replied: The difference occurs because it is impossible to determine directly the extent of oxygen adsorption with precision. Also, the pulse method is inherently less accurate because no carbon dioxide is adsorbed at the beginning of the experiment. Thus the moments observed in the pulse experiment do not differ as much from those which would be observed with no reaction as in the case of introduction of a step change.

A. Sárkány (*Inst. of Isotopes, Budapest*) asked: How are your response waves for $^{13}CO_2$ tracing influenced by the reactor itself? What is the profile of the response waves if there is no reaction?

J. Happel (*Columbia Univ., New York*) replied: The reactor system includes a volume of dead space which is designated β in the equations presented. If there were no reaction, the equations in the paper would reduce to the form required for continuous introduction and removal of feed and tracer to a stirred tank of volume β.

Separation of Rate Processes for Isotopic Exchange between Hydrogen and Liquid Water in Packed Columns

JOHN P. BUTLER, JAMES DEN HARTOG, JOHN W. GOODALE, and JOHN H. ROLSTON*

Atomic Energy of Canada Limited, Chalk River Nuclear Laboratories, Chalk River, Ontario, Canada, K0J 1J0

ABSTRACT Wetproofed platinum catalysts in packed columns promote isotopic exchange between counter-current streams of hydrogen saturated with water vapour and liquid water. The net rate of deuterium transfer from isotopically enriched hydrogen has been measured and separated into two rate processes involving the transfer of deuterium from hydrogen to water vapour and from water vapour to liquid. These are compared with independent measurements of the two rate processes to test the two-step successive exchange model for trickle bed reactors.

The separated transfer rates are independent of bed height and characterize the deuterium concentrations of each stream along the length of the bed. The dependences of the transfer rates upon hydrogen and liquid flow, hydrogen pressure, platinum loading and the effect of dilution of the hydrophobic catalyst with inert hydrophilic packing are reported. The results indicate a third process may be important in the transfer of deuterium between hydrogen and liquid water.

INTRODUCTION

The advantages of producing heavy water, D_2O, from natural water by a dual temperature chemical exchange process involving hydrogen and liquid water have been recognized for many years.[1] However, an efficient rate of transfer of deuterium between hydrogen and the liquid water is required to develop an economic enrichment process employing multi-stage cascades in which liquid water flows counter-current to the hydrogen stream. Such a process would be an attractive alternative to the existing bithermal exchange process between hydrogen sulphide and water. Although effective catalysts for isotopic exchange between hydrogen and water vapour have been known for many years, they are characterized by extremely low rates of exchange in the presence of liquid water.[2] An appreciation of these requirements led W. H. Stevens[3] to attempt to preserve the high vapour phase catalytic activity of a supported platinum catalyst in contact with liquid water through the application of a silicone polymer to wetproof the catalyst. Subsequently, the concept of providing an improved hydrophobic environment for platinum catalysts was extended to the use of porous plastics such as polytetrafluoroethylene.[4] Yet another approach to preparing an effective catalyst relies heavily on techniques used in the preparation of hydrogen electrodes for fuel cells.[5] Typically, such electrodes consist of a porous layer composed of platinum black, or of platinum supported on carbon, and Teflon fibrils which act both as a binder and as a wetproofing agent. Such composites, when deposited on a variety of column packings, produce very effective catalysts for isotopic exchange between hydrogen and liquid water.

As a first approximation, the overall rate of transfer of a deuterium tracer between streams of hydrogen and liquid water over such wetproofed catalysts can be considered in terms of two transfer steps. The first corresponds to the catalytic rate of transfer of deuterium from an enriched hydrogen stream to water vapour, R_1 (g atom D/s) and the second, R_2, corresponds to the transfer rate from water vapour to liquid as shown in eqns. (1) and (2).

$$HD + H_2O \text{ (vap.)} \underset{R_{-1}}{\overset{R_1}{\rightleftharpoons}} HDO \text{ (vap.)} + H_2 \tag{1}$$

$$HDO \text{ (vap.)} + H_2O \text{ (liq.)} \underset{R_{-2}}{\overset{R_2}{\rightleftharpoons}} HDO \text{ (liq.)} + H_2O \text{ (vap.)} \tag{2}$$

An analysis of the trickle bed in these terms can be made from a knowledge of the molar flows of the three streams and of their deuterium concentrations at the inlet and outlet of the reactor. It is the ability of the wetproofed catalyst to promote these reactions efficiently and to couple them closely in space that raises the prospects for a deuterium separation process based upon hydrogen–water exchange.

EXPERIMENTAL

(a) *Description of the Apparatus*

A schematic diagram of the exchange column (diameter 2·5 cm) and associated apparatus used for measurements of the exchange activity of a catalyst in a trickle bed reactor is shown in Fig. 1. In the reactor, liquid water is directed downward through the catalyst bed and hydrogen gas, saturated with water vapour, moves upward through

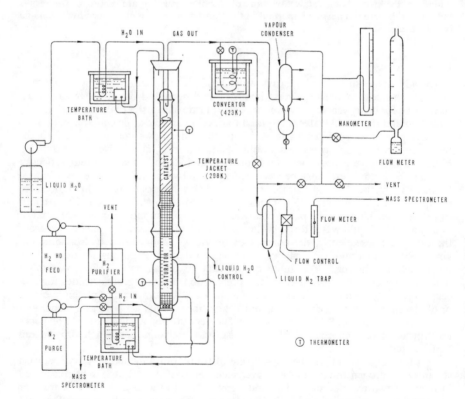

FIGURE 1   Schematic diagram of exchange column and associated apparatus.

the column. The catalyst bed was placed on top of a 45 cm bed of untreated 6·1 mm diameter alumina spheres similar to those used to support the catalyst. This section was maintained at the temperature of the catalyst and served as the final humidifying section and flow distributor for the gas stream entering the reactor. The lower 30 cm of the saturator was independently thermostatted to provide the heat for the initial humidification. This arrangement ensured that the hydrogen gas entering the bottom of the catalyst column was saturated with water vapour which was at isotopic equilibrium with the liquid water flowing from the bottom of the catalyst bed.

Considerable care was taken to prevent isotopic exchange between water vapour and liquid water above the catalyst bed before the deuterium content of the water vapour was measured. A hooked 3 mm diameter tube served as a spray trap and all tubing was heated to prevent condensation of water vapour. Valves directed samples of effluent gas stream through either a liquid-nitrogen trap to the mass spectrometer or through a high temperature converter at 423 K. The converter consisted of a coil of 9 mm Pyrex tubing filled with wetproofed platinum catalyst and served to bring the hydrogen and water vapour into isotopic equilibrium. The deuterium concentration of water vapour could then be calculated from measurements of the deuterium content of the hydrogen stream before and after passage through the converter.

Large diameter (15 cm) soap film flow meters were used to monitor the hydrogen flows. The deuterium atom fractions of the water-free hydrogen streams were measured to a precision of 0·1 ppm in a Thomson–Houston 202 mass spectrometer modified at CRNL[6] to permit on-line deuterium analyses at near natural abundance levels. Individual analyses were routinely performed at intervals of about 1 min and were recorded automatically on paper tape. The spectrometer was calibrated periodically against hydrogen standards of known deuterium content to compensate for spectrometer drift and to maintain the high accuracy necessary for calculations of the two transfer rates.

Isotopically enriched hydrogen (10^6 D/D + H $\sim$ 300) was purified by passage through a hydrogen purifier (Johnson-Matthey HP-50) and fed to the bottom of the saturator unit. Liquid water at natural deuterium abundance, was fed by a positive displacement diaphragm pump through either a 3 mm diameter tube or a perforated plastic ring distributor to the top of the catalyst column.

(b) *Kinetic Model of the Exchange reaction*

Since an appreciable isotope effect exists for exchange between hydrogen and water (eqn. (1)), the net rate of deuterium transfer to water vapour in a unit volume of the reactor T_R (g atom D/s cm³) is equated to the difference of the forward, R_1, and reverse, R_{-1}, rates weighted in terms of the probability of their occurrence.[7]

$$T_R = R_1 n(1 - v) - R_{-1} v(1 - n) \qquad (3)$$

Following Cohen[8] the probability factors are written in terms of the deuterium atom fractions (D/D + H) of the hydrogen, n, vapour, v, and liquid, N, streams. An analogous equation can be written for the net rate of transfer of deuterium from the vapour to the liquid.

$$T_D = R_2 v (1 - N) - R_{-2} N(1 - v) \qquad (4)$$

At low deuterium levels these equations reduce to

$$T_R = R_1 n - R_{-1} v \qquad (5a)$$

$$T_D = R_2 v - R_{-2} N \qquad (5b)$$

749

At equilibrium $T_R = 0 = T_D$, hence it follows that

$$\alpha_R = \frac{R_1}{R_{-1}} = \frac{v_e}{n_e} \text{ and } \alpha_D = \frac{R_2}{R_{-2}} = \frac{N_e}{v_e} \qquad \text{(6a) and (6b)}$$

Variables subscripted with e denote concentrations at equilibrium and α_R and α_D are respectively the separation factors between hydrogen and water vapour[9] and between water vapour and liquid water.[10] Elimination of the forward rates gives

$$T_R = R_{-1}(\alpha_R n - v) = \rho k_R \eta(\alpha_R n - v) \qquad (7a)$$

$$T_D = R_{-2}(\alpha_D v - N) = \rho k_D(\alpha_D v - N) \qquad (7b)$$

In applying these equations to the trickle bed reactor, it is assumed that there exists a uniform distribution of the process streams and of their deuterium concentrations across the column diameter. The reverse rates of transfer per unit volume of packed bed (g atom D/s cm³) have been expressed in terms of transfer rates per gram of catalyst, k_R or k_D, the packing density, ρ, of catalyst in the column (g/cm³ bed) and for the catalytic rate an overall effectiveness factor, η. The latter varies between 0 and 1 and is defined as the ratio of the observed rate over the catalyst spheres to that observed in the absence of any effects of catalyst morphology.

The differential equations governing the steady state transfer of an isotope in a column of height z(cm) with counter-current flows of the liquid and gas phases are given by

$$\frac{l}{A} \times \frac{dn}{dz} = -T_R \qquad (8a)$$

$$\frac{V}{A} \times \frac{dv}{dz} = T_R - T_D \qquad (8b)$$

$$\frac{L}{A} \times \frac{dN}{dz} = -T_D \qquad (8c)$$

These equations have been solved by Palibroda[11] for columns of cross-section A (cm²) where l, V, and L represent the molar flows (mol/s) of hydrogen, water vapour and liquid through the column.

(c) *Treatment of the Data*

To obtain the deuterium transfer rates $\rho k_R \eta$ and ρk_D, a knowledge of the inlet and outlet deuterium concentrations in all three process streams is required. Natural water is fed to the top of the column with a constant deuterium atom fraction, N_t, of 144 ppm. Measurement of the deuterium atom fraction of the hydrogen at the bottom of column, n_b, the top of the column, n_t and after passage through the converter, n_c, are sufficient to define the concentration in the remaining process streams. Thus v_t, the deuterium atom fraction of the water vapour at the top of the bed can be calculated from the equation

$$v_t = \frac{n_C(1 + \alpha_C H) - n_t}{H} \qquad (9)$$

The humidity, H, is defined as the molar ratio of water vapour to hydrogen in the gas stream and α_C is the hydrogen–water vapour separation factor at the temperature of the converter.[9] This relationship is a valid approximation for low deuterium levels.

750

Since the water vapour entering the column is in isotopic equilibrium with the liquid from the catalyst bed, it follows that

$$v_b = N_b/\alpha_D \tag{10}$$

where the variables subscripted with b denote the deuterium concentrations at the bottom of the bed, $z = 0$. A deuterium mass balance over the column permits the deuterium atom fraction of the effluent liquid to be calculated from the equation

$$N_b = \left[l\left(n_t - n_b\right) + Vv_t - LN_t \right] \bigg/ \left(\frac{V}{\alpha_D} - L\right) \tag{11}$$

Thus in a given experimental run it was necessary to determine only the flow rates of the three process streams and the deuterium concentration of the hydrogen stream at three specified sampling points.

RESULTS

The separation of the overall rate of deuterium transfer in the trickle bed reactors has been made on the basis of the kinetic treatment of successive exchange reactions described by Palibroda. The transfer coefficients $\rho k_R \eta$ and ρk_D (g atoms D/s cm^3) corresponding to the reverse transfer rates of the mono-deuteriated species in eqns. (1) and (2) were calculated with the aid of a short computer program which matched the computed deuterium atom fractions of the three streams at the top of the column against those determined experimentally. The dependence of the two separated rates upon hydrogen flow at 298 K and 0·1 MPa is shown in Fig. 2. The data were obtained for beds containing 100% catalyst spheres and mixed beds containing catalyst randomly diluted with an equal volume of inert hydrophilic alumina spheres. Also shown are independent measurements of the separated rates, $\rho k_R \eta$ and ρk_D which were made respectively in the absence of a liquid phase[12] and in the absence of catalyst.[13]

All transfer rates increase as a power of the hydrogen flow rate. The transfer rate $\rho k_R \eta$ shows an exponent of 0·02 which is independent of catalyst dilution and is similar to the direct measurements of the rate in the absence of the liquid stream. In contrast ρk_D in the trickle bed reactor shows a flow exponent which varies from 0·27 to 0·48 and depends strongly upon the composition of the catalyst bed. It should also be noted that $\rho k_R \eta$ is roughly five times greater than ρk_D in the 100% catalyst column whereas they are nearly equal in mixed meds. The addition of inert hydrophilic material to form a mixed bed substantially increases ρk_D whereas $\rho k_R \eta$ decreases in proportion to the amount of catalyst.

The transfer rates depend directly upon (D/D + H) measurements of three hydrogen streams (n_b, n_t and n_o) and each measurement has an accuracy of $\pm$ 0·3 ppm. As a result, the transfer rates under our operating conditions have a precision of $\pm$ 8%. Although measured rates in a given run show good reproducibity with hydrogen flow, variations in the rate of about $\pm$ 25% are observed in duplicate runs using different samples of catalyst. These variations are considered to result primarily from changes in the water flow distribution through the hydrophobic bed. Although the absolute values of transfer rates vary somewhat, the power dependence of the rates on hydrogen flow are reproducible to within $\pm$ 0·04.

The dependence of the separated rates upon the column height was studied in mixed beds and typical results at 298 K and 0·10 MPa are summarized in Table 1. The data were obtained during continuous column operation over 24 h on beds of freshly packed catalyst. Within the reproducibility of the measurements, $\pm$ 25%, the separated rates $\rho k_R \eta$ and ρk_D are independent of bed height and therefore they can be used to calculate the changes in the deuterium concentration of the various streams along the length of a column.

751

Butler, den Hartog, Goodale, Rolston

Variation of the liquid flow rate appears to have little effect on either of the separated transfer rates for beds of 100% catalyst as shown in Fig. 3. The liquid tends to flow as a

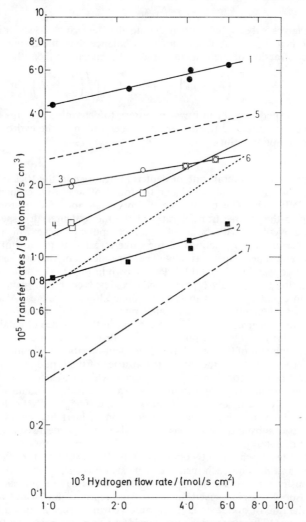

FIGURE 2 Dependence of the separated rates on hydrogen flow rate. Liquid flow: 9×10^{-3} mol/s cm², Column length: 10 cm, Temperature: 298 K, Pressure: 0.10 MPa. (1) $\rho k_R \eta$ for 100% catalyst, (2) ρk_D for 100% catalyst, (3) $\rho k_R \eta$ for 50% catalyst, (4) ρk_D for 50% catalyst, (5) $\rho k_R \eta$ in absence of liquid phase, (6) ρk_D for 50% hydrophobic spheres and 50% hydrophilic spheres, (7) ρk_D for 100% hydrophobic spheres.

smooth film down the column wall with very little of the liquid penetrating into the centre portion of the hydrophobic catalyst bed. Such test columns might more accurately be described as "wetted wall" rather than "trickle bed" reactors.

The absence of a significant effect of liquid flow rate upon the separated rates was

Table 1

Dependence of separated rates on height for mixed catalyst beds

Hydrogen flow: $2{\cdot}7 \times 10^{-3}$ mol/s cm²
Liquid flow: 9×10^{-3} mol/s cm²

Bed height l/(cm)	10^5 Transfer rates/(g atom D/s cm³)	
	$\rho k_R \eta$	ρk_D
5·0	2·1	2·6
5·0	2·0	2·4
10·0	2·8	1·8
30·0	2·3	1·9
30·0	1·7	3·0
75·0	2·7	2·5
Average	2·3	2·3

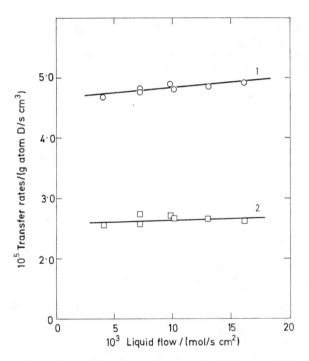

FIGURE 3   Dependence of the separated rates on liquid flow rate. Hydrogen flow: $2{\cdot}5 \times 10^{-3}$ mol/s cm², Column length: 15 cm, Temperature: 298 K, Pressure: 0·10 MPa. (1) $\rho k_R \eta$, (2) ρk_D.

753

also observed in mixed beds. However, in these beds, much more of the liquid was directed into the centre of the bed by the more easily wetted alumina spheres.

The separated rates summarized in Table 2 show an inverse square root dependence on hydrogen pressures up to 0·2 MPa. A similar effect was noted for $\rho k_R \eta$ measured on catalyst spheres in the absence of the liquid phase.[12]

Table 2

Effect of pressure on separated rates at 298 K

Hydrogen flow: $2·0 \times 10^{-3}$ mol/s cm²
Liquid flow: $8·8 \times 10^{-3}$ mol/s cm²
Column length: 20 cm

H₂ Pressure /(MPa)	10^5 Transfer rates/(g atom D/s cm³)	
	$\rho k_R \eta$	ρk_D
0·101	2·19	1·14
0·146	1·83	0·97
0·191	1·57	0·83

Catalysts with platinum loadings between 0·02 and 0·30 wt % were prepared from known amounts of a 10% platinized carbon powder. The separated transfer rates are plotted in Fig. 4 against the platinum metal area. The metal areas of the finished catalysts were calculated from the hydrogen uptake of the platinized carbon at 298 K and the

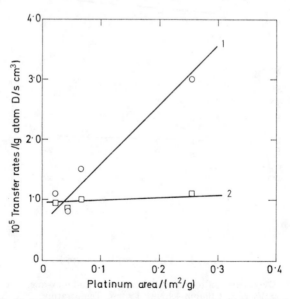

FIGURE 4 Dependence of the separated rates on platinum area. Hydrogen flow: $3·0 \times 10^{-3}$ mol/s cm², Liquid flow: $9·2 \times 10^{-3}$ mol/s cm², Column height: 15 cm, Temperature: 298 K, Pressure: 0·10 MPa. (1) $\rho k_R \eta$, (2) ρk_D.

weight of powder on each catalyst. Each atom of absorbed hydrogen was assumed to occupy an area of 8.83×10^{-16} cm^2. The transfer rate $\rho k_R \eta$ increases with the available metal area while ρk_D is essentially constant.

DISCUSSION

The data in Table 1 and Fig. 2 show that the transfer rate, ρk_D, is of the same order of magnitude as that of $\rho k_R \eta$. Thus neither transfer step can be considered to be rate limiting and changes in the rate of either reaction will influence the overall transfer of deuterium between hydrogen and liquid water. However, since the reactions are coupled in series the full effect of a change in one of the rate coefficients will not produce a change of the same magnitude in the overall exchange rate.

The separated transfer coefficient $\rho k_R \eta$ defines the exchange rate between the isotopically enriched water vapour and hydrogen at 100% relative humidity. At present it is not possible to relate it directly to the inherent catalytic exchange rate over a sphere of catalyst. Indeed, Fig. 2 shows that $\rho k_R \eta$ is not independent of hydrogen flow rate so that bulk gas phase mass transfer effects are operative in the column. Because of these effects, the observed inverse square root dependence of $\rho k_R \eta$ on hydrogen pressure is unlikely to represent the true kinetic order of the surface exchange reaction. In the absence of the liquid phase at very high hydrogen flow rates ($> 2 \times 10^{-2}$ mol/s cm^2), it is possible to obtain values of $\rho k_R \eta$ which are independent of hydrogen flow and which represent the inherent catalytic activity of the whole spheres for the vapour phase exchange reaction.[12] These high flows of hydrogen, however, could not be studied in the trickle bed reactor because they exceeded the flooding velocities of of the packings.

From Fig. 4 it is evident that exchange between hydrogen and water vapour is indeed a catalytically controlled process since higher transfer coefficients, $\rho k_R \eta$, are observed for higher platinum loadings. The measured rates of this reaction are nearly proportional to the platinum metal areas, indicating that the observed rates are not yet limited by pore diffusion. In contrast the rate of the vapour–liquid transfer reaction, ρk_D, is independent of platinum loading which is consistent with the non-catalytic nature of this transfer step.

Both transfer coefficients, $\rho k_R \eta$ and ρk_D, increase slightly with liquid flow (Fig. 3) in columns composed of either 100% catalyst or mixed beds. This can be attributed to an increase in the water holdup at the higher liquid flows giving rise to larger interfacial areas for vapour–liquid transfer. Similar behaviour is observed for the rates of vaporization of liquids into a carrier gas in packed columns.[14]

The data in Fig. 2 show that both separated transfer rates, $\rho k_R \eta$ and ρk_D, are about a factor of two larger than values obtained by direct measurements. Further, ρk_D is markedly affected by the composition of the catalyst bed. For beds containing 100% catalyst, this transfer rate increases with gas flow to the 0·27 power; while rates for mixed beds are higher and have a 0·48 power dependence on the gas flow. Direct measurements of ρk_D over alumina spheres in the absence of catalyst are 20 to 60% lower and have a much steeper slope, 0·64, than the separated rates from the trickle bed reactor. The higher power dependence obtained in the direct measurements of ρk_D is identical to that normally found for gas phase mass transfer rates in packed columns[15] and for rates of vaporization of a liquid into a carrier gas.[14] These discrepancies suggest some deficiencies in the application of the two parameter model of Palibroda.

The kinetic model assumes that there is a uniform radial distribution of deuterium in the three process streams and that back-mixing of the gas phase can be neglected. These assumptions seem reasonable in that $\rho k_R \eta$ and ρk_D are independent of column height (Table 1). Since the slope and magnitude of ρk_D measured in the trickle bed are strongly dependent upon the dilution of the catalyst, it is tempting to relate the above discrepancies to a non-uniform liquid distribution. However, direct measurements of ρk_D in beds of either hydrophobic spheres or an equal mixture of hydrophobic and hydrophilic spheres, have the same power dependence in spite of radically different liquid flow patterns. It therefore seems unlikely that the lower slopes of ρk_D separated from the trickle column are the result of poor liquid distribution. A more plausible explanation for these discrepancies is to assume that a third process, independent of the water vapour, is involved in the transfer of deuterium from the hydrogen to the liquid water. Neglect of such a process could account for both the lower slope of ρk_D and the higher magnitude of both separated rates.

ACKNOWLEDGEMENTS. We wish to thank J. F. Atherly for assistance in writing the computer program and R. W. Jones and M. W. D. James for maintaining and calibrating the spectrometer.

REFERENCES

[1] M. Benedict, *Progress in Nuclear Energy Series IV, Technology and Engineering*, eds. R. Hurst and S. McLain, (McGraw-Hill, New York, 1956), Ch. 1, p. 52.

[2] *Production of Heavy Water, National Nuclear Energy Series*, (McGraw-Hill, New York, 1955) eds., G. M. Murphy, H. C. Urey, and I. Kirshenbaum, Ch. 2, p. 16.

[3] W. H. Stevens, *Canadian Patent No.* 907,292, (1972).

[4] J. H. Rolston, W. H. Stevens, J. P. Butler, and J. den Hartog, *Canadian Patent No.* 941,134, (1974).

[5] R. G. Halemen, W. P. Colman, S. H. Langer, and W. A. Barber, *Fuel Cell Systems, Advances in Chemistry Series No.* 47, (Amer. Chem. Soc., Washington, 1965), p. 106.

[6] W. M. Thurston, *Rev. Sci. Instr.*, 1970, **41**, 963; 1971, **42**, 700.

[7] L. Melander, *Isotope Effects on Reaction Rates* (Ronald Press, New York, 1960), Ch. 3, p. 60.

[8] K. Cohen, *Theory of Isotopic Separation as Applied to the Large, Scale Production of* U^{235}, ed. G. M. Murphy, (McGraw-Hill, New York, 1951), Ch. 7, p. 128.

[9] E. Cerrai, C. Marchetti, R. Renzoni, L. Roseo, M. Silvestri, and S. Villani, *Chem. Eng. Progress Series No.* 50, 1954, 271.

[10] M. Majoube, *J. Chim. phys.*, 1971, **68**, 1423.

[11] N. Palibroda, *Z. Naturforsch.*, 1966, **21a**, 715.

[12] J. H. Rolston, G. A. Smith, and D. E. Clegg, to be published.

[13] J. P. Butler and J. den Hartog, to be published.

[14] A. E. Surosky and B. F. Dodge, *Ind. Eng. Chem.*, 1960, **42**, 1112.

[15] H. L. Shulman, C. F. Ullrich, A. Z. Proulx, and J. O. Zimmerman, *A.I.Ch.E. Journal*, 1955, **1**, 253.

DISCUSSION

A. Farkas (*Media, Pa.*) asked: Is the deuterium exchange rate sufficiently high for the utilization of your system in the production of heavy water according to the dual-temperature exchange process? What type of additional reaction or process do you think is needed to explain the discrepancies mentioned? Would it be a process involving mass transfer or diffusion limitation?

J. H. Rolston (*AEC, Chalk River*) replied: The overall rate of deuterium transfer over the catalysts described would be too low by a factor of between 3 and 5 for use in a dual-temperature exchange process. In addition, catalyst stability under hot tower operating conditions is uncertain. However, these catalysts are sufficiently active for use in a monothermal process based on the hydrogen liquid–water isotopic exchange reaction. Since our use of a two-parameter model gives transfer rates which are higher than direct measurements of these rates, the additional process transferring deuterium must be one that does not involve the intermediary water vapour. The independent measurements of the catalytic rate $\rho k_R \eta$ were obtained in the absence of a liquid flow. Hence, we expect this estimate of the transfer rate to be larger than the rate extracted from the trickle bed reactor where there are obviously additional resistances. An interesting possibility for the additional transfer process involves the transfer of deuterium between dissolved hydrogen and liquid water.

C. McGreavy (*Leeds Univ.*) said: It has been suggested that an explanation of the enhancement of the experimentally observed rates over the separated rates might be the result of transfer of the hydrogen isotope in the liquid. Since this is an additional resistance, it might be expected that the disagreement would be even greater. Perhaps another physical effect is more probable, *e.g.* the dependence of the rate on the interfacial areas. Is it possible the models have not made appropriate allowance for the change in surface areas arising from varying coverage of the catalyst or liquid distribution in the column?

J. H. Rolston (*AEC, Chalk River*) replied: We are suggesting that any transfer which occurs by a process which does not involve transfer through the vapour phase has been neglected in the two parameter model. Thus, if some transfer occurs by a third process operating in the trickle bed, it would decrease the magnitude of the two transfer rates $\rho k_R \eta$ and ρk_D separated from the column data. Such a step represents an alternative path for transfer rather than an additional resistance.

Your point concerning the importance of interfacial areas appropriate to the respective transfer rates expressed per unit volume of catalyst bed is well taken. Our estimate of the catalytic rate $\rho k_R \eta$ was made under conditions where no liquid was present and we have not attempted to include any effect arising from coverage of platinum catalyst by multilayer films of liquid water.

R. W. Coughlin (*Lehigh Univ., Bethlehem, Pa.*) asked: Have you measured the contact angle of water on the wet-proofed catalyst? Related phenomena are of importance for fuel cell electrodes where the proper hydrophilic–hydrophobic balance is necessary to assure the juxtaposition of gas, liquid electrolyte, and solid electrode. Related wet-proofed catalysts have been described in recent years in the patent literature for liquid-phase oxidation using air and solid catalysts.

J. H. Rolston (*AEC, Chalk River*) replied: We have measured contact angles for platinized carbon–Teflon layers deposited on glass slides. These are well above 100 degrees at 298 K, indicating that no wetting by bulk liquid water occurs.

There is indeed a considerable similarity between the formulation of our platinized carbon–Teflon catalysts and those of fuel cell electrodes. The use of such wet-proofed catalysts for catalytic oxidation of aqueous liquid streams has not been attempted by our group, but it is an interesting application and may prove of general use in the clean-up of liquid streams providing catalyst poisoning is not too severe.

S. P. S. Andrew (*ICI, Billingham*) asked: Is there any evidence for capillary water moving from the inside of the catalyst to the main water stream? For instance, is much water found inside the catalyst pellets after use?

J. H. Rolston (*AEC, Chalk River*) replied: Movement of water from the inside of porous platinized alumina pellets has been observed, but this is much slower than the transfer of deuterium between hydrogen and liquid water. Catalysts described in this paper contain essentially no entrapped water since they have been prepared by depositing the catalytically active platinum–carbon–Teflon layer on non-porous ceramic spheres.

Self-oscillations in the Catalytic Rate of Heterogeneous Hydrogen Oxidation on Nickel and Platinum

Vladimir D. Belyaev[†], Marina M. Slin'ko*, and Mikhail G. Slin'ko[†]

A. V. Topchiev Institute of Petrochemical Synthesis, Moscow, U.S.S.R.
[†]Institute of Catalysis, Novosibirsk, 630090, U.S.S.R.

ABSTRACT A study of self-oscillation in the heterogeneous catalytic reaction rate due to hydrogen oxidation on nickel and platinum has been performed. Investigations were carried out in a flow reactor under isothermal conditions with metallic wires or foils as catalysts. During the reaction process on nickel and platinum foils, the contact potential difference was measured using the vibrating condenser method. The waveform and frequency of oscillations in the reaction rate were similar to oscillations in the contact potential difference. Self-oscillations were caused by the catalytic mechanism of interaction between hydrogen and oxygen and not by thermokinetic or hydrodynamic effects. A mathematical model for self-oscillations is suggested, and model parameters have been determined on the basis of experimental data.

INTRODUCTION

The periodic variations in the catalytic reaction rate in homogeneous and enzymic systems have long been known. In a series of studies[1-3] the results of these investigations have been summarized and some possible models have been considered. Recently self-oscillations in heterogeneous catalytic reaction rates have also been discovered. Wicke and co-workers[4-6] investigated a self-oscillating regime for oxidation of carbon monoxide on platinum, Dauchot[7] also studied self-oscillations in the oxidation rate of carbon monoxide on platinum. The authors had various ideas on the occurrence of self-oscillations; however, no mechanism was given, nor any kinetic model for heterogeneous reactions. In refs. 8—10 oscillations in the heterogeneous catalytic rate of hydrogen oxidation on nickel were discussed. It was suggested that the occurrence of reaction rate self-oscillations was due to the change in catalytic properties of the nickel surface caused by adsorbed intermediates. In the present paper the data on self-oscillations in heterogeneous catalytic interaction between hydrogen and oxygen on nickel and platinum are presented and some possible mechanisms are discussed.

EXPERIMENTAL

In the present work an electrothermographic method was used to determine the reaction rates. The method for continuous measurements of variations in the work function during the reaction was used to monitor the state of the catalyst surface. These methods require application of catalysts in the form of fine wires, foils, or plates. The following catalysts were used:

(i) Nickel catalyst:
  foil of dimensions $600 \times 1 \times 0.0025$ mm with a geometric surface area of 12 cm²; plate of dimensions $18 \times 10 \times 0.1$ mm with a geometric surface area of 3·6 cm².

(ii) Platinum catalyst:
  wire with 0·05 mm diameter, 1460 mm in length, a geometric surface area 2·29 cm² plate of dimensions $11 \times 40 \times 0.05$ mm with a geometric surface area of 8·8 cm².

Experiments were carried out in a flow reactor. For purification, hydrogen and nitrogen were passed through columns containing a nickel–chromium catalyst, silica, and molecular sieves. The dosage of small amounts of a stoicheiometric hydrogen–oxygen mixture was carried out using an electrolyser. After mixing, the initial mixture was passed into the reference cell of a thermal conductivity detector, then to the reactor, and finally to the measuring cell of the detector. Continuous analysis of the gas-phase composition was performed. The hydrogen content was determined through the change in the thermal conductivity of the mixture. The oxygen concentration was measured using an MX-6202 mass spectrometer. The reaction rate was determined by measuring the difference between hydrogen or oxygen concentrations at the inlet and outlet of the reactor. The reaction rate on nickel foil was also measured using the electrothermographic method. The method consists of measuring the rate of heat evolution, which is proportional to the reaction rate. Quantitative measurements were carried out using a thermoregulator to maintain constant temperature of the foil (with accuracy of $\pm 0.02°C$) by varying the electrical power supplied. The rate was determined through the difference between the electrical power supplied before and during the reaction. This method was also used for a platinum wire to provide qualitative observations of the reaction rate.

To monitor the state of the catalyst surface during the reaction, the contact potential difference (CPD) was measured using the vibrating condenser method.[11] A molybdenum plate sealed into glass, with a surface area of 0.5 cm², was used as a reference electrode. Variations in CPD were recorded by a self-recording potentiometer. Simultaneously with CPD measurements, the gas-phase composition was also determined by the above-mentioned methods.

RESULTS

Nickel foil. Experimental conditions: the temperature was varied from 150 to 300°C, hydrogen and oxygen pressures varied from 20 to 760 torr and from 0 to 10 torr, respectively. The ratio between the reactor volume and the catalyst surface area was 0.417 cm. The rate of the gas flow was varied from 2 to 25 cm³/s at room temperature. Nitrogen, argon, and helium were used as inert gases.

Self-oscillations in the hydrogen oxidation rate on nickel foil were detected in the presence of an excess of hydrogen at oxygen concentration of about 0.1%. The amplitude of oscillations increases with oxygen concentration in the reactor, then passes through a maximum and falls to zero. The amplitude of oscillations increases with temperature. The maximum recorded amplitude attains 40% of the average rate for a period. The frequency of oscillations varies from 0.5 to 10 min⁻¹ with increase in temperature at hydrogen pressure equal to 1 atm. The plot of the period of oscillations against temperature is given in Fig. 1, curve (A). The region of oxygen concentrations where oscillations exist becomes narrower and a shift toward lower concentrations is observed as hydrogen pressure is decreased. At hydrogen pressures from 20 to 150 torr the frequency of oscillations increases with hydrogen pressure. When hydrogen pressure is less than 20 torr oscillations are not observed.

Platinum wire. Experimental conditions: temperature range 135—210°C; initial hydrogen concentrations 0.210—3.48 vol %; initial oxygen concentrations 0.105—68 vol %. Reactor volume/catalyst surface area ratio was 5.33 cm. The rate of gas flow was 2.86 cm³/s. The catalyst was first of all heated in a stream of oxygen to 450°C and then subjected to alternating oxidation–reduction. Before each experiment the catalyst was heated to the required temperature in a stream of nitrogen containing 21.5% oxygen.

Self-oscillations in the reaction rate were observed at a wire temperature higher than 145°C and initial oxygen concentration equal to 21.5 vol % and hydrogen concentration in the reactor higher than 0.5 vol %. A decrease in the amplitude and in the period of oscillations was observed as the temperature was increased (Fig. 1, curve B). When the catalyst temperature was 167 or 178°C and the initial hydrogen concentration was

2·16 vol % the maximum amplitudes were reached at oxygen concentration equal to 21·5 vol % and attained 18% of the average overall reaction rate. A decrease of less than 10% and an increase of more than 45% in oxgyen concentration results in the complete disappearance of self-oscillations.

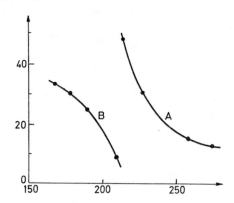

FIGURE 1 Plot of the period of reaction rate oscillations against temperature for nickel (A) and platinum (B). (Abscissa: temperature; °C; ordinate; period, s).

Platinum plate. Experimental conditions: temperature range 75—150°C; initial hydrogen concentration in the mixture was 0·210–3 vol %; initial oxygen concentration was 0·105—40 vol %. The ratio between the reactor volume and the catalyst surface area was 1·14 cm. The rate of gas flow passed through the reactor was 2·86 cm³/s. The platinum foil was purified by heating to 600–800°C in oxygen. Before each experiment the catalyst was heated to the required temperature in a mixture of nitrogen and 21·5% oxygen.

Self-oscillations in the reaction rate were observed within the temperature range 85—130°C.

There are upper and lower boundaries for the hydrogen concentration regions, dependent on temperature, where reaction rate self-oscillations appear. Initially the range of hydrogen concentration increases with temperature until a definite maximum is reached, and then starts to decrease. The amplitude of oscillations is characterized by similar behaviour. The period of self-oscillations decreases as the temperature is increased. When the temperature is constant, reaction rate self-oscillations appear at a certain hydrogen to oxygen ratio in the reactor. The amplitude and the period of oscillations first increase with hydrogen concentration in the reactor, pass through a maximum, and then decrease. As oxygen concentration is increased at 85°C, the amplitude and the frequency of oscillations increase until their complete disappearance. A shift of the range of self-oscillations toward high hydrogen concentrations is observed for higher oxygen concentrations.

Contact potential difference. Simultaneously with the analyses of oxygen and hydrogen content in the reactor during hydrogen oxidation on nickel and platinum, respectively, we measured the change in contact potential difference (CPD). Fig. 2 shows oscillations of the contact potential difference observed on a nickel plate at temperatures above 200°C simultaneously with reaction rate self-oscillations. The amplitude of CPD oscillations reached 0·25 v. Figure 3 shows the self-oscillations of the contact potential difference on a platinum plate. They were observed simultaneously with reaction rate self-oscillations at 110°C. It can be seen from Figs. 2 and 3 that CPD and concentration oscillations have the same waveform and frequency, and have no phase shift. The amplitude and frequency of CPD oscillations on a nickel plate increase with temperature and depend only slightly on the rate of gas flow passed through the reactor. The behaviour of CPD oscillations observed for a platinum plate is similar to that of concentration oscillations described in the earlier section

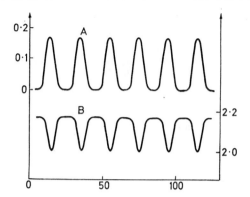

FIGURE 2   Oscillations of contact potential difference (A) and reaction rate (B) on nickel catalyst at 240°C. (Abscissa: time, s; left ordinate: change in contact potential difference, v; right ordinate: change in the reaction rate $\times$ 10^7, mole/s).

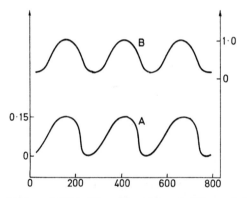

FIGURE 3   Oscillation of CPD (A) and reaction rate (B) on platinum at 110°C. (Abscissa: time, s; left and right ordinates are variations in CPD, v and reaction rate $\times$ 10^7, mole/s, respectively).

DISCUSSION

Experimental work in the field of self-oscillations is hindered to a great extent by a high parametric sensitivity of the dynamic properties of the system to the experimental conditions. In the present study it was not possible to observe continuously the composition and concentration of surface intermediates. Therefore, it was difficult to construct any detailed mathematical model. However, some principal questions can be solved.

Self-oscillating systems are widely distributed in physics, engineering, and biology. They are widely applied in radio engineering, mechanics, and in the control of technical devices. The theory for self-oscillations is highly developed. Any self-oscillating system may be considered to consist of four elements: a constant (non-oscillating) source of energy or matter; an oscillating system; a mechanism regulating the feed from the source of energy or matter; and the feedback between the oscillating system and a regulator to control the dosage of matter and energy.

If we consider a catalyst and a reaction medium as a single heterogeneous catalytic system then under certain conditions it may contain all four of the elements typical of a self-oscillating system. A non-oscillating source of matter is the constant flow of reactants entering the reactor. The oscillating system is represented by a complex of adsorbed species whose concentrations periodically vary with time (the catalyst together with adsorbed substances forms a single system). The catalyst controls the accumulation of adsorbed substances produced through adsorption and reaction, and thus plays the role of the regulator. The influence of reactants on the catalyst's properties can be regarded as a feedback in a self-oscillating heterogeneous system.

Self-oscillations in the reaction rate of hydrogen oxidation on nickel and platinum have some common characteristics. In both cases the range of occurrence of self-oscillations depends on the catalyst temperature and on the composition of the gaseous mixture. When the concentration of an excessive component of the gaseous mixture is increased, upper and lower boundaries for the range of occurrence of oscillations are shifted toward higher concentrations of another component. For nickel and platinum, similar dependences of the amplitude of oscillations on the concentration of a deficient reactant were observed. These dependences are represented by curves with the maximum obtained at a certain concentration of the deficient reactant. The frequency of self-oscillations increases with temperature or with pressure of an excessive component (hydrogen for nickel catalyst, oxygen for platinum catalyst).

The difference observed in the temperature range where self-oscillations appear on these catalysts is due to the higher activity of platinum toward hydrogen oxidation. This difference is observed not only for various metals but also for various forms of a metal. Moreover, during prolonged work with the same catalyst a shift in the temperature range may also be observed. This may be caused by variations in catalytic activity when passing from one sample to another or by changes in a particular sample resulting from prolonged use. A change in the concentration range where self-oscillations occur may be due to the influence of reactants on the properties of the nickel and platinum. Self-oscillations in the reaction rate on nickel foil were found in the presence of an excess of hydrogen at 180°C in the region where a first-order reaction changed into a zero-order one. On a platinum plate, a region of nonlinear kinetic behaviour was found in the presence of an excess of oxygen, in which the reaction rate was inhibited by hydrogen. In this region self-oscillations in the reaction rate were observed on platinum catalysts. Under similar conditions we tested palladium wire with diameter 0·1 mm. Within the ranges of temperature and hydrogen and oxygen pressures studied, only linear kinetic dependences were observed, no oscillations in the reaction rate being detected. This implies that the presence of a strong nonlinear change in the properties of the metal, caused by variations in the gas-phase composition and in the amount of adsorbed and dissolved reactants in the catalyst, is needed for the occurrence of reaction rate oscillations.

The above-mentioned data on the existence of self-oscillations at a constant catalyst temperature, on critical conditions for the appearance and disappearance of oscillations with variations in temperature and reaction mixture composition, as well as the dependence of the characteristics on these parameters, indicate a chemical nature for the phenomenon observed. The high sensitivity of the characteristics of the oscillations to variations in catalytic activities, and dependences on time of catalyst pretreatment, also constitute evidence in favour of the chemical nature of the oscillations. The results of specially designed experiments show that self-oscillations are caused neither by hydrodynamic effects not by any peculiarities of the apparatus. The change in heat conductivity or heat capacity of the gaseous mixture by substitution of one inert gas for another did not lead to any change in the amplitude, frequency, or waveform of the oscillations. These data, together with observations of oscillations at a constant temperature, indicate that self-oscillations in the reaction rate are not thermokinetic. One of the main pieces of evidence for the chemical nature of the oscillations is the self-oscillations of CPD. They have a waveform, frequency, and phase which are also characteristic of self-oscillations in the reaction rate, and they reflect periodic changes in the surface composition. Unfortunately, a quantitative interpretation of the CPD oscillations observed presents difficulties because of the complicated dependence of the electron work function on the amount of oxygen and hydrogen adsorbed on nickel and platinum. However, the absence of the dependence of CPD oscillations on the flow rate proves the idea that oscillations of concentrations in the gaseous phase are due to the oscillations of surface compounds. Thus, for the mathematical description of the process we will consider only variations in the degrees of metal surface coverage, and as a first approximation assume that the concentration of substances in the gaseous phase is constant.

THE MATHEMATICAL MODEL

The mathematical model in its simplest form must depict the main properties of the process: adsorption of hydrogen and oxygen, and interaction of adsorbed substances without giving all the details of the mechanism.

$$H_2 + 2[M] \underset{k_{-1}}{\overset{k_1}{\rightleftharpoons}} 2[M-H]$$

$$O_2 + 2[M] \underset{k_{-2}}{\overset{k_2}{\rightleftharpoons}} 2[M-O] \tag{1}$$

$$[M-O] + 2[M-H] \overset{k_3}{\longrightarrow} H_2O + 3[M]$$

If the rate constants of the several stages do not depend on the surface coverage of the catalyst with hydrogen and oxygen, then the system is a non-oscillating one since it lacks an essential element, *i.e.* a feedback. The concept of "feedback" may be described physically as an effect of an oscillating system on the regulator. Therefore, we expect that within the concentration range observed the catalyst's properties vary by virtue of being affected by adsorbed substances.[12] The rate constants of the given stages depend on the surface concentrations of oxygen and hydrogen. Let us consider the simplest case when the third-stage constant varies with coverage. We assume that the activation energy of the reaction increases linearly with the degree of surface coverage with oxygen or hydrogen for nickel and platinum catalysts, respectively. For nickel catalysts

$$E_3 = E_{30} + \alpha\theta_0 \tag{2}$$

where E_3 is the third stage activation energy, θ_0 is the degree of surface coverage with oxygen, and α is a coefficient. Then $k_3 = k_{30} \exp(-\mu\theta_0)$, where $\mu = \alpha/RT$.

763

Proceeding from these assumptions the following mathematical model may be obtained:

$$\frac{d\theta_H}{dt} = k_1 P_{H_2}(1 - \theta_0 - \theta_H)^2 - k_{-1}\theta_H^2 - 2k_{30}\theta_0\theta_H^2 \exp(-\mu\theta_0) = F(\theta_0, \theta_H)$$

$$\frac{d\theta_0}{dt} = k_2 P_{0_2}(1 - \theta_0 - \theta_H)^2 - k_{-2}\theta_0^2 - k_{30}\theta_0\theta_H^2 \exp(-\mu\theta_0) = Q(\theta_0, \theta_H) \qquad (4)$$

where θ_H is the degree of surface coverage with hydrogen; P_{H2} and P_{02} are hydrogen and oxygen pressures.

The necessary conditions for the occurrence of oscillations in the system of two non-linear equations are:

Static condition:

$$\Delta = \frac{\partial F(\theta_0, \theta_H)}{\partial \theta_H} \times \frac{\partial Q(\theta_0, \theta_H)}{\partial \theta_0} - \frac{\partial F(\theta_0, \theta_H)}{\partial \theta_0} \times \frac{\partial Q(\theta_0, \theta_H)}{\partial \theta_H} > 0 \qquad (5)$$

Dynamic condition:

$$\sigma = \frac{\partial F(\theta_0, \theta_H)}{\partial \theta_H} + \frac{\partial Q(\theta_0, \theta_H)}{\partial \theta_0} > 0 \qquad (6)$$

The static condition can be explained physically, since when the concentration of surface oxygen is varied, the change in the rate of O_2 removal due to the reaction must be greater than the rate of O_2 accumulation on the surface due to adsorption.

If
$$k_{-1}(\mu - 1)^2 < 2k_{-2} \qquad (7)$$

then condition (5) is fulfilled and the only steady state exists within the whole range of variations of hydrogen and oxygen surface concentrations. If conditions (5) and (6) are fulfilled simultaneously (when $\mu > 30$) then the only steady state becomes an unstable steady state around which a steady limit cycle appears. In this case the value of the stationary oxygen coverage is $\theta_{os} > 1/\mu$. The model describes correctly the type of variations in amplitudes and critical conditions with variations in the composition of the reaction mixture. The mechanism for the occurrence of self-oscillations in the system may be considered as the following. When the amount of adsorbed oxygen on the surface of a nickel catalyst is increased, the reaction rate decreases and the surface becomes covered with oxygen and hydrogen. According to condition (7) a quicker relative decrease in oxygen concentration on the catalyst surface takes place. The value of $\exp(-\mu\theta_0)$ reaches a critical value with a certain value of θ_0; a rapid reaction between oxygen and hydrogen takes place, and the surface becomes almost completely free of oxygen. Then there follows the range where adsorption of oxygen, which hinders the reaction, takes place. This is the last process of the cycle observed. A limitation of this model is that it does not describe slow oscillations with a frequency of less than 1 s^{-1}, and is very sensitive to variations in parameters.

To describe "slow" oscillations with periods of more than 10 s it is necessary to take into account processes which are slower than the reaction rate. These are processes such as solution in a metal of hydrogen and oxygen, which affect the metal's catalytic properties by decreasing or increasing the rate constants of individual stages. Thus, the presence of low-frequency oscillations indicates that deeper catalyst layers participate in the catalytic reaction.

REFERENCES

1   I. Higgins, *Ind. and Eng. Chem.*, 1967, **59**, 19.
2   H. J. Degn, *Chem. Educ.*, 1972, **49**, 302.
3   A. M. Zhabotinskii, *Concentration oscillations*, [In Russian], (Nauka, Moscow, 1974).
4   H. Bensch, P. Fieguth, and E. Wicke, *1st Internat. Symp. on Chem. Reaction Eng.* (Washington, 1970), p. 615.
5   H. Bensch, P. Fieguth, and E. Wicke, *Chem. Eng. Tech.*, 1972, **44**, 445.
6   E. Wicke, *Chem. Eng. Tech.*, 1974, **46**, 365.
7   J. Dauchot and J. Van Cakenberghe, *Nature Phys. Science*, 1973, **246**, 61.
8   V. D. Belyaev, M. M. Slin'ko, V. I. Timoshenko, and M. G. Slin'ko, *Kinetika i Kataliz*, 1973, **14**, 810.
9   V. D. Belyaev, M. G. Slin'ko, and V. I. Timoshenko, *Kinetika i Kataliz*, 1975, **16**, 555.
10 V. D. Belyaev, M. M. Slin'ko, M. G. Slin'ko, and V. I. Timoshenko, *Doklady Akad. Nauk S.S.S.R.* 1974, **214**, 1098.
11 V. A. Lyando, Yu. A. Alabuzhev, L. A. Sazonov, and I. S. Sazonova, *Kinetika i Kataliz*, 1962, **3**, 794.
12 G. K. Boreskov, *Zhur. fiz. Khim.*, 1958, **32**, 2739; 1959, **33**, 1969.

DISCUSSION

A. Jones (*Health and Safety Exec., Sheffield*) said: We have observed oscillations in a number of simple gas-phase oxidation reactions over platinum and have studied recently those occurring during the oxidation of carbon monoxide. These oscillations we have established as isothermal. In general, we have found the phenomena difficult to reproduce. An exception, however, is the reproducible oscillatory behaviour that

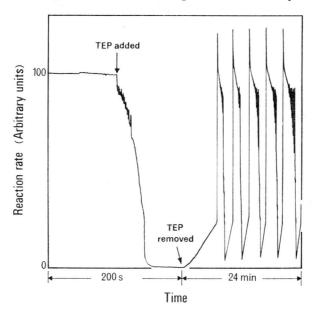

FIGURE   Influence of triethylphosphate (TEP) on the rate of oxidation of carbon monoxide over a platinum wire catalyst at 673 K in a flow reactor. Reactant mixture: 2·5 kN m^{-2} of carbon monoxide, 20·0 kN m^{-2} of oxygen, 78·8 kN m^{-2} of nitrogen. Added TEP pressure, 2×10^{-3} kN m^{-2}. Flow rate 200 ml min^{-2} through a 0·5 cm bore reactor.

can be *induced* by the deliberate poisoning of the catalyst. The Figure shows typical oscillations induced by the temporary addition to the reaction stream of triethylphosphate. The oscillations can occur over the whole range of reaction rates and persist for a great many oscillations. The periods of the oscillations can vary, depending on conditions, from a few seconds to several minutes. Our results thus suggest that at least some oscillatory behaviour is connected with contamination of the catalyst surface.

O. V. Krylov (*Inst. Chem. Phys., Moscow*) said: We have studied several catalytic oxidations in which we have observed auto-oscillation phenomena. Examples include cyclohexane oxidation on NaX zeolite, propylene oxidation on solid solutions CoO–MgO, and various hydrocarbon oxidations on cobalt-containing spinels. In all cases auto-oscillation had no thermokinetic basis. In some, we observed such auto-oscillations even when a pulse method was used. Thus each period of oscillation can be divided into several pulses. One explanation of these phenomena is the slow generation of some structure on the surface, which then undergoes rapid transformation into an active structure which allows rapid catalytic reaction and a new period of transformation.

M. Slin'ko (*Karpov Inst. of Phys. Chem., Moscow*) in reply to the two previous speakers, said: A knowledge of the kinetic laws of heterogeneous catalytic reactions under non-stationary conditions (*i.e.* involving non-stationary concentrations of intermediates on the catalyst), have become sufficient for solving the important practical problems which have arisen during recent years. The problems include processes in fluidized bed reactors, cyclic processes, and the study of catalysts for emission control. Self-oscillating regimes give an opportunity for investigations of rate changes in the non-stationary state of the catalyst during any desirable period of time. These self-oscillating regimes not only elucidate the state of the system at a particular moment, but also the evolution of the catalytic system as a whole. The amplitude, the frequency, the phase shift, and critical conditions for the appearance and disappearance of the self-oscillations are the general characteristics of the system, which are functions of the separate rates for the various steps of the complicated mechanism. That is why self-oscillating regimes are so interesting for the theory of catalysis. Experimental work in the field of self-oscillations is hindered by a high parametric sensitivity of the dynamic properties of the system to the state of the catalyst surface. In this work it was impossible to measure continuously the concentration of surface intermediates. Therefore, it was difficult to construct any detailed mathematical model. However, some important questions can in principle be resolved.

1. The behaviour of catalytic reactions far from equilibrium can be examined; this differs significantly from the behaviour of the system near to equilibrium, where self-oscillations are impossible.

2. The non-linear effects of the concentration of adsorbed species on catalyst properties during reaction can be investigated.

3. The presence of synchronous self-oscillations at different catalyst surface sites indicates that the model for an active metal involving non-local electronic properties is the most realistic.

4. The observation of oscillations of low frequency suggests that some sub-surface layers of metal take part in reaction.

P. A. Sermon (*Brunel Univ., Uxbridge*) said: In Figure 2 the oscillations in contact potential difference and in reaction rates for nickel at 240 °C are out of phase, whereas those for platinum at 110 °C (Figure 3) are in phase. Does this reflect differences in the catalytic properties of these metals rather than the difference in experimental temperature?

M. Slin'ko (*Karpov Inst. of Phys. Chem., Moscow*) replied: The difference in the phase shift for nickel and platinum catalysts is determined by different dependence of the work function as a result of the variations of the catalyst surface layer content. This difference depends primarily on the chemical properties of the metal.

S. P. S. Andrew (*ICI, Billingham*) said: The explanation of the oscillations given by the authors is the periodic filling up and then the emptying of a capacity, namely the quantity of adsorbed molecules. Does the period of the oscillation agree with the figure obtained by dividing this capacity by the average specific reaction rate?

M. Slin'ko (*Karpov Inst. of Phys. Chem., Moscow*) replied: The period and the shape of self-oscillations, which may change greatly, are functions of the feedback, *i.e.* the influence of the adsorbed or solved substances on the catalytic properties of metal. The period of self-oscillation is not equal to the ratio of the quantity of adsorbed molecules to the average specific reaction rate, and depends on the rate constants of the elementary stages.

Analysis of the Stability of Bifunctional Catalysts

R. A. AL-SAMADI AND W. J. THOMAS*

School of Chemical Engineering, University of Bath, Bath, Avon

ABSTRACT Because of the overall endothermicity of petroleum reforming reactions in which aromatics are produced, no attention has been given to catalyst instability resulting from temperature runaway. However, there remains the question of whether local instabilities can be generated by the exothermic hydrogenation reactions promoted *in situ*.

A steady state analysis of the differential mass and energy balance equations, representing the behaviour of systems based on both experimental and hypothetical kinetic models, suggests that the phenomenon of steady state multiplicity is only displayed for extreme conditions and when the catalyst is composed of a discrete mixture of a hydrogenation catalyst and an isomerisation catalyst. When the two catalyst components are compounded in a single pellet, the resulting catalyst particle is stabilised for a wide range of operating conditions. A dynamic analysis of the system of equations reveals, however, that the time required for a compounded catalyst to recover from a perturbation is relatively large.

INTRODUCTION

The use of bifunctional catalysts for promoting petroleum reforming reactions, in addition to providing the usual range of problems associated with reaction mechanisms[1] and the formulation of mathematical models representing catalyst pellet behaviour[2-4] and reactor performance,[5,6] raises the question of whether local instabilities in catalyst pellets can be generated by the exothermic hydrogenation reactions generally occurring as a constituent part of an overall endothermic process. In this paper we consider, specifically, the aromatisation of methylcyclopentane in the presence of a bifunctional catalyst utilised in two distinct forms, and for both of which experimental data have been published.[5,6] One preparation consisted of a discrete physical admixture of pellets of platinum dispersed and supported on inert silica–alumina together with pellets of fluorinated silica–alumina, while a second preparation was composed of particles compounded from platinum dispersed on a fluorinated silica–alumina support. In the former preparation the two active components were present as a discrete physical mixture of separate particles while the latter preparation existed in the form of composite particles containing both catalyst components.

The object of the present work is to utilise kinetic models already suggested,[5,6] incorporate them into material and heat balance equations for the pellet, and from iterative solutions of the heat balance equation examine whether or not multiple steady states exist for the pellet temperature. Such analyses should then reveal whether there are any conditions of catalyst composition which would lead to unstable operation. A brief excursion into the realms of dynamic analysis also provides some preliminary indication that the time required for composite pellets to recover from a temperature perturbation is a function of the catalyst composition.

Stability criteria

For heuristic purposes we initially consider a first order irreversible reaction which is taken to represent either an endothermic reaction in the presence of the supported platinum dehydrogenation catalyst or an exothermic reaction over the fluorinated silica–alumina catalyst. The usual steady state material balance equation for the reactant, representing chemical reaction and diffusion within a spherical pellet of the catalyst in question, is

$$D_e \frac{\partial}{\partial R} \left\{ R^2 \frac{\partial c}{\partial R} \right\} - kc = 0 \tag{1}$$

where k is the appropriate first order chemical rate constant. Eqn (1) acknowledges that diffusion, the mode of intraparticle mass transfer, is significant within the pellet. In order to include the possiblity of interparticle mass transfer resistance, Robin boundary conditions[7] are ascribed; thus

$$\frac{\partial c}{\partial R} = 0 \text{ at } R = 0 \tag{2}$$

and

$$D_e \frac{\partial c}{\partial R} = h_D (c_f - c) \text{ at } R = R_o \tag{3}$$

the driving force between fluid and pellet being the corresponding reactant concentration difference. In contrast to this situation the principal resistance to heat transfer will be within the relatively stagnant fluid boundary layer bathing the pellet.[8-10] Hence the pellet itself, having a fairly good thermal conductivity, will remain virtually isothermal and the appropriate steady state heat balance for the pellet will be

$$\frac{3h}{R_o}(T_f - T_p) = -\eta(-\Delta H) kc_0 = \frac{3h_D}{R_o}(c_0 - c_f)(-\Delta H) \tag{4}$$

where T_p is the uniform pellet temperature and η the pellet effectiveness factor with respect to the reactant. Eqn (1), coupled to boundary conditions (2) and (3), is sufficient to determine the reactant flux at the pellet surface and consequently the magnitude of the effectiveness factor. Furthermore the surface concentration will then be uniquely defined in terms of the prevailing fluid concentration and hence eqn (4) provides an estimate of the rise (or fall) in temperature of the fluid as a result of the heat released (or absorbed) by chemical reaction. Solution of eqn (1), satisfying boundary conditions (2) and (3), together with eqn (4) therefore leads to

$$(T_f - T_p) = -(-\Delta H) \frac{h_D}{h} \left\{ \frac{2(\lambda R_o \coth \lambda R_o - 1)}{Sh + 2(\lambda R_o \coth \lambda R_o - 1)} \right\} c_f \tag{5}$$

where

$$\lambda = -\left(\frac{A \exp[-E/R_g T_p]}{D_e} \right) \quad \text{and} \quad Sh = \frac{2R_o h_D}{D_e}. \tag{6}$$

For convenience we will refer to the uniform pellet temperature attained by pellets promoting an endothermic reaction as T_X and to that reached by pellets promoting an exothermic reaction as T_Y. In Fig. 1, the curve labelled 1 represents the form of eqn (5) when the reaction promoted is exothermic while curve 2 corresponds to the situation when an endothermic reaction is catalysed. The diagonal line merely describes a case when the reactant fluid concentration is insufficient to sustain any reaction. Multiple steady state solutions will exist between two bounds, such as T_f' and T_f'' where the gradient dT_f/dT_p vanishes. From eqn (5)

$$\frac{dT_f}{dT_p} = 1 - (-\Delta H)\alpha\beta \tag{7}$$

where

$$\alpha = \frac{h_D \, Sh \, E \, \lambda R_o}{R_g T_p^2 h \, \{2(\lambda R_o \coth \lambda R_o - 1) + Sh\}^2} \, c_f \tag{8}$$

and

$$\beta = \lambda R_o(1 - \coth^2 \lambda R_o) + \coth \lambda R_o. \tag{9}$$

It is a simple matter to demonstrate that both α and β are always positive for all values of T_p and so dT_f/dT_p can only vanish when the reaction prevailing is exothermic (*i.e.* when $(-\Delta H) > 0$).

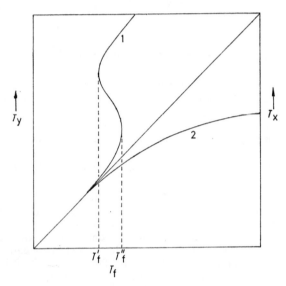

FIGURE 1   Sketch of possible solutions to eqn (5) depicting a heat balance for pellets.
Curve 1 represents exothermic conditions, curve 2 represents endothermic conditions.

In order to obtain a numerical basis for a comparison between the stability range of a pellet in which a simple first order irreversible reaction occurs and pellets in which more complex reactions proceed, eqn (5) is solved for the fluid temperature in terms of the uniform pellet temperature for a dual set of parameters (corresponding to exothermic and endothermic reaction). Thus eqn (5) is solved directly for T_f over a wide range of pellet temperatures T_x in the case of exothermic reaction. On the other hand, in the case of endothermic reaction, it is necessary to solve for T_x by an iterative procedure. Fig. 2 illustrates the results obtained for the typical sets of parameters listed in the figure caption. Inspection of this family of curves indicates that pellets promoting these hypothetical reactions would be stable for a wide range of fluid temperatures and reactant concentrations. There is, however, the possibility that conditions of relatively low temperature and high reactant concentration, which may occur during start up procedures, are likely to cause instability in particles catalysing exothermic reactions. Such a possibility is revealed by the curve labelled A in Fig. 2. The discontinuity in this curve simply implies that solution of eqn (5) for some pellet temperatures produces physically unattainable fluid temperatures. It is evident, however, that there may well exist some fluid condition which might produce instability, and further investigation is justified. It will also be clear from what follows that there is a strong resemblance between families of curves describing fluid and pellet temperature relations for simple first order kinetics and more complicated kinetic schemes, therefore offering the advantages of considerable model reduction when conditions for pellet stability are being sought.

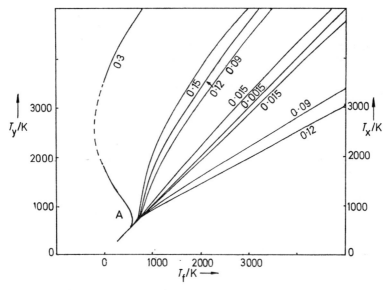

FIGURE 2 Numerical solutions to eqn (5) depicting the heat balance for pellets in which a simple reaction $A \rightarrow B$ occurs. Figures on curves denote fluid concentrations in kg mol m^{-3}. Values of parameters are: $A_1 = 1 \cdot 13 \times 10^3$ s^{-1}, $E_1 = 4 \cdot 5 \times 10^4$ kJ (kg mol)$^{-1}$, $\Delta H_X = 1 \cdot 058 \times 10^5$ kJ (kg mol)$^{-1}$, $\Delta H_Y = -6 \cdot 28 \times 10^4$ kJ (kg mol)$^{-1}$, $R_0 = 2 \times 10^{-3}$ m, $D_e = 4 \cdot 0 \times 10^{-7}$ m^2 s^{-1}, $h_D = 4 \cdot 0 \times 10^{-2}$ m s^{-1} and $h = 4 \cdot 19 \times 10^{-2}$ kJ m^{-2} s^{-1} K^{-1}.

Catalyst pellet models

Prior to an examination of catalyst pellet stability it is necessary to outline briefly the nature of the pellet model and the kinetic schemes adopted for incorporation within the material and energy balance equations. All catalyst pellets referred to are supposed to have spherical geometry and, to avoid excessive complications, the diffusivity of each of the reacting components and products is characterised by a single effective diffusivity D_e. This is justified on the grounds that, with the exception of hydrogen which was present in excess during the kinetic experiments,[5] the molecular weights of all the reacting components lies between 78 and 84.

The kinetic model (Scheme I) suggested by Jenkins and Thomas[5] for the aromatisation of methylcyclopentane over a mixture of separate discrete pellets of a hydrogenation –dehydrogenation catalyst (catalyst component X) and an isomerisation catalyst (component Y), the details of which are summarised in Table 1, is adopted for writing the material and energy balance equations describing chemical reaction, diffusion, and heat transfer for each type of pellet. Details of these equations have already been given in a separate publication[6] and the kinetic scheme (I) has been discussed elsewhere[5]. The experimentally determined kinetic parameters are summarised in Table 1. Expressing the concentration of reacting species in terms of the dimensionless dependent variable $\gamma_j = c_j/c_{jo}$ and the radius as $\rho = R/R_o$, the differential equations describing chemical reaction and diffusion within the pellet X or Y become

$$\frac{\mathrm{d}^2\gamma_j}{\mathrm{d}\rho^2} + \frac{2}{\rho}\frac{\mathrm{d}\gamma_j}{\mathrm{d}\rho} + \frac{R_o{}^2}{D_e}\sum_i\left\{A_i\exp\left(-\frac{E_i}{E_1\tau}\right)\right\}\gamma_i = 0 \tag{10}$$

the summation sign implying the appropriate algebraic sum of dimensionless kinetic terms signifying the rate of consumption or production of reacting component i within the pellet. The equations describing the heat balance for the pellets may be reduced to algebraic equations, however, since computations utilising a fully distributed parameter model of the pellet[6] indicate that the pellet behaves as an isothermal entity, the temperature rise (or fall) occurring in the stagnant gas film surrounding the pellet. Thus equations such as

$$(\tau_f - \tau_p) = -\left\{\frac{h_D c_{jo}(-\Delta H_1)\,R_g}{hE_1}\right\}\left(\frac{\Delta H_1}{\Delta H_1}\right)\sum_j(\gamma_{jf} - \gamma_{jo}), \tag{11}$$

where τ is the dimensionless temperature $\tau = TR_g/E_1$, are coupled with eqn (10). With boundary conditions similar to eqn (2) and (3), rendered dimensionless for each of the reacting components j, the pellets are now fully defined in terms of the model described.

If the above system of equations is solved without reference to conditions within a reactor in which the catalyst pellets are packed, the solutions presented are entirely arbitrary and somewhat artificial. Advantage was therefore taken of computations reported[6] which provide numerical solutions for reactors packed with mixtures of pellets X (containing the metallic platinum component) and Y (the acidic type fluorinated silica–alumina) in the optimum proportions necessary to maximise the yield of the desired product benzene. To compute the concentration and temperature profiles within a reactor operated in any given manner, the mass and heat fluxes to and from pellets within the system must be ascertained. These are found from the equations

$$N_{jX} = \frac{3(1-e)\varepsilon\,c_{jo}D_e}{R_o{}^2}\left(\frac{\partial\gamma_i}{\partial\rho}\right)_{\rho=1} \quad\text{and}\quad Q_X = \frac{3(1-e)h\,E_1}{R_oR_g}(\tau_f - \tau_X) \tag{12}$$

for X type pellets, and

$$N_{jY} = \frac{3(1-e)(1-\varepsilon)c_{jo}D_e}{R_o{}^2}\left(\frac{\partial\gamma_i}{\partial\rho}\right)_{\rho=1} \quad\text{and}\quad Q_Y = \frac{3(1-e)hE_1}{R_oR_g}(\tau_f - \tau_Y) \tag{13}$$

for Y type pellets, where ε is a quantity defining the volume fraction of X type particles in the mixture of particles at a position along the reactor where the fluid concentrations are c_{jo}.

Turning now to the composite pellets in which both catalyst components are present in the same particle, mass and energy balance equations are similar to eqn (10) and (11) but now ε (in this case the fraction of active surface of component X) necessarily appears in the appropriate kinetic terms of eqn (10) as ε or $(1 - \varepsilon)$ depending on whether the particular kinetic term arises from catalysis by the X or Y catalytic component. The flux equations are similar to eqn (12) and (13) but the factors ε and $(1 - \varepsilon)$ must now be excluded. The kinetic scheme (II) for the composite pellets is summarised in Table 1.

Both of the above systems of equations enable the calculation of concentrations and temperature for the pellets in terms of fluid conditions which, in turn, are determined

Table 1
Kinetic models

Scheme I (separate pellets)

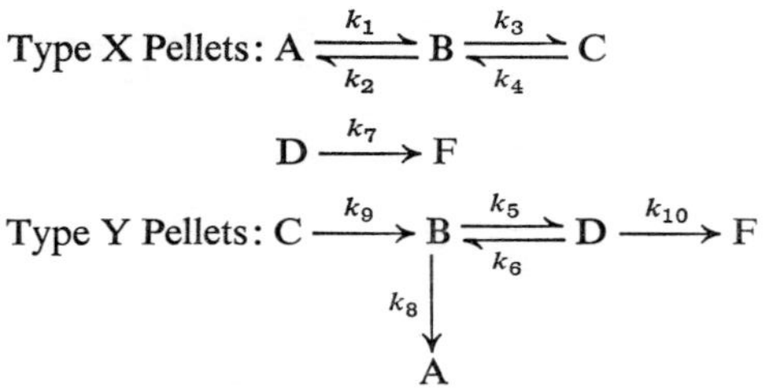

Type X Pellets: $A \underset{k_2}{\overset{k_1}{\rightleftharpoons}} B \underset{k_4}{\overset{k_3}{\rightleftharpoons}} C$

$D \xrightarrow{k_7} F$

Type Y Pellets: $C \underset{k_6}{\overset{k_5}{\rightleftharpoons}} D \xrightarrow{k_{10}} F$

$B \underset{k_8}{\overset{k_9}{\rightleftharpoons}} A$

Scheme II (compounded pellets)

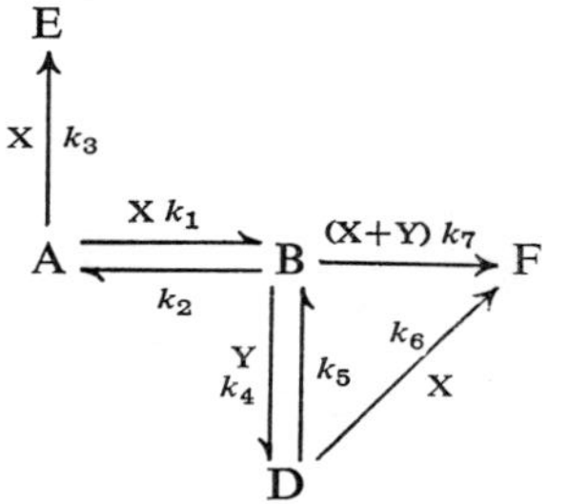

$$
\begin{array}{c}
E \\
\uparrow x\ k_3 \\
A \underset{k_2}{\overset{x\,k_1}{\rightleftharpoons}} B \xrightarrow{(\alpha X + Y)\,k_7} F \\
\quad Y\ k_4 \updownarrow\ \quad k_5 \searrow \nwarrow k_6\ \ x \\
\quad\quad D
\end{array}
$$

Reaction	Scheme I			Scheme II		
	A /s^{-1}	$^\dagger\Delta H$ /kJ (kg mol)$^{-1}$	$E \times 10^{-4}$ /kJ (kg mol)$^{-1}$	A /s^{-1}	$^\dagger\Delta H$ /kJ (kg mol)$^{-1}$	$E \times 10^{-4}$ /kJ (kg mol)$^{-1}$
1	1.07×10^3	1.06×10^5	5.25	5.85×10^{10}	1.05×10^5	16.7
2	2.39×10^2		4.19	63.4		3.76
3	1.13×10^3	6.28×10^4	4.48	1.86×10^{12}	-4.95×10^4	20.5
4	2.34×10^2		3.70	5.70	4.69×10^3	3.6
5	4.83×10^3	4.69×10^3	4.06	1.52×10^2		4.1
6	1.39×10^4		5.15	1.39×10^{10}	9.54×10^4	14.4
7	9.31×10^3	9.54×10^4	4.06	2.25×10^{13}	1.00×10^5	20.0
8	1.75×10^4	-1.06×10^5	5.17			
9	4.49×10^4	-6.28×10^4	5.69			
10	2.27×10^3	9.54×10^4	7.19			

† Enthalpies are listed as net values for equilibrium steps.

773

by the position of the catalyst pellet in the reactor. To complete the description, therefore, simple one dimensional reactor equations describing concentration and temperature conditions are written:

$$euc_{jo}\frac{\partial \gamma_{jf}}{\partial z} = -N_j, \text{ for A, B. } \ldots \text{ F} \tag{14}$$

and

$$eu\rho_f c_p \frac{\partial \tau_f}{\partial z} = -\frac{R_g Q}{E_1} + \frac{4h_w}{d_t}(\tau_h - \tau_f). \tag{15}$$

The mode of operation of the reactor may be defined at will merely by ascribing an appropriate numerical value to the reactor wall heat transfer coefficient h_w. It is assumed that the temperature of the heating fluid τ_h is constant along the reactor length. The reactor eqn (14) and (15) are subject to the inlet conditions (at $z = 0$) $\gamma_j = \gamma_{jo}$, $\tau_f = \tau_{fo}$, both of which must be specified.

Stability of bifunctional catalyst pellets

Solution of eqn (10) and (11) for the system of mixed pellets provides a means of computing the pellet temperatures as a function of the fluid temperature. The reactor conditions chosen, which ultimately determine the fluid conditions within the reactor [since eqn (14) and (15) are inextricably coupled with (10) and (11)], were those which approximate to the behaviour of an industrial reformer. Thus non-isothermal operation, in which heat is supplied to the reactor in quantities sufficient to maintain the catalyst pellet temperature at a level not far removed from the value at the reactor inlet, was considered to be a reasonable simulation of practice.

Eqn (10) may be reduced to a set of linear ordinary differential equations by the substitution $f_j = \gamma_{j}\rho$ and may be represented

$$\ddot{\mathbf{f}} = \mathbf{B} \cdot \mathbf{f} \tag{16}$$

where $\ddot{\mathbf{f}}$ is a column matrix of second order differential coefficients, $\mathbf{f}$ a column matrix of the new vectors f_j and $\mathbf{B}$ is a square matrix of the coefficients B_j associated with each f_j. The system of equations may thus be solved numerically by seeking the eigenvalues and eigenvectors of the matrix eqn (16).[11] Once these equations have yielded the concentration profiles within the catalyst pellet for the given fluid conditions at the reactor entrance, equations such as (11) are iteratively solved for T_X and T_Y. Subsequently the reactor eqn (14) and (15) are invoked and solved using the Runge–Kutta–Merson algorithm.

For the purpose of investigating particle stability, therefore, both the pellet and reactor equations were simultaneously solved, as described, to obtain a typical range of fluid conditions. Although inlet conditions were specified, calculations were performed for a range of constant catalyst composition between the extremes zero and unity and included the optimum composition which, for non-isothermal operating conditions, proved to be in the region of 0·5 with the metal loading used for the experimental kinetic determinations.[5,6] Having obtained sets of reactor fluid conditions by means of such numerical computations, eqn (11) was then solved iteratively for a wide range of fluid concentrations corresponding to the optimum pellet composition (*ca.* 0·5). Fig. 3 gives the pellet temperatures T_X and T_Y in terms of the fluid temperature T_f for a range of fluid concentration c_f. The similarity between this family of curves and those shown in Fig. 2 for the simple kinetic scheme is quite remarkable and suggests that little would be lost, when attempting to predict pellet instability, by reducing a complicated kinetic scheme to one containing a single rate controlling step. In fact, as is evident from Fig. 3, the curves lying to the right hand side of the diagonal line, which describe heat transfer and chemical reaction over pellets promoting endothermic conversion, demonstrate the uniqueness of the steady state: each of these curves represents a monotonically increasing function of the fluid temperature. The effect of increasing the fluid concentration of reactant is to increase the gradient dT_t/dT_x. On the other hand increasing the fluid concentration causes the temperature of pellets promoting exothermic reactions to increase sharply, and eventually the phenomenon of steady state multiplicity is possible. For the system investigated, however, such multiplicity does not reveal itself until the

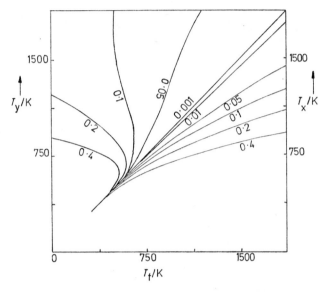

FIGURE 3 Numerical solutions to eqn (11) depicting the heat balance for discrete pellets catalysing the reforming of methylcyclopentane (A). Figures on curves denote fluid concentrations of reactant A in kg mol m^{-3}. Values of parameters are: $c_{Bf} = 0.3 \times c_{Af}$, $c_{Cf} = c_{Df} = 0.1 \times c_{Af}$, $R_o = 2 \times 10^{-3}$m, $D_e = 4.0 \times 10^{-7}$ m^2 s^{-2} $h_D = 4.0 \times 10^{-2}$ m s^{-1} and $h = 4.19 \times 10^{-2}$ kJ m^2 s^{-1} K^{-1}. Kinetic and thermo-dynamic quantities used are given in Table 1.

fluid concentration of the intermediate B (methylcyclopentene) exceeds 0.02 kg mol m^{-3}, and which concentration represents extreme conditions for reforming operations. This observation was verified by data obtained from a numerical study of the performance of reactors packed with various pellet mixture compositions. Temperature runaway only occurs for a discrete mixture of catalysts when the composition is as high as $\varepsilon = 0.99$ and in practice this means in the virtual absence of the isomerisation catalyst. In the limit ($\varepsilon = 1$) when the reactor is packed only with the hydrogenation catalyst (promoting the irreversible exothermic hydrogenation of components C and D), escalation in pellet temperature is therefore enhanced.

We conclude from the above analysis that only for extreme conditions of operation, when excess hydrogenation occurs, is particle instability likely to be important. The superior performance of composite pellets, however, renders even this event unlikely since heat released by the exothermic cracking reaction (A → E in Scheme II of Table 1) is immediately utilised to enhance the overall endothermic reactions occurring in the same pellet.

Since composite pellets are used in practice, a dynamic analysis of scheme II was undertaken by including in eqn (11) a term representing the accumulation of heat in the solid. No such term was included in the corresponding set of eqn (10) since the mass capacity of the catalyst pellet is so much smaller than its thermal capacity. The modified system of equations, now containing the unsteady state heat balance, was solved for step perturbations in fluid temperature and concentration. The technique used is simply an extension of numerical methods already outlined and reported elsewhere.[11] Fig. 4 is a typical record of the results obtained and clearly shows that for a perturbation of about 25 K or a 5% change in reactant concentration, the time required to reach the new steady state is over 350 s. Response to perturbations within reactors should be more rapid than this if process control is to be effective. A comprehensive numerical

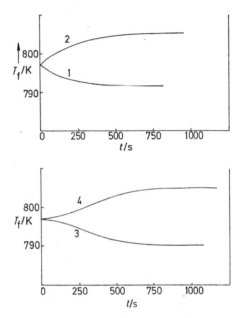

FIGURE 4   Transient responses to temperature and concentration perturbations in reactor packed with composite pellets.
Curve 1, 5% decrease in inlet concentration; curve 2, 5% increase in inlet concentration; curve 3, 25 K decrease in inlet temperature; curve 4, 25 K increase in inlet temperature.

investigation shows that the relaxation time is increased to values of 600 s as $\varepsilon \to 1$. Practical operation of reforming reactors should therefore take account of the long times required for recovery from disturbances.

NOMENCLATURE

A	Arrhenius constant
A	methylcyclopentane
B	methylcyclopentene
B	matrix of coefficients B_j

C	methylcyclopentadiene
c	concentration
$\bar{c}_p$	specific heat of fluid
D	cyclohexene
D_e	effective diffusivity
d_t	reactor tube diameter
E	cracked products
E_1	activation energy corresponding to reaction step A → B
e	reactor bed voidage
F	benzene
f	vector defining the product of dimensionless concentration and radius
ΔH	enthalpy change of reaction
h	fluid to solid heat transfer coefficient
h_D	fluid to solid mass transfer coefficient
h_w	reactor wall heat transfer coefficient
k	pseudo first order chemical rate constant
N	mass flux
Q	heat flux
R	pellet radius
R_g	universal gas constant
u	fluid velocity
X	hydrogenation catalyst
Y	isomerisation catalyst
z	position coordinate along reactor
β	defined by eqn (8)
α	defined by eqn (9)
γ	dimensionless concentration
ε	volume fraction of pellets of type X (for pellet mixtures) or active surface are fraction of component X (for composite pellets)
λ	defined by eqn (6)
ρ	dimensionless radius
τ	dimensionless temperature

Subscripts

o	conditions at reactor inlet and particle periphery
f	conditions within fluid
j	index indicating reaction component
p	conditions within pellet
X	conditions specifically within pellets of X
Y	conditions specifically within pellets of type Y

REFERENCES

[1] P. B. Weisz and E. W. Swegler, *J. Phys. Chem.*, 1955, **59**, 823.
[2] P. B. Weisz, *2nd Internat. Congr. Catalysis*, (Technip, Paris, 1961), p. 937
[3] D. J. Gunn and W. J. Thomas, *Chem. Eng. Sci.*, 1965, **20**, 89.
[4] W. J. Thomas and R. M. Wood, *Chem. Eng. Sci.*, 1967, **22**, 963.
[5] B. G. M. Jenkins and W. J. Thomas, *Canad. J. Chem. Eng.*, 1970, **48**, 179.
[6] R. A. Al-Samadi, P. R. Luckett, and W. J. Thomas, *Adv. Chem. Series No. 133* (*Chem. Reactn. Eng. II*), (Amer. Chem. Soc., Washington, 1974), p. 316.
[7] R. Aris, *The Mathematical Theory of Diffusion and Reaction in Permeable Catalysis, Vol. 1, The Theory of the Steady State* (Clarendon Press, Oxford, 1975), p. 58.
[8] C. McGreavy and J. H. Thornton, *Canad. J. Chem. Eng.*, 1970, **48**, 187.
[9] J. Beek, *A.I.Ch.E.J.*, 1961, **7**, 337.
[10] R. A. Al-Samadi, *Ph.D. Thesis* (University of Bath, 1975), p. 38 *et seq.*
[11] L. Fox, *Numerical Solution of Ordinary and Partial Differential Equations* (Pergamon Press, London and New York, 1962).

DISCUSSION

H. Heinemann (*Mobil, Princeton*) said: Instability of pellets is not necessarily as important as stability of active sites. The temperature at the active sites can sometimes differ by several hundred degrees, from that measured in the pellets. How does this effect your statements, and is the time at this high temperature (in exothermic reactions) sufficient to cause sintering? I would make a plea for measurements of rate of heat dissipation from active sites to the substrate.

W. J. Thomas (*Univ. of Bath*) replied: The local temperature rise generated by reaction at an active site is, of course, a very important factor in determining reactivity. Excessive local temperature rise may, under certain circumstances, lead to sintering of the active surface with consequent loss in catalytic activity. However, techniques for estimating transient local temperature rises are insufficiently developed for reliable assessments to be made. The thermal properties of many catalyst pellets are such that heat generated locally is quickly dispersed throughout the pellet. Measurements demonstrating that only small temperatures rises occur within a pellet catalysing a hydrocarbon oxidation have recently been made.[1]

When heat is dispersed rapidly, the temperature rise of the pellet as a whole is determined by the nature of the reactions occurring and the mass and heat transfer properties of the system. If heat is not dispersed rapidly, a local transient temperature rise will appear. As the last section of the paper demonstrates, conditions may be such that a temperature perturbation takes a relatively long time to subside.

S. P. S. Andrew (*ICI, Billingham*) asked: Would Prof. Thomas agree that significant deviations in average catalyst particle temperature from the gas stream local temperature can only occur when the reaction is either highly endothermic or exothermic and furthermore when the reaction rate is so high as to be approaching external gas–film mass-transfer limitation?

W. J. Thomas (*Univ. of Bath*) replied: Extremely exothermic or endothermic chemical reactions will give rise to temperature differences between the porous solid catalyst particles in which the reaction occurs and the surrounding gas because there is usually a resistance to heat transfer from particle to fluid caused by the relatively stagnant boundary layer of gas adjacent to the particle. If, in addition, the chemical reaction rate is so high that mass transfer of reactants and products through the gas film limits the overall rate, then transfer of heat to and from the particle is exacerbated.

J. B. Butt (*Northwestern Univ., Evanston*) said: I would like to second the comment made earlier by Dr. Heinemann concerning the importance of, and our ignorance of, possible differences in metal crystallite temperature and average catalyst temperature. The latter can be measured; the former cannot. One means of determining experimentally these differences is to 'calibrate' the crystallite temperature using a thermally neutral reaction of known kinetics; this may then be used as an indirect temperature probe when run competitively with an exothermic reaction. The requirements are that the competitive reaction experiment must not perturb the kinetics of either reaction run independently, and both reactions must occur on the metal crystallites only.

Unfortunately, I have not met anyone who is aware of a reaction-catalyst system fulfilling these requirements.

[1] J. Corrie, Ph.D. Thesis, Univ. London (1972).

The Kinetics of the Oxidation of Naphthalene over a Molten Vanadium Pentoxide–Potassium Sulphate Catalyst

PETER VINCENT BUTT and C. N. KENNEY*

Department of Chemical Engineering, Cambridge, England

ABSTRACT   The kinetics of the oxidation of naphthalene over an unsupported molten V_2O_5 catalyst and a solid potassium sulphate promoted V_2O_5 catalyst, below and above the catalyst melting point (433°C), are described. It is found that although kinetic equations based on a reduction–oxidation process are largely adequate to correlate the data they do not account for the mechanism. The rate of naphthalene oxidation falls with increasing temperature between 400°C and 433°C and the selectivity for phthalic anhydride formation decreases. It is concluded that the oxidised and reduced forms of the catalyst involve forms of vanadium which lie between quadrivalent and quinquevalent, that the catalyst probably undergoes a structural change near 400°C, that the selectivity of the catalyst declines with increasing reduction, and that the reaction probably takes place on the melt surface in contrast to the diffusion controlled reaction associated with sulphur dioxide oxidation in related catalyst systems.

INTRODUCTION

The recognition that molten salts are active catalysts for a number of reactions has stimulated growing interest in their chemistry and practical applications in recent years. We have discussed the kinetic behaviour of some inorganic oxidation reactions previously[1-3] but to date little related work has been reported on organic systems. The oxidation of naphthalene is discussed here. A catalyst of promoted vanadium pentoxide was employed, being of particular interest since its melting point lies within the operational temperature range of industrial processes (350—500°C) and it is likely that at or near the 'hot-spot' in industrial reactors, molten phases are present. Apart from the work of Loftus[4] who studied the oxidation of o-xylene by bubbling it with air through a melt of vanadium pentoxide and potassium sulphate, there have been no other reports of attempts to obtain the oxidation kinetics of a hydrocarbon using a molten salt catalyst. The principal aim of the work was to ascertain whether an unsupported molten catalyst would be active for naphthalene oxidation and it was hoped to throw light on the mechanism by which the reaction occurs above and below its melting point. Also it is of interest to know whether a molten catalyst acts as a reaction medium in which reaction takes place between dissolved gases followed by desorption of reaction products, or whether reaction occurs on the melt surface between adsorbed gas molecules. It is not possible to distinguish with certainty between these two mechanisms by a kinetic study alone, but some evidence may be obtained in favour of one or other mechanism.

The criteria governing the selection of the catalyst were that it should contain vanadium pentoxide and melt below 500°C. A eutectic of V_2O_5–KVO_3 is known (m.p. 389°C) which is reported to exhibit appreciable foaming.[4] The $K_2S_2O_7$–V_2O_5 system used for sulphur dioxide oxidation is an obvious candidate and is potentially attractive since it offers a wide range of operating temperatures (390—500°C) but it was rejected because of the appreciable sulphur trioxide equilibrium pressure which exists over the melt above 400°C. Quite apart from the detrimental effects on the analytical system, sulphur trioxide is a powerful homogeneous oxidising agent, making it difficult to separate the effects of the catalyst as such. It was consequently decided to employ a catalyst containing 39% of vanadium pentoxide and 61% K_2SO_4, by weight. This is close in composition to some industrial catalysts and also forms a sharp relatively low-melting-point eutectic (433°C). This mixture was used throughout the work since even small changes in composition will raise the melting point to temperatures where the homogeneous gas phase oxidation is no longer negligible. The results reported here summarise data on four different types of experiment *i.e.*, kinetic measurements on the oxidation of naphthalene over molten and over solid catalyst, together with reduction and oxidation studies using only one gaseous reactant.

EXPERIMENTAL

The same basic equipment was employed for all four types of study. Essentially integral rate measurements were made using a well mixed reactor, constructed of stainless steel. This was a vertically mounted hollow cylinder, internal diameter 5 cm, length 18 cm. Band heaters were fixed round the outside of the vessel and gave accurate temperature control ($\pm$ 0·25°C). The melt was stirred with a small paddle mounted just below the melt surface. It was established that good mixing, with very little disturbance of the surface, occurred at stirrer speeds below 150 rpm. The stirrer shaft passed through the centre of the reactor top, and a gastight seal was made with a gland using an asbestos packing treated with oil and graphite: during use this packing deteriorated and was replaced at intervals. The catalyst, *ca.* 40 g, of surface area 15 cm² was contained in small disposable Pyrex beakers which fitted into the reactor base: in stainless-steel containers there was a significant tendency of the melt to creep up the walls.

Naphthalene was added to the inlet gas stream passing some or all of a nitrogen/oxygen stream through a vaporiser containing liquid naphthalene. Gas chromatography was used to analyse both reactants and products, sampling being done with a heated pneumatic valve. Condensible organics, naphthalene, maleic anhydride, 1,4-naphthaquinone, and phthalic anhydride were separated in *ca.* 10 min on a five-foot Diatomite column held at 180°C and were measured using a flame ionisation detector coupled to a digital integrator. Oxygen, nitrogen, and carbon monoxide were separated on a 6 ft column at 25°C containing 5A molecular sieve and measured with a hot-wire thermal conductivity dector. Water and carbon dioxide were determined chemically. Changes in oxygen concentration were also followed with a Servomex oxygen analyser, coupled to a digital voltmeter. The reaction rate was found to be proportional to melt surface area but independent of the mass of the melt, gas flow-rate, and also stirrer speed, provided the melt surface remained undisturbed.

RESULTS

1. *Product distribution*

In all the runs with the molten catalyst over the temperature range 440—470°C, naphthalene conversions were between 18% and 32%. Of the naphthalene converted, approximately 11% formed phthalic anhydride and 14% formed 1,4-naphthaquinone, the remainder going to carbon dioxide and water. The selectivity of the molten catalyst was essentially independent of both temperature and reactant concentrations over the entire ranges studied. This finding is in contrast to the results of other workers studying the oxidation of naphthalene using solid supported catalysts containing vanadium pentoxide at lower temperatures. Ross[5] employed a catalyst of V_2O_5–SnO_2 on alumina in the temperature range 330—422°C and found that selectivity was independent of reactant concentrations but dependent on temperature. It could well be that the selectivity of the molten catalyst does change with temperature, but because the range investigated was not large, 440—470°C, any selectivity changes were too small to be detected. On the other hand, Roiter *et al.*,[6] using a supported catalyst of V_2O_5–K_2SO_4

on silica at temperatures from 300—360°C found selectivity to be independent of temperature but dependent on reactant concentrations.

It was found from plots of the rate of naphthalene disappearance and product formation against reactant concentrations, that the rate of reaction has a strong positive dependence on temperature and naphthalene concentration. Data for the dependence of the rate of naphthalene disappearance on oxygen concentration showed an almost linear relation with respect to oxygen partial pressure. Table 1 summarises the results

Table 1

Rate constants for solid catalysts (S) and molten catalysts (M)

Temp (°C)	$k_1 \times 10^4$ l/cm² s	$k_2/B \times 10^5$ l/cm² s	$k_2 \times 10^4$ l/cm² s	Activation energy associated with $k_1 = 13 \pm 1\cdot5$ kcal/g mol
300(S)	0·81	0·279	1·57	
320(S)	1·22	0·684	3·83	Activation energy
340(S)	1·94	1·43	8·39	associated with $k_2 = 29\cdot3 \pm 1\cdot5$ kcal/g mol
440(M)	4·79	4·25	4·06	Activation energy
450(M)	5·69	5·19	4·96	associated with
460(M)	6·73	6·3	6·02	$k_1 = 17\cdot6$ kcal/g mol
470(M)	7·92	7·61	7·26	Activation energy associated with $k_2 = 20\cdot5$ kcal/g mol

obtained in the study of the overall kinetics of naphthalene oxidation using the molten catalyst.

Mars and Van Krevelen,[7] Shelstad *et al.*,[8] and Ross[5] all fitted their data on the oxidation of naphthalene over solid supported catalysts to the following rate expression.

$$R = \frac{k_1 C_N k_2 C_O}{k_2 C_O + B k_1 C_N} \tag{1}$$

R = Rate of naphthalene disappearance
k_1 and k_2 = Naphthalene and oxygen rate constants
C_N and C_O = Naphthalene and oxygen concentrations
B = Molecules oxygen required per molecule of naphthalene reacted

This equation is derived from a steady-state analysis of the two-step redox mechanism.

Naphthalene + active catalyst $\xrightarrow{k_1}$ products + deactivated catalyst

Deactivated + gaseous oxygen $\xrightarrow{k_2}$ active catalyst.

Depending on the concentrations of naphthalene and oxygen and the values of the rate constants, equation (1) will yield apparent reaction orders of zero to one for both oxygen and naphthalene. In view of the observed variable dependence of reaction rate on naphthalene concentration, this equation represents a suitable starting point for correlation of the data.

The number of molecules of oxygen required for the oxidation of one molecule of naphthalene, B, can be calculated for each run from:

$$B = \frac{\text{Oxygen consumed (g mol/min)}}{\text{Naphthalene reacted (g mol/min)}}$$

Since the selectivity to phthalic anhydride remained essentially constant (10·6%) over the entire ranges of reactant concentrations and temperature investigated, the average value of B was taken from the results of all the runs and proved to be 9·55. This figure can be compared with that resulting had all the naphthalene been converted into carbon dioxide and water, *i.e.* $B = 12$. A knowledge of B allowed the calculation of the rate constants, k_2. Values of k_1, k_2/B, and k_2 calculated from a nonlinear least squares analysis are presented in Table 1. The accuracy of these calculated constants is probably ± 5%.

In the related studies on the unsupported solid catalyst the same apparatus was used with the same range of naphthalene and oxygen concentrations. Molten catalyst was poured into a hot Pyrex vessel in a furnace and then allowed to cool. The reaction products over the solid catalyst were the same as those from reaction catalysed by the melt *i.e.*, 1,4-naphthaquinone, phthalic·anhydride, carbon dioxide, and water. Maleic anhydride was detected but never in measurable amounts. However, in contrast to the melt studies, the distribution of these products varied considerably between 300 and 425°C. Two types of behaviour were observed: (a) From 400—425°C, the product distribution was essentially constant and similar to that observed with the molten catalyst. The selectivities to phthalic anhydride and 1,4-naphthaquinone were approximately 11% and 15%, respectively. (b) As the temperature was decreased from 400 to 380°C a dramatic increase in reaction rate occurred as shown in Figure 1. The selectivity towards phthalic anhydride showed a marked improvement compared with the observed at 400°C and above. As temperature in the range 300—380°C was decreased, selectivity to phthalic anhydride increased, for example, at 380°C approximately 50% of the naphthalene oxidised formed phthalic anhydride compared with over 70% at 300°C. The oxidation of phthalic anhydride to its decomposition products was not examined.

All the points at a particular temperature lie on a straight line which passes through the origin, suggesting that for a given temperature, selectivity is constant and, hence, independent of reactant concentrations. The optimum temperature for phthalic anhydride production was approximately 360°C. The data could also be satisfactorily correlated between 300 and 370°C by the same equation as for the molten catalyst. The reaction order with respect to oxygen is close to one: the order with respect to naphthalene falls from one to zero as the naphthalene concentration is increased. Different activation energies are obtained. Typical results are given in Table 1.

2. *Reduction experiments*

The main products from the oxidation of naphthalene concentration range (4—21 × 10^{-5} g mol/l), using the molten catalyst in the absence of oxygen, were carbon dioxide and water. 1,4-Naphthaquinone was the only other product detected, and accounted for approximately 34% of the naphthalene converted. The selectivity of the oxidation using the melt seemed to be essentially independent of both temperature and naphthalene concentration, but such a conclusion is tentative. The selectivity of the oxidation with the solid catalyst in the temperature range 400—425°C was similar to that observed with the melt. However, as soon as the temperature was lowered to 380°C, phthalic anhydride was detected in addition to the above mentioned products. Although carbon dioxide and water remained the main reaction products, as temperature decreased in the range 300—380°C, phthalic anhydride accounted for an increasing proportion of the naphthalene converted, mainly at the expense of the products of complete combustion;

e.g. at 380°C, initial selectivity to phthalic anhydride, 19·9%; to 1,4-naphthaquinone, 11·7%; at 300°C initial selectivity to phthalic anhydride, 38·6%; to 1,4-naphthaquinone, 13·3%. In view of the scant experimental data obtained for the solid catalyst, no exact conclusions can be drawn regarding the concentration- and temperature-dependence of the selectivity between 300 and 380°C, but the rates under comparable conditions were 0·1 to 0·15 times those with oxygen present. The initial rates of the reduction reaction decreased with increase in temperature in the range 400—425°C. Moreover at 380°C,

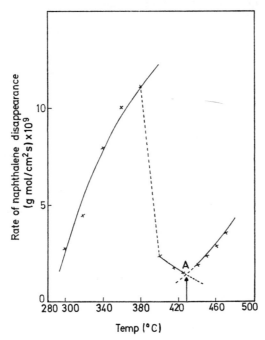

FIGURE 1   The activity of the catalyst as a function of temperature. The arrow marks the melting point.

the initial rate of naphthalene disappearance showed a dramatic increase to almost three times that at 400°C. Thereafter as the temperature was further reduced, so the initial reaction rates increased. Rates of naphthalene disappearance fell more rapidly with time than in the melt experiments. Estimates of reaction order and activation energy were not derived. However, some relevant calculations can be made assuming a linear decay in naphthalene consumption rate and constant selectivity. Thus if the reduction of V_2O_5 is assumed to form V_2O_4 at 380°C, the depth of the reduced layer is approximately $8·0 \times 10^{-2}$ cm when the reaction ceases. Also, if the only available oxygen was a surface monolayer the $1·25 \times 10^{-8}$ g mol of atomic oxygen (assuming 10^{15} sites/cm^2) would be consumed in approximately $3·5 \times 10^{-4}$ min. instead of the estimated time of 150 min. Clearly an appreciable transfer of oxygen from bulk to surface can occur.

3. Re-oxidation experiments

When the catalyst components are fused together in air some oxygen is evolved, equivalent to the formation of $V_2O_{4·92}$: this figure agrees closely with an earlier result of Boreskov.[11] More oxygen is evolved on heating in a nitrogen stream (equivalent to

$V_2O_{4\cdot 85}$) but the catalyst readily returns to the more highly oxidized form ($V_2O_{4\cdot 92}$) on exposure to air. This labile oxygen cannot, on thermodynamic grounds, arise from the simple vanadium oxides and is presumably associated with V_2O_5–K_2SO_4 complexes.

The activation energy was determined for the re-oxidation of the partially reduced molten catalyst. The value of 10 ± 2 kcal/g mol, agrees well with the value obtained by Holroyd and Kenney.[9] $9\cdot 7 \pm 2\cdot 5$ kcal/g mol, for the oxidation of partially reduced V_2O_5–$K_2S_2O_7$ melts. They found the rate of oxygen absorption to be proportional to the quadrivalent vanadium content of the melt and to the oxygen partial pressure to the one-half power; their rate constants in the same units are an order of magnitude larger than those determined in this work. Nonetheless the data appear to be reasonably well accounted for in terms of a surface reaction between quadrivalent vanadium and gaseous oxygen.

If it is supposed naphthalene oxidation may be considered exclusively as a reduction–reoxidation process, then it is possible to extract reaction rate constants from the following highly simplified rate expressions:

$$R_{RED} \quad\quad = k_{red} (V_2O_5)(C_N) : R_{OX} = k_{ox} (V_2O_4)(C_o) \tag{2}$$

where,

R_{RED}	= Rate of reduction, ($B \times$ rate of naphthalene disappearance, g mol/cm^2. s).
(V_2O_5)	= mole fraction of vanadium pentoxide initially in the melt determined by volumetric analysis.
R_{OX}	= Initial rate of oxygen consumption during re-oxidation (g mol/cm^2 s).
(V_2O_4)	= mole fraction of quadrivalent vanadium in the melt, determined from difference between initial V_2O_5 and oxygen removed in naphthalene oxidation.
(C_N) and C_o)	= Naphthalene and oxygen concentrations (g mol/l).

Thus, at 460°C we have,

$$k_{red} \quad\quad = 1\cdot 0 \times 10^{-5} \text{ l/cm}^2 \text{ s}$$
$$k_{ox} \quad\quad = 3\cdot 8 \times 10^{-4} \text{ l/cm}^2 \text{ s.}$$

The rate of re-oxidation is thus approximately four times faster than the rate of catalyst reduction (rate of oxygen removal from the catalyst during reduction). The rate constants derived here for the separate catalyst reduction and oxidation steps are lower than those obtained from kinetic measurements of oxidation rate using equation (1). Further, $k_{ox} > k_{red}$ in contrast to the common inference that, in the presence of oxygen, catalyst reoxidation is the slower step. However k_{ox} and k_{red} were obtained with catalyst studied far from steady-state conditions, and the low rate of oxidation of naphthalene, together with the poor selectivity, in the absence of oxygen, suggests that oxygen adsorption on the catalyst is required both to form the optimum catalyst surface structure, and to oxidise naphthalene.

DISCUSSION

There has been considerable speculation regarding the role of vanadium pentoxide oxygen in oxidation, but no clear picture has yet emerged. It is known that solid vanadium pentoxide is an 'n' type semiconductor and, as such, contains lattice defects, and there is evidence that the addition of a promoting agent such as potassium sulphate increases the number of lattice defects which enhance the catalytic activity. Furthermore, vanadium pentoxide catalysts have been shown to be reduced to a certain extent during oxidation reactions, and it is thought that the activity of the catalyst depends on the ratio of V^{IV} to V^V.[10] V_2O_5, $V_2O_{4\cdot34}$, V_2O_4, and V_2O_3 have been detected in a catalyst initially charged as vanadium pentoxide after use in the oxidation of o-xylene and it has been suggested that the active portion of the catalyst was confined to $V_2O_{4\cdot34}$ (V_6O_{13}) and V_2O_5 and to intermediate structures between the two. Our findings are consistent with much earlier work, but they differ in two respects from a recent investigation by Andreikov *et al.*[13] who used a pulse technique with the solid catalyst. Our absolute reaction rates (mol/cm²s) are much higher, but this discrepancy can be accounted for if their stated surface areas relate to the support rather than active catalyst. More important is their assertion that the rate of the overall reaction is the same as the rate of the separate reduction and oxidation steps. However, their technique permits measurements at smaller perturbations from steady operating conditions than here, where we calculate a monolayer of oxygen will be removed in 2×10^{-2}s. Also a different catalyst composition was employed.

The reaction may take place on the surface of the melt, or homogeneously, or in a diffusion layer at the gas–melt interface. Using the scaled particle theory of liquids,[14] minute solubilities of naphthalene are predicted (10^{-12} g mol/cm³ atm) as expected given the energy required to form cavities in the melt. This suggests reaction on the melt surface. Such arguments do not apply to oxygen transport which we have discussed elsewhere in relation to sulphur dioxide in melts.[9] Even in the solid catalyst there appears to be substantial mobility of oxygen species.

The investigation has demonstrated that highly significant structural changes may occur in a catalyst over a relatively narrow temperature range. It has been confirmed that the attractively simple V^{IV} and V^V redox mechanism is not adequate to explain hydrocarbon oxidation over promoted vanadium pentoxide even when the kinetic predictions of such a mechanism are closely followed over a solid and by a molten catalyst. Increased amounts of quadrivalent vanadium arising with reaction in the absence of oxygen diminish the selectivity towards phthalic anhydride formation. The ease with which the catalyst will take up additional oxygen suggests a relatively highly oxygenated labile complex is necessary to complete the formation of phthalic anhydride from naphthaquinone. Further speculation about the detailed mechanism and catalyst structure is unlikely to be productive until more data are obtained. In particular the roles of "lattice" and "adsorbed" oxygen must still be regarded as unresolved.

ACKNOWLEDGEMENT. The award of an SRC Studentship to P.V.B. during this investigation is gratefully acknowledged.

REFERENCES

1 A. R. Glueck and C. N. Kenney, *Chem. Eng. Sci.*, 1968, **23**, 1257.
2 D. M. Ruthven and C. N. Kenney, *Chem. Eng. Sci.*, 1968, **23**, 981.
3 C. N. Kenney, *Catalysis Rev.-Sci. Eng.*, 1975, **11**, 197.
4 J. Loftus, *D.Sc. Thesis*, M.I.T., 1963.
5 G. L. Ross, *Ph.D. Thesis*, Edinburgh University, 1970.
6 V. A. Roiter, V. P. Ushakova, G. P. Kornelchuk, and J. G. Skorbilina, *Kinetika i Kataliz*, 1961, **2**, 94.
7 P. Mars and D. W. Van Krevelen, *Chem. Eng. Sci., Special Supplement*, 1954, **3**, 41.
8 K. A. Shelstad, J. Downie, and W. F. Graydon, *Canad. J. Chem. Eng.*, 1960, **38**, 102.
9 F. P. B. Holroyd and C. N. Kenney, *Chem. Eng. Sci.*, 1971, **26**, 1971.
10 G. I. Simard, I. F. Steger, R. J. Arnott, and L. A. Siegel, *Ind. Eng. Chem.*, 1955, **47**, 1424.
11 G. K. Boreskov, V. V. Illarianov, R. P. Ozerov, and E. V. Kildisheva, *Zhur. obshchei Khim.*, 1954, **24**, 23.
12 L. Ya. Margolis, *Adv. Catalysis*, 1963, **14**, 429.
13 E. I. Andreikov, Yu. A. Sveshnikov, and N. D. Ruskyanova, *Kinetika i Kataliz*, 1974, **15**, 1207.
14 A. K. K. Lee and E. F. Johnson, *Ind. Eng. Chem. Fundamentals*, 1969, **8**, 726.

DISCUSSION

M. Baerns (*Ruhr-Univ., Bochum*) said: With respect to your oxidation experiments carried out in the absence of gaseous oxygen I should like to mention some preliminary results on but-1-ene oxidation for preparation of maleic anhydride which were obtained by periodic operation of a fixed bed reactor (*i.e.* first air and secondly a mixture of but-1-ene and nitrogen were passed over the catalyst). This procedure was repeated over a period of several hours, the cycle times ranging from 10 to 60 s. The activity was not markedly affected by the absence of gaseous oxygen as compared to stationary operation (*i.e.* passing a mixture of but-1-ene and air continuously over the catalyst). However, the selectivity for maleic anhydride formation decreased significantly in favour of the formation of carbon monoxide and dioxide. It seems to me that this is caused by the lower oxidation state of the catalyst achieved during a certain time of the reaction. Would you comment on these results in the light of your work?

H. Schaeffer (*Degussa, Frankfurt*) said: We have investigated the related reaction of oxidation of benzene to maleic anhydride.[1] Pure solid vanadium oxides in the range of composition $VO_{2.5}$ to $VO_{1.5}$ were studied under non-stationary conditions using microcatalytic techniques. The results were as follows. First, in the presence or absence of molecular oxygen the maximum specific activity for benzene oxidation was observed with $VO_{2.0}$. With oxygen in the reaction mixture, benzene reacted 75–100 times faster in the presence of $VO_{2.0}$ than in the presence of $VO_{2.5}$. Secondly, the reaction of benzene is slower in the absence of oxygen than in its presence. This ratio of rates is increased if oxygen is pre-adsorbed. Thirdly, in the presence of oxygen no great differences in the oxidation rate for maleic anhydride were observed in the range of catalyst composition $VO_{2.5}$ to $VO_{2.0}$. Consequently $VO_{2.0}$ showed not only the maximum activity but also the maximum selectivity for maleic anhydride formation in the presence of oxygen. Lastly, VO_2 has, to a first approximation, a rutile structure. Rutile itself is inactive for benzene oxidation but rutile doped with 5% VO_2 is very active (unpublished results). The latter observation reminds me of the fact that a series of industrially developed vanadium oxide catalysts for the oxidation of benzene and *o*-xylene contain large amounts of titania.

C. N. Kenney (*Cambridge Univ.*) in reply to the two previous speakers said: The above comments support the view that the activity and particularly the selectivity of the catalyst are dependent on the $V^{IV}:V^V$ ratio. The simple reduction–oxidation mechanism involving a single type of oxygen cannot easily account for these findings. It may be that gaseous oxygen is required both to oxidize the (chemisorbed) hydrocarbon and to provide 'lattice' oxygen so that the catalyst surface has the appropriate composition and structures for selective oxidation. In the absence of gaseous oxygen the hydrocarbon would react less selectively with the lattice oxygen.

S. P. S. Andrew (*ICI, Billingham*) asked: Is the reaction velocity effect reversible as a function of temperature? Does the measured velocity traverse the same path in Figure 1 when the temperature is decreased as when it is increased?

P. Mars (*Twente Univ. of Tech.*) said: The increase in activity accompanying solidification of the catalyst amounts to a factor of 10 to 15. Could this not be caused by the formation of cracks in the solidifying catalyst? In the case where the order with respect to oxygen is close to unity, the value of the rate coefficient of the step which is hardly influencing the rate (the reduction step) cannot be determined accurately. Could the differences between the activation energies calculated for the solid and those for the molten catalyst (Table 1) be related to this fact?

C. N. Kenney (*Cambridge Univ.*) in reply to the two previous speakers said: It was difficult to study the effect of cooling the catalyst through its melting point since the

[1] H. Schaefer, *Ber. Bunsengesellschaft phys. Chem.*, 1967, **71**, 222.

melt expanded on solidification and broke the glass container. I agree with Prof. Mars' comments. We are examining the behaviour of the melt on a porous catalyst support to see whether crack formation is important but capillary effects may complicate any conclusions.

H. Heinemann (*Mobil, Princeton*) said: In reactions catalysed by molten salts, gas–liquid contact is often difficult to control because of bubble size and aggregation. Heat transfer is subject to the same problems. How do you differentiate between catalytic and thermal reactions? The viscosity of non-catalytic molten salts is probably different from that of catalytic ones and thus will give different bubble size and different heat transfer. Will this make it difficult to have dependable blank runs?

C. N. Kenney (*Cambridge Univ.*) replied: The homogeneous oxidation occurs at a quite different rate and gives largely carbon dioxide and water. We limited our measurements to temperatures below 753 K because we found that above this temperature homogeneous reaction occurred in the reactor, with no catalyst, at an appreciable rate. Below 743 K there is little reaction in the absence of catalyst.

M. Schiavello (*Univ. of Rome*) (communicated): It is well known that vanadium pentoxide is easily reduced under vacuum and by treatment in hydrogen. It is also known that the oxygen deficiencies can be accommodated by a mechanism of crystallographic shear formation. According to the amount of oxygen removed, a series of well-defined non-stoicheiometric phases can be formed. On the other hand, it is likely that the mechanism is in part reversible and that adsorbed oxygen can restore the original structure. It is then evident that various oxygen ions are present in the phases formed by reduction of vanadium pentoxide. There is a possibility that such a mechanism can operate also in the vanadium pentoxide–potassium sulphate catalyst. The authors state that an appreciable transfer of oxygen from bulk to surface can occur. Can you comment on this being an indication in favour of a shear mechanism?

C. N. Kenney (*Cambridge Univ.*) replied: Prof. Schiavello's remarks apply to the behaviour of pure vanadium pentoxide. Although it is convenient to discuss reduction–oxidation mechanisms in terms of the vanadium : oxygen ratio, the work of Boreskov and others suggests that when large proportions of potassium sulphate are present, complex compounds, possibly polymeric, are formed which can fairly readily acquire and lose oxygen. It would be interesting and helpful if some of the newer physical techniques were used to identify these. Consequently there seems no need to consider shear mechanisms until the species present have been identified.

The Measurement and Control of Moving Reaction Zones in Fixed-bed Reactors

ALI A. GHANDI and LESTER S. KERSHENBAUM*

Department of Chemical Engineering and Chemical Technology,
Imperial College, London S.W.7.

ABSTRACT   It is well known that exothermic reactions taking place in fixed-bed reactors can exhibit temperature profiles which slowly traverse the bed, either upstream or downstream, while the reactor is otherwise operating under steady- or near steady-state conditions. In an earlier theoretical study, we had demonstrated that in systems with decaying catalysts, a reaction zone moving in a direction counter to that of the flow of fluid can be close to the optimal operating method for that system. The current work is an experimental study of the moving reaction zones and methods of controlling the reaction zone velocity to achieve some objective. The effect of various controls on the velocity of propagation and the position of the ultimate steady-state (if any), have been studied. The chemical systems studied to date include the oxidation of carbon monoxide on a copper oxide catalyst, and the oxidation of carbon deposited on silica (*i.e.* the regeneration of a cracking catalyst). Propagation velocity has been measured with reasonable accuracy and reproducibility, and correlated as a function of feed conditions and initial conditions.

INTRODUCTION

In recent years several workers have noted that propagating temperature profiles can be observed in fixed-bed tubular reactors in which strongly exothermic reactions are taking place.[1-4] Depending upon the feed and controls, the profiles can propagate in either a forward or reverse direction and can lead to a stable steady-state or to a blow-out.

There has also been a substantial amount of work seeking to explain the above phenomenon by postulating various models of such systems.[5-8] These models differ considerably by the inclusion or omission of various contributing factors, including gas and solid conductivity, radiation, dispersion, the kinetics of the reaction, *etc.*, but are still able to predict the observed behaviour of such systems. A rather complete analysis by Eigenberger[6] concluded that the foremost influence is the conduction of heat in the solid phase, and that the transient propagation of the reaction zone is primarily a transition from one state to another stable steady-state. The number of possible stable steady-states in a given configuration is also discussed. By allowing for heat conduction in the solid phase, the infinite number of steady-states predicted by early two-phase models[9] is reduced to two or three, depending upon the exact configuration and the boundary conditions chosen.

In a recent theoretical study,[10] we had demonstrated that in certain systems with decaying catalysts, a reaction zone moving in a direction counter to that of the flow of fluid can be close to the optimum operating method for that system. In that work, the complete time- and space-dependent optimal control policy was generated for a tubular reactor subject to several catalyst decay laws. The solutions revealed that a critical parameter is the ratio of the activation energy for the main reaction to that for the catalyst decay reaction. For values of this ratio less than one, the optimum temperature profiles represent a low temperature reaction zone that occupies the entire active bed at all times. However, where this ratio is greater than one, the profiles represent a zone of sharply rising temperature which propagates with time along the length of the reactor.

The similarity of the above results to the phenomenon of propagating reaction zones,

788

indicated that further experimental study was warranted in the area of moving reaction zones. Special attention should be paid to measurement of the reaction zone velocity and methods of controlling this velocity, when such control is desirable.

Both experiment and theory have shown that effective control variables include the composition and flow rate of the feed stream and, in certain cases, the feed temperature. The current work seeks to determine the effectiveness of these controls for specific systems.

EXPERIMENTAL SYSTEM

A schematic diagram of the experimental system is shown in Figure 1. The reactions studied experimentally were the catalytic oxidation of carbon monoxide to carbon dioxide over a cupric oxide catalyst and the regeneration of the fouled cracking catalyst.

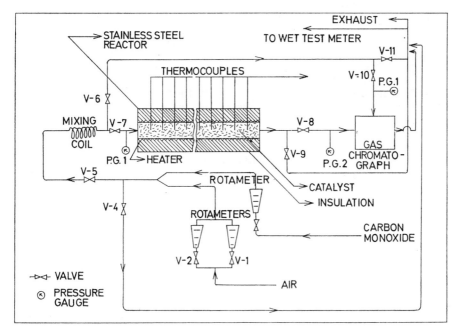

FIGURE 1   Schematic of the experimental system.

During the regeneration of cracking catalyst the deposit of carbon that causes the deactivation of the catalyst is removed through the oxidation to CO and CO_2. The feed flowrate and composition are used to control the zone temperature and propagation rate.

The reactor is operated with reverse propagation of the reaction zone, that is with the reaction zone moving upstream towards the entrance of the reactor. The materials used for the oxidation of carbon monoxide are 99·5% pure CO and dry air, with the catalyst in the form of thin cylindrical wires (0·5 mm diameter × 2 mm length). Only results referring to the oxidation of CO will be discussed in this paper.

The reactor was constructed of 18/8 stainless steel, approximately 1 cm internal diameter, and 60 cm in length. Several layers of ceramic fibre insulation were provided to minimize heat losses. Stainless-steel sheathed Chromel–Alumel thermocouples

(1 mm diameter) were placed along the axis of the reactor at uniform intervals of 2·5 cm. Four electrical heating tapes were wrapped around the reactor, providing four zones each of 15 cm. The heating tape on the exit section was used for the initial imposition of a temperature profile prior to each run. The purpose of the heaters was to heat up the reactor for the regeneration of the catalyst after each carbon monoxide oxidation run.

Analyses were performed on a Servomex gas chromatograph. The samples were taken automatically from the reactor effluent and from the inlet stream at suitable intervals. Temperature profile readings were taken automatically at intervals of 1 min on a Solartron Data Logger, with the output punched on paper-tape.

In order to determine the kinetic data for the oxidation of CO for subsequent use in modelling studies, a set of preliminary steady-state isothermal experimental runs were made in a smaller reactor.

These kinetic data were also used to compare the experimental transient temperature profiles with those predicted by our theoretical models.[11]

The rate expression for the oxidation can be reasonably well represented by a power law expression,

$$r = k_0 \exp\left(-E/RT\right) C_A C_B$$

The runs utilized a high air to CO ratio (approximately 1 % to 2 % CO) to minimize thermal effects, and keep the runs near isothermal.

The frequency factor and activation energy were found to be $3·165 \times 10^6$ m³/kmol s and $3·6 \times 10^4$ kJ/kmol, respectively. These results were used in the theoretical predictions shown in Figure 7.

Discussion of Results

Typical temperature profiles at various times are shown in Figures 2—4 for the oxidation of CO on copper oxide, for both constant and variable feed conditions. Only selected data points are illustrated to avoid confusion.

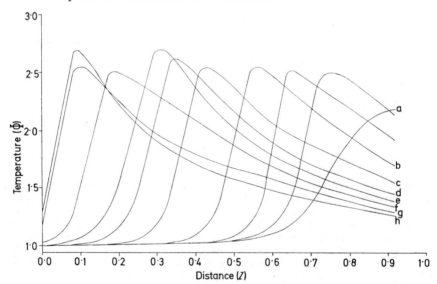

FIGURE 2   Temperature profiles with constant feed conditions. Time: (a) 0; (b) 2 h; (c) 3 h; (d) 4 h; (e) 5 h; (f) 6 h; (g) 7 h; (h) 8 h.
$C_{Ao} = 8·267 \times 10^{-3}$ kmol/m³; $u = 0·256$ m/s.

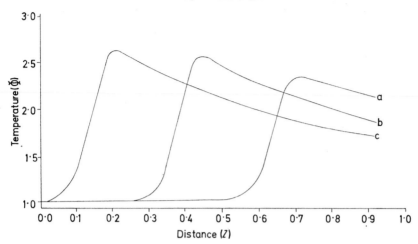

FIGURE 3 Temperature profiles with variations of concentration keeping velocity constant.
(a) Time = 15 min, C_{Ao} = 5·08 × 10^{-3} kmol/m³.
(b) Time = 50 min, C_{Ao} = 6·0 × 10^{-3} kmol/m³.
(c) Time = 80 min, C_{Ao} = 6·47 × 10^{-3} kmol/m³.
First increase of concentration after 15 min.
Second increase of concentration after 50 min.

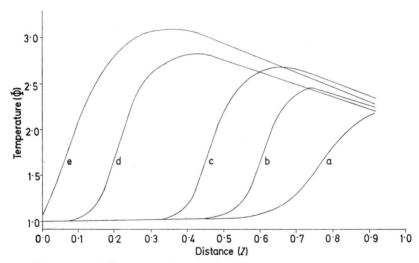

FIGURE 4 Temperature profiles with perturbation in velocity after 20, 40, and 80 min, respectively keeping concentration constant.
(a) Time = 0
(b) Time = 20 min, W = 0·79 × 10^{-4} m/s, u = 0·652 m/s.
(c) Time = 40 min, W = 0·625 × 10^{-4} m/s, u = 1·12 m/s.
(d) Time = 80 min, W = 0·3125 × 10^{-4} m/s, u = 1·305 m/s.
(e) Time = 120 min.
$$C_{Ao} = 5·08 \times 10^{-3} \text{ kmol/m}^3.$$

791

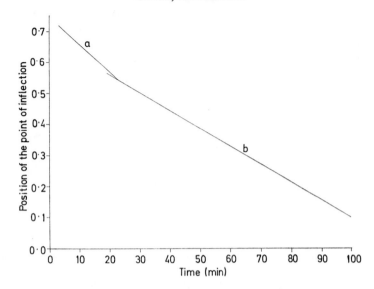

FIGURE 5   Position of the point of inflection against time.
Slope of a, $W = 0.879 \times 10^{-4}$ m/s, $u = 0.466$ m/s;
Slope of b, $W = 0.577 \times 10^{-4}$ m/s, $u = 0.512$ m/s.

Since the geometry of the profile varies very slowly with time, it is assumed that the velocity of the point of inflection characterizes the velocity of the profile as a whole.

In order to find the point of inflection, a third order polynomial fit was made for the region in question on each profile, and the inflection point found from the resulting constants.

The average velocity of propagation is measured directly from the slope of the plot of the position at the point of inflection against time as shown in Figure 5. It was found that this method of determining the velocity of propagation was much more accurate and reproducible than a simple differencing technique. The propagation velocities are indicated on the graphs of variable feed runs, shown in Figures 3 and 4.

The data were then tested against the semi-empirical correlations proposed by Vortmeyer and Jahnel.[5] In that work it was found that, over a wide range of conditions, the velocity of propagation, W could be correlated in the form

$$W = \alpha u^{0.77} C_A^{0.5} - \beta u$$

where u is the feed velocity (m/s), C_A is the concentration (kmol/m^3), β is the ratio of gas to solid thermal conductivity, and α is essentially an empirical constant. For the oxidation of CO over CuO, $\beta \approx 3.2 \times 10^{-4}$.

All previous work has been done at pseudo-steady-state operating conditions in which the feed conditions were kept constant for long periods of time. For purposes of control, however, it is important to know whether alterations in the feed velocity will quickly lead to a corresponding change in the velocity of propagation according to the above correlation. It is quite conceivable, that a rapid change in feed conditions could alter the behaviour of the system so that it obeyed the above correlation only after a long period of time, if at all. A prime example of such a perturbation is at start up where blow-out is very possible under conditions which would normally lead to propagation. This is discussed in further detail in the next section.

For other than start-up conditions, it was found that the system responds to changes in feed conditions within several thermal time constants of the perturbation. This is shown typically in Figure 5 where the break point in the curve occurs approximately 2 min after the disturbance. The new velocity of propagation then satisfies the correlation quoted above. A typical plot of velocity of propagation as a function of feed velocity is shown in Figure 6.

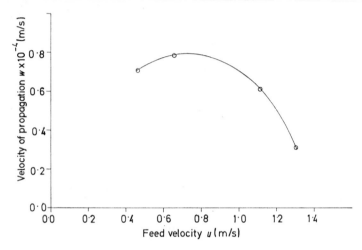

FIGURE 6   Velocity of propagation for selected values of feed velocity.
$C_{Ao} = 5.08 \times 10^{-3}$ kmol/m³.

The experimental results compared quite well with those predicted by earlier theoretical studies.[10,11] In those studies, the physical system is described by two coupled non-linear parabolic partial differential equations. These were solved by quasi-linearization and application of a Crank Nicholson finite difference algorithm.

A comparison of the theoretical model with the experimental results is shown in Figure 7. The discrepancy between the two may be attributed to the use of such a simple

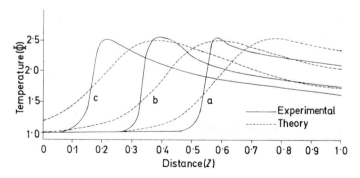

FIGURE 7   Time: (a) 20 min; (b) 50 min; (c) 80 min.
$C_{Ao} = 5.39 \times 10^{-3}$ kmol/m³;
$u = 0.59$ m/s.

model which neglects radiative energy transport and differences between catalyst and bulk gas temperatures. However the point of inflection of the profiles obtained both by the model and by the experiment coincide, and yield the same velocity of propagation.

CONCLUDING REMARKS

The experimental results indicated an effective reverse propagation of the reaction zone along the reactor bed. The effect upon various perturbations in the system parameters were examined. The propagation of quasi-steady-state profiles was investigated with different values of feed velocity (u), and feed concentration (C_A). Variation of the initial temperature profiles upon propagation velocity (W) has also been studied.

In summary, perturbation in the initial temperature profile has only a temporary effect on the propagation rate. The hotter zone with its higher temperature gradient initially has a higher velocity of propagation, but later reverts to the standard velocity, as might be expected.

High inlet velocities increase the rate at which fuel is fed to the reaction zone, raising the rate of heat output, the peak temperature, and velocity of propagation, but also raising the sensible heat load on the system. As the inlet velocity is increased further, the velocity of propagation starts to decrease; for even higher inlet velocity the reaction zone breaks down completely causing blow-out.

A point worthy of note is that during the initial period of operation, a low inlet velocity must be used, so that the zone can be well established. The increased thermal load caused by the proximity of the reactor boundary makes it more difficult to initiate propagation. Hence, if a high feed rate is used, blow-out will result. Once propagation is obtained, and the reaction zone has moved away from the reactor boundary higher feed rates can be used without difficulty.

At higher feed concentration the rate of energy formation is increased; however, the resistance to conduction within the reactor bed allows only a fraction of this extra heat to be dissipated by an increase in the velocity of propagation. Therefore, the remainder is stored to raise the peak temperature of the reaction zone. Figure 3 illustrates the rise of peak temperature and increase in the velocity of propagation as the inlet concentration is increased.

Except for the special circumstances associated with start-up there appears to be a unique value of the velocity of propagation for each set of feed conditions with only a short transient encountered upon a change in these controls.

LIST OF SYMBOLS

C_A	= Concentration of CO	$(kmol/m^3)$
C_{Ao}	= Concentration of CO in the reactor feed	$(kmol/m^3)$
C_B	= Concentration of oxygen	$(kmol/m^3)$
E	= Activation energy of the reaction	$(kJ/kmol)$
k_o	= Pre-exponential frequency factor	$(m^3/kmol\ s)$
l	= Distance along the reactor axis	(m)
L	= Total length of the reactor	(m)
r	= Rate of reaction	$(kmol/m^3\ s)$
T	= Temperature of the reactor	(K)
T_o	= Temperature of the feed	(K)
ϕ	= T/T_o = Dimensionless temperature	
u	= Feed velocity	(m/s)
W	= Velocity of propagation	(m/s)
Z	= l/L = Dimensionless distance	

REFERENCES

[1] E. Wicke and D. Vortmeyer, *Z. Elektrochem.*, 1959, **63**, 145.
[2] R. L. Reed, D. W. Reed, and J. H. Tracht, *Pet. Trans. AIME*, 1960, **219**, 99.

[3] G. Padberg and E. Wicke, *Chem. Eng. Sci.*, 1967, **22**, 1035.
[4] P. Fieguth and E. Wicke, *Chem.-Ing.-Tech.*, 1971, **43**, 604.
[5] D. Vortmeyer and W. Jahnel, *Chem. Eng. Sci.*, 1972, **27**, 1485.
[6] G. Eigenberger, *Chem. Eng. Sci.*, 1972, **27**, 1907.
[7] J. M. Berty, J. H. Bricker, S. W. Clark, R. D. Dean, and T. J. McGovern, *Proc. 5th European Symp. Chem. React. Engineering*, Amsterdam, 1972, B8–27.
[8] H. K. Rhee and N. R. Amundson, *Ind. Eng. Chem. Fundamentals*, 1974, **13**, 1.
[9] S. L. Liu and N. R. Amundson, *Ind. Eng. Chem. Fundamentals*, 1962, **1**, 200.
[10] R. G. Earp and L. S. Kershenbaum, *Chem. Eng. Sci.*, 1975, **30**, 35.
[11] R. G. Earp, Ph.D. Thesis, University of London, 1973.

DISCUSSION

S. P. S. Andrew (*ICI, Billingham*) said: The reactor is shown horizontal in the author's Figure 1. Were any measurements made of the velocity of propagation with the reactor vertical both for up-flow and down-flow?

L. Kershenbaum (*Imperial College, London*) replied: No measurements were made with a vertical reactor in this study, although such measurements had been made on previous occasions. Are you concerned that some effects may have been caused by natural convection? The reactor was operated at atmospheric pressure and I would not expect convection to contribute greatly under these conditions.

J. D. Potts (*Sun Oil Co., Philadelphia*) said: In our current research programme on ammoxidation, we always employ vertical fixed bed reactors. They have a centrally located axial thermocouple well in which is placed a $\frac{1}{32}''$ sliding thermocouple. Temperature readout is on a digital pyrometer. With this system a temperature profile can easily be established with as many points of reference as desired. The so-called 'hot spot' is readily located and its movement quickly followed. We have experienced a phenomenon similar to that described by the authors with respect to perturbations in the reactant velocities. From a practical point of view, it is not difficult to control a 'blow-out' if one uses diluted and/or stratified reaction beds which are widely described in the literature.

P. G. Menon (*Indian Petrochem. Corp., Baroda*) said: In 1961 Froment and Bischoff[1] gave an elegant mathematical treatment of temperature profiles and rate profiles for a fixed catalyst bed under the conditions of a rapidly-fouling exothermic catalytic reaction. They predicted that these profiles should exhibit maxima, the loci of which traverse through the catalyst bed as a function of process time.

Such moving temperature profiles and rate profiles have been measured by us[2,3] during the oxidation of hydrogen sulphide on active carbon catalysts, where the sulphur deposited rapidly covers the surface thereby deactivating the catalyst. We have also studied the dependence of the velocity of these profiles on initial reactor temperature and reactant concentrations. For instance, the velocity of the maximum in the profile, and hence the effective life of the catalyst bed (as determined from the break-through point of hydrogen sulphide), depends strongly on the initial temperature of the catalyst bed.[4]

The work reported in the present paper is in fairly good agreement with the Froment–Bischoff prediction and our own earlier experimental results.

[1] G. F. Froment and K. B. Bischoff, *Chem. Eng. Sci.*, 1961, **18**, 161.
[2] P. G. Menon and R. Sreeramamurthy, *J. Catalysis*, 1967, **8**, 95.
[3] P. G. Menon, R. Sreeramamurthy, and P. S. Murti, *Chem. Eng. Sci.*, 1972, **27**, 641
[4] R. Sreeramamurthy and P. G. Menon, *J. Catalysis*, 1975, **37**, 287.

A Deactivation Reactor for Catalyst Screening and Evaluation

JOHANNES V. JENSEN and ESMOND J. NEWSON

Inst. for Kemiteknik, Lab. for Energiteknik, Technical University of Denmark, 2800 Lyngby, Denmark.

ABSTRACT A deactivation reactor was designed and operated under essentially isothermal conditions for a reaction system typified by its highly exothermal nature, *i.e.* auto exhaust oxidation. The reactor facilitated the simultaneous deactivation of different catalysts in various sizes and shapes in the temperature range 300—600°C. Catalyst activity as a function of time was monitored by separate measurements in a recirculation reactor. Sulphur poisoning of the transition-metal oxide catalysts accounted for most of the observed deactivation but other mechanisms were also apparent. A reversible poisoning mechanism involving sulphate formation was suggested. An estimate of the active surface areas was made and compared to the total surface area.

INTRODUCTION

In catalyst screening and deactivation studies, several different requirements must be met simultaneously so that catalyst performance can be conveniently and fundamentally assessed. For convenience in catalyst formulation and preparation, only small quantities of catalyst should be required and it should be possible to test several catalysts simultaneously. An assessment of the fundamental kinetic parameters is of considerable value if it is known that the catalysts were all exposed to the same well-defined conditions of temperature, pressure, and concentration for the same length of time in the same reactor.

An approximation to some of these conditions has been described.[1] Differential thermal analysis was used as a microactivity test for measuring the low-temperature activity of auto exhaust catalysts. Individual tests of small catalyst quantities, both fresh and aged, were rapidly performed, but no attempt was made to quantify kinetic parameters. This method has recently been extended to include studies of reaction kinetics and catalyst decay.[2] A review of conventional laboratory reactors concludes that all reactor types considered have·some limitations particularly if catalyst decay is present.[3] An unconventional reactor type is the single pellet diffusion reactor[4] where well-defined conditions of temperature and concentration were established so that fundamental kinetic and diffusion parameters were evaluated. However, the adaptability of this technique to the simultaneous evaluation of several catalysts in various commercial forms is difficult to envisage. A multitube converter for simultaneous aging of catalyst samples followed by determination of fundamental kinetic parameters in a separate reactor, led to model discrimination for deactivation.[5] Since the deactivation took place in integral reactors, the kinetics of deactivation would be difficult to determine.

The purpose of this paper is to describe a deactivation reactor for simultaneous evaluation of several small catalyst samples in practically any shape or form. The reactor is used only for purposes of deactivation under well-defined conditions of temperature and concentration. Fundamental kinetic parameters on both fresh and spent catalysts are determined by measurements in a conventional recirculation reactor. Theoretical inferences on the mechanism for deactivation are suggested by independent measurements and references to the literature.

796

EXPERIMENTAL

Reactor procedures required observation of the deactivation process by intermittent kinetic measurements in the recirculation reactor. The experimental assembly is shown on Figure 1. The recirculation reactor for the kinetic data is of conventional design,

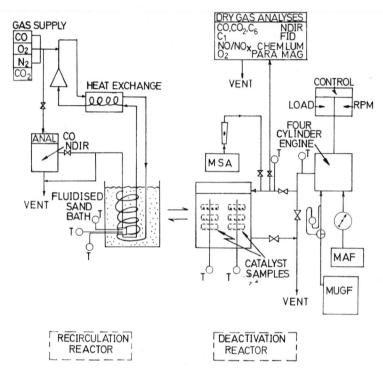

FIGURE 1   Experimental assembly for kinetic and deactivation measurements.

T	thermocouple.
NDIR	non-dispersive infrared.
FID	flame ionisation detector.
CHEM LUM	chemiluminescene.
PARA MAG	paramagnetic resonance.
MSA	measured secondary air.
MUGF	measured unleaded gasoline flow.
MAF	measured air flow.

with a fluidised sand bath providing essentially isothermal conditions. A gas mixture consisting of carbon monoxide, carbon dioxide, and air was added to the recirculation loop so that the kinetics of carbon monoxide oxidation were measured. Analytical data were obtained from an infrared analyser.

The deactivation reactor was used to study the decline in activity of non-noble metal, auto exhaust oxidation catalysts. A conventional four-cylinder engine using lead-free gasoline provided the necessary conditions for deactivation. The addition of secondary air ensured an overall oxidising atmosphere in the reactor. The engine was operated in a constant driving mode so that well-defined conditions of temperature and concentration were obtained. Analytical data for the various gas species in the exhaust were obtained by various methods. CO, CO_2, and C_6H_{12} were analysed by nondispersive infrared (NDIR), hydrocarbons as C_1 were analysed by flame ionisation (FID), NO/NO_x by chemiluminescence (CHEM LUM), oxygen by paramagnetic resonance (PARA MAG). Twelve samples of catalysts of various sizes and shapes were located in the baskets inside the deactivation reactor. Each basket was about 2 cm in height and 3 cm in diameter. Three different catalysts were used. Two of them were of the same type, copper oxides in alumina, prepared by coprecipitation techniques. The third was a commercial copper–chromium oxide in alumina.

RESULTS

Chemical and physical analyses and kinetic parameters of fresh catalysts are shown on Table 1.

Table 1

Chemical and physical analyses and kinetic parameters of fresh catalysts

	Catalyst 410	designation CC9
Chemical		
Copper, wt %	12·9	12·0
Chromium, wt %	—	4·7
Aluminium, wt %	40·0	36·9
Physical		
Surface area (BET), m^2/g	84	60
Pore volume (mercury), cm^3/g	0·435	0·234
Particle density, g/cm^3	1·26	1·95
Sieve size range, cm	0·05–0·1	0·05–0·1
Kinetic		
Pre-exponential factor, mol/g cat/s	$1·27 \times 10^3$	$1·43 \times 10^4$
Activation energy, cal/g mol	$23·9 \times 10^3$	$20·1 \times 10^3$

Preliminary work with the deactivation reactor showed that temperature profiles were non-uniform at about 600°C. Figure 2 shows temperature profiles parallel to the vertical axis in the three quadrants (K1—K3) of the deactivation reactor. Profiles are compared with different catalyst weights in the baskets, *i.e.* 2 g at 2 h and 3 h, and about 1 g at 10·5 h. In addition, the location of different particle sizes of the same catalyst were also changed.

Particle sizes ranged from about 0·05 to 0·4 cm diameter, but only the small size range 0·05—0·1 cm is considered in this paper.

Details of conditions in the deactivation reactor for runs A—E are shown on Table 2. Temperatures from 300—600°C were used and run lengths were from 20—117 h. Exhaust (dry) gas composition before the reactor and before secondary air addition, is also shown together with calculated oxygen content after secondary air addition. Oxygen is clearly in excess (4—5%) for the desired oxidation reactions (CO: 1%, C_1: 1000 pm).

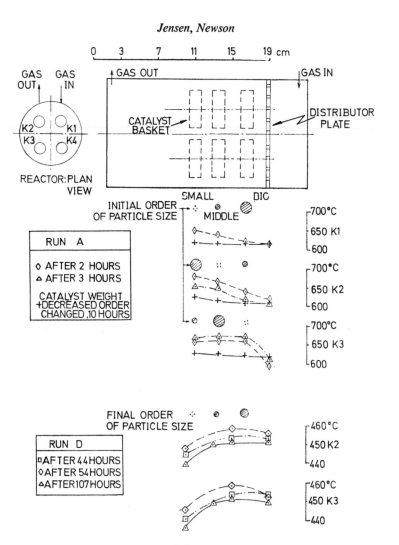

FIGURE 2   Deactivation reactor temperature profiles. K1 Quadrant 1.

Combination of the results from the recirculation reactor and the deactivation reactor gives activity data as Arrhenius plots with time and deactivation temperature as parameters. Data for different catalysts, CC9 and 410, are shown on Figure 3. The relative activity of fresh and aged catalysts for CO oxidation in the recirculation reactor is shown as a function of time in Figure 4. Activity comparisons were made at a temperature of 250°C so that comparison between intrinsic activities was essentially being made.

DISCUSSION

By passing all the exhaust gases from the four-cylinder engine through the deactivation reactor containing a relatively small quantity of catalyst, it was intended that isothermal, differential reactor operation should be approximated. As shown by the temperature profiles in Figure 2, isothermal operation was not observed initially in the

Table 2

Deactivation reactor experiments with unleaded gasoline

Run	Temperature (°C)	Run length (h)	CO %	CO$_2$ %	THC ppm	NO$_x$ ppm	O$_2$ %	Reactor O$_2$ %
			\multicolumn{5}{c}{Exhaust gas composition}			
A	600–650	20	2–3	—	600	—	0	0·5
B	450	40	0–1	13	1000	1100	1–3	5
C	300	51	0–1	13	1250	750	1	4
D	450	117	0–1	13	1250	1250	1–2	4
E	550	31	0–1	13	1000	1750	1–2	5

Gasoline and Crank-case oil properties

| | Gasoline | | Crank-case oil | |
	73–5	74–3	Fresh	Used
Density at 15°C g/cm^3	0·775	0·748		
Research octane number	94·5	92·5		
Lead content		50 ppb		
Sulphur content	0·1 ppm	0·1 ppm	1·5%	1·76%

temperature range 600—650°C. The deviation was particularly associated with the small particle sizes—profile K3, at 2 and 3 h. A clear particle size dependence was also observed when comparing profiles K3, K2 and K1, at 2 and 3 h. By estimating the interphase transfer rate of CO using the well known correlation[6]

$$\varepsilon j_D = \frac{0·357}{Re^{0·359}} \tag{1}$$

in the Reynolds number (Re) range $3 < Re < 2000$, adiabatic temperature increases of 4°C and 66°C were calculated for "big" and "small" particles respectively. This agrees reasonably well with the observed 10° and 75°C temperature rises in profile K3 and K1 at 2 h. By decreasing the catalyst weight to 1 g and changing the order of catalyst particle sizes from reactor inlet to big–middle–small, essentially isothermal profiles were measured at 10·5 h. This however is quite fortuitous as temperature increases one-half those observed are to be expected, if interphase transfer was controlling the reaction rate. As the profiles at 450°C show, heat losses from the reactor occur, noticeably beginning where the "small" particles are located. It is this combination of heat production and heat losses which give the impressive isothermal profiles at 600°C, Figure 2, 10·5 h.

Figures 3 and 4 ,show the results of combining the operation of the recirculation and deactivation reactors. Catalyst comparison can be made for both activity and stability. Figure 3 shows that CC9 is initially more active than 410. The rate constant for CO oxidation is derived from the relation

$$r_w = k \frac{C^n}{C_{oo}{}^n} \text{ mol g}^{-1} \text{ s}^{-1} \tag{2}$$

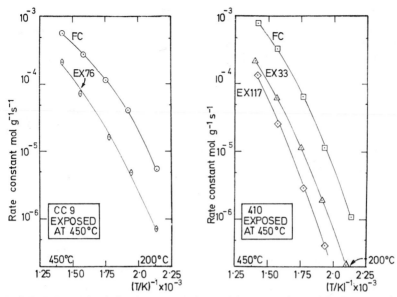

FIGURE 3   Recirculation reactor results: activity comparison of fresh and aged
catalyst for auto exhaust oxidation.
FC   fresh catalyst.
EX 76   exposed at 450°C, 76 h
EX 33   exposed at 450°C, 33 h
EX 117   exposed at 450°C, 117 h

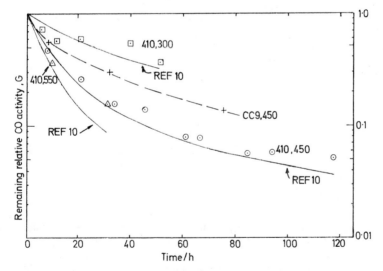

FIGURE 4   Deactivation reactor results: stability comparison of catalysts for auto
exhaust oxidation.
410, 300, 410 catalyst exposed at 300°C
410, 450, 410 catalyst exposed at 450°C
410, 550, 410 catalyst exposed at 550°C
CC9, 450, CC9 catalyst exposed at 450°C

where k has the usual Arrhenius temperature dependence. Figure 4 shows that CC9 is more stable than 410 at 450°C on subsequent exposure to auto exhaust under oxidising conditions. Remaining relative activity is defined as

$$G = \frac{a}{a_0} = \frac{a_0 - a_p}{a_0} \tag{3}$$

The shape of the deactivation curves is indicative of "fast" followed by "slow" deactivation which suggests that for the catalysts studied, CO oxidation takes place at "fast" and "slow" sites. The selectivity of SO_2 poisoning on active sites for CO oxidation by leached copper chromite catalyst has been shown to take place in two steps *i.e.*, "fast" and "slow" adsorption.

An impurity poisoning mechanism due to sulphur dioxide was initially suggested in this study by DTG and sulphur analyses of copper oxide catalysts exposed to auto exhaust in an integral reactor.[8] A weight loss from the catalyst was observed between 650 to 800°C, the former value corresponding to the decomposition temperature of copper sulphate. Transition-metal oxide catalysts for oxidation are reported to be particularly susceptible to sulphur oxides poisoning, leading to sulphate formation.[7,9] A sintering mechanism is not considered plausible to describe deactivation in Figure 4 due to the relatively low temperatures ($\leqslant 550$°C) of operation and the fact that BET measurements on fresh and aged catalyst showed little difference in surface areas.

Analyses of the unleaded gasoline for sulphur showed surprisingly low sulphur contents of 0·1 ppm, Table 2. In contrast, the crank-case oil sulphur content was rather high at 1·7—1·8 weight per cent. Assuming an oil consumption of 1 l per 5000 km and utilising the measured air/fuel ratios, which were essentially stoicheiometric, a value of 5 ppm SO_2 can be calculated for the exhaust gas. This corresponds to an equivalent 40 ppm sulphur in the unleaded gasoline.

In an earlier paper,[10] it was admitted that a simple irreversible, poisoning mechanism could not completely describe the temperature dependence of activity with time in the range 300—550°C, Figure 4. The data at 550°C compared to 450°C would particularly point to a reversible poisoning mechanism due to copper sulphate formation especially since significant quantities of sulphur have been found in the aged catalysts, Figure 5. The suggested mechanism is therefore

$$[CuO] + \begin{bmatrix} SO_2 + \tfrac{1}{2} O_2 \\ \Updownarrow \\ SO_3 \end{bmatrix} \rightleftharpoons CuSO_4 \tag{4}$$

Under oxidising conditions and at low temperatures, both the oxidation of SO_2 and the formation of $CuSO_4$ are thermodynamically favoured. Another, similar, equation for sulphate formation from alumina·can also be written.

The data in Figure 5 for CC9 are to be expected if sulphur poisoning is the cause of catalyst deactivation, since the CO oxidation activity was measured under diffusion-free conditions. However, the data for 410 suggests another temperature dependent poisoning mechanism in addition to sulphate formation. This could possibly be copper removal by volatile copper chloride formation. In run A at about 600°C, it was observed by microprobe analyses of spent catalysts that the copper and chloride contents were reduced significantly after 20 h operation. This mechanism of copper oxide catalyst deactivation by chloride volatilisation has been observed previously.[11] However, catalyst activity in run A remained constant with time but this surprising result could be due to reducing conditions that prevailed in this first run of the series.

The sulphur content of aged catalyst 410 after 117 h of operation at 450°C can be used to estimate the active surface area for CO oxidation. This catalyst contained

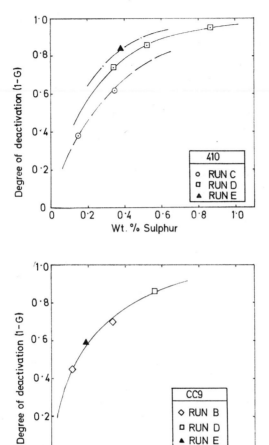

FIGURE 5 Effect of sulphur deposition on catalyst activities.

0·87% weight sulphur corresponding to a measured activity loss of 95%. Using a value of about 2Å for the radius of a sulphur dioxide molecule,[12] and assuming all the sulphur is adsorbed as a poison on the active surface for CO oxidation, then the following ratio can be calculated

$$\frac{\text{active surface}}{\text{total BET surface}} \approx \frac{22}{84} \approx 0.25 \qquad (5)$$

This value represents a maximum since sulphate formation can also take place with alumina. It should also be noted that area for a sulphur dioxide molecule on the surface was assumed to cover $12Å^2/SO_2$ compared to a value of $26Å^2/SO_2$ used in the literature.[13]

ACKNOWLEDGEMENTS. The work of the analytical department Haldor Topsøe A/S is specially acknowledged.

NOMENCLATURE

a	active surface of catalyst, $cm^2\ cm^{-3}$
a_o	active surface of fresh catalyst, $cm^2\ cm^{-3}$
a_p	poisoned surface, $cm^2\ cm^{-3}$
C	concentration of main reactant in bulk phase, $mol\ cm^{-3}$
C_{oo}	concentration selected for normalisation, $mol\ cm^{-3}$
d_p	particle diameter, cm
D	coefficient of molecular diffusion $cm^2\ s^{-1}$
G	remaining relative activity for CO oxidation, eqn (2)
G_M	molal velocity of fluid, $mol\ s^{-1}\ cm^{-2}$
j_D	mass transfer factor $= \dfrac{k_g P}{G_M} Sc^{2/3}$
k_g	mass transfer coefficient, $mol\ s^{-1}\ cm^{-2}\ atm^{-1}$
k	apparent rate constant, $mol\ g\ cat.^{-1}\ s^{-1}$
n	order of reaction
P	total pressure, atm.
r_w	reaction rate, $mol\ g\ cat.^{-1}\ s^{-1}$
Re	Reynolds number $= \dfrac{d_p G}{\mu}$
Sc	Schmidt number $= \dfrac{\mu}{\rho D}$
μ	viscosity, $g\ cm^{-1}\ s^{-1}$
ρ	density, $g\ cm^{-3}$
ε	bed voidage

REFERENCES

[1] E. W. Carlson, P. Chu, and F. G. Dwyer, 166th Natl. ACS Meeting, Chicago, Div. Pet. Chem. Preprints, 1973, **18**, No. 3, 467.
[2] B. Wedding and R. Farrauto, Ind. and Eng. Chem., Proc. Des. Dev., 1974, **13**, 45.
[3] V. W. Weekman, Jr., AIChEJ, 1974, **20**, No. 5, 833.
[4] L. L. Hegedus and E. E. Petersen, Ind. and Eng. Chem., Fundamentals, 1972, **11**, 549.
[5] F. G. Dwyer and C. R. Morgan, 164th Natl. ACS Meeting, Boston, Div. Pet. Chem. Preprints, 1972, **17**, No. 4, H9.
[6] C. N. Satterfield, Mass Transfer in Heterogeneous Catalysis, (MIT Press, Cambridge, Mass., 1970), vol. 2, p. 82.
[7] R. J. Farrauto B. Wedding, J. Catalysis, 1973, **33**, 249.
[8] K. A. Simonsen, Inst. Silikatindustri, Tech. Univ. Denmark, personal communication, Nov., 1973.
[9] N. A. Fishel, R. K. Lee, and F. C. Wilhelm, 165th ACS Natl. Meeting, Dallas, Div. IEC Preprints, April, 1973.
[10] J. Jensen, E. Newson, and J. Villadsen, paper submitted to 4th Internat. Symp. Chem. Reaction Eng., Heidelberg, April 1976.
[11] J. F. Roth, Ind. and Eng. Chem., Prod. Res. Dev., 1971, **10**, 381.
[12] Gmelins Handbuch, **9**, Teil B1, 206 S B.
[13] Y. Yao and J. T. Kummer, J. Catalysis, 1973, **28**, 124.

DISCUSSION

P. Davies (*ICI, Billingham*) asked: How did the authors succeed in ensuring the same gas flow (or the same space velocity) through each of the catalyst beds? Would not the first catalysts in the system remove the poisons such as phosphates, sulphur, *etc.*, from the exhaust gas and hence lead to wrong conclusions on the deactivation of the following samples?

E. J. Newson (*Tech. Univ. of Denmark, Lyngby*) replied: We have assumed that the gas distributor in the deactivation reactor (Figure 1), is sufficient to cause the same gas flow through each of the catalyst beds. We checked this by measuring the temperature in three of the four quadrants (one was blocked). Figure 2 shows axial temperature profiles K2 and K3 for Run D. Isothermal operation in respective quadrants was considered indicative of good gas distribution.

With respect to your second question, we would agree that the situation you envisage could arise, but in our system the VHSV was about $3 \times 10^6 \ h^{-1}$, so we consider that differential operation with sufficient poison for all catalysts was achieved.

S. P. S. Andrew (*ICI, Billingham*) said: I would agree that adsorbed sulphur dioxide is a reversible temporary poison but am surprised to note that the author also places sulphate ion in this category.

E. J. Newson (*Tech. Univ. of Denmark, Lyngby*) replied: Sulphate is placed in the reversible temporary poison category because of the effect of temperature on the decomposition of $CuSO_4$. At about 650 °C, copper sulphate decomposes to the oxide. Reference 9 in the paper particularly considers sulphate ion poisoning and its reversibility with temperature.

We also now have some sulphur analyses for fresh and spent catalyst 406 which was run with 410 and CC9. 406 Catalyst run for the *same* time at two *different* temperatures 450 °C (run B) and 600–650 °C (run A) accumulated about 2000 and 200 p.p.m. sulphur respectively. This shows that the higher temperature kept sulphur (presumably as sulphate ion) off the catalyst.

The Role of Lattice Oxygen in Selective and Complete Oxidation of Propylene over Binary Oxide Catalysts

GEORGE W. KEULKS* and L. DAVID KRENZKE

Department of Chemistry and Laboratory for Surface Studies, The University of Wisconsin–
Milwaukee, Milwaukee, Wisconsin 53201, U.S.A.

ABSTRACT   The use of isotopic oxygen in an investigation of the catalytic oxidation of propylene
has shown that lattice oxygen participates in the formation of both selective and complete
combustion products. There appeared to be no distinction between the oxide ion which is
incorporated into the acraldehyde and the oxide ion which is incorporated into the carbon
dioxide. By using the extent of lattice oxygen participation and catalyst reduction data to
calculate relative diffusion constants, it was concluded that for various binary oxide catalysts
the catalytic activity for both selective and complete combustion is directly related to the rate
of diffusion of the oxide ion through the lattice.

INTRODUCTION

The selective oxidation of propylene to acraldehyde over various oxide catalysts
has been studied extensively by a number of workers.[1] These studies have firmly established that the activation of the hydrocarbon produces an allylic intermediate. However,
the form of active oxygen responsible for partial and complete oxidation is still not
completely resolved.

Studies using isotopic oxygen have provided strong evidence for the participation of
lattice oxygen. Keulks[2] observed that during the oxidation of propylene with $^{18}O_2$
over a bismuth molybdate catalyst, only 2—2·5% of the oxygen atoms in
the acraldehyde and carbon dioxide produced were ^{18}O. A similar result was reported
by Wragg et al.[3] The lack of extensive incorporation of ^{18}O into the reaction products implies the participation of bulk oxide ions in the oxidation reaction. Moreover,
the diffusion of oxygen from the surface into the bulk and from the bulk back to the
surface must be rapid.

A recent study by Otsubo,[4] using isotopically labelled bismuth molybdate catalysts
as well as $^{18}O_2$, lends support to the concept of extensive lattice oxygen participation.
The authors prepared catalysts with the ^{18}O concentration in the molybdenum layers
and with the ^{18}O concentration in the bismuth layers. Their results indicate that propylene is oxidized by a Bi_2O_2 layer and bulk oxygen migration occurs from the MoO_2
layers to the Bi_2O_2 layers. The migration of oxide ions in the bulk produces surface
anion vacancies in the MoO_2 layers which serve as adsorption sites for gas-phase
oxidation.

The studies using isotopic oxygen mentioned above were conducted in recirculation
reactors. While such a reactor is convenient for such isotopic tracer studies, there is the
possibility that the data obtained do not represent the behaviour of the catalyst in its
steady state. In order to circumvent this difficulty, we decided to investigate the oxidation of propylene with $^{18}O_2$ over various binary oxide catalysts in a flow reactor under
steady-state conditions. We felt that this approach would provide greater insight into
the role of lattice oxygen in the selective and complete oxidation of propylene.

EXPERIMENTAL
Catalyst preparation

The γ-phase bismuth molybdate, Bi_2MoO_6, was prepared by the method of Batist *et al.*[5] *via* a slurry reaction between bismuth nitrate and molybdic acid. The product was filtered, dried, and calcined in air for 2 h at 500°C.

The α-phase bismuth molybdate, $Bi_2Mo_3O_{12}$, was prepared according to Keulks *et al.*[6] by the coprecipitation of ammonium paramolybdate and bismuth nitrate in an acidic solution (pH = 1·5). The precipitate was filtered after aging 20 h in the final solution then dried, calcined in air for 5 h at 200°C, and calcined again for 15 h at 460°C.

The ferric molybdate, $Fe_2(MoO_4)_3$, was prepared by the method of Kerr *et al.*[7] *via* the coprecipitation of ferric nitrate and ammonium paramolybdate. The precipitate was dried and calcined at 500°C for 2 h.

The α- or violet-phase of cobalt molybdate, $CoMoO_4$, was prepared by coprecipitation of cobalt nitrate and ammonium paramolybdate. The precipitate was washed repeatedly, dried, and calcined for 5 h at 500°C.

The β- or green-phase of cobalt molybdate, $CoMoO_4$, was obtained by grinding the α-phase. X-Ray diffraction analysis indicated that this procedure gave approximately a 1:1 mixture of the α and β phases.

The reagents used for catalyst preparation were Mallinckrodt AR grade ammonium paramolybdate, Fisher Certified bismuth nitrate, Mallinckrodt AR grade molybdic acid, Mallinckrodt AR grade ferric nitrate, and Mallinckrodt AR grade cobalt nitrate.

The primary purpose of this work was to determine the extent of lattice oxygen participation in selective and complete oxidation by using oxygen-18 as a tracer. In order to make a direct comparison between the catalysts with regard to the rate of oxygen-18 incorporation into the reaction products, we decided to have each catalyst charge contain the same amount of lattice oxygen (1,000 μmoles). Table 1 gives the amount of each catalyst used in the experiments and its BET surface area.

Table 1

Catalyst data

Catalyst	Surface Area (m²/g)	Amount of Catalyst Containing 1,000 μmoles of O^{2-}, g
Bi_2MoO_5	2·1	0·10
$Bi_2Mo_3O_{12}$	1·6	0·075
$Fe_2(MoO_4)_3$	1·2	0·05
α–$CoMoO_4$ (violet)	13·8	0·055
β–$CoMoO_4$ (green)	13·8	0·055

Apparatus

The experiments were run in a single-pass, integral flow reactor at atmospheric pressure. The reactor consisted of 8 mm Pyrex tubing as a preheat volume and catalyst chamber. To minimize the possibility of a homogeneous reaction in the postcatalytic zone, a section of 1 mm capillary tubing was used below the catalyst bed. The reactor was heated by a tubular furnace, controlled by a Thermoelectric R-100 temperature controller. The reaction temperature was monitored by a 1/16 inch sheathed Type K thermocouple inserted into the catalyst bed.

The feed gases, oxygen (Airco, 99·6%), propylene (Mathesen, C.P. grade), and helium diluent (Airco, 99·99%) were used without further purification. The individual gas flows were controlled by Whitey micrometering valves and stabilized against downstream pressure changes by a Moore differential flow controller (63BU-L). The oxygen-18 (91% ^{18}O, 0·5% ^{17}O), obtained from Yeda Research and Development Company, Rehovoth, Israel, was added to the feed stream at the reactor inlet using a syringe pump and a 10 cm³ syringe. During the $^{18}O_2$ addition, the $^{16}O_2$ flow was shut off at the reactor inlet using a syringe valve. The flow rate of $^{16}O_2$ and $^{18}O_2$ (*ca.* 1·8 cm³/min) were matched so that the oxygen flow remained constant throughout the entire experiment. This configuration allowed a rapid changeover from $^{16}O_2$ to $^{18}O_2$ in the feed gas and maintained the previously established steady-state conditions.

The product distribution was determined by an in-line gas chromatograph using a Porapak T column (1/4 inch × 6 ft) at 120°C and a thermal conductivity detector. The peak areas were obtained using an Infotronics digital integrator, and the actual quantity of each component was calculated using the peak area, the detector response factor, and the thermal response factors of Dietz.[8]

The oxygen-18 concentration in the gas-phase oxygen, carbon dioxide, and acraldehyde was determined by a Bendix time-of-flight mass spectrometer coupled to the reactor exit through a Granville-Phillips automatic pressure controller. During the oxygen-18 experiment, the mass range of 25 to 65 was scanned every 30 s. The percentage of oxygen-18 in O_2, CO_2, and C_3H_4O was obtained by measuring peak heights at the appropriate mass numbers. These data were then plotted as a function of time. At the end of the oxygen-18 addition, the flow of oxygen-16 was resumed and the effluent was again continuously analysed by mass spectrometry. These data were also plotted against time and showed the decay of the oxygen-18 content in CO_2 and C_3H_4O.

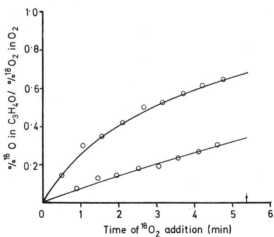

FIGURE 1 $^{18}O_2$ Incorporation into acraldehyde at 430°C. Bottom, Bi_2MoO_6; top, $Bi_2Mo_3O_{12}$; Arrow indicates end of $^{18}O_2$ addition.

RESULTS

The feed gas used for the activity tests and isotopic experiments contained 0·1 atm. of C_3H_6, 0·09 atm. of O_2, and 0·81 atm. of He. The total flow rate was 20 cm³ (STP)/min. This composition was chosen to give reasonable conversion levels and to match the

Table 2
Activity data

Catalyst	Temperature, °C	Conversion, %	Selectivity to Acraldehyde, %	Rate of Formation (μmoles/m² min)			E_a (kJ/mole)	
				C_3H_4O	CO_2	C_3H_6O	C_3H_4O	CO_2
Bi_2MoO_6 (0·10g)	400	20	78	61·4	29·9			
	415	31	85	112·0	40·3			77·0
	430	34	83	117·0	54·3			
*Bi_2MoO_6ᵃ (0·05g)	400	7·9	85	55·9	29·6			
	415	12·2	87	88·9	39·5		126	74·5
	430	18·9	90	148·0	52·6			
$Bi_2Mo_3O_{12}$ (0·075g)	400	8·0	70	46·2	19·8			
	415	10·7	85	72·7	26·4		114	70·3
	430	16·2	87	113·0	33·1			
$Fe_2(MoO_4)_3$ (0·05g)	400	1·2	trace	trace	16·5	trace		
	415	1·4	trace	trace	25·1	trace		79·5
	430	1·8	trace	trace	31·1	trace		
α-$CoMoO_4$ (violet) (0·055g)	400	2·1	trace	trace	4·8			
	415	2·5	trace	trace	7·3			111·0
	430	3·7	trace	trace	11·7			
β-$CoMoO_4$ (green) (0·055g)	400	2·1	trace	trace	4·1			
	415	3·1	trace	trace	6·2			110·0
	430	3·7	trace	trace	9·9			

ᵃ A half charge was used because the full charge was too active to yield reliable activity data.

809

$^{16}O_2$ flow with the $^{18}O_2$ flow from the syringe pump. Before any data were collected, the catalyst was lined out at 430°C for 4—5 h. During this time, several g.c. analyses were run to verify that steady-state conditions had been reached. The activity data were obtained at 400, 415, and 430°C, while the isotopic experiments were run at 430°C and all tests were run at steady-state flow conditions. The activity data are summarized in Table 2.

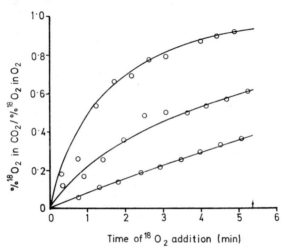

FIGURE 2 ^{18}O Incorporation into carbon dioxide at 430°C. Bottom, Bi_2MoO_6; middle, $Bi_2Mo_3O_{12}$; top, α–$CoMoO_4$; Arrow indicates end of $^{18}O_2$ addition.

^{18}O Experiments

Oxygen-18 experiments were run on three catalysts, Bi_2MoO_6, $Bi_2Mo_3O_{12}$, and α-$CoMoO_4$ (violet). The β-$CoMoO_4$ was not used because its activity was identical to that of the α-form, and $Fe_2(MoO_4)_3$ was eliminated because its activity was too low. Figs. 1 and 2 show the incorporation of ^{18}O into acraldehyde and carbon dioxide during the addition cycle. Using these data and the following equation, the amount of lattice oxygen which participated in acraldehyde and carbon dioxide formation can be calculated:

$$\%^{18}O \text{ in product} = (1 - e^{-FT/V}) \times 100,$$

where $\%^{18}O$ in product is read directly from Fig. 1 or 2;

$F =$ flow of oxygen into the catalyst, which equals the flow of oxygen out of catalyst as oxygenated products, in μg atoms/min;
$T =$ time in min; and
$V =$ amount of catalyst oxygen participating in μg atoms.

This equation was solved for V at various times (1, 2, 3, 4, and 5 min) and the values averaged. The results are summarized in Table 3.

Discussion

As expected, the bismuth molybdate catalysts were the most active and selective of the catalysts tested. The activation energies and product distributions are consistent with published data.[9-11] It should be noted, however, that the low activity of cobalt

Table 3

Extent of lattice oxygen participation

Catalyst containing 1,000⁻ moles of O^{2-}	Flow of O into catalyst, μg atom/min	μmoles of catalyst oxygen participating in the formation of[a]		Number of oxygen layers involved[b]	
		CO_2	C_3H_4O	CO_2	C_3H_4O
Bi_2MoO_6	79·4	890	1,000	254	286
$Bi_2Mo_3O_{12}$	37·9	182	150	91	75
α–$CoMoO_4$ (violet)	25·8	45	—	3·6	—

[a] The decay data was analysed by using a similar method and gave results in good agreement with the reported data.
[b] Assuming 1×10^{15} oxygen atoms/cm².

811

and iron molybdate is probably due to the higher GHSV used in this study compared with that used by other workers.[10,11] A trace of acetone and a trace of acraldehyde were observed with iron molybdate and cobalt molybdate, respectively.

In the experiments using isotopic oxygen, we did not observe heterophase exchange, homophase exchange, or heterophase equilibration of oxygen with the catalyst. The concentration of $^{18}O_2$ in the reactor effluent remained within $\pm$ 2% of the $^{18}O_2$ concentration in the feed gas. Therefore, the effect of these isotopic exchange reactions is negligible under our reaction conditions.

Moreover, the exchange of oxygen through CO_2 appeared to be negligible. If this exchange reaction were operative, we would have observed a change in the $^{18}O_2/^{16}O_2$ ratio in the effluent and the ratio of $[C^{16}O^{18}O]^2/[C^{16}O_2][C^{18}O_2]$ would have approached the value of 4. As noted above, the $^{18}O_2/^{16}O_2$ ratio remained constant and the $[C^{16}O^{18}O]^2/[C^{16}O_2][C^{18}O_2]$ ratio was always significantly larger than 4. While we did not investigate the isotopic exchange between $H_2^{18}O$ and the catalyst, Wragg *et al.*[12] have shown that oxygen-18 exchange through H_2O would not affect the interpretation of the experimental results obtained for propylene oxidation.

From Table 3, we can conclude that the source of oxygen for incorporation into the products for propylene oxidation over α- and γ-bismuth molybdate and α-cobalt molybdate is lattice oxygen. Furthermore, there appears to be no distinction between the lattice oxygen incorporated into CO_2 and that incorporated into acraldehyde. This suggests that selectivity is determined by a specific interaction between propylene and an oxygen species at a surface site and not by the crystallographic position of the oxide ion in the bulk.

The activity, on the other hand, appears to be directly related to the degree of participation of lattice oxygen (see Tables 2 and 3). This implies that there should be a correlation between catalytic activity and bulk oxygen mobility. To test this hypothesis, we calculated the relative diffusion constants of O^{2-} for α-$CoMoO_4$, $Bi_2Mo_3O_{12}$, and Bi_2MoO_4 using Fick's law, *i.e.*,

$$J = - D \left(\frac{dc}{dx} \right) dydz,$$

where J = the flux of oxide ions from the catalyst surface through an area $dydz$;
D = the diffusion constant; and

$\dfrac{dc}{dx}$ = the concentration gradient in the x-direction.

Values of J were obtained from Table 2 using the data at 430°C by means of the equation,

$J(\mu g$ atoms/min m^2) = -3(rate of CO_2 formation) $-$ 2(rate of C_3H_4O formation)

which takes into account the oxygen flux in the H_2O also formed.

Calculation of the concentration gradient requires a value for the degree of reduction, dc. This value cannot be determined from the steady-state activity data. Instead, we obtained an estimate of the values for dc by using the relative rates of reduction by H_2 of α-$CoMoO_4$, $Bi_2Mo_3O_{12}$, and Bi_2MoO_6. At 430°C under constant hydrogen pressure, we determined the relative rates of reduction to be 0·5:12·4:19 for α-$CoMoO_4$, $Bi_2Mo_3O_{12}$, and Bi_2MoO_6, respectively.

For a value of dx, the depth of reduction, we chose to use the number of oxygen layers which participate in the reaction. For the bismuth molybdate catalysts, the average of the number of layers participating in both CO_2 and acraldehyde formation was used. However, for Bi_2MoO_6, this average represents essentially all of the layers in the catalytic mass. Therefore, this value was halved in order to obtain the number of layers beneath any given surface.

In Fig. 3, we have plotted the flux, J, which represents the catalytic activity for the formation of the oxygen-containing products as a function of the relative diffusion

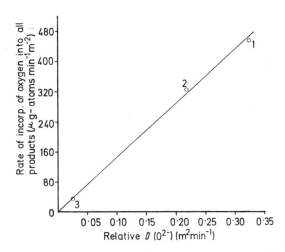

FIGURE 3 Total catalyst activity against relative diffusion constant of O^{2-} at 430°C $1 = Bi_2MoO_6$; $2 = Bi_2Mo_3O_{12}$; $3 = \alpha-CoMoO_4$.

constants for the oxide ion. The plot is linear and shows that the catalytic activity is directly proportional to the mobility of lattice oxygen. Consequently, we have concluded that catalytic activity for both partial and complete oxidation is controlled by the diffusion of oxide ions from the bulk to the surface.

The results obtained in this investigation are consistent with mechanistic suggestions which have appeared in the literature. It is relatively well established that the formation of acraldehyde involves the activation of the propylene *via* an allylic species.[1,13] This allylic intermediate then quickly reacts with catalyst oxygen to yield acraldehyde. However, the isotopic oxygen experiments have shown that catalyst oxygen is also incorporated into CO_2. One possible explanation is that acraldehyde or its surface precursor is strongly chemisorbed on the surface of catalysts of low selectivity. The strong chemisorption allows the surface species to undergo subsequent reactions involving lattice oxygen. Another possibility is that there is an activation of the oxide ion to form a highly reactive O^- anion radical *via* charge transfer to the cation. This process has been described in detail by Kazanskii.[14] As suggested by Haber,[13] this activation of oxygen leads to complete combustion. It is conceivable that both of these factors become important with catalysts of low selectivity for acraldehyde, and additional experiments are needed to examine their relative importance.

ACKNOWLEDGEMENTS. The authors thank Mr. H. Asada, who obtained the reduction data. Acknowledgement is also made to the Donors of the Petroleum Research Fund, administered by the American Society, for support of this research.

REFERENCES

[1] (a) L. Ya. Margolis, *Catalysis Rev.*, 1973, **8**, 241; (b) R. J. Sampson and D. Shooter, *Oxidation and Combustion Rev.* 1965, **1**, 223; (c) H. H. Voge and L. R. Adams, *Adv. Catalysis*, 1967, **17**, 151; (d) W. M. H. Sachtler, *Catalysis Rev.*, 1970, **4**, 27; (e) D. J. Hucknall, *Selective Oxidation of Hydrocarbons* (Academic Press, New York and London, 1974), Ch. 3. p. 23.
[2] G. W. Keulks, *J. Catalysis*, 1970, **19**, 232.
[3] R. D. Wragg, P. R. Ashmore, and J. A. Hockey, *J. Catalysis*, 1971, **22**, 49.
[4] T. Otsubo, H. Miura, Y. Morikawa, and T. Shirasaki, *J. Catalysis*, 1975, **36**, 240.
[5] Ph. A. Batist, J. F. H. Bouwens, and G. C. A. Schuit, *J. Catalysis*, 1972, **25**, 1.
[6] G. W. Keulks, J. L. Hall, C. Daniel, and K. Suzuki, *J. Catalysis*, 1974, **34**, 79.
[7] P. F. Kerr, A. W. Thomas, and A. M. Langer, *Amer. Mineral.*, 1963, **48**, 14.
[8] W. A. Dietz, *J. Gas Chromatog.*, 1967, **5**, 68.
[9] G. W. Keulks, M. P. Rosynek, and C. Daniel, *Ind. Eng. Chem. Prod. Res. Develop.*, 1971, **10**, 138.
[10] N. Pernicone, *J. Less-Common Metals*, 1974, **36**, 289.
[11] Y. Moro-oka, S. Tan, and A. Ozaki, *J. Catalysis*, 1968, **12**, 291.
[12] R. D. Wragg, P. G. Ashmore, and J. A. Hockey, *J. Catalysis*, 1973, **28**, 337.
[13] J. Haber, *Z. Chem.*, 1973, **13**, 241.
[14] V. B. Kazanskii, *Kinetika i Kataliz*, 1973, **14**, 95.

DISCUSSION

J. Haber (*Inst. Catalysis, Krakow*) said: I fully agree that a catalyst, to be active in selective oxidation, must show a high mobility of lattice oxygen. However, I have doubts concerning the significance of the correlation between activity and the value of the diffusion constant as calculated from reduction experiments. Suppose that some other factor determines the activity. A catalyst which is much more active will obviously give off much more oxygen than a catalyst which is less active. As it is, the lattice oxygen which participates in the selective oxidation, the amount of oxygen used, must always correlate with the conversion of the reactant irrespective of the factor determining the rate; however, this correlation is artificial.

G. W. Keulks (*Univ. of Wisconsin, Milwaukee*) replied: We agree that if high mobility of lattice oxygen is assumed to be important in selective oxidation, then a correlation with lattice oxygen participation is expected. Unfortunately, some workers do not agree with the assumption, and we wanted to show that it is valid under steady-state conditions. What was surprising was that we were able to correlate total activity (rate of oxygen incorporation into all products) with the diffusion constant for catalysts having widely differing selectivities. This implies that the oxygen source for carbon dioxide formation is also the bulk under our conditions.

S.-J. Teichner (*Univ. Claude Bernard, Lyon*) said: The problem of the participation of lattice oxygen or adsorbed oxygen in the partial or total oxidation of olefins is not, in my opinion, definitely set for the range of catalysts used in this reaction. We have, for instance, experimental evidence that, in the partial oxidation of isobutene into methylacraldehyde (60%) and acetone (40%) with a 100% selectivity for this partial oxidation,[1] oxygen which participates in this reaction is adsorbed oxygen and not lattice oxygen. The catalyst is NiO on alumina, prepared[2] by thermal decomposition of nickel hydroaluminate $Al_2O_3.2NiO.6H_2O$.

G. W. Keulks (*Univ. of Wisconsin, Milwaukee*) replied: I agree. What we are suggesting is a little more restrictive, namely, that for selective oxidation catalysts which produce acraldehyde by a stepwise mechanism *via* an allylic intermediate, lattice oxygen

[1] A. Muller, F. Juillet, and S. J. Teichner, *Bull. Soc. chim. France*, 1976, 1356, 1361.
[2] A. Merlin and S. J. Teichner, *Compt. rend.*, 1953, **236**, 1892.

is the source of the reactive oxygen. We do not suggest that this is the only mechanism for selective oxidation reactions. In fact, we have suggested a hydroperoxide mechanism to explain some of the results we obtained in our investigations of heterogeneous–homogeneous reactions.

A. Ozaki (*Tokyo Inst. of Tech.*) said: Work by Otsubo *et al.*, mentioned in the text, deals with reduction by hydrogen of isotopically labelled bismuth molybdates; their work on the reaction of propylene is soon to be published, and I would like to introduce their results. The concentration of labelled acraldehyde formed on differently labelled bismuth molybdates varies as shown in the Figure. Labelled acraldehyde

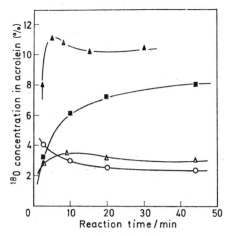

FIGURE Time course of ^{18}O concentration in acrolein produced by the oxidation of propylene. (—◦—) γ-Bi$_2$^{18}O$_3$. MoO$_3$(^{18}O 4·5%); (—▲—) γ-Bi$_2$O$_3$. Mo^{18}O$_3$ (11%); (—△—) γ-Bi$_2$O$_3$. Mo^{18}O$_3$ (4·5%); (—■—) oxidation of propylene with gaseous ^{18}O$_2$.

increases with time when the bismuth oxide layer is enriched in ^{18}O, demonstrating that the oxygen in acraldehyde comes from the bismuth oxide layer.

The result shown in Figure 3 of the text appears convincing, but the correlation might be superficial. Miura *et al.* oxidized carbon monoxide with $^{18}O_2$ over various molybdate catalysts.[3] (This work was published in Japanese, so you might not be aware of it.) Their results were consistent with yours in so far as the three catalysts used here are concerned. However, they attribute the high mobility of oxygen to the layered structure because lanthanum molybdate, which has the layered structure, exhibited a high mobility in spite of showing very low catalytic activity. There can be effects due to differences in the reactant used; however, your conclusion can be tested by the reaction on lanthanum molybdate.

G. W. Keulks (*Univ. of Wisconsin, Milwaukee*) replied: Thank you for outlining this work; we were aware of it, a translation having been made by one of my Japanese co-workers. It is difficult to compare oxidation of carbon monoxide with that of hydrocarbons. In the case of lanthanum molybdate, we would predict that propylene is not converted into a π-allylic species. Consequently, even though lattice oxygen mobility is great, the catalyst would be inactive because it would be unable to activate the hydrocarbon. The same explanation may hold true for carbon monoxide oxidation.

[3] H. Miura, Y. Morikawa, and T. Shirasaki, *Nippon Kagaku Kaishi*, 1975, 1875.

M. Schiavello (*Univ. of Rome*) said: According to your results lattice oxygen ions participate in both partial and total oxidation. From such results you infer that selectivity is determined by a specific interaction between propene and an oxygen species at a surface site and not by the crystallographic position of the oxide ion in the bulk. Since there are various types of lattice oxygen ion in the catalysts studied, how can you exclude the possibility that one of these is responsible for partial oxidation and that others are responsible for total oxidation? The structural arrangements in which the oxygen ions are located can be of importance.

G. W. Keulks (*Univ. of Wisconsin, Milwaukee*) replied: We agree that structural properties of catalysts and bulk metal–oxygen bond strengths are important in determining the characteristics of oxygen mobility. However, what we have inferred from our experiments is that the origin of the reactive oxygen for both selective and total oxidation is the same. The origin of this oxygen is the lattice.

These results also indicate that the carbon dioxide is being formed almost exclusively by a consecutive reaction of acraldehyde and not by a parallel route from propylene. The parallel route would most likely involve an adsorbed oxygen species. It should be noted, though, that these experiments were conducted at a relatively high temperature. It may be that the interaction of an adsorbed oxygen species is less important (or of little importance) at the high temperature.

This interpretation is consistent with the results we reported on heterogeneous–homogeneous reactions. We found such reactions occurring at low temperatures but not at high temperatures. We suggested that they were initiated by a surface peroxide intermediate formed by the interaction of a π-allylic species and adsorbed oxygen.

Thus, we would expect the parallel reaction pathway to carbon dioxide to become more important at lower temperatures. We plan to test this possibility. A recent paper by Sancier *et al.* suggests that adsorbed oxygen species are reactive at lower temperatures.

Finally, some of our more recent work suggests that the oxygen species involved in the formation of acraldehyde is different from that involved in the formation of water. It appears that these different oxygen species are associated with different cations.

T. G. Alkazov (*Pet. and Chem. Inst. Baku*) said: An important conclusion of this paper is that activities of the catalysts investigated depend on the mobility of lattice oxygen. The diffusion constants of oxygen have been obtained by using the relative rates of reduction by hydrogen. We have also studied the reducibility of different molybdates, but we used other reducing agents (*e.g* C_3H_6) as well as hydrogen. The relative reducibility of molybdates was also compared with their activities for olefin oxidation. The relative reducibilities of molybdates depend on the agents used for reduction, and the best correlation between the activities for propylene oxidation and the reducibilities of molybdates was obtained when propylene was used as the catalyst reductant.

G. W. Keulks (*Univ. of Wisconsin, Milwaukee*) replied: We have found that the reducibility of various molybdates, and in particular the extent of reduction obtained, depends on the agent used for reduction. However, we have not attempted to compare the reducibility of catalysts with their activity. Instead, we have used only the relative initial rates of reduction to estimate the concentration gradient. Our results indicate that the relative initial rates of reduction are essentially the same when propylene and hydrogen are used as reductant.

K. van der Wiele (*Akzo Zout Chemie, Hengelo*) said: The rapid exchange of gas-phase oxygen absorbed by the catalyst with all oxygen anions of the bulk is probably connected with the reduced state of the catalyst under steady-state conditions. I therefore expect that at higher oxygen:hydrocarbon ratios (lower degrees of reduction) internal oxygen transfer will be slower and more gas-phase oxygen will appear directly in the reaction products. Do you agree, and have you any experimental evidence concerning this point?

G. W. Keulks (*Univ. of Wisconsin, Milwaukee*) replied: We plan to examine the effect of the propylene:oxygen ratio on the incorporation of ^{18}O into the products. However, our mechanisms would lead us to expect that temperature would exert a greater effect than the propylene:oxygen ratio on the total amount of bulk oxygen which participates in the reaction.

M. F. Portela: (*Inst. Superior Tecnico, Lisbon*) said: Niwa and Murakami, using a technique of periodical and separate admission of but-1-ene and oxygen over bismuth molybdate, found that more than 95% of butadiene and but-2-enes were formed during but-1-ene admission, and that 80% of carbon dioxide was formed during oxygen admission.[4] I have injected pulses of pure but-1-ene over bismuth molybdate in a micro-reactor and noted formation of butadiene and isomers. After such an operation, the injection of pulses of oxygen produced only carbon dioxide, no traces of hydrocarbons were observed. Such results suggest the formation of carbon dioxide through the attack on strongly chemisorbed olefin by gaseous oxygen. How do you reconcile such facts with your results.

G. W. Keulks (*Univ. of Wisconsin, Milwaukee*) replied: We have observed similar results using propylene in a pulse reactor. However, the production of carbon dioxide was probably due to the removal of a carbonaceous residue from the catalyst. This was evident from the mass balance calculations because, for the early pulses, not all of the carbon was recovered.

Another problem is that selectivity determined using a pulse reactor is sometimes at variance with the value obtained under steady-state conditions. This is particularly true when there is product inhibition, which is frequently observed for butadiene. Consequently, under conditions which you describe, some carbon dioxide production may be a result of the further oxidation of the adsorbed product by gaseous oxygen.

In general, we have found that the results of oxidation reactions in pulse reactors must be treated with caution. This is one of the factors which prompted us to undertake this study under steady-state conditions.

W. J. Muizebelt (*Akzo, Arnhem*) said: We have made observations using ESCA on multicomponent catalysts used in a flow reactor under steady state conditions for isobutene oxidation in the presence of excess oxygen. It was found that molybdenum was reduced to a considerable extent at the surface (up to 50%). Bismuth was reduced to a lesser degree. I think these observations are generally in agreement with your experiments on pure bismuth molybdates; that is, lattice oxygen is utilized even in the presence of gas phase oxygen.

G. W. Keulks (*Univ. of Wisconsin, Milwaukee*) replied: I agree but I have one concern. It is known that bismuth molybdate catalysts provide a finite vapour pressure of oxygen. Under vacuum conditions, this leads to a reduction of the catalysts, and I wonder if some of the reduction which you observe by ESCA could be due to the vacuum environment in the ESCA unit.

M. V. Mathieu (*CNRS, Villeurbanne*) said: Dr. Coudurier and I wish to comment on the nature of the metal–oxygen bond. Some years ago, Trifiro and co-workers said that selective oxidation catalysts contain metal–oxygen double bonds characterized by a high frequency band near $1000\ cm^{-1}$. This is an oversimplification. We made a comprehensive study of the i.r. spectra of the oxides V_2O_5, WO_3, MoO_3, CuO and Bi_2O_3, taking into account their structure. We used the high frequency band to estimate the bond order of the shortest metal–oxygen bond. These bond orders are between 1·5 (V_2O_5) and 1·1 (CuO or Bi_2O_3). It is far from being a double bond, and CuO is known to be a good catalyst for propene oxidation to acraldehyde.

4 M. Niwa and Y. Murakami, *J. Catalysis*, 1972; **26**, 359; *ibid*. 1972, **27**, 26.

But this reaction goes by a redox process. So we studied the i.r. spectra of our oxides after reduction by propene. We found a good correlation between the selectivity:activity ratio for selective propene oxidation and the ratio of the value of the force constant for the shortest metal–oxygen bond before reduction to that observed after reduction. The smaller the relative change, the better the selectivity:activity ratio. A small change of the strength of the metal–oxygen bond must be favourable either for the initial reduction or for the final reoxidation of the catalyst surface.

Prof. Keulks thanked Dr. Mathieu for his comment.

G. K. Boreskov (*Inst. Catalysis, Novosibirsk*) said: Prof. Keulks has made interesting measurements but I do not quite understand his interpretation of them. In my opinion the rate of propylene oxidation cannot be controlled in the steady-state condition by diffusion of oxygen ions from the bulk to the surface. The most difficult step of this reaction is oxygen removal from the catalyst surface by an oxidized substance and the rate of this process is independent of oxygen diffusion in the bulk of a solid catalyst. I suppose that the only conclusion from Prof. Keulks' results on oxygen isotope distribution is that the mobility of bulk oxygen is high. However, this mobility is not connected directly with the mechanism of the catalytic reaction.

The correlation observed between the diffusion coefficients (D) and the reaction rates (r) may be explained by their dependence on oxygen binding energy (q).

From the linear dependence

$$E_D - \log D = A + \alpha q$$
$$E_r - \log r = B + \beta q$$

where
E_D = diffusion activation energy,
E_r = reaction activation energy,
q, A, B, α, β = constants

we always obtain $\log r = B - \beta/\alpha\, A + \beta/\alpha \log D$.

(Comment received too late for author to reply. Ed.)

The Nature of the Active Sites on Olefin Oxidation Catalysts

IKUYA MATSUURA

Department of Inorganic Chemistry and Catalysis, University of Technology,
Eindhoven, The Netherlands.

ABSTRACT The adsorption equilibria of but-1-ene and butadiene have been studied on the following series of catalysts: α-Fe_2O_3, Fe_3O_4, $Fe_4Bi_2O_9$, $FeSbO_4$, $FeAsO_4$, and $FePO_4$: and also Sb_2O_4/SnO_2.

On $FeSbO_4$, $FeAsO_4$, and Sb_2O_4/SnO_2 which are selective oxidation catalysts, there is a weak and reversible adsorption of but-1-ene. However, but-1-ene was found to be adsorbed strongly on α-Fe_2O_3, Fe_3O_4, and $Fe_4Bi_2O_9$ which are non-selective oxidation catalysts. On $FePO_4$, with no activity, the adsorption was very weak.

Similar types of adsorption sites for but-1-ene and butadiene on USb_3O_{10} were observed on the selective oxidation catalysts. From these results a surface reaction site model is suggested which refers to the (110) surface plane of rutile and is very similar to that given earlier for USb_3O_{10}. Furthermore, olefin adsorption characteristics are discussed in connection with Mössbauer parameters for Fe^{3+} ions in these catalysts.

INTRODUCTION

Earlier studies by the author on the adsorption of olefins, butadiene, acraldehyde, and ammonia on selective oxidation catalysts such as Bi_2MoO_6[1] and USb_3O_{10}[2] led to a suggestion that the active site consists of a combination of two types of adsorption sites. One of these, named the A-site is visualized as a reactive surface oxygen ion able to adsorb butadiene, acraldehyde, and ammonia by a strong activated adsorption. The other site, named the B-site, adsorbs olefins by a fast, weak and dissociative adsorption, one of the surface intermediates in the case of olefins being an allylic species. The reaction site combines the two, an olefin being adsorbed on a B-site, the allylic intermediate moving to an A-site for oxidation and the reaction product being released after a subsequent transfer to another B-site. Arguments were given to ascribe A-sites to oxygen anions connected to Bi^{3+} (or Sb^{5+}) cations, the B-sites to anion vacancies connected with Mo^{6+} (or U^{6+}) cations.

This concept has been tested on the following catalyst series: α-Fe_2O_3, Fe_3O_4, $Fe_4Bi_2O_9$, $FeSbO_4$, $FeAsO_4$, and $FePO_4$; and also Sb_2O_4/SnO_2. The problem of olefin adsorption was studied in connection with Mössbauer parameters for Fe^{3+} ions in these series.

EXPERIMENTAL

Preparation: In all cases the first step consisted of precipitating single hydroxides or oxide hydrates separately. To obtain the mixed oxides, freshly precipitated mixtures of hydroxides were slurried under constant stirring in boiling water for 8 h (mixing ratio Fe or Sn/other metal = 0·8). Sometimes colour changes were observed, brown to white for P/Fe and As/Fe, brown to green for Sb/Fe after processing. The solids were filtered, washed, and calcined: 8 h/750°C for Fe, Sb/Fe, Bi/Fe, and Sn/Sb; 4 h/500°C for As/Fe and P/Fe. Fe_3O_4 was prepared from α-Fe_2O_3 by hydrogen reduction in the presence of water vapour for 8 h at 400°C.

X-Ray patterns: Diffraction patterns obtained were: single Fe oxide → α-Fe_2O_3, reduces Fe oxide → Fe_3O_4, Fe/P → $FePO_4$, Fe/As → $FeAsO_4$, Fe/Sb → $FeSbO_4$, Fe/Bi → $Fe_4Bi_2O_9$. No separate Sb oxide could be observed for Sn/Sb.

Adsorption: The adsorption of but-1-ene and butadiene was measured over a range of pressures from 10^{-2} to 10^3 N/m^2 and a range of temperatures from 0°C to 250°C as described earlier (Matsuura *et al.*[1]).

Catalytic activity: Activity and selectivity of the various samples were measured for the oxidation of but-1-ene in a flow system at atmospheric pressure. Flow rate = 0·1 m^3/kg catalyst, ratio of reactants: C_4H_8:O_2:He = 1:1·5:4. Temp. 300—500°C.

Mössbauer spectra: Mössbauer spectra of Fe were measured by Dr. A. M. van der Kraan (IRI-Delft) at 25°C on a constant acceleration spectrometer with a ^{57}Co–Rh source. Isomer shifts are given relative to the NBS reference material Na_2Fe $(CN)_5$,NO, $2H_2O$ at 25°C.

RESULTS

Catalysts α-Fe_2O_3, Fe_3O_4 and $Fe_4Bi_2O_9$ are very active for the oxidation of but-1-ene, but they produce only CO_2, CO, and H_2O. $FeSbO_4$ and $FeAsO_4$ are less active but very selective for the butene to butadiene oxidation. It is noteworthy that the selectivity depends on the Fe/Sb or Fe/As ratio, an excess of Sb(As) being necessary for good selectivity. $FePO_4$ only catalyses double bond isomerization. The selective oxidation is first order in but-1-ene above 450°C but below 400°C is zero order in butene and oxygen with slight inhibition by butadiene, similar to that observed for USb_3O_{10}[3] and Bi_2MoO_6[4]. Apparent activation energies are given in Table 1.

Table 1

Adsorption and oxidation data

| Catalyst | Adsorbate | Adsorption Data | | | Oxidation of C_4H_8 | |
		Ads. No ($\times 10^{17}$/m^2)	Q_{ads} (kJ/mol.)	$-\Delta S_{ads}$ (J/mol. K)	E_H	E_L (kJ/mol.)
$FeSbO_4$	C_4H_8	9·0	49·6	131·0	50·4	60·9
	C_4H_6(W)	8·5	42·8	121·0		
	(S)	4·0	97·9	128·1		
$FeAsO_4$	C_4H_8	8·4	42·4	100·8	65·1	80·9
	C_4H_6(W)	8·4	42·0	93·2		
	(S)	3·8	101·6	119·7		
Sb_2O_4/SnO_2	C_4H_8	8·0	59·6	144·8	27·3	69·3
	C_4H_6(W)	7·3	58·0	147·4		
	(S)	4·0	92·8	146·2		
USb_3O_{10}[2]	C_4H_8	1·5	66·4	134·4	33·6	76·5
	C_4H_6(W)	1·5	62·2	135·7		
	(S)	0·7	97·9	143·6		

Adsorption characteristics for the various catalysts were as follows: but-1-ene and butadiene adsorb strongly and irreversibly on α-Fe_2O_3. On Fe_3O_4 and $Fe_4Bi_2O_9$ the adsorption is reversible and dissociative with relatively high heats of adsorption. Distinction between strong and weak butadiene adsorption was impossible. However, on $FeSbO_4$, $FeAsO_4$, and Sb_2O_4/SnO_2 the butadiene adsorption provided clear evidence for a strong and a weak type of adsorption, while on these catalysts but-1-ene adsorption is weak. On $FePO_4$ only a very weak adsorption of but-1-ene could be detected (Tables 1 and 2). The number of sites for weak and strong butadiene adsorption are in the

Table 2

Mössbauer parameters, adsorption and oxidation data

Catalyst	Mössbauer Parameter		Adsorption of C_4H_8			Oxidation of C_4H_8
	IS (mm/s)	QS (mm/s)	Ads. No. ($\times 10^{17}/m^2$)	Q_{ads} (kJ/mol.)	$-\Delta S_{ads}$ (J/mol. K)	
α-Fe$_2$O$_3$	0·47	0·21	13·2	Strong, irreversible		
Fe$_3$O$_4$	0·67 octa / 0·26 tetra	— / —	5·1	86·1	215·8	non selective
Fe$_4$Bi$_2$O$_9$	0·64 octa / 0·44 tetra	0·44 / 0·87	5·1	71·8	220·5	
FeSbO$_4$	0·62	0·72	9·0	49·6	161·7	selective
FeAsO$_4$	0·60	0·67	8·4	42·4	108·0	
FePO$_4$	0·45	0·68	1·9	25·2	63·0	only isomerization

ratio 2:1. The number of sites for weak but-1-ene and butadiene adsorption are usually very similar, as observed earlier for USb_3O_{10} and Bi_2MoO_6.

As to the connection between catalytic properties and adsorptive properties, it is a striking observation that on passing from high heats of but-1-ene adsorption to lower values the catalysis changes from highly active but nonselective *via* less active but selective to zero activity. It is only in the range of selective catalysts that the phenomenon of the two types of adsorption for butadiene becomes apparent.

Mössbauer spectra of the ferric oxide sample are those known from α-Fe_2O_3 *i.e.* an isomer shift of 0·47 and the formation of a hyperfine line structure. Fe_3O_4 as produced here shows two magnetically split 6-line patterns. One hyperfine line structure is due to tetrahedral Fe^{3+} ions, the other arises from octahedral Fe^{2+} and Fe^{3+} ions in $[Fe^{3+} (Fe^{2+}Fe^{3+}O_4)]$.

The spectra of $Fe_4Bi_2O_9$ consist of two quadrupolar doublets in good agreement with Bokov *et al.*[5] Those of $FeSbO_4$ and $FeAsO_4$ consist of only one quadrupolar doublet and are closely similar. The spectrum of $FePO_4$ consists of a superposition of at least two different quadrupolar doublets, indicating two inequivalent iron states.

Mössbauer parameters are given in Table 2.

DISCUSSION

In Fig. 1 a plot is given of the entropies of adsorption against the heats of adsorption of but-1-ene on various oxidation catalysts: a single linear relation suffices to cover all

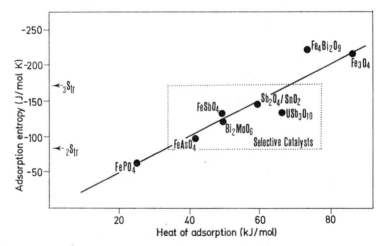

FIGURE 1   Relation of adsorption entropy to heat of adsorption for but-1-ene.

data available. The relation is also interesting because it leads to a classification of catalysts. Catalysts with high heats of adsorption and completely immobile surface species (see value of $_3S_{tr}$ calculated for a complete loss of all translational degrees of freedom) are very active but nonselective. Low heats of adsorption coupled with extensive surface mobility are not active.

Selective catalysts are systems that bind but-1-ene moderately strongly, allowing also some mobility to the surface species.

Table 2 shows that the Mössbauer parameters for Fe differ considerably from sample to sample. All samples active in oxidation and showing reversible adsorption contain Fe^{3+} in an octahedral environment. It is seen from the Table and Fig. 2 that for these systems *viz.* Fe_3O_4, $Fe_4Bi_2O_9$, $FeSbO_4$, and $FeAsO_4$ the Q_s and I_s values are interrelated in an approximately linear fashion, Q_s increasing and I_s decreasing in this direction indicating an increasing distortion of the oxygen octahedron. Plotting heats of adsorption against I_s shows a linear decrease with decreasing chemical shift (Fig. 2). This suggests that B-sites are connected with bulk octahedral Fe^{3+}. If B-sites are taken to be surface Fe^{3+} cations in tetragonal pyramidal configuration with an oxygen vacancy, an allylic species might be presumed to occupy the vacancy. Stronger distortion in the

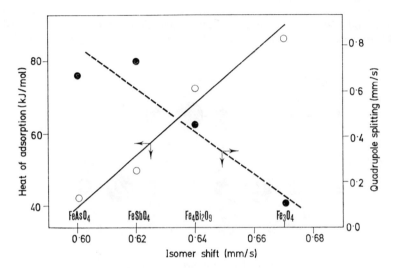

FIGURE 2 Correlation between heats of adsorption of but-1-ene and Mössbauer parameters for catalysts with octahedral Fe^{3+}.

bulk would then accompany weaker bonding. At the surface A-site adsorption of butadiene must occur on another type of site since it occurs next to another type of butadiene adsorption that is closely similar to the but-1-ene adsorption.

The author suggested earlier,[6,2] that an A-site is an oxygen atom bonded to a Bi or Sb cation on the surface. However, pure Sb_2O_4 does not adsorb butadiene strongly. So the site must acquire its properties by being adjacent to one or more B-sites. It is noteworthy that the ratio of numbers of B- to A-sites is 2, just as found earlier for other systems. A model for the reactive site cluster for selective catalysts would then be:

$$\begin{array}{ccccccc} & & O & & & & \\ O Fe^{3+} & O & Sb^{5+} & O & Fe^{3+} & O & \qquad I \\ O & & O & & O & & \end{array}$$

A non-selective site in this representation is then:

$$\begin{array}{ccccccc} & & O & & & & \\ O & Fe^{3+} & O & Fe^{3+} & O & Fe^{3+} & O \qquad II \\ & O & & O & & O & \end{array}$$

823

It should be observed in this connection that the actual site concentration on selective catalysts is low (Tables 1 and 2). This is what should be expected, since a high concentration of (I) should lead to the occurrence of (II). Presumably site (I) is separated from its neighbour by patches of inactive sites (III).

$$\bigcirc\ Sb^{3+}\ \bigcirc\ \overset{\displaystyle\bigcirc}{Sb^{5+}}\ \bigcirc\ Sb^{3+}\ \bigcirc \qquad \text{III}$$

The site model could be elaborated in a two dimensional model from the knowledge that $FeSbO_4$, $FeAsO_4$ and Sb_2O_4 have the rutile structure. Following Jones and Hockey[7]

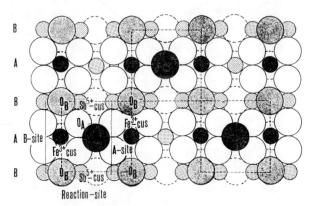

FIGURE 3   The (110) surface plane of $FeSbO_4$.
(a) Active site position.

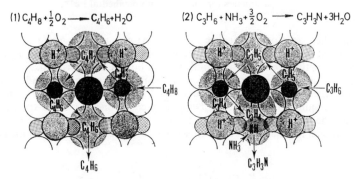

(b) Scheme of suggested surface reaction.

(1) $C_4H_8 + \frac{1}{2}O_2 = C_4H_6 + H_2O$
(2) $C_3H_6 + NH_3 + \frac{3}{2}O_2 = C_3H_3N + 3H_2O$

and Parfitt *et al.*[8] we choose the (110) plane as the surface plane and thus construct Fig. 3. All Fe cations are in one row with some Fe replaced by Sb, all other Sb cations in the next row. An active and selective site would then be as shown in Fig. 3a.

The reaction mechanism suggested is similar to the given earlier. C_4H_8 is adsorbed dissociatively on one Fe^{3+} site, the allyl moves to a vacancy next to the O_A on Sb^{5+}

where it donates two electrons and finally moves to the second B-site with loss of a second H-atom. The two H-atoms form H_2O with O_A. Mobility of the adsorbed species is therefore necessary. The formation of acrylonitrile is again similar to the model suggested earlier and given in the second part of Fig. 3b.

ACKNOWLEDGEMENT. The author thanks Dr. A. M. van der Kraan (Inter Universitair Reactor Instituut, Delft) for the Mössbauer spectra and for his kind assistance in their interpretation.

REFERENCES

[1] I. Matsuura and G. C. A. Schuit, *J. Catalysis*, 1971, **20**, 19.
[2] I. Matsuura, *J. Catalysis*, 1974, **35**, 452.
[3] Th. G. J. Simons, P. N. Houtman, and G. C. A. Schuit, *J. Catalysis*, 1971, **23**, 1.
[4] Ph. A. Batist, H. J. Prette, and G. C. A. Schuit, *J. Catalysis*, 1971, **15**, 267.
[5] V. A. Bokov, G. V. Norikov, V. A. Trukhtanov, and S. I. Yushchuk, *Sov. Phys. Solid State*, 1970, **11**, 2324.
[6] I. Matsuura and G. C. A. Schuit, *J. Catalysis*, 1972, **25**, 314.
[7] P. Jones and J. A. Hockey, *Trans. Faraday Soc.*, 1971, **67**, 2679.
[8] P. Jackson and G. D. Parfitt, *Trans. Faraday Soc.*, 1971, **67**, 2469.

DISCUSSION

J. Haber (*Inst. Catalysis, Krakow*) said: You identify the B-sites, at which butene molecules adsorb, as iron ions. As the function of iron in oxysalts such as antimonates, arsenates, or molybdates corresponds to the function of bismuth in the analogous oxysalts of bismuth, an obvious conclusion would be that in bismuth molybdate catalysts the B-sites should be identified with bismuth ions. Such a conclusion agrees with the mechanism of olefins oxidation which we advanced some years ago, and which has since been confirmed by many experiments.

When discussing the relation between adsorption and catalysis we must always remember that the adsorption sites operating at low temperature may be responsible for a quite different type of adsorption from those involved in catalytic reaction at high temperatures. We have measured the adsorption of propylene on a number of oxides and we have found that the number of adsorption sites in the temperature range 20–150 °C is of the same order of magnitude as that quoted in your paper for butene, although the investigated samples differed considerably in their activity. Also the values of the heat of adsorption were comparable.

It seems to me therefore that in the temperature range 20–200 °C we measure a different type of adsorption than that relevant to oxidation. Thus, the presence of such adsorption might be a necessary condition for the catalyst to be active, but it is certainly not a sufficient condition, and it may not correlate directly with activity.

I. Matsuura (*Tech. Univ. of Eindhoven*) replied: We have observed that bifunctional oxides such as bismuth molybdate, uranium antimonate and iron antimonate possess two different types of site of which one is an olefin adsorption site (the B-site) and the other is an oxidation site (the A-site). The heat of adsorption of but-1-ene is related to the quadrupole splitting and the isomer shift of Fe^{3+} ion (present work). Fe^{3+}-cations are olefin adsorption sites, and therefore the B-site may be formed by transition metals. However, we consider that the passage of the allylic intermediate, from the B-site to the A-site (Bi in Bi_2MoO_6 or Sb in USb_3O_{10} or $FeSbO_4$) is important for oxidation (see Figure 3b).

L. J. R. Vandamme (*Catholic Univ. of Louvain*) asked: Would you comment on the likely differences in behaviour between various oxides, and do you expect to observe a shift in selectivity over the range of oxides that you used? In particular, do you expect that only the phosphate could catalyse butene oxidation?

You show that weak adsorption leads only to isomerization. I wonder whether this is true for each butene, in view of our observation of the selective adsorption of *cis*-but-2-ene on zeolites.[1]

I. Matsuura (*Tech. Univ. of Eindhoven*) replied: Butene adsorption on selective oxidation catalysts leads to the formation of allylic intermediates. If allylic intermediates are not formed oxidation may not proceed. The acidic properties of iron phosphate are such that isomerization will occur *via* carbonium ion intermediates.

F. S. Stone (*Univ. of Bath*) said: My comment relates to the evaluation of the apparently precise Q_{ads} and ΔS_{ads} values in Table 1 and their significance for the correlations advanced later in the paper. I am concerned about the extent to which there is thermodynamic reversibility in the chemisorbed layers described here, a requirement which must obtain if meaningful values of Q_{ads} and ΔS_{ads} are to be extracted from adsorption isotherms. What is your criterion of reversibility? Do you regard the heats which you have evaluated as being independent of coverage?

Secondly, I note that you refer to the temperature range 0—250 °C for your adsorption measurements. I would have thought that isomerization of but-1-ene at the high temperature end of this range might be affecting your evaluation of Q_{ads} and ΔS_{ads}.

I. Matsuura (*Tech. Univ. of Eindhoven*) replied: The adsorption of butene and propene over the selective catalysts gave reasonable reversible isotherms. The surface of the catalysts might be energetically heterogeneous. However, if the heat of adsorption is low, as we observed ($Q < 80$ kJ mol^{-1}), the heterogeneity might be not so extensive. Therefore we have not observed a dependence of the heat on coverage.

The adsorption of but-1-ene has been studied at temperatures lower than 100 °C. The isomerization of but-1-ene over bismuth–molybdate is known to be very slow under these conditions (Tamaru *et al.*) hence the effects of isomerization on the isotherms could be neglected.

[1] P. A. Jacobs, L. J. Declerck, L. J. R. Vandamme, and J. B. Uytterhoeven, *JCS Faraday I*, 1975, **71**, 1545.

Electron Spectroscopic Studies of the Transformations of Hydrocarbon Molecules at Surfaces of Oxidation Catalysts

Jerzy Haber,* Wiesław Marczewski, Jerzy Stoch, and Leonard Ungier

Research Laboratories of Catalysis and Surface Chemistry,
Polish Academy of Sciences, 30–060 Kraków, Poland

ABSTRACT   The consecutive transformations aldehyde → acid → CO_2 + H_2O on selective oxidation catalysts $CoMoO_4$, MoO_3, and MoO_2 have been studied by the XPS technique at 77—773 K. On heating acraldehyde, the transformation of vinylic carbon atoms into paraffinic was observed. Simultaneously, the carbonyl group was transformed into carboxyl. This indicates that acraldehyde is bonded to the surface by means of the formation of a carboxyl group structure. On desorption, these species undergo decarboxylation, leaving the hydrocarbon part at the surface in form of a polymer. Studies of propionic acid showed that reduction of the surface favours decarboxylation of the acid molecule, which does not occur on oxidised $CoMoO_4$.

INTRODUCTION

Some years ago a multicentre model of the surface of selective oxidation catalysts was advanced,[1-3] identifying in molybdates the transition metal cations as active centres for generation of allyl radicals, which is the first step in the selective oxidation of olefins. It has been postulated that Mo–O polyhedra act as the second type of centres, capable of inserting oxygen into organic molecules by nucleophilic attack on allyl species to form an unsaturated aldehyde. The mechanism of the insertion of the second oxygen to form an unsaturated acid and the centres responsible for this step remain unidentified. It seemed thus of interest to exploit the potentiality of X-ray photoelectron spectroscopy for obtaining information on the properties of surface layers and to follow by this technique the course of the transformations aldehyde → acid → CO_2 + H_2O at the surface of selective oxidation catalysts $CoMoO_4$, MoO_3, and MoO_2. Acraldehyde was chosen as the reactant, and comparison with the behaviour of carboxylic groups was made by studying propionic acid instead of acrylic acid in order to avoid polymerisation.

EXPERIMENTAL

A Vacuum Generators ESCA-3 apparatus has been used with X-ray excitation from Al (1486·6 eV). Specimens of MoO_3, MoO_2, and $CoMoO_4$, each in the form of finely ground powder, were deposited from acetone suspension on to the sample holder. The Au $4f_{7/2}$ peak was taken as the reference peak, assuming its binding energy (B.E.) value to be 84·0 eV. The estimated accuracy of the determination of B.E. values was ±0·1 eV. The construction of the probe enabled the temperature of the sample to be varied from 77 K to 773 K. Before each adsorption run the sample of the catalyst was outgassed at 393 K for 1 h at 10^{-8} Torr.

The experimental procedure was as follows. After outgassing, the sample was cooled to 77 K and exposed at this temperature to 0·1 Torr of the given reactant in the preparation chamber until the Mo $3d$ and O $1s$ peaks of the catalyst disappeared, indicating total coverage of the surface with an adsorbed layer sufficiently thick to give a spectrum, which could be considered as characteristic of the adsorbate. The sample was then

827

heated in the preparation chamber consecutively to higher and higher temperatures, the sample being every time outgassed at the given temperature and the spectrum recorded.

RESULTS

1. MoO₃ *as catalyst*

Figures 1 and 2 show results of a series of experiments, in which the reactivities of acraldehyde and propionic acid at the surface of MoO_3 were studied, respectively. The

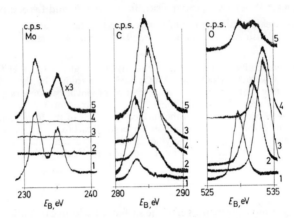

FIGURE 1 Spectra of Mo 3d, C 1s, and O 1s electrons in the course of transformations of acraldehyde on MoO_3. 1, initial sample of MoO_3; 2, after condensation of acraldehyde at 77 K; 3,4,5, after desorption at 123 K, 200 K, and 298 K, respectively.

B.E. values of Mo 3d electrons in the initial sample of MoO_3 are 231·9 and 235 eV, and that of O 1s electrons is 529·7 eV, in agreement with our earlier studies[4] and the data quoted in the literature.[5,6] They show excellent reproducibility.

Changes of peak positions and intensities observed in the course of adsorption at liquid-nitrogen temperature and subsequent heating are summarised in Table 1. The fact that the positions of the Mo 3d doublet and the lattice O 1s peak in samples covered with chemisorbed species are identical with those in the initial samples indicates that the observed shifts are due to chemical transformations and not to charging of the surface. The charge effects are more difficult to estimate in the case of thick adsorbed layers, when peaks due to the underlying solid are not visible. In such cases, however, information on the effect of charging may be obtained from comparison of the peaks due to one adsorbate on different catalysts. Identical positions of the C 1s and O 1s peaks of propionic acid condensed on MoO_3, as well as on $CoMoO_4$, rule out the influence of charging. The interpretation of the spectra of acraldehyde is more complex, but here also the important chemical shifts may easily be recognised.

2. MoO₂ *as catalyst*

a. Transformations of CO₂

In order to investigate the influence of the reduction of the MoO_3 surface on transformations of chemisorbed carboxyl groups, a series of experiments was carried out, in which CO_2 or propionic acid were adsorbed on MoO_2. The results of the experiments in which CO_2 together with a small amount of H_2O was contacted with the surface are

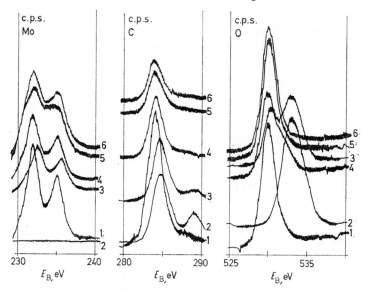

FIGURE 2 Spectra of Mo $3d$, C $1s$, and O $1s$ electrons in the course of transformation of propionic acid on MoO_3. 1, initial sample; 2, after condensation of propionic acid at 77 K; 3, 4, after desorption at 223 K, and 298 K respectively; 5, after outgassing at 573 K for 1 h; 6, after exposure to 0.1 Torr of O_2 at 573 K for 3 h.

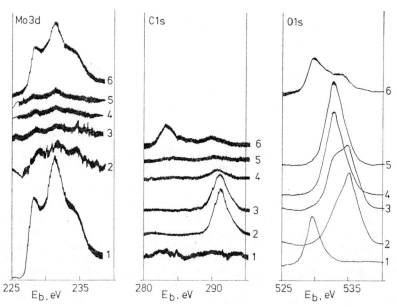

FIGURE 3 Spectra of Mo $3d$, C $1s$, and O $1s$ electrons of CO_2 and H_2O adsorbed on MoO_2. 1, initial sample; 2, after adsorption of CO_2 at 77 K; 3,4,5, after consecutive desorptions on heating; 6, after desorption at 298 K.

Table 1

Transformations of acraldehyde and propionic acid at the surface of MoO_3

T, K	Mo 3d		C 1s		O 1s	
			Initial sample			
77	231·9	235·0	283·2		529·9	
			Adsorption of acraldehyde			
			–C=C–	>C=O	>C=O	
77	—	—	283·1	286	532	
123	—	—	—	285·3		533·4
				↓ i		↓ d
200	—	—		285·1		533·3
				↓ n		
298	231·9	235·0		284·5	529·9	532·4
			Adsorption of propionic acid			
			–C–C–	–COO		–COO
77	—	—	284·9	289·0		533·0
			↓ n	↓ d		↓ d
123	—	—	284·6	289·0		532·9
			↓ n	↓ d		↓ d
223	232·6	235·6	284·6	289·0	530·4	532·8
			↓ n	↓ d	↓ i	↓ d
298	231·9	235·0	284·0	—	529·9	shoulder
			↓ d		↓ i	
573	232·0	235·1	284·0		530·0	—
	230·8	233·8				
543 in O_2	232·0	235·1	284·0		530·0	
			↓ d		↓ i	

i, intensity increases; *d*, intensity decreases; *n*, no change.

shown in Fig. 3. The position of the Mo 3d doublet at 228·2 and 231·4 eV is character-istic for MoO_2.[4] The presence of the shoulder in the higher B.E. range corresponding to the Mo $3d_{3/2}$ peak of Mo^{6+} indicates that the surface also contains a certain amount of Mo^{6+} ions.

In Fig. 3, curve 2 represents the spectrum of the condensed adsorbate. We can thus conclude that the peaks at 291·2 and 534·9 eV correspond to C 1s and O 1s levels in condensed CO_2. After removal of the condensed CO_2, the carbon peaks disappeared completely, whereas an O 1s peak of high intensity at 532·5 eV became visible (spectra 4 and 5). Its position is identical with that of the O 1s peak in ice. Further heating resulted in gradual removal of condensed water, which eventually disappeared, and the spectrum of the catalyst as well as that of chemisorbed species became visible (spectrum 6). It is composed of two C 1s peaks at 283·2 and 290·2 eV, and three O 1s peaks at 529·4, 531·3 and 533·5 eV. Since the positions of the C 1s peak at 290·2 eV and the O 1s peak at 533·5 eV are identical with those we observed on recording the spectrum of dehydrated sodium carbonate, we assume that they may be interpreted as belonging to surface carbonate groups. The C 1s peak at 283·2 eV is due to the carbonaceous deposit formed at the surface. The O $1s$ peak at 529·5 eV represents the lattice oxygen (*cf.* spectrum 1 of the initial sample) and that at 531·3 eV should be identified as belonging to the surface hydroxyl groups, in agreement with data quoted by Robert *et al.*[7]

b. Transformations of propionic acid

Results of the experiments with propionic acid were qualitatively similar to those obtained with MoO_3; there were, however, important quantitative differences. The changes in the intensity of the different peaks are shown in Fig. 4. We would draw

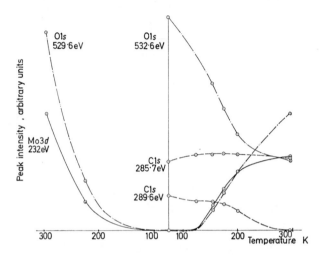

FIGURE 4   Changes of peak intensities as a function of the temperature of adsorption and desorption of propionic acid on MoO_2. a, adsorption on cooling in propionic acid vapour; b, desorption on heating *in vacuo*.

particular attention to the behaviour of two C $1s$ peaks: that at 285·8 eV belonging to the ethyl carbon, and the one at 289·8 eV related to the carboxyl carbon. The intensity of the former remains practically unchanged on heating, whereas that of the latter decreases practically to zero. A parallel decrease in the intensity of the oxygen peak at 532·6 eV is observed. Simultaneously the intensity of the Mo $3d$ doublet and of the lattice oxygen at 529·9 eV increases, indicating that the coverage of the surface by the adsorbed layer decreases.

3. $CoMoO_4$ *as catalyst*

The behaviour of acraldehyde and propionic acid at the surface of $CoMoO_4$ is summarised in Table 2. The positions of Mo $3d$, O $1s$, and carbon residue peaks in $CoMoO_4$ are shifted in comparison with these in MoO_3 by 3·4—3·5 eV to higher B.E. values.

DISCUSSION

Let us first discuss the behaviour of propionic acid on MoO_3. It may be seen from Fig. 2 that after adsorption at liquid-nitrogen temperature the surface is at first covered by molecules of propionic acid, two kinds of carbon atoms in the ratio 2:1 being visible in the spectrum. The effective charge on the carbon atom in the carboxylic group is more positive than the charge on atoms of the ethyl group, and the B.E. value of C $1s$ electrons is higher. Because of the hydrogen bonding between the carboxylic groups,

Table 2

Transformations of acraldehyde and propionic acid at the surface of CoMoO$_4$

T, K	Binding Energy, eV			
	Mo 3d		C 1s	O 1s
			Initial sample	
77	235·4 238·4		287·6	533·2
			Adsorption of acrolein	
		–C=C–	>C=O	>C=O
77	— —	285·1	288	532·8
		↓ d		↓ n
123	— —	285·2	—	532·3
		↘ i		↘ d
220	234·5 237·5	285·9	—	533·2
		↘ n	lattice	↘ d
298	234·5 237·5	286·8	532·6	534·3
		↓ n	↓ i	↓ d
593 15m	234·5 237·5	286·6	532·6	534·3
		↓ n	↓ i	
593 105m	234·5 237·5	286·8	532·6	shoulder
523		$\mid n$	$\mid n$	
in O$_2$	234·6 237·6	286·8	532·6	—
			Initial sample	
77	235·3 238·4		287·6	533·2
			Adsorption of propionic acid	
77	— —	285·0	289·1	532·8
298	235·3 238·4	—	288	— 533·2

i, intensity increases; d, intensity decreases; n, no change.

both oxygens in such a group are equivalent and only one peak due to oxygen appears. On heating, the C 1s and O 1s peaks of the carboxylic group decrease in intensity and finally disappear completely, whereas that of the ethyl-group carbon changes to a much smaller degree. This may be explained by assuming that heating results in two parallel processes: desorption of the acid molecules and their decarboxylation

$$CH_3\text{–}CH_2\text{–}CO_2H \rightarrow CO_2 + 2H \text{ (ads.)} + CH_3\text{–}CH_2\text{–(ads.)}$$

leaving at the surface the chemisorbed ethyl groups. They can be taken off from the surface in the form of oxidised species on outgassing at 573 K, removing lattice oxygen and leaving reduced molybdenum ions, as indicated by the appearance of the Mo 3d doublet at 230·8 and 233·8 eV, which may be assigned to Mo^{5+} ions. On exposing the sample to oxygen, the surface is totally reoxidised.

The decarboxylation of acid molecules is much more pronounced on reduced surfaces, as indicated by the experiments with MoO$_2$. In this case, heating of the adsorbed layer results in decarboxylation without any changes in intensity of the ethyl-group carbon peak at 285·7 eV (Fig. 4). It may thus be concluded that there is practically no desorption of the acid molecules, all of them undergoing decarboxylation and leaving at the surface the chemisorbed ethyl groups. These are not removed by outgassing, which indicates that they are much more strongly bonded to the surface than in the case of MoO_3 and probably undergo polymerisation forming a carbonaceous layer. This is consistent with the fact that on MoO_2 the C $1s$ B.E. values of chemisorbed acid molecules are by 0·6 eV higher than on MoO_3, suggesting the interaction of the hydrocarbon chain with the surface. Some information about the mode in which the carboxylic groups are bonded to the surface may be obtained from the observation that the ratio of the intensities of lattice oxygen to Mo $3d$ peaks reappearing on decarboxylation of the acid is much smaller than in initial sample or after total decarboxylation. This could be taken as a hint that lattice oxygen ions are involved in bonding or—what is equivalent—that the carboxylic group is bonded through its oxygens, which become incorporated into the lattice sites of the surface. The results discussed above thus lead to the conclusion that reduction of the surface favours the decarboxylation of the acid molecule.

After adsorption of acraldehyde on MoO_3 at liquid-nitrogen temperature (Fig. 1) the surface is covered with acraldehyde molecules, the spectrum being composed of the C $1s$ peak of the vinyl carbons as well as the C $1s$ and O $1s$ peaks of carbonyl group. On heating, two processes may be clearly distinguished. At first the two carbon peaks disappear and in their stead only one peak of high intensity appears at a B.E. value 285·3 eV, similar to that of C $1s$ electrons of paraffinic chain carbons. This may be explained by assuming that polymerisation of acraldehyde took place and the vinyl carbons disappeared as a result of the formation of paraffinic chains. A second process must apparently be responsible for the shift of the B.E. value of oxygen to 533·4 eV. This value is near to that of oxygen in a carboxyl group (*cf.* Figs. 1 and 2) and may be taken as a hint that the aldehyde group of acraldehyde has reacted with a lattice surface oxygen ion. In such a case we would also expect the appearance of a C $1s$ peak of carbon atom linked to two oxygen atoms in the carboxyl group and characterised by B.E. values of about 289·0 eV found for propionic acid. This peak is not visible in the spectrum, which may be explained by taking into account the fact that the intensity of the carbon peak due to a carboxyl group is much smaller than that of its oxygen peak, and may be masked by the broad peak of polymerised chains. On further heating, the intensity of the C $1s$ peaks of polymerised chains remained practically unchanged, whereas the intensity of the carboxyl oxygen peak rapidly decreased, indicating that carboxyl groups were removed from the surface as a result of decarboxylation. This result may be understood in view of the fact that the transformation of aldehyde into carboxyl group entails the reduction of the surface, the situation attained being thus similar to that described for the surface of MoO_2.

The behaviour of acraldehyde on a $CoMoO_4$ catalyst is practically identical to that in the case of MoO_3. On heating, the O $1s$ peak of acraldehyde shifts at first to a value of 534·3 eV. This value is higher than that observed in propionic acid; however, as in the case of acraldehyde on $CoMoO_4$ the O $1s$ peak was also shifted to a higher value, the appearance of this peak may be taken as an indication that the aldehyde group has reacted with surface oxygen to form a carboxyl group. On further heating, the intensity of this O $1s$ peak decreases, and it disappears on outgassing at 600 K, whereas the intensity of the C $1s$ peak assigned to the hydrocarbon portion of the molecule remains practically unchanged. This indicates that, as in the case of the MoO_3 catalyst, extensive decarboxylation takes place with the formation of a carbonaceous deposit. On the other hand, experiments with propionic acid showed that it is desorbed from the surface of $CoMoO_4$ without decomposition. These results may be rationalised when it is taken into account that acraldehyde is oxidised to acid, thus reducing the surface.

This conclusion is consistent with the results of studies on adsorption and reduction of $CoMoO_4$ with acraldehyde. It was found by Kozłowska and Słoczyński[8] that at 713 K no adsorption of acraldehyde on $CoMoO_4$ is observed in the presence of oxygen in the gas phase, whereas in the absence of oxygen considerable amounts of acraldehyde are adsorbed.

Kozłowska and Słoczyński also measured the rate of reduction of $CoMoO_4$ by acraldehyde and noticed that this process is accompanied by the formation of a carbonaceous deposit. They found that the ratio of the rate of reduction to the rate of "coke" formation was constant, irrespective of the temperature and degree of reduction. This may be explained if we assume that abstraction of lattice oxygen by acraldehyde and formation of the deposit strongly bonded to the surface occur in one and the same elementary reaction.

We may thus formulate a hypothesis that on reduced surfaces of molybdate catalysts acraldehyde is strongly bonded due to the formation of a carboxyl group structure with both oxygen atoms incorporated into the surface layer of the catalyst. On desorption these species undergo decarboxylation, leaving at the surface the hydrocarbon portion of the molecule, which may easily polymerise with formation of a carbonaceous deposit. Conversely, at the oxidised surface, acraldehyde is probably bonded through its aldehyde carbon to one lattice oxygen ion, this species being easily desorbed from the surface, as indicated by the weaker adsorption of propionic acid at oxidised surfaces.

REFERENCES

1 J. Haber and B. Grzybowska, *J. Catalysis*, 1973, **28**, 489.
2 J. Haber, *Z. Chem.*, 1973, **13**, 241.
3 K. German, B. Grzybowska, and J. Haber, *Bull. Acad. polon. Sci., Ser. Sci. chim.*, 1973, **21**, 319.
4 J. Haber, W. Marczewski, J. Stoch, and L. Ungier, *Ber. Bunsenges. phys. Chem.*, 1975, **79**, 970.
5 A. Cimino and B. A. de Angelis, *J. Catalysis*, 1975, **36**, 11.
6 E. L. Aptekar, M. G. Chudinov, A. M. Alekseev, and O. V. Krylov, *Reaction Kinetics and Catalysis Letters*, 1974, **1**, 493.
7 T. Robert, M. Bartel, and G. Offergeld, *Surface Sci.*, 1972, **33**, 123.
8 J. Haber, A. Kozlowska, and J. Sloczynski, *Bull. Acad. polon. Sci., Ser. Sci. chim.*, in press.

DISCUSSION

P. Canesson (*Catholic Univ. of Louvain*) said: I think it is rather speculative to discuss the failure of ethyl groups to desorb following the adsorption of propionic acid. In particular, Figure 2 shows that the initial C $1s$ peak from carbon contamination is very important and is situated at practically the same value of binding energy as the peak which you assign to ethyl groups. What evidence do you have that the C $1s$ line obtained after outgassing the MoO_3 surface at 573 K is due to ethyl groups and not to carbon contamination originating, for example, from pump oil? Have you any other evidence concerning the nature of these ethyl groups?

It is difficult to understand why the O $1s$ line shift to 533·4 eV, for MoO_3, on adsorption of acraldehyde, can be attributed to the formation of carboxyl groups. If such a situation is created a C $1s$ peak should appear at 289 eV with an intensity equal to nearly one-quarter that of the O $1s$ peak (because the ratio of relative line intensities[1] for C $1s$ and O $1s$ is roughly $1:2$). Moreover, the C $1s$ line intensity is less perturbed than that for O $1s$ because of the presence of a superficial layer on the solid.

R. W. Joyner (*Univ. of Bradford*) said: For propionic acid on MoO_3 the C $1s$ binding energy for —C—C— decreases by 0·9 eV, to 284·0 eV, while other binding energies remain constant. How does Prof. Haber interpret this interesting chemical

1 V. I. Nefedov, N. P. Sergushin, I. M. Band, and M. B. Trzhaskovskaya, *J. Electr. Spectroscopy*, 1973, **2**, 383.

shift? In the case of acraldehyde on $CoMoO_4$ (Table 2) the C $1s$ and O $1s$ values both shift by 1.6 eV between 123 K and 298 K. Does Prof. Haber agree that this shift is probably due to changes in surface charging?

F. Bozon-Verduraz (*Univ. of Paris VI*) said: My first question concerns the arguments you use to rule out any influence of the charging effect on your XPS spectra. Would you explain why the presence of chemisorbed species on the surface can prevent the occurrence of any charge effect?

Secondly, as Dr. Canesson comments, the level of carbon contamination is especially high (see Figure 2) and this may lead to dubious conclusions regarding the assignment of lines in the 284–285 eV range. In addition, the carbon contamination peaks may be shifted or broadened by the charge effect, which may render the interpretation difficult.

Y. Kubokawa (*Univ. of Osaka*) said: We have used the i.r. technique to investigate the transformation of adsorbed acraldehyde to carboxylate ions at the surface of a $SnO-MoO_3$ catalyst—a catalyst essentially the same as that used by Prof. Haber. However, carboxylate ion formation occurs at much higher temperatures (about 423 K). I would expect, therefore, that the conditions in the ESCA apparatus may affect your conclusions.

J. Haber (*Inst. Catalysis, Krakow*), in reply to the four previous speakers, said: Two observations indicate that there is no charging effect interference. First, the binding energy values of carbons and oxygens in adsorbate molecules are, for 77 K, independent of the solid studied and are identical to those observed after condensation of the adsorbate directly onto the sample holder. Secondly, the binding energy values of Mo $3d$ and lattice O $1s$ electrons reappearing in the course of adsorption experiments are identical with those measured for outgassed samples before the experiments.

The peak for residual carbon could be easily subtracted from the carbon peak due to the deposit formed by decomposition of propionic acid. This is so because the rate of increase of intensity of the residual carbon peak was determined in a blank experiment and the peak was found to be one order of magnitude smaller than that due to adsorption. In order to show it the curve 1 in Figure 2b had to be registered at much higher gain.

The hypothesis that the carboxylate group is formed at the surface as the result of the transformation of acraldehyde is based on the observation that the position of the oxygen peak shifts to 533.4 eV (*i.e.* a value near to that observed on adsorption of propionic acid), whereas the position of the lattice oxygen peak at 529.9 eV remains practically unchanged indicating that no charging effects interfere and that the observed changes are true chemical shifts. The absence of a carboxylate carbon peak is explained in the paper.

The small shift of C $1s$ binding energy for —C—C— on MoO_3 on heating (Figure 1) may be explained by condensation of the radicals into larger polymer macromolecules, this process affecting the bonds with the surface and increasing the negative charge on carbon atoms. As regards the shift of C $1s$ in —C=C— and O $1s$ in —C=O on heating between 123 K and 298 K being equal, it is rather a coincidence and not charging, because simultaneously the positions of Mo $3d$ peaks remain exactly at the value characteristic for pentavalent ions. The interpretation of these shifts is given in terms of two simultaneous processes: condensation of vinyl groups and formation of surface carboxylate groups.

The possibility of the influence of X-radiation in the spectrometer on the observed surface phenomena must always be taken into account. To check it we carried out experiments in which the sample with the adsorbate was kept in the spectrometer for a longer time and many spectra were registered consecutively. However, no changes of the spectra were observed after the first change due to adsorption. Thus we believe that the irradiation does not affect our results.

^{57}Fe and ^{121}Sb Mössbauer Study of Mixed Iron and Antimony Oxide Catalysts for Ammoxidation of Propene

H. Kriegsmann, G. Öhlmann, J. Scheve,* and F.-J. Ulrich

Zentralinstitut für physikalische Chemie der AdW der DDR, DDR–1199 Berlin-Adlershof, German Democratic Republic

ABSTRACT Only by the use of Mössbauer spectrometry is it possible to observe the formation of compounds between Fe_2O_3 and Sb_2O_5 or Sb_2O_3. $Fe_2Sb_2O_7$ or $FeSbO_4$ are believed to be the catalytically active part of the catalyst. Correlations of the quadrupole splitting of ^{57}Fe with acrylonitrile (ACN) yield show volcano shaped curves. These results, and the fact that δ_{is} ^{57}Fe varies proportionally with ΔE_Q ^{57}Fe, mean that a medium strength of lattice distortion or an optimal change of the M–O bond are responsible for a good selective oxidation catalyst. The effect of lattice distortion was confirmed by doping $Fe_2Sb_2O_7$ with cations of different radii.

The ratio of the two valence states of iron and antimony varies with the concentration of the oxides. The highest catalytic activity at a ratio of 1 shows that electron transfer is necessary for activating the reacting molecules.

Introduction

Because X-ray diffraction measurements and i.r. spectrometry had generally failed, Mössbauer spectrometry was used to obtain information concerning the structure of our samples. This method has had considerable success with all the catalysts, with or without supports. The mixtures of Fe_2O_3 with Sb_2O_5 or Sb_2O_3 also contained 10 or 25% by weight of SiO_2.

In particular, the combination of the γ-resonance spectra of the two catalytically active ions in the mixed oxide catalyst enabled us to determine valence state, s-electron density, and crystal-field distortion. Only these parameters permitted us to explain the catalytic activity of the catalysts during the ammoxidation of propene.

Experimental

Preparation of compounds

The samples were prepared by mixing strong acid solutions of iron nitrate and antimony chloride, followed by precipitation with ammonia or dilution with water. The slurry was mixed with 10 or 25% by weight of silica, dried by evaporation or spraying, and the powder thus obtained was fired for 8 h at 500°C.

The catalytic activities of the samples were determined in fixed or fluid bed reactors. The feed consisted of air, propene, and ammonia in the ratio 10:1:1. The products were determined by gas chromatography and titration.

Mössbauer spectra

The spectra for the mixed oxide catalysts were obtained by means of the Mössbauer effect using the 14·4 keV γ-rays of ^{57}Fe and the 37·2 keV γ-rays of ^{121}Sb. Spectra were recorded with a constant acceleration drive unit, used in connection with a multichannel analyser operating in the time-mode. The sources used were a 10 mCi ^{57}Co(Pd) for ^{57}Fe spectra, obtained from New England Nuclear, and a $\sim$200 μCi ^{121m}SnO$_2$ for ^{121}Sb spectra. The detector for both methods was a thin NaI(Tl) scintillation crystal. The window of the single channel analyser was set on the escape peak of the 37·2 keV ^{121}Sb γ-ray.

836

The velocity scale of all spectra was calibrated from the known energies of the six lines of the Mössbauer spectrum of a metallic iron foil absorber using a ^{57}Co source. The isomer shifts δ_{is} are reported for the ^{57}Fe spectra related to sodium nitrosylferricyanide, and for the ^{121}Sb spectra related to the ^{121m}SnO$_2$ source.

Absorbers were prepared by mixing the powdered samples with powdered polyethylene; pressed into thin discs, they were placed in a holder. All the samples for ^{57}Fe spectra contained 10 mg Fe/cm^2 while those for ^{121}Sb spectra contained about 15 mg Sb/cm^2. The measurements were carried out on ^{57}Fe nuclei at 300 K and on ^{121}Sb nuclei continuously at 80 K. The ^{121}Sb source was not cooled and remained at room temperature (300 K). The ^{57}Fe spectra were analysed by means of a curve-fitting programme,[1] the ^{121}Sb spectra were analysed manually.

RESULTS

We investigated catalysts with Sb/Fe ratios of 10:1, 5:1, 3:1, 1:1, 1:2, and 1:5, respectively. All the Mössbauer spectra of ^{57}Fe obtained at $T = 300$ K showed a doublet. Only at an Sb/Fe ratio of 1:5 (and in some cases at 1:1) an additional sextet appeared typical for α-Fe$_2$O$_3$. The doublet at $T = 300$ K is caused by paramagnetic Fe^{3+} ions. We assume these ions to be formed by reaction of Fe$_2$O$_3$ with Sb$_2$O$_4$ or Sb$_2$O$_5$, respectively, yielding Fe$_2$Sb$_2$O$_7$ or FeSbO$_4$. This was confirmed by comparing the spectra of the pure compounds. The δ_{is} and the quadrupole splitting (ΔE_Q) at room temperature are typical for Fe^{3+} (high spin) in octahedral co-ordination and relatively high distortion. The sextet at 80 K and at a high Fe$_2$O$_3$ concentration is due to superparamagnetic α-Fe$_2$O$_3$ particles. All catalysts with a Sb/Fe ratio of 5:1 and one of the spray dried (3:1) showed a superposed second doublet originated by Fe^{2+} ions. The concentration of Fe^{2+} was estimated assuming the lattice vibrations to be nearly equal. The amount of Fe^{2+} is increased by testing the catalyst and the Fe^{3+} portion is decreased. Figure 1 shows that the quadrupole splitting ΔE_Q increased with decreasing concentration of Fe$_2$O$_3$. Because the isomer shift increases too, we can say that the s-electron

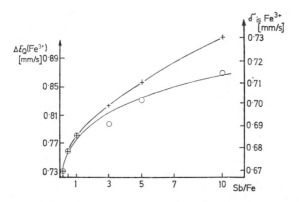

FIGURE 1   The change of quadrupole splitting (ΔE_Q) and of isomer shift (δ_{is}) of ^{57}Fe^{3+} with increasing Sb/Fe ratio.

density at the ^{57}Fe nucleus decreases. That means that the M–O bond becomes more covalent and the lattice distortion increases. All the ^{121}Sb spectra are single peaks except those of Sb$_2$O$_4$ compounds. The peaks of some spectra consist of eight lines caused by the ^{121}Sb quadrupole splitting.[2] These lines are not resolved under experimental conditions because their widths are of the same order of magnitude as the value of ΔE_Q.

The Sb spectra of the catalysts show two completely resolved peaks, one observed at a velocity associated with the $+3$ state and the other with the $+5$ state. It should also be noted that the Sb^{III} peak has a δ_{is} (-14 to -13.2 mm/s) differing from the experimental value for pure Sb_2O_3 (-12 ± 0.2 mm/s), but the Sb^V δ_{is} ($+0.1$ to -0.6 mm/s) lies close to the experimental value for pure Sb_2O_5 (0.2 ± 0.2 mm/s).

The difference in δ_{is} for the Sb^{III} implies that the bonding of Sb^{3+} in Sb_2O_4/Fe_2O_3 is different from that in pure Sb_2O_3.

On account of the distance between the two peaks the method is well suited for demonstrating the simultaneous presence of Sb^{III} and Sb^V because the absorption peaks for these valence states are well separated.[3]

Because Sb^{III} is not present in either of the samples the δ_{is} and the peak-width Γ of Sb^V are used for comparing the different catalysts. The change of the isomer shift of Sb^V with increasing Sb/Fe ratio is illustrated in Fig. 2. The *s*-electron density at the

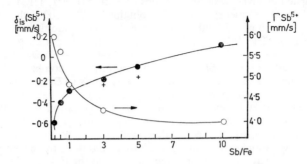

FIGURE 2   The change of isomer shift (δ_{is}) and of peak width (Γ) of $^{121}Sb^{5+}$ with increasing Sb/Fe ratio.

Sb^V nucleus decreases with increasing content of antimony oxide. The peak-width of the Sb^V spectrum is assumed to give information concerning the electric field gradient, *i.e.* the lattice distortion. Because of the spherosymmetric charge cloud density, no quadrupole splitting is normally observable. The small peak broadening effect we found is due to the deviation from cubic symmetry. It originates from the charges of the neighbouring lattice sites. The effect of the Sb/Fe ratio on the peak broadening of Sb^V is shown in Fig. 2. The lattice distortion becomes smaller as the Sb/Fe ratio increases.

In order to establish the existence of the quadrupole splitting we prepared pure $Fe_2Sb_2O_7$ and doped this compound with different cations by impregnation. The quadrupole splitting of the Fe^{3+} increases with decreasing ionic radius (Fig. 3). If we compare the quotient of charge and ionic radius we obtain a function indicating that the greater this quotient is the greater is the quadrupole splitting (Fig. 3, open circle). Only a few supported catalysts show *X*-ray deflection spectra. The *d*-values and relative intensities can be ascribed to both $Fe_2Sb_2O_7$ and $FeSbO_4$. These compounds have almost identical *X*-ray spectra.[5] The conclusion is confirmed by electron diffraction, and by comparison of the Mössbauer data of the supported catalysts with those of $Fe_2Sb_2O_7$. The ^{57}Fe doublet, corrected by measurements at 80 K which eliminate the superparamagnetic contribution, can only be derived from a compound other than Fe_2O_3.

DISCUSSION

According to the Mössbauer spectra all samples contained Fe^{3+} at octahedral sites, and a new compound, an iron antimonate, is formed. We prepared four spray-dried catalysts in addition to the six slurry evaporated catalysts with Sb/Fe ratios of 10:1,

Kriegsmann, Öhlmann, Scheve, Ulrich

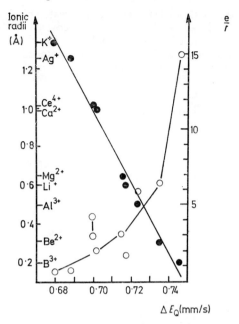

FIGURE 3   Effect of ionic radii and quotient *e/r* of doping elements on the quadrupole splitting of $^{57}Fe^{3+}$ in $Fe_2Sb_2O_7$.

5:1, 3:1, 1:1, 1:2, and 1:5. Three of them had an Sb/Fe ratio of 5:1 and one an Sb/Fe ratio of 3:1. We checked the effect of the Sb valence by preparing a slurry-evaporated 1:1 catalyst with Sb_2O_5 as starting material. The other evaporated catalysts were made from mixtures of antimony and ferric salt solutions. Thus, 11 different mixed antimony–iron oxide catalysts were investigated. The yield of acrylonitrile (ACN) is plotted in

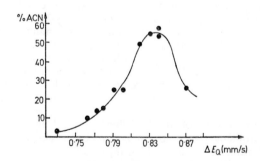

FIGURE 4   Acrylonitrile yield (mole %) against quadrupole splitting of $^{57}Fe^{3+}$.

Fig. 4. The ACN yield increases at first with increasing ΔE_Q, and after reaching a maximum it decreases. All four catalysts at the maximum have an Sb/Fe ratio of 5:1, spraydried or evaporated. This can be explained if we assume ΔE_Q to reflect the lattice distortion. Then this volcano shaped curve means that a moderately distorted lattice or a

839

moderate change of the M–O bond strength has the greatest catalytic effect for selective oxidation. Because the Fe–O bond becomes more covalent with increasing δ_{Is} and ΔE_Q the bond changes from Fe^{III}–O to Fe^{II}–O. In order to check this effect we investigated the influence of the ionic radii of different elements used for doping on the ACN yield. Figure 5 demonstrates that a medium radius, *i.e.* a medium quadrupole splitting

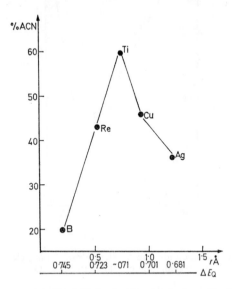

FIGURE 5   Change of ACN yield (mole %) with ionic radii of doping elements.

or a medium lattice distortion, increases the selective oxidation towards a maximum value. This effect can also be deduced from the linear increase of the quadrupole splitting with decreasing ionic radius of the doping element (Fig. 3).

If we compare the quadrupole splitting of Sb^V with the catalytic parameters, we get the same picture (Fig. 6). At a medium value of Γ, assumed equal to ΔE_Q, ACN yield and propene conversion are high. Because according to Fig. 2 Γ decreases, if the isomer shift is less negative, we explain the curves of Fig. 6 by assuming that the covalency of the Sb^V–O bond of the Sb_2O_5 should increase slightly towards an Sb^{III}–O bond. The medium effect is seen for the four catalysts with an Sb/Fe ratio of 5:1, known from the correlation of ^{57}Fe ΔE_Q with ACN yield (Fig. 4). By means of this the accuracy of our measurements is proved. In all plots we use the quadrupole splitting ΔE_Q and not the isomer shift δ_{Is}, because the first one is more exactly measured and needs no correction to a standard.

Summarizing the results, we obtain the following model. By annealing mixtures of Fe_2O_3 and Sb_2O_5 we obtained new compounds with Fe at octahedral sites. The covalency of the M–O bond increases in both oxides, thus strengthening the M–O bond. This corresponds to our results in determining the starting temperature of the reduction in hydrogen.[4] While Fe_2O_3 and Sb_2O_5 begin to be reduced at 400°C or 415°C, respectively, $FeSbO_4$ began to be reduced only at 515°C. The best catalysts with Sb/Fe ratios of 5:1 showed this effect between 470 and 500°C, less active catalysts are reduced at higher or lower temperatures, thus supporting the above mentioned idea. This means, the main effect is to change the M–O bond to a medium value between the two oxidation states of Fe^{II} and Fe^{III}, or Sb^{III} and Sb^V, respectively.

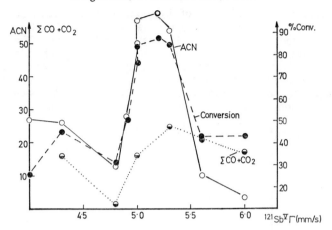

FIGURE 6   Dependence of ACN yield, propene conversion, and sum of CO and CO₂ (all in mole %) from the peak width (Γ) of $^{121}Sb^{5+}$.

If this is true, both valency states are expected to occur in the good catalysts. The presence of Sb^{III} is known from the description of our results. In Fig. 7 it is proved that the greatest amount of ACN is yielded at the highest Sb^{III} content. The amount of Sb^{III} is increased during the reaction and tends to an Sb^{III}/Sb^V ratio of 1, that is, to Sb_2O_4 (*i.e.* Sb_2O_3, Sb_2O_5) or $Fe_2Sb_2O_7$, respectively.

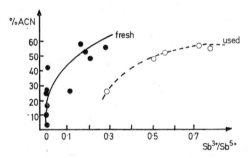

FIGURE 7   ACN yield (mole %) against Sb^{3+}/Sb^{5+} ratio.

The Mössbauer spectra of the four catalysts on the top of the volcano-shaped curve (Fig. 4) have two doublets. One is caused by Fe^{3+} ions, but the second undoubtedly arises from Fe^{2+}. The relation between the Fe^{3+}/Fe^{2+} ratio and the ACN yield is plotted in Fig. 8. Selective oxidation is increased, if the amount of Fe^{3+} decreases. The catalytic reaction reduces the Fe^{3+} content. The more nearly the Fe^{3+}/Fe^{2+} ratio becomes equal to one, the closer the ACN yield approaches a maximum value. This effect confirms the model, *i.e.* means that a medium M–O bond strength or reducibility is responsible for a good oxidation catalyst.

This is valid also for selective oxidation, because the proportion of the total oxidation shows the same relation to the Mössbauer parameters, shown in Fig. 6. The fact that in the best Fe_2O_3–Sb_2O_5 catalysts the cations have two valence states and the tendency towards a ratio of 1:1 between them during the reaction confirms the importance of a reduction–oxidation mechanism and of electron transfer. Summarizing all the results, we can conclude that by annealing coprecipitated mixtures of Fe_2O_3 and Sb_2O_5 or Sb_2O_3 a compound is formed with Fe^{3+} ions on octahedral sites. If we distort this oxygen octahedron by a growing excess of Sb ions or by doping with cations which have a high quotient of charge to ionic radius, the catalytic activity increases at first because the M–O bond strength is decreased. But at a certain value this effect becomes too great, the reoxidation rate becomes too slow, and catalytic activity decreases. The same is true for the antimony oxide. The quotient of charge and ionic radius means the greater

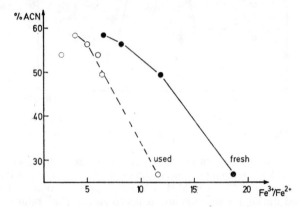

FIGURE 8   ACN yield (mole %) against Fe^{3+}/Fe^{2+} ratio.

the charge at a constant distance, the more the oxygen ion is displaced relative to the central cation of the octahedron. The octahedral configuration is confirmed for Fe^{3+} by e.s.r. measurements. For two of the best catalysts, we obtained peaks with a definable g-factor of $2·08 \pm 0·05$ or $2·18 \pm 0·05$, respectively. The width of these resonance peaks is 470 Oe for unused catalysts, and narrows during reaction to 130 Oe. $Fe_2Sb_2O_7$ has a resonance with a g-factor of $2·0 \pm 0·1$ which is 2300 Oe wide. All values are obtained at 300K. Because α-Fe_2O_3 gives no e.p.r., and Sb ions are not paramagnetic, these peaks originated from Fe^{3+}. The change of the peak during reaction confirmed the change of the valence state during reaction (Figs. 7 and 8), because it must be interpreted by crystal field variation.[5]

ACKNOWLEDGEMENTS.   The authors are indebted to Dr. A. Yu. Alexandrov and Dr. Meisel for assistance in our work.

REFERENCES

[1] W. Brückner and E. Wieser, *Exp. Techn. Phys.*, 1971, **19**, 31.
[2] S. L. Ruby, G. M. Kalvius, *et al.*, *Phys. Rev.*, 1966, **148**, 176.
[3] F.-J. Ulrich, W. Meisel, J. Scheve, and A. Yu. Alexandrov, *Proc. Conf. Mössbauer Spectrometry*, Dresden, 1971 (Physik. Gesellschaft der DDR, Berlin, 1971), Vol. 2, p. 542.
[4] J. Freiberg *et al.*, *Izvest. Bulg. Acad. Sci. Chem. Ser.*, 1973, **4**, 455.
[5] These and all other data are taken from F.-J. Ulrich, *Thesis*, Berlin, 1975.

DISCUSSION

G. Manara (*Snamprogetti, Milan*) said: We have studied the catalytic activity of iron–antimony catalysts for propene oxidation in the absence of gaseous oxygen[1] and agree

[1] V. Fattore, Z. A. Fuhrman, G. Manara, and B. Notari, *J. Catalysis*, 1975, **37**, 223.

that, by adding an excess of antimony to the iron, the catalytic activity is improved, but that too great a concentration of antimony leads to a decrease in activity. Other workers have shown that, as the Sb:Fe ratio is varied, a change occurs at Sb:Fe $\sim$ 2:1 in (i) the activation energy for isotopic oxygen exchange, (ii) the work function, and (iii) the electrical conductance. Maximum selectivity in butene dehydrogenation and in propene oxidation is also observed at Sb:Fe $\sim$ 2:1. Furthermore, the patent literature indicates that the same ratio is preferred in the preparation of industrial catalysts. Thus I would ask whether Mössbauer spectroscopy can suggest an interpretation for the behaviour of catalysts having Sb:Fe ratios near to 2:1.

G. Ohlmann (*Zentralinst. für Phys. Chem. der AdW., Berlin, DDR*) replied: We did not study the reactions that you mention so I cannot give a definite answer. However, the Mössbauer parameter did not reveal any special behaviour near Sb:Fe $\sim$ 2:1. The fact that the most favourable lattice distortion for a maximum yield of acrylonitrile lies at Sb:Fe $\sim$ 5:1 does not necessarily mean that this ratio has to be the same for other reactions too.

I. Matsuura (*Tech. Univ. of Eindhoven*) enquired whether antimony present in excess might become substituted for iron, and so decrease the number of iron sites. Prof. Öhlmann replied that catalysts had been examined by X-ray reflection spectroscopy and by i.r. spectroscopy and, whilst not all catalysts had given spectra, those spectra that had been obtained were consistent with the compounds and surface structures mentioned in the paper.

Association Effects in Hydrogenolysis and Hydrogenation over Bimetallic Supported Catalysts

J. P. BRUNELLE,† R. E. MONTARNAL,* and A. A. SUGIER

Institut Français du Pétrole, 1 et 4, avenue de Bois Préau, 92500 Rueil-Malmaison, France, and †Rhone Poulenc, 182–184, Av. A. Briand, 92 Anthony, France

ABSTRACT   Synergistic effects have been observed for bimetallic supported iridium–copper and iridium–rhenium species, prepared by an ion-exchange technique on alumina. Copper causes a drastic decrease in the turn-over of n-pentane hydrogenolysis, and an increase in isomerization and cyclization selectivity. A smaller decrease was observed for benzene hydrogenation. For iridium–rhenium associations, the activity passes *via* a maximum for both reactions; the variation of depth of hydrogenolysis gives further proof of metal interactions. Such interactions reflect the existence of bimetallic crystallites.

Synergistic effects for benzene hydrogenation lead us to invoke the electronic factor for both reactions, without excluding the geometrical factor for hydrogenolysis.

An interpretation of the electronic effect is suggested, based upon the "softness" characteristics of the metals. A modification of "softness" by the ligand effect of additives explains why the addition of copper to iridium modifies the properties of the latter so as to resemble platinum or palladium, and why the addition of rhenium to iridium causes the latter to become more like osmium.

INTRODUCTION

In a previous study, we showed that for supported rhenium and the six noble Group VIII metals, the turn-over number in n-pentane hydrogenolysis or benzene hydrogenation varied in a roughly parallel manner when the nature of the metal was changed.[1] The variation was found to be greater for the former type of reaction: around 10^9 as against 10^2 for the latter. This difference reflects the greater influence of the nature of the metal on the demanding activation of paraffins, than on the non-demanding activation of unsatureds.

Numerous papers have been published on the "alloy" effect, in bimetallic catalysts, and particularly with Group VIII and IB metal associations, for these two types of reactions.[2-8] In the present study of bimetallic associations, we chose to start from iridium, and to add to it not only a Group IB metal (Cu) but also a metal poorer in *d* electrons (Re). We used a well defined exchange technique[9] for impregnation, to see if it was efficient in inducing association effects. The modifications that were hoped for in the activity were interesting from a fundamental point of view, so as to locate the association in comparison with pure metals and to see if a parallelism was maintained in the evolution of the two reactions. From a practical point of view, research on iridium could be justified by its interest for catalytic reforming.

EXPERIMENTAL

Catalyst preparation

The homogeneous deposition of active elements was achieved by an exchange technique on a large-area alumina carrier. It was possible thus to obtain a high dispersion of the metals. We tried to work with a constant number of active sites. For this reason, the Ir–Cu association was prepared with an increasing proportion of copper added to the same amount of iridium. On the other hand, for the Ir–Re association, the total amount of metals was kept constant.

Iridium–copper associations

Using a nitric acid–chloroiridic acid solution, a homogeneous deposit of 0·25% weight of iridium was achieved by an anionic exchange. After filtration, washing and drying at 120°C, the solid was calcined at 500°C during 3 h, to regenerate hydroxyl sites on the alumina. Cationic exchange with a cuprammonium nitrate solution was then performed to deposit up to 1·25% of the Group Iʙ metal. Filtration, washing, drying, and calcination were performed as before, followed by reduction during 6 h at 500°C in a stream of hydrogen.

Iridium–rhenium associations

The metal catalyst (0·6% by weight total metal) was prepared by a single anionic exchange, by using an aqueous solution of chloroiridic and perrhenic acids. After the same operations as before, the reduction was conducted at sufficiently high temperature (550°C) to reduce the rhenium compound.

Characterization of the catalysts

The homogeneity of the metal distribution was controlled by microprobe analysis. The specific surface areas of the reduced metals were determined by classical chemisorption techniques. For Ir–Cu associations, chemisorption was performed by a dynamic method.

Hydrogen and carbon monoxide, respectively, characterize superficial iridium and total metallic sites. The stoicheiometry H/Ir was chosen equal to 1. For Ir–Re associations, hydrogen chemisorption was performed by the conventional static technique, using the stoicheiometry of 1 H for one metallic site. Of course, we must bear in mind the uncertainties, that have been well described, in such nonetheless indispensable characterizations.[10]

Apparatus

The classical apparatus described elsewhere[11] was used under dynamic and differential conditions. All the experimentation was done at atmospheric pressure with the following conditions: for n-pentane hydrogenolysis, 100 to 400°C, $H_2/n\text{-}C_5 = 10$, helium diluent (pressure of 0·5 bar); for benzene hydrogenation, 10 to 200°C, $H_2/C_6H_6 = 40$. Phillips "pure grade" n-pentane was distilled to reduce the impurities (isopentane and cyclopentane) to less than 0·1%. "Pure grade" benzene was purified over sodium powder in an argon atmosphere. The final quantities of sulphur were less than 0·5 ppm for both hydrocarbons. Analyses were performed by gas chromatography. In the case of n-pentane activation, isomerization into isopentane and cyclization into cyclopentane could be observed as well as hydrogenolysis. Moreover, the depth of hydrogenolysis can be deduced from the distribution of light products. We chose to characterize this depth numerically by the following equation:

$$P = (C_1 + C_2) - (C_3 + C_4)/(C_1 + C_2 + C_3 + C_4) \qquad (1)$$

C_i, being the molar fraction of the paraffin with i atoms. P goes from zero for stepwise to 1 for quite deep hydrogenolysis.

RESULTS

Figure 1D shows that the number of superficial iridium atoms decreases only slightly with the atomic copper fraction and that the total number of superficial metallic atoms increases roughly linearly with the wt % Cu. The application of the classical formula: $\phi = 5/\rho S_{\text{metal}}$ gives a crystallite size of about 40Å in the case of monometallic iridium. Figure 1A shows that, for the Ir–Re association, the total number of superficial metallic atoms remains appreciably constant, except for rhenium-rich associations. In any case,

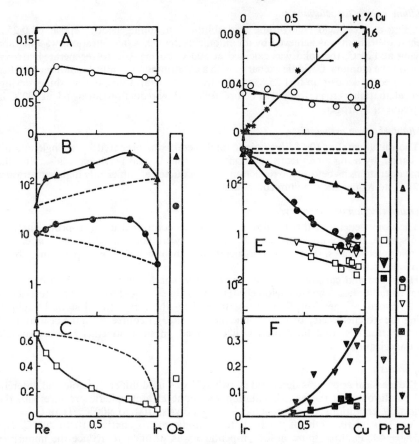

FIGURE 1   left. Variations in properties of rhenium–iridium alumina catalysts with the atomic fraction of iridium. Comparison with osmium–alumina catalysts.

A. Hydrogen chemisorption, in cm³/g catalyst (○).

B. Turn-over measured at 30°C for benzene hydrogenation (▲) and 185°C for n-pentane hydrogenolysis (◉) in mole/h, site.

C. Depth of n-pentane hydrogenolysis at 185°C, expressed by Equation (1) (□).
   right. Variations in properties of iridium–copper alumina catalysts with the atomic fraction of copper. Comparison with platinum and palladium silica catalysts.

D. Hydrogen chemisorption, in cm³/g catalyst (○). Carbon monoxide chemisorption variation with wt % copper (*).

E. Turn-over measured at 100°C for benzene hydrogenation (▲) and 250°C for pentane hydrogenolysis (◉), isomerization (▽) and cyclization (□).

F. Isomerization (▼) and cyclization (■) selectivities.

such measurements will give access to the activities in terms of turn-over from the measurement of superficial active sites. Preliminary experimentation showed the inactivity of the carrier under our test conditions. The absence of external and internal diffusional limitations was verified by varying the linear flow-rate of gases and the sizes of the alumina pellets. In the first moments of hydrogenolysis, the well known self-poisoning phenomenon appears. So experimental measurements were undertaken after 5 h, the time necessary to attain the steady state. The great variations observed in activity made it necessary to operate at different temperatures so as to remain within the range of low conversions (0·05 to 5%). Therefore extrapolations were sometimes made to compare activities at the same temperature. Such extrapolations appear valid, at least formally, due to the linear variation of the rate in the Arrhenius co-ordinates.

Iridium–copper associations

Figure 1E shows that the turn-over of n-pentane hydrogenolysis decreases drastically with the addition of copper. For the atomic fraction (α) of 0·5, the activity is reduced by a factor of 10^3. Isomerization and cyclization also decrease, but their selectivity clearly increases. Such an evolution leads toward the catalytic properties of platinum of the same dispersion as is illustrated by Figs. 1E and 1F. Pure crystallite size effects were studied for comparison; increase of iridium crystallite sizes from 20 to 150Å only produces a decrease of a factor of 10 in the n-pentane hydrogenolysis turn-over number.

Figure 1E also shows a decrease in the turn-over of benzene hydrogenation, but not such a drastic one. For $\alpha = 0·5$, a reduction by only a factor of 10 is observed. Such an evolution leads, in this case, toward the properties of palladium rather than platinum.

Very similar results were also obtained in hydrogenolysis and hydrogenation over silica supported iridium–gold associations,[12] in accordance with published results.[13]

Iridium–rhenium associations

Both for hydrogenolysis and hydrogenation, a maximum turn-over value appears in Figure 1B, when going from rhenium to iridium. In Figure 1C is also plotted the evolution of the depth of hydrogenolysis. It decreases when the atomic fraction of iridium increase, as also illustrated by the following values:

	C_1	C_2	C_3	C_4	P
Ir	10	43	42	5	0·06
Re/(Ir + Re) = 0·54	27	34	28	11	0·22
Re	64	19	10	7	0·66

In the Figure, the broken lines represent the theoretical values obtained on the hypothesis of the simple addition of the performance relative to each metal, without any interaction between them, and also assuming surface and bulk alloy compositions to be equivalent.

DISCUSSION

Modifying effects of quite a different extent appear depending on the metal added to iridium; much more drastic effects are observed with copper than with rhenium. Another characteristic is their selectivity; copper leads to a negative and rhenium to a positive synergistic effect. Moreover, a parallelism in the evolution of both reactions must be underlined. We can now ask ourselves what are the causes of these synergistic effects?

It is first possible to eliminate any intervention of the carrier acidity due to chlorine. For experiments performed with Ir–Cu associations at relatively high temperature (350°C) chlorine was fortunately eliminated by hydrolysis during impregnation with the basic copper solution. For iridium alone or iridium–rhenium associations, experimentation carried out with pent-1-ene on alumina chlorinated to a much greater extent showed quite negligible activity at the temperature of the tests. Secondly, metal–metal bifunctional effects due to the presence of monometallic crystallites of each metal, can hardly explain the decrease in activity with the addition of copper to iridium.

Moreover, we found that if copper was deposited at the periphery of a pellet homogeneously impregnated with iridium, the turn-over was characteristic of pure iridium. For Ir–Re bimetallic catalysts, bifunctionality is not absolutely excluded but does not seem very probable if we consider the coherence of interpretation presented below on the basis of association effects.

Two types of well known association effects must be considered: geometric and electronic effects.[14,15]

For n-paraffin activation, geometric effects may have a noticeable influence if we consider that, as has been demonstrated for platinum[16,17] the chemisorption and transformation of the hydrocarbon involve several metallic sites. For benzene hydrogenation, on the other hand, it is generally assumed that a geometric effect does not appear.[18]

Considering that association effects have been observed here for hydrogenation as well as for hydrogenolysis, it appears that the electronic effect must be considered. Our interpretation will be based on a synthetic presentation, which we will now describe, of specific characteristics of each type of activation.

n-Pentane activation requires the use of a relatively high temperature (150 to 300°C) to allow for dissociative chemisorption. Two or three mechanisms have been invoked and even proven for platinum.[19,20] An important point for our interpretation is that we can classify metals in an order of increasing depth of hydrogenolysis. From our own results[1] this is:

$$\text{Pd, Pt} \leqslant \text{Ir, Rh,} < \text{Ru} < \text{Os} < \text{Re}.$$

We can assume that such a classification also implies the increasing order of the metal–hydrocarbon bond strength. From a more synthetic point of view, it seems to correspond fairly well to a decrease in the "softness" of the metal considered, according to the concepts of Pearson.[21] Indeed, transition metals are so considered as soft acids, but their "softness" increases with their polarizability, and, on this basis, considering all Group VIII metals, an inverse correlation has been suggested between their "softness" and the extent to which they cause paraffin hydrogenolysis.[22]

Our previous results[1] also show that a maximum value for the turn-over of hydrogenolysis appears around Rh, Ru, Os, for which the strength of chemisorption would be optimal. When going from these metals towards Pt or towards Re, the decrease in activity is then, respectively, correlated with a decrease or increase in chemisorption strength. Maximum isomerizing and cyclizing activity is obtained for platinum *i.e.*, for weaker chemisorption, permitting the rearrangement of the molecule without carbon–carbon bond breaking, as already emphasized by some authors.[6]

Benzene hydrogenation can be achieved at much lower temperature (here 30°C) and involves charge transfer and chemisorption, which may be different from the previous case (σ-bonding and π-bonding by retroco-ordination for benzene, and σ-bonding for hydrogen). If metals are considered in the same order as before, a maximum in their activity again appears around Ru, Rh, Os.[1] Activity variation appears much less than for hydrogenolysis, and, mainly experimental arguments to determine in what direction chemisorption strength increases are not so easy to handle. As a working hypothesis, we will consider that, since benzene is a "soft" base, it will be more strongly bonded with softer metals *i.e.*, those on the platinum "side". So, from Ru, Os, metals, the decrease and increase in activity will be respectively, connected with the increase in bonding toward Pt, and the decrease in bonding toward Re. It is also possible and even necessary to consider hydrogen chemisorption. Although several kinds of bonding exist for hydrogen we can consider, as a working hypothesis, that bonding would be optimal around Ru, Os; weaker toward Pt, and stronger toward Re, on the basis of an electron transfer, preferentially in the direction of the metal, for this covalent bond.

Interpretation concerning iridium–copper associations
n-Pentane activation

Copper, which is inactive when alone, can first operate on iridium as an inert diluent leading, as already emphasized for Ni–Cu, Ni–Ag, Pt–Au associations,[14,23] to a preferential reduction of hydrogenolysis, which, generally needs the biggest "ensemble" of active sites. Secondly it can operate as an electron donor, thus increasing the "softness" of iridium, weakening the metal–n-pentane bond strength, and thus decreasing the turn-over value. Selectivity in isomerization and cyclization is increased by the evolution toward the bond strength characteristics of platinum. Beyond this rather general interpretation, we can seek to specify the mechanism of copper intervention. For monometallic iridium, crystallite size is about 40Å, and, if we consider that Ir–Cu associations are formed by such iridium crystallites, surrounded by copper crystallites, this can hardly explain the profound influence of copper up to a high fraction of this additive. The geometric effect would be only an edge effect, and the electronic effect would be ineffective for so great a distance between Ir and·Cu. So it seems necessary to invoke an "interpenetration" between iridium and copper although Ir and Cu are said to be practically immiscible at temperatures lower than 1000 K.[24] Such a possibility has been presented for small crystallites on the basis of thermodynamic arguments,[25] which have been contested by other authors, who prefer to invoke kinetic arguments.[26] A recent publication emphasizes that the thermodynamics lead to the segregation of the most volatile components of the alloy on the lower co-ordination sites, either with or without a miscibility gap, with the kinetics, of course, governing the degree of evolution.[27] Mention must also be made of the possibility of superficial enrichment with the metal (here, iridium) more strongly bonded with reactants.[28,29] Finally, through this complexity, the necessary interpenetration between Ir and Cu appears possible. In our case, the sensibly constant value of chemisorbed hydrogen at least shows that iridium is not "covered" by copper, which is consistent with the interpenetration concept.

Benzene hydrogenation

The geometric effect of copper is nil or very weak. By its ligand effect (at a short distance), copper increases the retroco-ordination toward benzene, and hence the strength of benzene bonding, which provides a contribution to the decrease in activity. The ligand donor effect would, on the other hand, weaken the hydrogen bond strength, but this again provides a coherent contribution to the decrease in activity.

The weaker influence of copper in hydrogenation as opposed to hydrogenolysis must be compared with the same weaker variation in this type of activation when the nature of the metal is varied. Such a variation can be explained by the sole influence, for this non-demanding reaction, of the electronic factor. We think it is important to consider the argument presented by Joice *et al.*[30] It consists in assuming that the chemisorption of an unsaturated hydrocarbon leads, by the interplay of σ–π bonding, to a profound electronic modification of the metallic site, which thus becomes less sensitive to electronic modification resulting from the use of another metal, or from the addition of a foreign ligand.

Interpretation concerning iridium–rhenium associations

For the n-pentane reaction, rhenium leads to a greater degree of hydrogenolysis than iridium, hence to a stronger chemisorption of the hydrocarbon. This appears consistent[31] with the positive hydrogen order observed in ethane hydrogenolysis on rhenium.[16] Quite a high hydrogen pressure would be required to cause the appearance of a negative hydrogen order by the displacement of firmly adsorbed hydrocarbon. It is diffiulct to determine here the direction of the intervention of the geometric factor, because both metals chemisorb the hydrocarbon. Such an effect is not excluded, but again our observation of association effects in hydrogenation leads us to consider electronic effects. Electronic effects can be explained by the progressive evolution of chemisorption strength, when going from Ir to Re, which pass through properties corresponding to Ru or Os. On this possibly rather naive basis, maximum activity would be nearer the intermediate metal, as the degree of interpenetration would be deeper. In our case, the maximum activity approaches that of osmium.

The variation in depth of hydrogenolysis is interesting to interpret. It decreases progressively from a high value for Re, to a low value for Ir, but the experimental result remains lower than the one given by the theoretical curve (in broken lines) corresponding to the absence of any interaction between the metals (Figure 1C). Such a difference gives further proof of association effects. For a 50/50 composition, the depth of hydrogenolysis is similar to that observed[1] with monometallic Os or Ru, a result consistent with electronic effects.

For benzene hydrogenation, iridium and rhenium are also "on each side" of more active metals, in the previous classification. The observation of a maximum when going from Ir to Re, is then, consistent with an electronic interaction.

REFERENCES

1. J. P. Brunelle, A. A. Sugier, and R. E. Montarnal, to be published.
2. E. G. Allison and G. C. Bond *Catalysis Rev.*, 1973, **7**, 233.
3. A. O'Cinneide and F. G. Gault, *J. Catalysis*, 1975, **37**, 311.
4. D. A. Dowden, *Proc. 5th Internat. Congr. Catalysis*, Miami Beach, 1972, paper 41.
5. W. G. Reman, A. H. Ali, and G. C. A. Schuit, *J. Catalysis*, 1971, **20**, 374.
6. A. Roberti, V. Ponec, and W. M. H. Sachtler, *J. Catalysis*, 1973, **28**, 381.
7. J. H. Sinfelt, *J. Catalysis*, 1973, **29**, 308.
8. J. K. A. Clarke, *Chem. Rev.*, 1975, **75**, 291.
9. J. P. Brunelle and A. A. Sugier, *Compt. rend.*, 1973, **276**, C, 1545.
10. J. R. Anderson, *Structure of Metallic Catalysts* (Academic Press New York, 1975).
11. J. P. Brunelle, A. A. Sugier, and J. F. Le Page, *J. Catalysis*, in press.
12. J. P. Brunelle, *Thèse*, Paris, 1976.
13. J. Plunkett and J. K. A. Clarke, *J. Catalysis*, 1974, **35**, 330.
14. V. Ponec, *Catalysis Rev. Sci. and Eng.*, 1975, **11**, 41.
15. J. R. Anderson, *Adv. Catalysis*, 1973, **23**, 1.
16. J. H. Sinfelt, *Adv. Catalysis* 1973, **23**, 91.
17. C. Kemball, *Catalysis Rev.*, 1971, **5**, 33.
18. P. C. Aben, J. C. Platteeuw, and B. Stouthamer, *Proc. 4th Internat. Congr. Catalysis*, Moscow, 1968, paper 31.
19. F. Garin and F. G. Gault, *J. Amer. Chem. Soc.*, 1975, **97**, 4466.
20. M. A. McKervey, J. J. Rooney, and N. G. Samman, *J. Catalysis*, 1973, **30**, 330.
21. R. G. Pearson, *Hard and Soft Acids and Bases* (Dowden Hutchinson and Ross, Pennsylvania, 1973).
22. R. Montarnal and G. Martino, *Séminaire franco-soviétique sur la catalyse*, (Kiev, September, 1974).
23. J. R. H. van Schaik, R. P. Dessing, and V. Ponec, *J. Catalysis*, 1975, **38**, 273.
24. H. Nowotny and A. Winkels, *Z. Physik*, 1939, **114**, 455.
25. D. F. Ollis, *J. Catalysis*, 1971, **23**, 131.
26. D. W. Hoffman, *J. Catalysis*, 1972, **27**, 374.
27. J. J. Burton, E. Hyman, and D. G. Fedak, *J. Catalysis*, 1975, **37**, 106.
28. R. L. Moss and D. H. Thomas, *J. Catalysis*, 1967, **8**, 151.
29. F. L. Williams and M. Boudart, *J. Catalysis*, 1973, **30**, 438.
30. B. J. Joice, J. J. Rooney, P. B. Wells, and G. R. Wilson, *Discuss. Faraday Soc.*, 1966, **41**, 223.
31. J. P. Boitiaux, G. Martino, and R. Montarnal, *Compt. rend.*, 1975, **280**, C, 1451.

DISCUSSION

V. Ponec (*Univ. of Leiden*) said: I should like to present some results which might provide an alternative explanation of the interesting work presented by Brunelle *et al.* Puddu[1] followed the hydrogenation of benzene and of cyclohexene on several platinum–gold alloys and obtained results summarized in the Figure. As you can see there is no

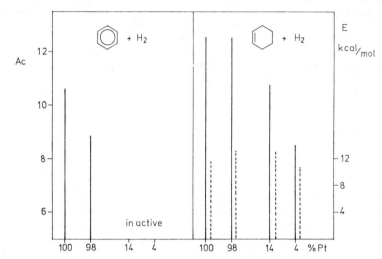

FIGURE   Activity (log rate under standard conditions) of different Pt-Au catalysts for benzene and cyclohexene hydrogenations (full lines). Activation energy of cyclohexene hydrogenation (dotted line).

change in the apparent activation energy of cyclohexene hydrogenation which would indicate a substantial change in the chemisorption properties (electronic structure) of individual Pt atoms. Also Stephan found that the thermal desorption of hydrogen does not indicate any sign of such changes due to alloying.[2] Therefore, we explained the dramatic decrease in the activity in benzene hydrogenation by the *lack of ensembles of active sites* of a size necessary for benzene activation and hydrogenation[1] (at. % Pt < 14).

Because the interaction of platinum with gold is relatively stronger than of iridium with copper, I wonder whether we have to expect changes in the electronic structure of iridium and corresponding variations in the chemisorption bond strength of benzene, when iridium is alloyed with copper. Do authors have any additional information on heats of adsorption, activation energies, UPS valence band structure, *etc.*, which support their statement on the role of the electronic factor in the reactions they followed?

Although copper is of low activity in most diluted alloys, its activity might not be negligible. This is the reason why alloying with gold suppresses the activity of platinum much more than alloying copper with nickel.

R. Montarnal (*Inst. Franç. du Petrole, Rueil-Malmaison*) replied: We have data on

[1] S. Puddu and V. Ponec, *Rec. Trav. chim.*, in press.
[2] J. J. Stephan, V. Ponec and W. M. H. Sachtler, *Surface Sci.*, 1975, **47**, 403.

the activation energies of benzene hydrogenation. However, we think that the activation energy (for aromatic and olefin hydrogenation) is not a sufficiently sensitive parameter to allow a distinction between geometric and electronic effects. For pure metals of Group VIII for example, their variation is quite small although it would be difficult to deny any electronic effect.

For the n-pentane hydrogenolysis we have found that the activation energy decreases from 36 kcal mol^{-1} for pure iridium, to 23 kcal mol^{-1} for an atomic fraction of copper of 0·5, and then increases to 33 kcal mol^{-1} for an atomic fraction of copper of 0·9. These results, which are in accordance with those you obtained for Ni–Cu associations,[3] are difficult to explain by geometric effects alone.

Concerning the differences between cyclohexene and benzene hydrogenation for platinum–copper associations, one can remark that the specificity of Group VIII metals is smaller for olefin hydrogenation than for benzene hydrogenation. Hence it seems normal that you find, as we do, an important association effect for benzene hydrogenation and a rather smaller effect for cyclohexene hydrogenation.

H. Charcosset (*CNRS, Villeurbanne*) said: The calcination in air seems to determine the final state of the catalysts. It may account for (i) the low value of the per cent dispersion of iridium (smaller iridium particles would probably be obtained if the exchanged catalysts had been directly reduced by hydrogen) and (ii) the formation of iridium–copper alloy particles. Mixed IrO_xCuO_y particles are probably being formed during calcination. Their subsequent reduction by hydrogen is practically instantaneous and leads therefore to alloy particles which are metastable at the moderate temperatures used in your experiments. Much less interaction would be observed between iridium and copper if a copper-exchanged iridium catalyst was reduced directly by hydrogen.

You propose that for certain co-precipitations of the iridium–rhenium bimetallic particles, the surface properties would be that of osmium. This hypothesis could be checked by the temperature programmed desorption of hydrogen and the measurement of the position of the metal–CO band, or by examination of bands detected by i.r. spectroscopy.

R. Montarnal (*Inst. Franç. du Petrole, Rueil-Malmaison*) replied: It is very likely that calcination is responsible both for crystallite growth and for the formation of mixed species, rapid reduction of which can lead to metastable bimetallic particles. We agree that the methods you suggest would bring further precision to our interpretation of bimetallic associations.

C. Brooks (*United Technologies Res. Center, East Hertford, Conn.*) said: My question concerns the characterization of these bimetallic catalysts. Would the authors amplify the discussion on 'homogeneity of metal distribution' as controlled by micro-probe analysis? For a bimetallic system, where the average crystallite size is small relative to the beam width resolution of the microprobe used, it is conceivable that an apparent homogeneity occurs for two metals due to signal averaging derived from two possible metal configurations: (i) a configuration where the metal crystallites are bimetallic alloy phases or (ii) a configuration where the metal crystallites are predominantly single metal crystallites. This latter configuration would be quite likely to occur if: (a) catalyst preparation consisted of consecutive additions of the metals with intermediate calcining treatments; (b) there was strong bonding of both metals to the support; or (c) surface mobility of both metals was of a low order during catalyst pretreatment. Are the catalyst systems studied here of such a nature that there is no ambiguity involved in interpreting micro-probe signal homogeneity as a measure of an alloy phase?

R. Montarnal (*Inst. Franç. du Petrole, Rueil-Malmaison*) replied: Since the incident beam

[3] V. Ponec and W. M. H. Sachtler, *Proc. 5th Internat. Congr. Catalysis*, Miami Beach, 1972, Vol. 1, p. 645.

area for microprobe analysis is about one square micron, we speak in this paper of macroscopic homogeneity of distribution of the two metals for a section of the impregnated alumina extrudate.

A. Sárkány (*Inst. of Isotopes, Budapest*) said: We have found a similar order for metal–hydrocarbon bond strength, *i.e.* the strength of interaction increases in the sequence Pt < Pd < Ru < Os < Re. This result is confirmed not only by the increasing degree of hydrogenolysis, but also by the observation that deuterium exchange and hydrogenolysis are separable to a large extent on platinum and palladium (the temperature gap is about two hundred Kelvin degrees) whereas on ruthenium and rhenium the two reactions occur simultaneously. It seems, therefore, that on platinum and palladium a hydrogen-rich hydrocarbon species exists, which may explain the superior isomerization ability of these metals. On the other hand we failed to detect isomerization on rhenium and ruthenium.

R. Montarnal (*Inst. Franç. du Petrole, Rueil-Malmaison*) replied: Your results confirm that reactions involving desorption of products having the same number of carbon atoms as the initial hydrocarbon (exchange or isomerization) are favoured as compared to hydrogenolysis over those metals that form weak chemisorption bonds with hydrocarbons.

K. Kochloefl (*Girdler Süd-Chemie, Munich*) said: Recently we have studied the hydrogenolysis of some alkylbenzenes, *e.g.* toluene, ethylbenzene, and isopropylbenzene, over palladium, platinum, and rhodium supported on various carriers. The activity of these metals increases in the order Pd < Pt < Rh, whereas the selectivity of hydrogenolysis decreases in the same order. These results agree well with yours obtained for n-pentane hydrogenolysis. However, we observed that the carriers used, their preparation and especially any thermal pretreatment employed can affect very strongly the activity and selectivity of supported noble metals. Hence, I miss, in your study, a detailed characterization (sort, basicity or acidity) of the carriers used.

R. Montarnal (*Inst. Franç. du Petrole, Rueil-Malmaison*) replied: Our catalysts were prepared using 1·2 mm diameter extrudates of pure γ-alumina of surface area 250 m² g^{-1} and pore volume 0·55 cm³ g^{-1}. Any acidity of this support is irrelevant under our experimental conditions.

V. Haensel (*Universal Oil Products, Des Plaines*) said: The concept of expression of depth of hydrocracking or hydrogenolysis of n-pentane in terms of the expression:

$$P = \frac{(C_1 + C_2) - (C_3 - C_4)}{(C_1 + C_2 + C_3 + C_4)}$$

is an interesting one. It may be applicable where demethylation by metals is the sole reaction and sequential in character, thus:

$$C_5 + H_2 \longrightarrow C_4 + C_1, \qquad C_4 + H_2 \longrightarrow C_3 + C_1,$$
$$C_3 + H_2 \longrightarrow C_2 + C_1, \qquad C_2 + H_2 \longrightarrow 2C_1.$$

However, it is not applicable to cases of supported catalysts, even at low temperatures, where the acidity of the catalyst plays an important role and n-pentane can yield C_3 and C_4 without formation of C_1. This occurs through the presence on the surface of the catalyst of alkyl carbonium ions (C_n^+) and olefins ($C^=$) which can combine to form an intermediate higher species which then decomposes:

$$2C_5 \longrightarrow 2C_5^+ \longrightarrow C_3^= + C_2^+ + C_5^+ \tag{1}$$
$$C_3^= + C_5^+ \longrightarrow (C_8^+) \longrightarrow C_4^= + C_4^+ \tag{2}$$

giving the overall reaction: $2C_5 \longrightarrow C_2 + 2C_4$.

Thus, C_4 is formed from C_5 without the formation of C_1 and, in order to make a correlation, the contribution of both the metal and acidity must be considered in the depth of hydrogenolysis expression. Even if the support by itself gives very little acidic reaction, the acidity can be enhanced when the metal component is added and the overall result may be a concerted reaction involving both the metal and support in promoting an acid type reaction.

R. Montarnal (*Inst. Franç. du Petrole, Rueil-Malmaison*) replied: Blank experiments with pent-1-ene on our γ-alumina carrier have shown the absence of carbonium ion reactions at this temperature on the carrier alone. However, it is not impossible that the nature of alumina surface might be modified by the deposition of precious metal. Our argument for the hydrogenolysis mechanism on the metal is based on the distribution of the light hydrocarbon products. On platinum, palladium, and iridium we invariably observe a good equality between C_1- and C_4-hydrocarbon yields and between C_2- and C_3-hydrocarbon yields at a given temperature, which is characteristic of stepwise hydrogenolysis.

Structure and Catalytic Properties of Bimetallic Reforming Catalysts

A. V. Ramaswamy, Paul Ratnasamy,* and S. Sivasanker

Indian Institute of Petroleum, Dehradun, India

and A. J. Leonard

Laboratoire de Physico-chimie Minerale et Catalyse, University of Louvain, Belgium

ABSTRACT   The structural and catalytic properties of Pt–Al$_2$O$_3$ and Pt–Ir–Al$_2$O$_3$ are compared. For both, the mechanism and relative rates of sintering of the metal crystallites in hydrogen are similar. Pt–Ir–Al$_2$O$_3$ is less active in dehydrogenation reactions compared to Pt–Al$_2$O$_3$. In presence of n-heptane, the deposition of "coke" is lower on the bimetallic catalyst under identical conditions. Pt–Ir–Al$_2$O$_3$ is more active in hydrogenolysis reactions. When the catalysts were presulphided at 523K, a selective, permanent deactivation of the hydrogenolysis activity is observed on Pt–Ir–Al$_2$O$_3$. It is suggested that the "dilution" of Pt by an element with a relatively lower dehydrogenation activity, viz., Ir, leads to lower surface concentrations of "coke" precursors on Pt–Ir–Al$_2$O$_3$ and hence to reduced fouling rates on it. However, it possesses a higher hydrogenolysis activity, thereby necessitating presulphidation to poison selectively sites responsible for that activity.

Introduction

Platinum catalysts had been dominating the field of naphtha reforming ever since their introduction in 1949 by U.O.P. Recently, both Pt–Re[1-4] and Pt–Ir[5,6] catalysts have been claimed to be superior to Pt catalysts. A particular advantage claimed for both these systems [4-6] is their outstanding stability with regard to product distribution-selectivity, enabling longer run (cycle) lengths at severity conditions similar to or even higher than those of Pt catalysts. Now, what are the distinguishing features in the structural and catalytic properties of Pt–Ir–Al$_2$O$_3$ and Pt–Re–Al$_2$O$_3$ compared to those of Pt–Al$_2$O$_3$ which account for their superior performance? The present study is an attempt to answer this question for the Pt–Ir–Al$_2$O$_3$ system.

The role, if any, of Ir in stabilizing the metal dispersion was investigated from the relative rates of sintering of the metal surface area of both the mono- and bi-metallic catalysts in H$_2$. The activities of the catalysts for hydrogenation–dehydrogenation, hydrogenolysis, and dehydrocyclization reactions as well as for "coke" deposition were compared. Since sulphur has recently been claimed[7] to be a useful tool for increasing the stable life of Pt catalysts, its influence on both Pt–Al$_2$O$_3$ and Pr–Ir–Al$_2$O$_3$ has also been studied. The results enable us (a) to account for the superiority of Pt–Ir–Al$_2$O$_3$ over Pt–Al$_2$O$_3$ in high severity reforming, and (b) to offer a plausible explanation for the difference in catalytic selectivities of Pt and Ir in hydrocarbon reactions.

Experimental

M1 is a platinum–alumina catalyst. M2 and M3 were obtained from M1 by sintering it in a flow of pure, dry hydrogen at 953K for 14·4 and 72·0 ks, respectively. B1 is a platinum–iridium–alumina catalyst. B2 to B5 were obtained by sintering B1 under similar conditions for 3·6, 14·4, 36·0, and 72·0 ks, respectively. A conventional high-vacuum adsorption system was used for determining the BET and the metal surface areas and ammonia adsorption isotherms. Both the experimental procedure and the

Table 1
Physicochemical Properties of Catalysts

Catalyst	H/M ratio	Metal surface area (MSA) m²/kg cat.	Metal Cryst. size $(\bar{d})$ nm	MSA/S_{BET} $\times 10^3$	NH_3ads. per 10 sq.nm	M–M_1 pm (Fig. 2)	Turn-over number	H/M on regeneration
M1	0·87	840	0·97	3·2	5·1	284	—	—
M2	0·24	230	3·54	1·2	5·6	284	—	—
M3	0·17	160	4·97	0·81	4·9	286	—	0·34
B1	0·64	690	1·32	2·8	3·9	282	47	—
B2	0·16	170	5·31	0·78	—	285	66	—
B3	0·19	200	4·49	0·90	5·8	284	35	—
B4	0·14	150	5·99	0·70	—	283	43	—
B5	0·12	130	7·08	0·60	4·5	284	40	0·43

Note:

1. The method of Benson and Boudart[9] was used for calculating the H/M (hydrogen/total metal) ratios.

2. The crystallite sizes were calculated from H/M ratios assuming cubic metal crystallites exposing five of their six faces to the adsorbing gas.

3. Ammonia adsorption was carried out at 573K and 1·33 kNm^{-2}. The results are expressed as the number of NH_3 molecules adsorbed per 10 sq.nm of the surface.

4. Turnover number is defined as the number of molecules of cyclohexane produced from benzene per exposed metal atom per kilosecond. Conditions: 378 ± 3K; benzene feed rate = 0·0193 mole s^{-1} kg^{-1}; Hydrogen/benzene mole ratio = 5·5.

5. The sintered catalysts were regenerated in flowing oxygen while raising the temperature from 303 to 773K over a period of about 7·2 ks and cooling in the same atmosphere.

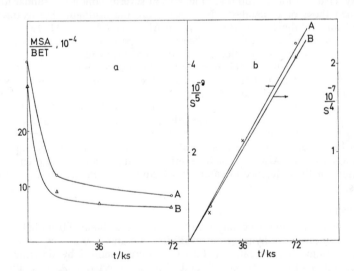

FIGURE 1a   Relative rates of sintering of M 1 (curve A) and B 1 (curve B), respectively

1b   Plot of $1/S^{n-1}$ against time for M 1 (A) and B 1 (B), respectively.

method of calculating the interatomic distances from X-ray scattering data have been detailed elsewhere.[8] A downflow, glass reactor was used for the hydrogenation of benzene, dehydrogenation of cyclohexane, and dehydrocyclization of n-heptane, all at 9.6×10^4 Nm^{-2} pressure. Care was taken to eliminate mass and heat transfer effects on the rate data.

RESULTS

Table 1 lists the physicochemical properties of the catalysts. In Fig. 1a, the ratio of metal surface to total BET area is plotted against the duration of sintering for both the mono- and bi-metallic catalysts. From the similar rates of fall of both the curves, it may be concluded that there are no significant differences in the rates of sintering of the metal crystallites in Pt–Al$_2$O$_3$ and Pt–Ir–Al$_2$O$_3$. In order to probe deeper into the mechansim of sintering, the kinetic data were fitted to the equation,[10]

$$- \, dS/dt = k \, S^n$$

where, S is the metal surface area, t is time and k and n are constants. Different values of n were tried by statistical regression and a value of $n = 6$ was found to fit most satisfactorily the data for catalysts M1 to M3, while the bimetallic samples gave the best fit for $n = 5$ (Fig. 1b). From an analysis of the sintering phenomena in supported metal crystallites, Ruckenstein and Pulvermacher[10] had postulated that when the values of the exponent n in the above equation fall between 3 and 8, the diffusion of the metal

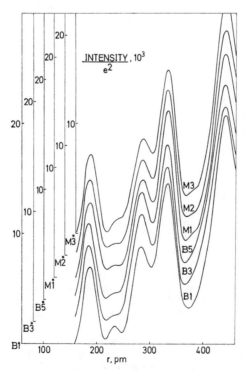

FIGURE 2   The radial electron density distribution curves; The code numbers of the catalysts are indicated in the curves.

crystallites on the alumina surface is the rate controlling step in the overall sintering process. Thus, on both Pt-Al₂O₃ and Pt-Ir-Al₂O₃, the mechanism of sintering is probably similar and the presence of Ir does not confer any additional stability against sintering to the metal crystallites under reduction conditions. This is not very surprising in view of the similar properties of the alumina surface in M1 and B1, like surface area, chlorine content, and acidity (Table 1). Others[11,12] have also reported values of n between 3 and 8 for the sintering of Pt-Al₂O₃ in hydrogen. On subsequent regeneration by an O₂-H₂ treatment, however, sintered Pt-Ir-Al₂O₃ regained higher values of the metal dispersion (Table 1).

The radial electron distribution curves obtained from the X-ray scattering data are shown in Fig. 2. The interatomic distance vector at 185 pm refers to the Al^IV-O (175 pm) and Al^VI-O (190 pm). The peak at 280 pm comprises the vectors Al^VI-Al^VI (279 pm) and M-M₁ (M = Pt, Ir; Pt-Pt₁ = 278, Ir-Ir₁ = 272, Pt-Ir = 274 pm). The resolution and accuracy of the r.e.d. method (about 5 pm) does not enable the distinction between Pt-Pt and Pt-Ir, or Ir-Ir vectors. The average value of M-M₁ for all the eight samples 283 ± 5 pm, Table 1) agrees very well with earlier values of Pt-Pt₁ in Pt-Al₂O₃ and Pt-SiO₂-Al₂O₃ catalysts[8] and also with the M-M₁ distances in the bulk metals. Hence, we conclude that the Pt-Pt(Ir) distances in our catalysts are similar, within the limitations of this method, to those for the bulk metals and that these values do not change on sintering.

Table 2

Dehydrogenation of Cyclohexane to Benzene on M1 and B1 catalysts

Reaction Parameters		Catalysts	
		M1	B1
Turnover number[a] at	498K	4·7	3·4
	523K	12·4	9·2
	548K	18·6	16·4
Apparent activation energy, kJ mole⁻¹		69·5	70·3
Pre-exponential factor, A, s⁻¹ × 10⁸		3·3	2·5

[a] Turnover number is defined as the number of benzene molecules formed from cyclohexane per second per exposed metal atom; Conditions: cyclohexane feed rate, 0·0239 mole s⁻¹ kg⁻¹; pressure, 9·6 × 10⁴ Nm⁻²; duration of run, 3·6 ks.

The rates of hydrogenation of benzene on Pt-Ir-Al₂O₃ catalysts was found to be zero-order with respect to the partial pressure of benzene under the reaction conditions employed. The hydrogenation activity expressed as the turn-over number is independent of the metal crystallite size (Table 1). From the relative rates of dehydrogenation of cyclohexane to benzene on M1 and B1 given in Table 2, it is seen that the latter is consistantly less active than the former. In the reactions of n-heptane (Table 3), the bimetallic catalysts are less active in dehydrocyclization and "coke" formation than Pt-Al₂O₃ but give rise to more hydrogenolysis products. The product yields were also more stable over B1 than on M1. Under the conditions given in Table 3, for example, the percentage conversion into toluene fell from 10·2 to 6·0 vol.% between 3·6 and 19·8 ks of reaction in the case of Pt-Al₂O₃. The corresponding values for Pt-Ir-Al₂O₃ were 4·3 and 3·9%, respectively. These data reveal the existence of a mechanistic and most probably a structural interaction between the Pt and Ir components.

Preliminary studies of the influence of sulphur on Pt-Ir-Al₂O₃ catalysts indicated that the catalytic activity was drastically attenuated when molecules containing sulphur were introduced along with the n-heptane feed at 793K. If, however, the catalysts were sulphided at a lower temperature (523K) prior to the n-heptane run, the product pattern, given in Table 4, was different. While the dehydrocyclization and hydrogenolysis activity of both M1 and B1 are initially suppressed by sulphur, the latter activity remains permanently suppressed even in the steady state in the case of catalyst B1 (1·5 and 1·7% or C_1–C_6 products).

Table 3

Reactions of n-heptane[a]: Products Distribution

Products (vol %)	Catalysts			
	M1	M2	B1	B3
n-Heptane conversion	38·4	39·0	36·0	38·1
Toluene	6·0	1·0	3·9	1·0
Total aromatics	12·0	5·2	6·6	5·1
C_1–C_5 products	16·0	20·0	22·0	24·4
C_6–C_7 products	10·4	13·8	7·4	8·6
Amount of 'coke' deposited,[b]				
during i) 19·8 ks	4·2	4·1	2·3	3·5
ii) 43·3 ks	9·0	—	5·7	—

[a] Conditions: $9·6 \times 10^4 Nm^{-2}$; 793 K; H_2/n-C_7 = 2·0; Feed rate = 0·017 mole $s^{-1}kg^{-1}$. The values are given as percentage conversions obtained, 19·8 ks after the start of the reaction, except where indicated.
[b] Percent by weight.

Table 4

Influence of Presulphidation on Product Distribution in Reaction of n-Heptane

Volume %

Catalyst	Conversion	Toluene	Total aromatics	C_1–C_6	C_7
M1	47·5(38·6)	4·5(2·6)	4·5(3·6)	15·0(0·7)	34·3(27·9)
B1	36·0(31·7)	3·6(0·1)	5·0(0·1)	1·5(1·7)	29·5(29·9)

Note:

1. The values refer to the steady state conditions (19·8 ks after the start of the reaction). Values in parantheses refer to initial (3·6 ks after commencing the n-heptane reaction) product distributions.

2. The catalysts were presulphided at 523 K, with n-heptane containing 3000 ppm of thiophen at $9·6 \times 10^4 Nm^{-2}$ pressure for 2·4 ks, at H_2/feed mole ratio of 1:1. n-Heptane test reaction was then carried out under the conditions given in Table 3.

DISCUSSION

At the low hydrogen pressures necessary to obtain good yields of high-octane gasoline, dehydrogenation reactions are favoured compared to hydrogenolysis or hydrocracking reactions.[13] It is also known that Pt is more active than Ir in C–H cleavage reactions (see Table 3). For example, the value of x in the species, C_2H_x formed by adsorbing C_2H_6 was found[14] to be zero for Pt but two for Ir. Thus, Pt catalysts due to their high dehydrogenative ability lead to excessive surface concentrations of olefinic intermediates and subsequent rapid catalyst fouling by "coking", as observed in practice. Hence the use of high hydrogen pressures with Pt–Al_2O_3 catalysts. In the case of Pt–Ir–Al_2O_3 catalyst, part of the surface Pt atoms are replaced by Ir. This replacement of Pt, which is highly active in C–H cleaveage, by a less reactive Ir leads to a "dilution" effect, as evidenced from the lower concentrations of benzene formed on Pt–Ir–Al$_2$O$_3$ (Table 2). In fact, the similar values of the apparent activation energy observed for both Pt–Al$_2$O$_3$ and Pt–Ir–Al$_2$O$_3$ suggest that the nature of the active centres and the reaction mechanism are probably similar on both the catalysts. The lower value of the pre-exponential factor for Pt–Ir–Al$_2$O$_3$, suggesting a lower *number* of active centres is also in accord with the above picture of "dilution" of surface Pt by Ir. Since the Pt–Ir–Al$_2$O$_3$ surface possesses a lower dehydrogenation capacity, the concentration of "coke" forming olefin precursors will be lower on its surface leading to lower fouling rates and hence better product-yield stability. The relatively higher selectivity of Pt–Ir–Al$_2$O$_3$ for the hydrogenolysis rather than the dehydrogenation reaction causes a lower yield of the liquid products during the reforming process. Presulphidation of the catalyst at lower temperatures eliminates this drawback (see Table 4). Apparently, sulphur compounds are adsorbed more strongly on Ir than on Pt.

What are the structural factors that cause the higher "coking" rate on Pt–Al$_2$O$_3$ and the lower "coking" but higher hydrogenolysis rates on Pt–Ir–Al$_2$O$_3$? These differences in selectivity are all the more striking in view of the closely similar structural properties of bulk Ir and Pt metals.[15] The M–M$_1$ distances, for example, are 272 and 278 pm for Ir and Pt, and the work functions are 521 and 508 kJ, respectively. Now, it may be recalled that X-ray scattering experiments[8,16] had revealed that in the top two surface layers of Pt crystallites in supported Pt–Al$_2$O$_3$ and Pt–SiO$_2$–Al$_2$O$_3$, the metal atom to metal vacancy stoicheiometric ratio corresponds to 1. Surface Pt atoms, thus, possess a low co-ordination number. Hence, both face as well as edge and corner atoms are structurally equivalent in that both present an almost isolated Pt atom relatively free of the electronic and steric influences of neighbouring atoms to the adsorbed molecules.[16] Let us now consider Ir and Pt atoms. Their ground state valence electronic structures are $5s^2p^6d^9s^0$ and $5s^2p^6d^9s^1$, respectively.[17] Their electron energy diagram[18] derived from the Thomas–Fermi–Dirac method reveals that the separation of the 5d and 6p energy levels increases from Ir to Pt. Thus, promotion of a 5d electron to the higher levels will be easier for Ir than for Pt. This, in turn, confers greater ability for Ir to accept electrons (from the adsorbed hydrocarbon molecules, for example) in the d shell.

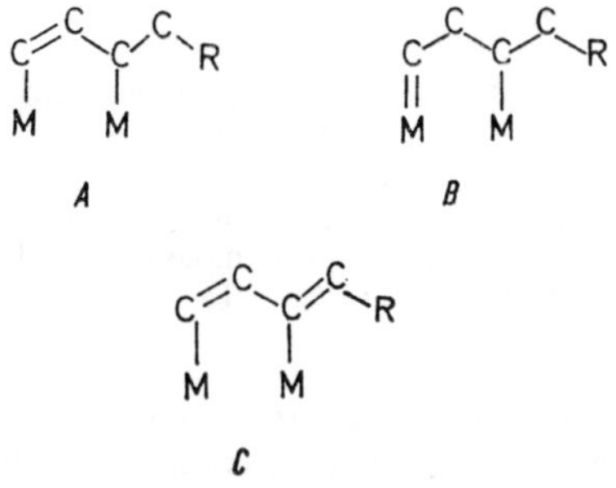

FIGURE 3   Adsorbed hydrocarbon species.

Applying these concepts to the catalytic phenomenon, it may be recalled that Anderson and Avery[19] had postulated the formation of 1,3-diadsorbed intermediates like A and B (Fig. 3) for the adsorption of saturated hydrocarbon molecules over Pt. The distinguishing feature of B is the presence of a multiple metal–carbon bond. On both

Pt and Ir, structures A and B are both present. Their relative preponderance will, however, differ. The presence of the double bond in B implies that the metal atom is able to accomodate more electrons from carbon in its d orbitals. As pointed out earlier this process is likely to be more favourable in the case of Ir. Pt, on the other hand, with a larger difference between its $5d$ and $6p$ energy levels, will form comparatively fewer metal–carbon multiple bonds and hence structure A will predominate over Pt. An additional factor favouring the formation of A over Pt and B over Ir is the matching of the C_1–C_3 distance with the M–M_1 distances. The C_1–C_3 distances are 263 and 252 pm in A and B, respectively. Due to the bond strain involved in fitting the C_1–C_3 distance to the M–M_1 distance, structure B will be energetically more stable over Ir with its shorter M–M_1 bond length compared to Pt. Now, since C–H cleavage occurs more readily on Pt, formation of intermediates like C (Fig. 3) from structures like A is also more likely over Pt than on Ir. Multiply-bonded polyolefins like C are the probable precursors to "coke". Intermediates like B on the other hand, (multiply bonded to the surface) are expected to lead to C–C cleavage and hence to hydrogenolysis products, as suggested by Anderson and Avery.[19] Differences in the catalytic selectivity of Pt–Al_2O_3 and Pt–Ir–Al_2O_3, thus, arise from the basic differences in the electronic properties of Pt and Ir.

ACKNOWLEDGMENT. We are grateful to the Director, Computer Centre, University of Louvain for the computing facilities. We thank Dr. S. K. Goyal and colleagues of the Microanalysis Lab. for "coke" determinations, and colleagues of the Analytical Physics Section for the gas chromatographic analyses. We are grateful to Drs. I. B. Gulati and K. K. Bhattacharyya for encouragement and support.

REFERENCES

[1] R. L. Jacobsen, H. E. Kluksdahl, C. S. McCoy, and R. W. Davis, *Preprints, API, Division of Refining*, May, 1969, 37–69

[2] M. J. Sterba, P. C. Weinert, A. G. Lickus, E. L. Pollitzer, and J. C. Hayes, *Oil & Gas J.*, 1968, **66**(53), 140.

[3] F. W. Kopf, W. C. Pfefferle, M. H. Dalson, W. A. Decker, and J. A. Nevison, *Preprints, API, Division of Refining*, May, 1969, 23–69.

[4] H. E. Kluksdahl, *U.S. Patent*, 3,415,737 (1968).

[5] B. Spurlock and R. L. Jacobsen, *U.S. Patent*, 3,507,781 (1970).

[6] J. H. Sinfelt, *U.S. Patent*, 3,835,034 (1974).

[7] J. C. Hayes, R. T. Mitsche, E. L. Pollitzer, and E. H. Homer, *Preprints, Div. Petrol Chem., Amer. Chem. Soc.*, 1974, **19**(2), 334; J. H. Sinfelt, *U.S. Patents*, 3,835,034 and 3,839,194 (1974); S. Sivasanker and A. V. Ramaswamy, *J. Catalysis*, 1975, **37**, 553.

[8] P. Ratnasamy, A. J. Leonard, L. Rodrigue, and J. J. Fripiat, *J. Catalysis*, 1973, **29**, 374; P. Ratnasamy and A. J. Leonard, *Catalysis Rev.*, 1972, **6**, 293.

[9] J. E. Bensen and M. Boudart, *J. Catalysis*, 1965, **4**, 704.

[10] E. Ruckenstein and B. Pulvermacher, *J. Catalysis*, 1973, **29**, 224.

[11] H. L. Gruber, *J. Phys. Chem.*, 1962, **66**, 48.

[12] T. R. Hughes, R. J. Houston, and R. P. Sieg, *Ind. Eng. Chem., Process Des. Dev.*, 1972, **1**, 96.

[13] E. F. Schwarzenbek, Paper presented at the Joint Conference of the Chemical Institute of Canada and the American Chemical Society, Toronto, Canada, May 24–29, 1970.

[14] J. H. Sinfelt, *Adv. Catalysis*, 1973, **23**, 91.

[15] R. C. Weast, *Handbook of Chemistry and Physics* (Chemical Rubber Co., Cleveland, Ohio, 50th Edn., 1970) pp. B 27 and B 37.

[16] P. Ratnasamy, *J. Catalysis*, 1973, **31**, 466.

[17] Ref. 15, p. B 2.

[18] L. Pauling, *The Nature of the Chemical Bond* (Cornell University Press, Ithaca, 3rd Edn., 1960) p. 56.

[19] J. R. Anderson and N. R. Avery, *J. Catalysis*, 1966, **5**, 446.

DISCUSSION

H. Charcosset (*CNRS, Villeurbanne*) asked: What is the composition of catalyst B1? Can you comment on the redispersion of the sintered catalysts? What mechanism would you suggest for this phenomenon?

M. Shelef (*Ford Motor Co., Dearborn*) said: What do you envisage as the process responsible for the regaining of metal surface area on heating in oxygen to 773 K? Is it redispersion of the metal or simply removal of some material blocking the metal sites? Redispersion by volatile oxides is hardly likely under your conditions.

P. Ratnasamy (*Indian Inst. of Petroleum, Dehradun*), in reply to the two previous speakers, said: The ratio of platinum to iridium in our bimetallic catalyst was about 10:1. Owing to the higher metal–oxygen bond energy of iridium compared to platinum, incorporation of oxygen atoms in the metallic clusters during oxidation in molecular oxygen will be favoured more in the case of the bimetallic catalyst. Removal of these oxygen atoms on subsequent reduction in hydrogen will then lead to more 'porous' metallic crystallites in the case of Pt–Ir–Al_2O_3 catalysts, and this accounts for the higher value of the dispersion. Blocking of metal sites responsible for hydrogen adsorption by carbon does not arise because there was no carbon deposit on the regenerated catalysts.

R. Maatman (*Dordt College, Sioux Center*) said: In 1971 we reported on the platinum–alumina catalysed dehydrogenation of cyclohexane.[1] Our rates and activation energy were very close to those that you report. Assuming that the reaction is a simple surface process, we calculated, using statistical mechanics, that only one-millionth of the surface is active. There is always the possibility that this is the correct conclusion; for example, it may be that the surface is poisoned very soon after the catalytic reaction begins. However, it seems more likely that Boudart's conclusion is correct, namely that the reaction is facile and that all surface atoms are active. We concluded that our original assumption (that the reaction is simple) was wrong and that it was likely that cyclohexene or cyclohexadiene is an intermediate. We have made calculations using your data and it seems that the same conclusions can be drawn concerning your system. In your opinion, is either cyclohexene or cyclohexadiene a *necessary* intermediate in the dehydrogenation of cyclohexane in your systems?

P. Ratnasamy (*Indian Inst. of Petroleum, Dehradun*) replied: We have not observed any cyclohexadiene in our products though, under some conditions, we have observed small quantities of cyclohexene.

P. Tétényi (*Inst. of Isotopes, Budapest*) said: It is rather surprising that the increase in the activity for hydrogenolysis was not followed by the growth of coke formation. This is explained by the authors in terms of a parallel decrease in dehydrogenation activity leading to lower production of olefins which are precursors for coke formation. I am not sure, however, that olefins are the only precursors of coke. It is possible that the highly dehydrogenated hydrocarbon residues which lead from one side to hydrogenolysis serve from the other side as precursors for coke production.

Comparing iridium and platinum with respect to hydrocarbon chemisorption we have observed an easier removal of chemisorbed hydrocarbons from platinum than from iridium. From this point of view, the coke formation on iridium should have been greater than that on platinum.

[1] R. W. Maatman, P. Mahaffy, P. Hoekstra, and C. Addink, *J. Catalysis*, 1971, **23**, 105

The Selectivity of Oxide Catalysts for the Alkylation of Phenols with Methanol

Kozo Tanabe* and Tadao Nishizaki

Department of Chemistry, Faculty of Science, Hokkaido University, Sapporo 060, Japan

ABSTRACT In order to elucidate the nature of catalysts which are highly selective for the alkylation at the *ortho*-position of phenols with methanol, the reaction kinetics, the infrared spectra of adsorbed phenol and *o*-cresol, and the acid–base properties of catalysts have been studied, using $SiO_2–Al_2O_3$ and MgO as typical solid acid and base catalysts. It was revealed that the *ortho*-selectivity was strongly controlled by the adsorbed states of phenol and *o*-cresol, which varied depending on the acid–base properties of the catalysts. The active sites necessary for the selective methylation at the *ortho*-position were concluded to consist of basic sites and relatively weak acid sites. In accordance with the conclusion, a new catalyst, $TiO_2–MgO$, which satisfies the requirements of the conclusion, was found to show a high selectivity.

Introduction

The alkylation of phenol with methanol is industrially important as a reaction to produce 2,6-xylenol, a monomer of a good heat-resisting poly-(2,6-dimethyl)phenylene oxide resin. MgO,[1] Ce_2O_3–MnO–MgO,[2] MgO–Al_2O_3,[3] and Fe_2O_3–ZnO[4] were recently reported to be highly selective for the formation of 2,6-xylenol. However, no reason for the high selectivity is given and no reaction mechanism has been suggested. In the present work, the factors controlling the selectivity and the reaction mechanism have been studied by the following experiments using mainly SiO_2–Al_2O_3 and MgO catalysts: (i) In addition to the reaction of phenol with methanol, the reaction of anisole, *o*-cresol, *o*-methylanisole, and 2,6-xylenol with methanol were studied. (ii) Infrared studies on adsorbed phenol and *o*-cresol were performed. (iii) The acidic and basic properties of the catalysts were measured.

Experimental

Catalysts and Reagents. SiO_2–Al_2O_3 (Nikki Chem. Co. N631L) was ground and sieved to 16–24 mesh and then calcined at 400°C for 3 h. MgO was obtained by calcining its basic carbonate (Kanto Chem. Co.) at 530°C for 3 h. TiO_2–MgO was prepared by grinding titanic acid and magnesium hydroxide with a small amount of water in a kneader and by calcining at 500°C for 3 h. Titanic acid and magnesium hydroxide were prepared respectively by hydrolysing titanium tetrachloride and magnesium chloride with ammonia water. The catalysts were stored in ampoules before use. Methanol, phenol, cresols (*o*-, *m*-, *p*-), methyl anisoles (*o*-, *m*-, *p*-), and xylenols (2,6-, 2,4-, 2,3-) were all guaranteed reagents of Wako Pure Chemicals Co.

Reaction Procedure. The reaction was carried out at 250–500°C by a conventional flow method. The reaction mixture (phenol or aromatic compounds/methanol, molar ratio = 1) was passed through 1–2 g of catalyst bed at an appropriate flow rate. The catalysts were recalcined at reaction temperatures before the reaction. The reaction products were trapped with ice and analysed by gas chromatography on a column containing 25% DC-550 on Shimalite. The catalytic activity and selectivity were expressed by (mole % of all alkylated products/mole % of supplied phenol) × 100 and (mole % of a particular product/mole % of all alkylated products) × 100, respectively.

Infrared Spectra and Acid–Base Properties. Phenol or *o*-cresol was adsorbed at room temperature on a disc of catalyst which had been evacuated at 400 or 500°C for 3 h in an i.r. *in situ* cell and evacuated at elevated temperatures for 1 h. Infrared spectra of the samples were measured at room temperature. The acidic properties were measured by titrating a catalyst suspended in benzene with n-butylamine, using various Hammett indicators,[5] and the basic properties were measured by titrating similarly with benzoic acid, using various nitroanilines and Bromothymol Blue as indicators.[5]

RESULTS

Infrared Spectra of Adsorbed Phenol and o-*Cresol.* When phenol or *o*-cresol was adsorbed on SiO_2–Al_2O_3, the absorption band (3350 or 3400 cm^{-1}) of OH group of phenol or *o*-cresol was not observed and the band (3751 cm^{-1}) of OH group on SiO_2–Al_2O_3 shifted to low frequency and became broad; the spectra for *o*-cresol are shown in Figure 1. This indicates that phenol and *o*-cresol are adsorbed as phenolate and methyl phenolate and that H^+ of OH of phenol or *o*-cresol reacts with O^{2-} on the surface to form a new OH. In the case of MgO, infrared studies also showed that phenolate and methyl phenolate were formed on the surface from phenol and *o*-cresol,

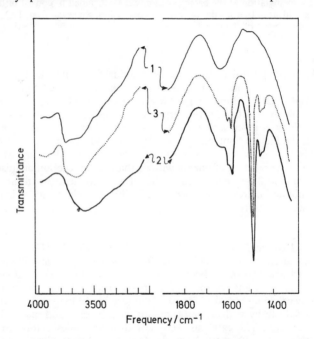

FIGURE 1   Infrared spectra of *o*-cresol adsorbed on SiO_2–Al_2O_3. 1: SiO_2–Al_2O_3 evacuated at 400°C for 3 h. 2: *o*-cresol was adsorbed at room temperature and evacuated at 100°C for 1 h. 3: after evacuating sample 2 at 250°C for 1 h.

respectively. The adsorbed species disappeared on addition of methanol followed by a brief evacuation. Since the methylated compounds desorbed on a brief evacuation, the species are suggested to be intermediates of the alkylation. On SiO_2–Al_2O_3, the ratio of the intensity of the band at 1496 cm^{-1} to that around 1600 cm^{-1} (both of the bands are assigned to the in-plane skeletal vibrations of benzene ring) was quite different from that of phenol or *o*-cresol in the liquid state. On MgO, however, the ratio was the same as that in the liquid state. These results indicate that there is an interaction between the benzene ring of phenolate or methyl phenolate and the catalyst surface for SiO_2–Al_2O_3, while no such interaction exists between them for MgO, as shown in Figure 2.

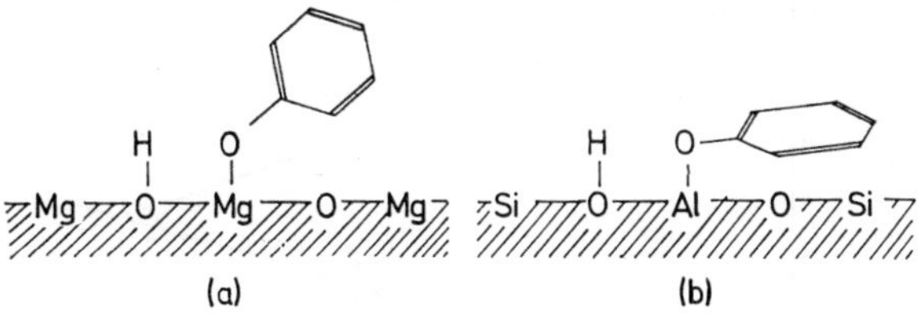

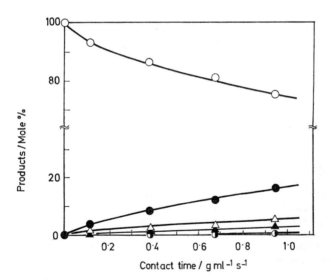

FIGURE 2 The adsorbed states of phenol. (a) MgO: no interaction between benzene ring and catalyst surface (b) SiO_2–Al_2O_3; interaction between benzene ring and catalyst surface.

Acidic and Basic Properties. The acidities of SiO_2–Al_2O_3 used were 0·13, 0·17, and 0·27 mmol g^{-1} at 3·3 $\geqslant$ acid strength (H_0) >—3·0, —3·0 $\geqslant$ H_0 > —8·2, and H_0 $\leqslant$ —8·2, respectively, the total acidity at H_0 $\leqslant$ 3·3 being 0·57 mmol g^{-1}. The basicities of MgO were 0·23, 0·13, and 0·05 mmol g^{-1} at 7·1 < basic strength (pK_{BH}) < 15·0, 15·0 < pK_{BH} < 18·4, and pK_{BH} > 18·4, respectively, the total basicity $(pK_{BH} = 7·1)$ being 0·41 mmol g^{-1}. TiO_2–MgO (molar ratio = 1) showed both acidic (0·2 mmol g^{-1} at 1·5 $\geqslant$ H_0 > 4·0) and basic properties (1·2 mmol g^{-1} at pK_{BH} $\geqslant$ 15·0).

Reactions of Various Compounds with Methanol. The reaction of phenol with methanol over SiO_2–Al_2O_3 formed anisole, *o*-, *m*-, and *p*-cresol and a small amount (1%) of *o*-, *m*-, and *p*-methylanisole as shown in Figure 3, while the reaction over MgO pro-

FIGURE 3 The reaction of phenol with methanol over SiO_2–Al_2O_3.
Reaction temperature; 273 $\pm$ 6°C
○; phenol, ●; anisole, △; *o*-cresol, ▲; *m*-, *p*-cresol, ◐; *o*-, *m*-, *p*-methyl anisole

duced *o*-cresol, 2,6-xylenol, and small amounts of *m*- and *p*-cresol as shown in Figure 4. Over SiO_2–Al_2O_3, anisole reacted with methanol to form mainly phenol and *o*-methyl-

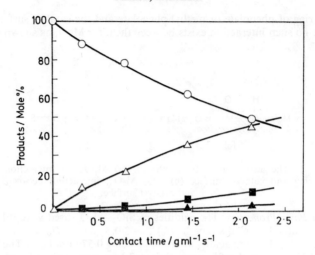

FIGURE 4   The reaction of phenol with methanol over MgO.
Reaction temperature; 496 ± 1°C
○; phenol, △; o-cresol, ▲; m-, p-cresol, ■; 2,6-xylenol

Table 1

Various reactions at 273 ± 6°C over SiO_2–Al_2O_3

Reactants	Anisole + methanol	o-Cresol + methanol	o-Methylanisole	2,6-Xylenol + methanol
Contact time (g cm^{-3} s^{-1})	0·8	0·8	0·8	*
Products (mole %)				
Phenol	11	1	0	0
Anisole	82	0	0	0
o-Methylanisole	3	11	86	1
m-,p-Methylanisole	1	0	0	0
o-Cresol	1	78	12	10
m-,p-Cresol	1	0	0	0
2,6-Xylenol	0	4	2	71
2,4-Xylenol	1	5	1	3
2,3-Xylenol	0	1	0	1
2,4,6-Trimethylphenol	0	0	0	14

* Catalyst: 1·52 g, flow rate: 0·182 cm³ min⁻¹, 2,6-xylenol/methanol (molar ratio) = 0·5.

anisole (Table 1) and the reaction of anisole alone gave almost the same results. Over MgO, the reaction of phenol with methanol formed 5·3 mole % of cresols at 450°C, whereas the reaction of anisole with methanol formed only 0·4 mole % of o-cresol at

Table 2
Various reactions over MgO (Contact time = 0·6 g cm⁻³ s⁻¹)

Reactants	Phenol + methanol	Anisole + methano	Anisole + methanol	2,6-Xylenol	2,6-Xylenol + methanol
Reaction temp. °C	452	453	505	496	496
Products (mole %)					
Benzene	0	0·2	11·0	0	0
Toluene	0	0	3·3	0	0
Anisole	0·3	97·1	43·2	0	0
Phenol	94·5	0·3	11·5	1·4	0
o-Methylanisole	0	0·7	2·0	0	0
o-Cresol	3·6	0·4	10·0	16·2	3·2
m-,p-Cresol	1·7	0	0·6	0·9	0
2,6-Xylenol	0	1·0	13·0	53·7	65·9
2,4-Xylenol	0	0·2	2·3	9·1	5·9
2,4,6-Trimethylphenol	0	0	3·2	18·7	25·0

almost the same temperature, but formed benzene and toluene at 505°C (Table 2). In the case of SiO₂–Al₂O₃, o-cresol reacted with methanol to form o-methylanisole, 2,6-, 2,4-, and 2,3-xylenol, and a small amount of phenol, but did not isomerize to m- or p-cresol (Table 1), while, in the case of MgO, the reaction of o-cresol with methanol gave 63% of 2,6-xylenol, 22% of phenol, 7% of 2,4,6-trimethylphenol, and 4% of m- and p-cresol. The reaction of o-methylanisole alone over SiO₂–Al₂O₃ gave o-cresol,

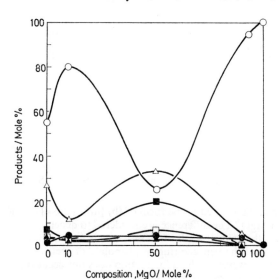

FIGURE 5   The alkylation of phenol with methanol over TiO₂–MgO of various compositions. Reaction temperature; 400°C.
○; phenol, ●; anisole, △; o-cresol, ▲; m-, p-cresol, ■; 2,6-xylenol, □; 2,4-xylenol

2,6-xylenol, and 2,4-xylenol (Table 1) and the reaction in the presence of methanol gave the same products, no isomerization to *m*- or *p*-methylanisole being observed. Over SiO_2–Al_2O_3, 2,6-xylenol reacted with methanol to form *o*-cresol, 2,4,6-trimethylphenol, 2,4-xylenol *etc.* (Table 1), and the reaction of 2,6-xylenol alone formed *o*-cresol by dealkylation. Over MgO, 2,6-xylenol formed mainly *o*-cresol and 2,4,6-trimethylphenol.

TiO$_2$–MgO (molar ratio = 1) showed a high *ortho*-selectivity (85 %) as well as a high activity, compared with TiO$_2$, MgO, and TiO$_2$–MgO of other compositions (Figure 5). At 400°C, MgO was completely inactive.

DISCUSSION

Factors Controlling Positional (o-, m-, p-) *Selectivity.* SiO$_2$–Al$_2$O$_3$ produced o-, m-, and p-isomers of cresol and xylenol, while MgO produced mainly the o-isomers. This can be considered to be due to the difference in the adsorbed states of phenol and o-cresol as shown in Figure 2, which are evidenced by infrared study. Since, in the case of SiO$_2$–Al$_2$O$_3$, the plane of the benzene ring of phenolate or methyl phenolate is close to the catalyst surface, any of the o-, m-, and p-positions can be attacked by a methyl cation formed from methanol. On the other hand, only the o-position can be methylated in the case of MgO, because the o-position is near to the catalyst surface and the m- and p-positions are located some distance away from the catalyst surface.

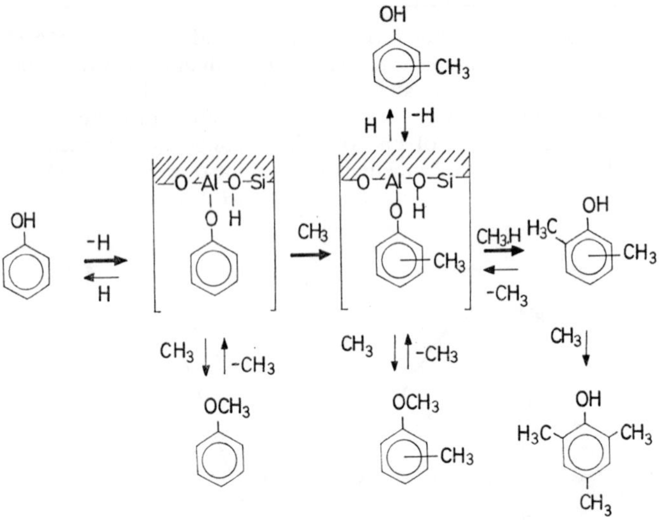

FIGURE 6   Reaction mechanism of alkylation over SiO$_2$–Al$_2$O$_3$

Then why is phenol adsorbed in the form of (b) [in Figure 4] on SiO$_2$–Al$_2$O$_3$ and in the form of (a) on MgO? The difference is considered to depend on the acid strength of the catalysts. Since the acid strength of SiO$_2$–Al$_2$O$_3$ is very high, the acid sites interact with the π-electrons of the benzene ring of phenolate or methyl phenolate, giving the adsorbed form (b). However, such an interaction does not occur on very weakly acidic MgO, and the adsorbed form (a) is produced.

At low reaction temperature (275°C), MgO, CaO,[6] SrO,[6] TiO$_2$,[7] ZnS,[7] Fe$_2$O$_3$–ZnO,[4] and MgO–Al$_2$O$_3$,[3] which have no strong acid sites, are inactive. Silica–alumina, Al$_2$O$_3$, and cation exchanged zeolites, which have strong acid sites, are active.[7] The higher the acid strength or acidity, the higher the activity, though the selectivity is low.[7] At high reaction temperature (450–500°C), strongly acidic catalysts showed very low selectivity. Basic solids such as MgO, CaO, SrO, MgO–Al$_2$O$_3$ *etc.* became active only at high temperature (450–500°C) and showed a high selectivity for *ortho*-position alkylation. Of MgO, CaO, and SrO, which are strongly basic, the weakly acidic MgO is the most active and selective.[7] The acid sites on TiO$_2$–MgO are so weak that they do not interact with a benzene ring, but are sufficiently strong to produce methyl cations. Thus, TiO$_2$–MgO, which contains such acid sites as well as strong basic sites, becomes active and shows high selectivity under moderate reaction conditions (400°C).

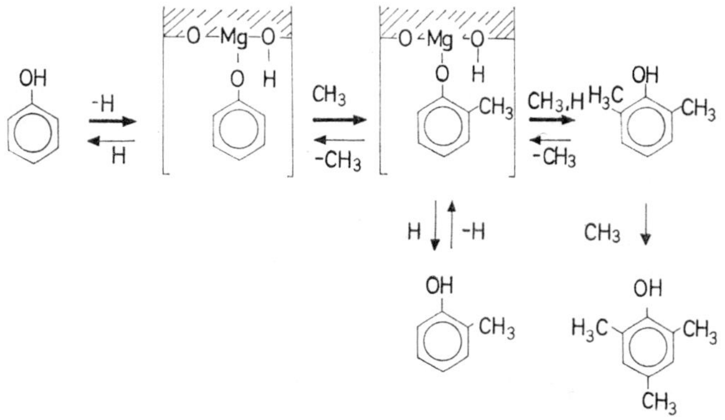

FIGURE 7   Reaction mechanism of alkylation over MgO

Reaction Mechanism. Reaction mechanisms for the alkylation over SiO$_2$–Al$_2$O$_3$ and MgO, shown in Figures 6 and 7, are suggested on the basis of kinetic and infrared studies described in the foregoing section. In the case of SiO$_2$–Al$_2$O$_3$, *o*-cresol is not directly formed from anisole, since there is no maximum in the time-variation curve of anisole formation, and the amount of *o*-cresol formed by the reaction of phenol with methanol is larger than that formed by the reaction of anisole alone. Thus, *o*-cresol is formed by the methylation of phenolate. Since *o*-cresol and 2,6-xylenol do not isomerize to *m*- and *p*-isomers, the distribution of the isomers is determined in the steps of methylation of phenolate and methyl phenolate. Therefore, the adsorbed state shown in (b) of Figure 2 becomes important.

In the case of MgO, anisole and methylanisole are not formed. However, it may be that anisole, once formed, changes instantly to *o*-cresol at high reaction temperature. This possibility is excluded by the facts that a larger amount of *o*-cresol is formed by the reaction of phenol with methanol than by the reaction of anisole alone, and that benzene and toluene are formed by the reaction of anisole with methanol, but not by the reaction of phenol with methanol at 500°C. As in the case of SiO$_2$–Al$_2$O$_3$, the steps of methylation of phenolate and methylphenolate are important and the adsorbed state shown in (a) of Figure 2 causes the formation mainly of *o*-cresol and 2,6-xylenol.

ACKNOWLEDGEMENTS. We are indebted to Professor H. Hattori for helpful discussions, and Mr. T. Sumiyoshi for his experiment on the alkylation over TiO$_2$–MgO.

REFERENCES

[1] Neth. Appl. 6506830, Nov. 30, 1965 (General Electric).
[2] S. Enomoto and M. Inoue, *Ann. Meeting, Catalysis Soc. Japan*, Sendai, 1968, Preprints, p. 3.
[3] Y. Fuduka, T. Nishizaki, and K. Tanabe, *Nippon Kagaku Zasshi*, 1972, 1754.
[4] T. Kotanigawa, M. Yamamoto, K. Shimokawa, and Y. Yoshida, *Bull. Chem. Soc. Japan*, 1971, **44**, 1961.
[5] K. Tanabe, *Solid Acids and Bases* (Kodansha, Tokyo; Academic Press, New York, London, 1970), Ch. 2 and 3.
[6] T. Nishizaki and K. Tanabe, *Shokubai*, 1973, **15**, 94.
[7] T. Nishizaki and K. Tanabe, *Shokubai*, 1972, **14**, 138.

DISCUSSION

J. Shabtai (*Weizmann Inst. of Science, Rehovot*) said: I should like to draw your attention to six papers, five of them published in *J. Org. Chem.* in 1968 and 1970, in which Klemm, Taylor, and I considered the mechanism of alumina-catalysed *ortho*-alkylation of phenolic compounds (*e.g.* naphthols, 5-indanol, and phenol itself), with methanol. The mechanism proposed is essentially the same as that discussed in your paper as far as the main steric aspects are concerned, *i.e.* we suggested a vertical orientation of the phenoxide species, which prevents *para*-alkylation and results in selective *ortho*-alkylation by the adsorbed methylating agent. However, there are important differences in mechanistic detail. First, the bond between the cationic site and the phenoxide oxygen cannot be a typical covalent bond as indicated in your scheme, but should be considerably polarized, especially when the bonding is to alumina. Secondly and more important, your suggestion that the alkylating agent derived from methanol is a methyl carbonium ion is highly unlikely. Rather, we have proposed that the reaction proceeds on an ensemble of acidic and basic sites by a concerted mechanism, involving electrophilic attack on the *ortho*-position of the phenoxide species by a surface methoxide group. Our molecular orbital calculation of superdelocalizabilities for electrophilic attack in phenoxide show that *O*-alkylation should be strongly favoured. The negligible extent of this type of alkylation in the presence of alumina indicates that the phenoxide oxygen is effectively shielded by the surface from such an attack, in agreement with our steric model.

K. Tanabe (*Hokkaido Univ.*) replied: We also have results obtained using alumina as catalyst. One difference between the catalytic activities of alumina and magnesia is that anisole is formed over alumina, but not over magnesia, due to the difference in the nature of the metal–oxygen bond in the adsorbed phenolate. Since the aluminium ion is more electronegative than the magnesium ion, the metal–oxygen bond of aluminum phenolate can be attacked both by a methyl cation and by a proton, whereas that of magnesium phenolate can be attacked only by a proton. We consider that the bond is less polarized for magnesia than for alumina. A second difference is that the *ortho*-selectivity is lower in the case of alumina, because of the higher acid strength of this oxide.

As to the formation of methyl cations, we speculate that methanol is adsorbed on acidic hydroxyl groups (formed by the reaction of surface oxygen anions with protons liberated by the dissociation of phenol) to form alkoxonium ions which easily produce methyl cations at 450 °C.

P. Pichat (*CNRS, Villeurbanne*) said: Your conclusions concerning the adsorption modes of phenols are based only on intensity changes in i.r. spectra and may thus be considered with some caution, since these intensity changes are not frequently used. They are difficult to explain and may arise from various effects, either in solutions or on solids. U.v. reflectance spectroscopy would be useful to confirm these adsorption modes. Did you obtain u.v. spectra of your samples?

K. Tanabe (*Hokkaido Univ.*) replied: We have not yet obtained u.v. reflectance spectra. However, since it is the change in intensity ratio of two bands that we observed, and since the intensity ratio changed over several solid acids but not over several solid base catalysts, our conclusion seems to be sound.

W. C. Conner (*Allied Chemical, Morristown*) said: The controversy concerning CH_3^+ or CH_3O attack on the phenolate surface ion may be answered by studying the reactions using isotopes. Mixtures of $CH_3^{18}OH$ with $^{13}CH_3OH$, and of $^{13}CH_3OH$ with CH_3OD should exchange differently by the two mechanisms (*i.e.* with and without phenol).

K. Tanabe (*Hokkaido Univ.*) replied: Since the reaction is an alkylation, CH_3O does not attack phenolate. In both cases, it is CH_3^+ or $CH_3\delta^+$ that attacks phenolate. Thus, the use of isotopes as you suggest does not distinguish between the two mechanisms.

Dual Functional Catalyst in Dehydroaromatization of Isobutene to Xylene

Masamichi Akimoto, Etsuro Echigoya,* Masayoshi Okada, and
Yoshio Tomatsu

Department of Chemical Engineering, Tokyo Institute of Technology, Ookayama,
Meguro-ku, Tokyo, Japan

ABSTRACT The effect of the carrier on vapour-phase dehydroaromatization of isobutene to xylene over a vanadia catalyst has been investigated with reference to the mechanism and the nature of the active site for the reaction. It has been shown that many acidic sites are formed on the catalyst supported on Al_2O_3 or SiO_2, in contrast to that on MgO or charcoal, and that these sites effect the skeletal isomerization of isobutene to n-butene, this latter participating in m- and o-xylene formation but not that of p-xylene. Furthermore, V^{3+} has been found to abstract a hydrogen atom from butenes and intermediate compounds. Thus it is concluded that formation of p-xylene proceeds via 2,5-dimethylhexa-1,5-diene over V^{3+} with no participation of acidic sites, in contrast to the formation of m- or o-xylene which does so via 2,4-dimethylhexa-1,5-diene or 3,4-dimethylhexa-1,5-diene over V^{3+} with participation of the acidic sites.

INTRODUCTION

Vapour-phase dehydrocyclization of linear hydrocarbons, first reported by Moldawski and Kamusher,[1] has been a very important process for the effective utilization of hydrocarbons in the field of petrochemistry, and many workers[2] have devoted themselves to the investigation of the reaction mechanism and of effective catalysts. In oxidative dehydrocyclization,[3] however, the selectivity is very low on account of the complete oxidation of hydrocarbons, with the exception of the reaction over oxides containing Bi_2O_3 in the absence of gaseous oxygen.[4] Thus the reaction has no industrial significance.

Recently, Csicsery[5] studied dehydroaromatization of lower hydrocarbons over Pt and Cr_2O_3 catalysts and discussed the effect of the nature of the alumina carrier on the distribution of the aromatics formed. However, the effect of changing the carrier on the distribution of the products and the mechanism for participation of acidic sites in the reaction have not yet been clarified in detail. We now report studies on the vapour-phase dehydroaromatization of isobutene to xylene over a vanadia catalyst, with reference to the effect of the carrier and the nature of the active site in relation to the mechanism of formation of the xylene isomers.

EXPERIMENTAL

Catalytic dehydroaromatization of isobutene was carried out using a conventional flow microreactor under atmospheric pressure. The reactor consisted of a 10 mm-i.d. quartz tube, 300 mm long, with a concentric thermowell. As reaction material, isobutene (purity 99% up) was used and nitrogen (diluent)was purified by passing through a silica-gel tower.

Catalysts were prepared by mixing either alumina sol, silica sol, titania gel, or magnesium hydroxide with an aqueous solution of ammonium metavanadate, evaporating the mixture to dryness on a water bath. The catalysts (32—60 mesh) were then calcined in a current of air at 650°C for 3 h and were reduced with hydrogen (60 NTP cm^3/min) at 550°C for 1 h before the reaction.

Akimoto, Echigoya, Okada, Tomatsu

The gaseous effluent from the reactor was analysed by gas chromatography. The acidity of the catalysts was measured by adsorption of pyridine on the catalyst, previously evacuated at 550°C, using a spring balance.

RESULTS AND DISCUSSION

Role of acidic sites in xylene formation

In the main reaction, 1 mole of *p*-xylene is formed from 2 moles of isobutene as shown below.

$$2 \; CH_2{=}C\overset{\displaystyle CH_3}{\underset{\displaystyle CH_3}{\big<}} \rightleftharpoons 3H_2 \; + \; CH_3{-}\!\!\bigcirc\!\!{-}CH_3 \quad (1)$$

In the reaction under atmospheric pressure, the equilibrium constant log K_P of reaction (1) is 4·021 (600 K), 4·787 (700 K), or 5·380 (800 K), and thus the equilibrium conversion of isobutene was calculated as 97·6% (600 K), 99·2% (700 K), and 100·0% (800 K). Hence, there is no barrier for the reaction based on the equilibrium.

First, the activity of various metal oxides (such as V_2O_5, Cr_2O_3, MoO_3, WO_3, Fe_2O_3, and Nb_2O_5) supported on alumina (M/Al = 5/95 atom/atom) was investigated at

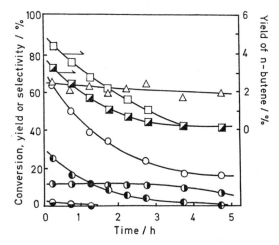

FIGURE 1 Change in the activity of V_2O_5–Al_2O_3 catalyst during the reaction of iso-butene.

Cat.: V_2O_5–Al_2O_3 (V/Al = 5/95 atom/atom), W/F/g cat.h g mol^{-1} = 40·0, reaction temp./°C = 550, feed: isobutene/N_2 = 50/50 v/v. ○: Conversion of isobutene, ◐: yield of *p*-xylene, ◑: yield of *m*-xylene, ◒: yield of *o*-xylene, △: selectivity to xylene, □: yield of but-2-ene, ◪: yield of but-1-ene.

550°C; V_2O_5 was found to be the best catalyst with a 27·4% yield of xylene and a selectivity of 54·5%. Furthermore, formation of isobutane was always observed, thus indicating that abstracted hydrogen atoms are accepted by isobutene. Therefore a detailed investigation of the V_2O_5 catalyst was made, and a catalyst of V/Al = 5/95 (atom/atom) was found to be the most active and selective for the reaction. Figure 1 shows the results of the reaction over V_2O_5–Al_2O_3 (V/Al = 5/95). Formation of *m*- and *o*-xylenes, especially the former, is considerable at the beginning of the reaction, and then decreases

873

gradually to a negligible value as compared with that of *p*-xylene. It is very instructive that the yield of n-butene produced by the skeletal isomerization of isobutene shows a tendency similar to that of *m*- and *o*-xylenes. On the other hand, the reaction over V_2O_5–MgO (V/Mg = 5/95) shows a very different result (Fig. 2): only *p*-xylene is formed, with no formation of *m*- or *o*-xylene, or n-butene.

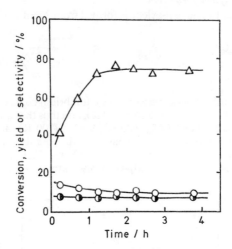

FIGURE 2 Change in the activity of V_2O_5–MgO catalyst during the reaction of isobutene.

Cat.: V_2O_5–MgO (V/Mg = 5/95 atom/atom), W/F/g cat.h g mol^{-1} = 40·0, reaction temp./C = 550, feed: isobutene/N_2 = 50/50 v/v. ○: Conversion of isobutene, ◑: yield of *p*-xylene, △: selectivity to xylene.

In a similar way, the activity and the selectivity of V_2O_5–SiO_2, V_2O_5–charcoal, and V_2O_5–TiO_2 were investigated, and the results are summarized in Table 1. Thus, based on the distribution of aromatics formed, these carriers can be broadly divided into

(A) Al_2O_3, SiO_2: *o*-, *m*-, and *p*-xylene and a small amount of benzene and toluene are formed;
(B) MgO, charcoal: only *p*-xylene is formed; and
(C) TiO_2: inactive.

It is worthwhile to note that formation of n-butene is observed during the reaction in the case of (A), in contrast to (B). As the isomerization of xylenes could hardly be observed over the V_2O_5–Al_2O_3 catalyst at 550°C (Table 2), the formation of *m*- and *o*-xylenes cannot be attributed to the isomerization of *p*-xylene formed, but must arise from different reaction paths. Thus, it seems very probable that n-butene formed by the skeletal isomerization of isobutene participates in *m*- and *o*-xylene formation.

In order to confirm this assumption, but-1-ene was introduced into the reactor during the reaction of isobutene over V_2O_5–MgO (Fig. 3). It is surprising that formation of *m*- and *o*-xylenes appears suddenly, with some decrease in the yield of *p*-xylene, on the introduction of but-1-ene. Furthermore, poisoning of the V_2O_5–Al_2O_3 with pyridine during the reaction of isobutene gave rise to a rapid decrease in the yields of *m*- and *o*-xylenes and n-butene, in contrast with that of *p*-xylene (Fig. 4). Thus, these results strongly suggest that n-butene participates in *m*- and *o*-xylene formation but not in

Table 1

Carrier effect on the reaction of isobutene[a]

Carrier		Conversion of isobutene/%	Selectivity/%					Yield/%	
			Benzene	Toluene	p-Xylene	m-Xylene	o-Xylene	Isobutane	n-Butene
(A)	Al₂O₃	49·4	0·6	3·2	22·7	31·8	trace	12·0	4·8
	SiO₂	25·1	0·0	4·0	53·4	5·2	0·0	6·5	3·5
(B)	MgO	10·8	0·0	0·0	73·2	0·0	0·0	3·0	0·8
	Charcoal[b]	44·7	0·0	2·9	36·9	trace	0·0	18·8	0·0
(C)	TiO₂	0·0	0·0	0·0	0·0	0·0	0·0	0·0	0·0

[a] Results obtained after 45 min; cat.: V/M = 5/95 atom/atom, M = Al,Si,Mg,Ti, W/F/g·cat.h g mol⁻¹ = 40.0 reaction temp./°C = 550, feed: isobutene/N₂ = 50/50 v/v.
[b] V₂O₅ content/wt% = 4·0 (impregnation method), W/F/g cat.h g mol⁻¹ = 25·0.

Table 2

Reaction of xylene or 2,5-dimethylhexa-1,5-diene[a]

Reactant	Conversion/%	Yield/%[b]			Conversion/%	Yield/%[c]		
		p-Xylene	m-Xylene	o-Xylene		p-Xylene	m-Xylene	o-Xylene
p-Xylene	27·5	72·5	1·6	0·2	13·1	86·9	trace	0·0
m-Xylene	20·6	1·6	79·4	1·6	10·9	0·6	89·1	0·3
2,5-Dimethyl-hexa-1,5-diene	100·0	66·7	9·2	3·4	100·0	84·9	3·3	2·9

[a] Cat.: V₂O₅-Al₂O₃ (V/Al = 5/95 atom/atom), W/F/g cat.h g mol⁻¹ = 40.0 reaction temp./°C = 550, feed: reactant/N₂ = 3/97 v/v.
[b,c] Results obtained after 135 or 255 min.

p-xylene formation, and that acidic sites on the surface are responsible for the skeletal isomerization of isobutene to n-butene.

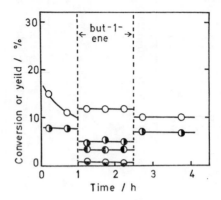

FIGURE 3 Effect of but-1-ene addition on the formation of *m*- and *o*-xylene during the reaction of isobutene.

Cat.: V_2O_5–MgO (V/Mg = 5/95 atom/atom), the reaction conditions are same as those of Figure 2, but but-1-ene injection is 8·3 v/v. ○: Conversion of isobutene, ◑: yield of *p*-xylene, ◐: yield of *m*-xylene, ◓: yield of *o*-xylene

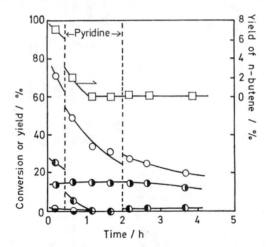

FIGURE 4 Poisoning of V_2O_5–Al_2O_3 catalyst with pyridine during the reaction of isobutene.

Cat.: V_2O_5–Al_2O_3 (V/Al = 5/95 atom/atom), the reaction conditions are same as those of Figure 1, but pyridine injection is 2·0 v/v. ○: Conversion of isobutene, ◑: yield of *p*-xylene, ◐: yield of *m*-xylene, ◓: yield of *o*-xylene, □: yield of n-butene.

The acidity of these catalysts was then determined using a spring balance and pyridine as adsorbate, and the results obtained are shown in Table 3. As expected from previous results, the order of the acidities is V_2O_5–Al_2O_3 > V_2O_5–SiO_2 ≫ V_2O_5–MgO, in accordance with the order in the selectivity with respect to *m*- and *o*-xylene formation.

The acidity of the V_2O_5–charcoal catalyst is expected to be of the same order as that of V_2O_5–MgO. Therefore, it is concluded that acidic sites on the surface participate in *m*- and *o*-xylene formation indirectly through the isomerization of isobutene to n-butene. The gradual decrease in the conversion over V_2O_5–Al_2O_3 and V_2O_5–SiO_2 with time is presumably to be attributed to the formation of carbonaceous materials on the surface, due to the acidic sites.

Table 3

Acidity of the catalysts

Catalyst	Pyridine adsorbed/mmol g cat.$^{-1}$	
	Before reduction	After reduction
V_2O_5–Al_2O_3[a]	0·844	0·781
V_2O_5–SiO_2	0·201	0·168
V_2O_5–MgO	0·032	0·020

[a] V/M = 5/95 atom/atom, M = Al,Si,Mg.

Nature of active sites for dehydrogenation.

X-Ray analysis revealed that the V_2O_5–Al_2O_3 is amorphous, with no peak due to V_2O_5, in contrast with the V_2O_5–SiO_2 which showed only a very broad peak at $2\theta = 20$—$30°$. On the other hand, the V_2O_5–MgO and the V_2O_5–TiO_2 were similarly shown to be periclase- or rutile-type solid solutions, with no peak due to V_2O_5. Thus, V_2O_5 seems to be finely dispersed in these carriers with the exception of the V_2O_5–SiO_2 catalyst.

In order to investigate the nature of the active sites for the dehydrogenation, the V_2O_5–MgO was poisoned with various pentenes during the dehydroaromatization of isobutene, and the strength of pentene adsorption on the sites was estimated from the decrease in *p*-xylene yield (Table 4). In this experiment, a part of the injected pentene

Table 4

Poisoning of V_2O_5–MgO catalyst with pentenes during the reaction of isobutene[a]

Yield of *p*-xylene/%	2-Methyl-but-2-ene	2-Methyl-but-1-ene	cis-Pent-2-ene	Pent-1-ene	3-Methyl-but-1-ene
Before injection, y_0	2·06	1·90	1·72	1·48	1·85
During injection, y	1·71	1·45	1·09	0·93	1·03
y/y_0	0·830	0·765	0·636	0·631	0·556

[a] Cat.: V_2O_5–MgO (V/Mg = 5/95 atom/atom), W/F/g cat.h g mol^{-1} = 33·3, reaction temp./°C = 575, feed: isobutene/N_2 = 50/50 v/v, pentene injection = 6·67 v/v.

(10—15%) was converted into isoprene or penta-1,3-diene by dehydrogenation, but no formation of products resulting from the combination of π-allyls from isobutene and pentene was observed. Moreover, the extent of isomerization of these pentenes was below 15%. The order of the strengths of adsorption is thus 3-methylbut-1-ene >

pent-1-ene > *cis*-pent-2-ene > 2-methylbut-1-ene > 2-methylbut-2-ene, which is in agreement with that on $AgNO_3$.[6] Hence, the active site for the dehydrogenation is expected to be a vanadium ion of low oxidation state, *e.g.* V^{2+} or V^{3+}. In order to determine the oxidation number of the vanadium ion on the surface, the V_2O_5–MgO catalys was evacuated at 300°C for 1 h and, at —100°C, the change in the amount of V^{4+} (h.f.s) before and after the introduction of oxygen (5 mmHg) was determined by means of e.s.r. using Mn^{2+} as an internal standard. The amount of V^{4+} increased to 123% of the initial value. However, the amount of V^{4+} increased to 320—340% of the initial value on the introduction of oxygen in the case of the catalyst previously reduced with hydrogen for 1 h at 550°C. This result gives strong evidence for the conclusion that a great deal of V^{3+} is formed over the surface by the reduction, this ion serving to abstract hydrogen from butene and intermediate compounds during the reaction of isobutene. In practice, the amount of V^{4+} formed by the reduction of these V_2O_5–Al_2O_3, V_2O_5–SiO_2, and V_2O_5–MgO catalysts with hydrogen at 550°C decreased to negligible values on reduction with a stream of isobutene at 550°C for 1 h. On the other hand, an inactive V_2O_5–TiO_2 catalyst was scarcely reduced with hydrogen or isobutene even at 550°C. These results also support the above conclusion.

Reaction scheme and rate-determining step

The following reaction scheme for xylene formation is suggested; it is similar to that suggested by Davis[7] who emphasized formation of toluene from n-heptane *via* the direct formation of a six-membered-ring structure.

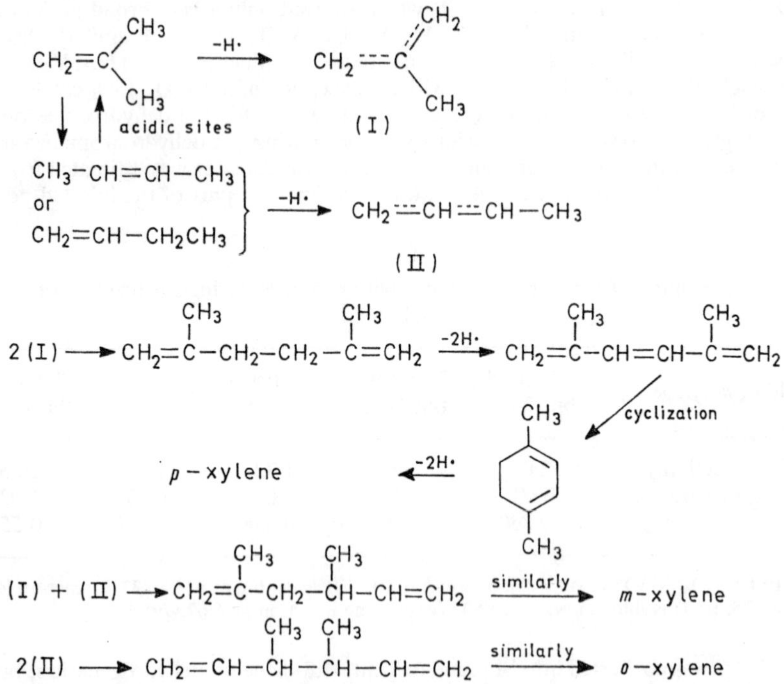

p-Xylene formation *via* 2,5-dimethylhexa-1,5-diene is apparent from the selective formation of *p*-xylene in the reaction of the hexadiene (Table 2). As intermediates for *m*- or *o*-xylene formation, 2-methylhepta-1,5-diene and octa-2,6-diene are also possible. But cyclization of 2-methylhepta-1,3,5-triene or octa-2,4,6-triene, followed by formation of a six-membered-ring structure seems unlikely to occur. It is thus concluded that formation of *p*-xylene from isobutene proceeds over V^{3+} with no assistance from the acidic sites, in contrast with that of *m*- or *o*-xylene which occurs from isobutene and n-butene or n-butene over V^{3+} with participation of the acidic sites. The reaction order with respect to isobutene and the overall activation energy for *p*-xylene formation over the V_2O_5–MgO catalyst were 1·1 and 7·5 kcal/g-mol, respectively. Hence, the rate-determining step for *p*-xylene formation is assumed to be a coupling of two isobutenes adsorbed by a process involving abstraction of a hydrogen atom, in accordance with the case of oxidative dehydroaromatization.[8]

REFERENCES

1. B. L. Moldawski and H. Kamusher, *Compt. rend. Acad. sci. U.R.S.S.*, 1936, 355.
2. R. C. Pitkethly and H. Steiner, *Trans. Faraday Soc.*, 1939, **35**, 979; H. Pines and C. T. Cher, *J. Org. Chem.*, 1961, **26**, 1057; H. Pines and S. M. Csicsery, *J. Catalysis*, 1962, **1**, 313; H. Pines, C. T. Goetskel, and S. M. Csicsery, *J. Org. Chem.*, 1963, **28**, 2713.
3. T. Sakamoto, M. Egashira, and T. Seiyama, *J. Catalysis*, 1970, **16**, 407; K. Ohdan, T. Ogawa, J. Umemura and K. Yamada, *Kogyō Kagaku Zasshi*, 1970, **73**, 842; T. Seiyama, M. Egashira, T. Sakamoto, and I. Aso, *J. Catalysis*, 1972, **24**, 76; D. L. Trimm and L. A. Doerr, *ibid.*, 1971, **23**, 49 and 1972, **26**, 1; T. Seiyama, T. Uda, I. Mochida, and M. Egashira, *ibid.*, 1974, **34**, 29.
4. H. E. Swift, J. E. Bozik, and J. A. Ondrey, *J. Catalysis*, 1971, **21**, 212.
5. S. M. Csicsery, *J. Catalysis*, 1970, **17**, 207 and 1970, **18**, 30.
6. R. J. Cvetanovic, F. J. Duncan, W. E. Falconer, and R. S. Irwin, *J. Amer. Chem. Soc.*, 1965, **87**, 1827.
7. B. H. Davis, *J. Catalysis*, 1973, **29**, 398.
8. T. Seiyama, T. Uda, I. Mochida, and M. Egashira, *J. Catalysis*, 1974, **34**, 29.

DISCUSSION

W. O. Haag (*Mobil, Princeton*) said: I should like to draw your attention to a related study[1] in which *p*-xylene was produced from 2,4,4-trimethylpent-1-ene, as a source of methylallyl radicals, and from 2,5-dimethylhexa-1,5-diene, in which MgO was used which acted as both a dehydrogenation and a hydrogen-transfer catalyst. One mole of hydrogen was produced per mole of *p*-xylene formed. Could you tell us the fate of the hydrogen that is removed from reactants and intermediates in your reactions over vanadia supported on magnesium oxide? In particular, is hydrogen gas produced, and if so, how much?

E. Echigoya (*Tokyo Inst. of Technology*) replied: In the formation of xylene from isobutene, three moles of hydrogen are evolved whereas two moles of hydrogen are evolved in the reaction of 2,5-dimethyl-1,5-hexadiene. As hydrogen was used as carrier gas for the chromatograph, the amount of hydrogen evolved was not determined. However, some abstracted hydrogen reacted with isobutene, as seen in the formation of isobutane (Table 1). On the other hand, hydrogenated products were minimal in the conversion of the hexadiene into xylene, showing that all of the hydrogen abstracted formed hydrogen gas.

A. Sárkány (*Inst. of Isotopes, Budapest*) said: I wish to discuss the V_2O_5–TiO_2 system, although this was not our best sample. We found that, in the temperature range of 923–973 K, V_2O_5 could be incorporated into the TiO_2 (anatase or rutile) lattice. During this process the V_2O_5 loses oxygen and simultaneously anatase transforms into rutile. The result is a rutile-type solid solution, $Ti_{1-x}V_xO_2$, in which vanadium is

1. W. O. Haag, *Ann. New York Acad. Sci.*, 1973, **213**, 288.

stabilized by the rutile matrix in the oxidation state of 4. I think this observation may give further evidence for the importance of V^{3+} in dehydroaromatization.

E. Echigoya (*Tokyo Inst. of Technology*) replied: V_2O_5–TiO_2 is transformed into rutile-type solid solution by the calcination at 923 K ,as you say. This was confirmed by X-ray analysis. This material was hardly reduced either by hydrogen or by isobutene and did not show any activity. Thus V^{4+} ion is not an active centre for this reaction, and our results agree with yours.

P. Ratnasamy (*Indian Inst. of Petroleum, Dehradun*) said: During an investigation of the acidity of WO_3–Al_2O_3 by the adsorption of ammonia,[2] we observed (in agreement with your results, Table 3), that there is a diminution in acidity on reduction at 773 K. Can you comment on the decrease in the amount of pyridine adsorbed after reduction of your supported V_2O_5?

E. Echigoya (*Tokyo Inst. of Technology*) replied: We did not investigate whether acidic sites over these catalysts have Lewis- or Brönsted-type nature. However, reduction of the catalysts gives rise to some decrease in the oxidation number of metal ions, which would cause a decrease in the strength of a Lewis acid. A similar effect would also be expected in the case of a Brönsted acid, due to a decrease in the oxidation number of metal ions present in the vicinity of the acidic sites. In addition to these effects, elimination of surface hydroxyl groups on reduction is also probable. The decrease in the amount of pyridine adsorbed on the catalysts following reduction can be attributed to the above effects.

J. W. Hightower (*Rice Univ., Houston*) asked: Have you considered the possibility that dehydrodimerization may actually involve an oxidative reduction mechanism with oxygen being supplied by the catalyst? The deep reduction of the catalyst could perhaps account for the observed decrease of the catalytic activity with time. If this were the case, a completely reduced catalyst should be inactive.

Also, is it possible that some of the reactions in the sequence may occur thermally through homogeneous reactions in the gas phase? For example, 1,4-dimethylcyclohexadiene may form p-xylene through such a process. Have you ever passed some of these materials through an empty reactor at these high temperatures to determine if such a process occurs?

E. Echigoya (*Tokyo Inst. of Technology*) replied: Benzene and toluene were produced in the initial stage of the reaction over an unreduced catalyst, and this was supposed to result from participation of lattice oxygen existing on the catalyst surface. Considerable yields of benzene and of toluene were produced in the presence of gaseous oxygen. This was attributed to the oxidation of the methyl group of hexadiene or hexatriene, followed by the decarboxylation. No demethylation of p-xylene in the presence of oxygen supports this. Fused unsupported V_2O_5 possessed considerable activity initially, but this decreased with reaction time and almost disappeared after several hours due to the deep reduction of the oxide. In contrast, V_2O_5 supported on magnesia or alumina was not completely reduced. The observation that these catalysts showed a constant activity after several hours suggests that the decrease in activity initially is attributable to the formation of carbonaceous materials on the surface.

Reaction of other materials proposed as intermediates, besides 2,5-dimethyl-hexa-1,5-diene, was not examined. Cyclization of 2,5-dimethyl-hexa-1,5-triene may occur thermally. However, this reaction was not carried out in an empty reactor. No product was obtained in an empty reactor in the case of isobutene. Furthermore, no intermediate was observed in the products during the catalytic reaction. Thus, the reactions of these intermediates were concluded to take place on the surface of the catalyst.

---

[2] S. Sivasankar, A. V. Ramaswamy, V. Savitha, A. J. Leonard, and P. Ratnasamy, *J. Catalysis* (submitted for publication).

Surface Composition and Catalytic Properties of Supported Mixed Metal Catalysts

M. Nakamura† and H. Wise*

Solid State Catalysis Laboratory, Stanford Research Institute, Menlo Park, California 94025, U.S.A.

ABSTRACT   Dispersed Pd/Au catalysts on a silica support were prepared in order to determine the effect of alloying on catalytic activity. For kinetic measurement the hydrogenation of acetone to propan-2-ol was employed. For each of supported mixed-metal catalysts the kinetic data were supplemented by measurement of the Pd surface area by "titrating" the number of oxygen adatoms with carbon monoxide. In addition the elemental surface composition of the crystallite was determined by Auger electron spectroscopy. Over the range from 0 to 75 atom percent gold, the activation energy for acetone by hydrogenation remains constant at 7.9 ± 0.4 kcal/mole, as compared to a value of 11.3 ± 0.5 kcal/mole for gold. However, the specific reaction rate goes through a maximum for a catalyst composition near 50 atom percent gold, a region in which the d-band holes of the t_{2g} sub-band of Pd are filled. Thus changes in activation entropy result from alloying of Pd with Au, which are interpretable in terms of ligand effects and bond formation with adsorbates.

INTRODUCTION

In spite of considerable effort devoted to the study of mixed-metal catalysts[1] the existence of an electronic or geometric factor has not been established unequivocally. Partially this lack of a correlation with solid-state properties—such as d-band filling due to alloy formation—or geometric effects—such as metal-atom ensembles—is due to the absence of three essential pieces of information on the same catalytic system, namely the surface composition of the mixed-metal crystallites, the number density of surface atoms of each of the metal components of the homogeneous metal mixture, and finally kinetic data on changes in reaction parameters with surface composition.

However with the advent of Auger electron spectroscopy (AES) the elemental composition of the surface is amenable to quantitative analysis. At the same time surface-area titrations (SAT) involving reaction specificity indigenous to only one surface component of a binary mixed-metal system offers a procedure for establishing the surface atom density of that component.

The silica-supported Pd/Au system was selected for our investigations. Over the entire composition range the metals form a series of homogeneous solid solutions and exhibit the physical properties of disordered alloys.[2] For the present study we explored a number of experimental procedures for the preparation of SiO_2-supported Pd/Au catalysts, and monitored catalyst structure by X-ray diffraction and AES. Catalytic activity at various metal compositions was evaluated in terms of the hydrogenation of acetone.

In a study[3] of the catalytic reaction between gaseous acetone and hydrogen on evaporated metal films of platinum, nickel, tungsten, iron, palladium, and gold, the formation of propan-2-ol as the major product was observed, with small amounts of propane being formed except in the case of palladium and gold. In several publications the same reaction was studied in the presence of alloy or mixed-metal catalysts such as copper-nickel,[4] nickel–cobalt,[5] palladium–molybdenum,[6] Raney platinum–iridium,[7] and Raney

† International Fellow 1974/76.

nickel activated with cobalt, copper, iron, and manganese.[8] In none of these studies was the existence of an alloy effect demonstrable.

Recently it has been found that Pd/Au alloy catalyst finely dispersed on supports can be prepared by reducing the mixed oxides of palladium and gold which are formed by treating the chlorides with alkali solutions.[9] The catalysts so prepared were used in the study of acetone hydrogenation.

EXPERIMENTAL DETAILS

1. *Catalyst preparation*

A number of experimental techniques were examined for the synthesis of silica-supported Pd/Au catalysts. Of these the most satisfactory procedure for preparation of supported, high-surface-area alloy catalysts involved precipitation of the metal components as hydroxides followed by hydrazine reduction.

In this method, 3 g of Cab-O-Sil powder was added to 20 cm^3 of an aqueous solution of sodium chlorapalladate and chloroauric acid at the concentrations required to attain the desired Pd/Au ratio. After evaporation to dryness in air at 373 K for 15 h, 10 cm^3 of sodium hydroxide solution (0·5 N) was added followed by an excess of aqueous hydrazine (5 vol %). The mixture was allowed to stand at room temperature for 24 h. The catalyst was washed thoroughly with water dried in air at 373 K for 15 h, and calcined at elevated temperatures. The degree of bulk alloying of the supported mixed metal catalysts was established from the X-ray diffraction spectra [(111) planes]. Calcination at 1000 K for 3 h in flowing helium resulted in the formation of alloys with bulk compositions corresponding to near nominal compositions as demonstrated by the appearance and location of a single peak in the diffraction spectrum (Table 1).

Table 1

Composition of silica supported
Pd/Au catalysts[a]

Composition (Au, atom %)		
Nominal	Bulk (X-ray)	Surface (AES)
0	0	0
25	28	25 ± 5
50	55	50 ± 8
75	76[b]	75 ± 10

[a] After calcination at 1000 K in He for 3 h.
[b] Additional small shoulder corresponding to 100 atom % Au.

2 *Surface composition by AES*

A detailed description of the apparatus and technique in the surface studies by AES has been given in ref. 10. In those measurements a series of alumina-supported Pd/Au catalysts was employed, supplemented for this study by a limited number of AES determinations on the SiO$_2$-supported samples. For quantitative elemental surface analysis the M_5–$N_{4,5}N_{4,5}$ transition of Pd (331 eV) and the $N_7O_{4,5}O_{4,5}$ transition of gold (73 eV) were selected. The escape depth of the Auger electrons associated with these energies are estimated[10] to be less than two atomic layers for Au and less than three atomic layers for Pd.

The absence of any composition difference between bulk and surface observed for Pd/Au microspheres (Figure 1) and Al_2O_3-supported Pd/Au catalysts[10] was found to apply to the SiO_2-supported samples as well. For example, a nominal 80 Pd/20 Au

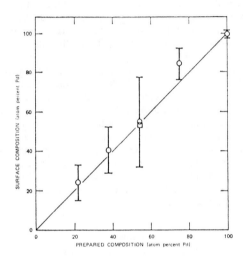

FIGURE 1   Surface composition as a function of bulk composition for supported Pd–Au catalysts.

(atom percent) on SiO_2 catalyst exhibited a surface composition of 79·5 Pd/20·5 Au (atom percent), well within the experimental error. On theoretical grounds[10] it was concluded that surface enrichment with respect to one of the metal components was not favoured for the Pd/Au system because of the nearly identical surface-free energies of the individual metals. Also our observations indicate that neither contact with the support material nor crystallite size causes any deviation in surface composition.

3 *Surface area measurement*

For the determination of the specific surface area of Pd in the mixed-metal catalysts we employed the surface reaction between oxygen adatoms and carbon monoxide, a process that is limited to transition-metal surface sites. This type of "surface-area titration" (SAT) was developed for supported Pt-catalysts[11] and has now been extended to dispersed Pd- and Pd/Au-catalysts. The principle of the SAT technique is based on the reaction

$$O(s) + 2CO(g) \rightarrow CO_2(g) + CO(s) \qquad (1)$$

where (s) refers to surface-adsorbed species and (g) to gaseous species. On the surface of Pd the oxidation of CO is known to occur quite rapidly in the presence of chemisorbed oxygen so that the formation of CO_2 can be used to determine quantitatively the number of Pd-sites once the degree of surface coverage of the metal with oxygen and carbon monoxide has been established. In the case of dispersed platinum on an alumina support the metal-to-oxygen atom ratio at maximum surface coverage was found to be $Pt/O = 2$. The same value was found to prevail for the CO-coverage following SAT (*i.e.*, $Pt/CO = 2$). As a result the ratio of the moles of CO consumed to CO_2 formed is $\Delta CO/CO_2 = 2$, in accordance with the stoicheiometry of reaction (1). One may interpret this result in terms of "bridge-bonded" oxygen and carbon monoxide species prevalent on the surface.

The same stoicheiometry was found to prevail with the dispersed Pd–SiO$_2$ sample (Table 2). However, for Pd–Au alloys account must be taken of changes in the bonding mode of CO and oxygen atom due to the presence of gold. As infrared studies have shown,[12,13] dilution of Group VIII atoms with Group IB atoms causes a modification in the bonding configuration of chemisorbed CO, *i.e.*, the alloying causes a decrease in the fraction of bridge-bonded CO and an increase of the fraction of linear CO. As a matter of fact when correction is made for the difference in surface composition of Pd–Ag on SiO$_2$ catalysts from that of the bulk,[10] the fraction of bridge-bonded CO to total chemisorbed CO is found to decrease linearly as the atomic fraction of surface silver atoms increases.[14] For the Pd–Au/SiO$_2$ system sorption studies with carbon monoxide and oxygen indicate[15] that (i) the configuration of oxygen adatoms is similar to that of CO admolecules, and (ii) the ratio of the number of surface palladium atoms (n_{Pd}) to that of oxygen adatoms (n_O) or CO admolecules is given by the values shown in Table 2. Based on this information the specific surface area of palladium for each alloy composition could be evaluated from the amount of CO_2 formed in the CO–O(s) titration (Table 2). Details of the CO titration for surface area measurement will be given in a forthcoming publication.[15]

Table 2

Surface area measurement for
(Pd/Au) SiO$_2$ catalysts[a]

Composition (atom %)		$\left(\dfrac{n_{Pd}}{n_O}\right)_s$	Pd Surface Area (m^2/g Pd)
Pd	Au		
100	0	2·0	21 ± 3
75	25	1·60	28 ± 3
50	50	1·34	24 ± 3
25	75	1·14	17 ± 3

Total metal loading (Pd + Au) = 5 weight per cent.

4 Catalytic studies

Catalytic activity was evaluated in a differential-flow reactor. The supported catalyst (20 mg) was placed in the Pyrex glass reactor, provided with a fitted glass disc for holding in place the powdered catalyst. The material was pretreated in a stream of hydrogen for 1 h at 425 K. Afterwards the temperature was set to the desired reaction temperature. The reactant feed was produced by passing the hydrogen stream through an acetone saturator maintained at 296 K. In this way, a mixture composed of 215 torr

Table 3

Hydrogenation of acetone on PdAu/SiO$_2$ catalysts[a]

Au (atom %)	Actual conversion of acetone into propan-2-ol (%)				Conversion number[b] (10^{-3} molecules $\cdot$ site^{-1} $\cdot$ s^{-1})				E (kcal/mole)	A $\left(\dfrac{\text{molecules}}{\text{site} \cdot \text{s}}\right)$
	423 K	448 K	473 K	498 K	423 K	448 K	473 K	498 K		
0	0·30	0·51	0·74	1·31	9	16	23	40	8·1 ± 0·4	1·8 × 10²
25	0·86	1·46	2·22	3·02	32	54	82	110	7·7 ± 0·4	2·4 × 10²
50	1·38	2·40	3·65	5·00	110	180	280	380	7·8 ± 0·4	1·1 × 10³
75	0·18	0·32	0·48	0·82	78	78	120	200	8·1 ± 0·4	8·7 × 10²
100	0·0054	0·012	0·022	0·040	—	—	—	—	11·3 ± 0·5	—

[a] Catalyst mass in each experiment = 0·020 g.
[b] Conversion number = number of molecules of propan-2-ol formed per Pd site per s.

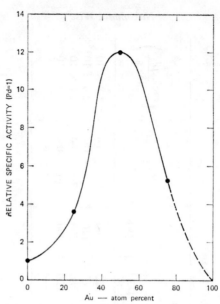

FIGURE 2   Relative change in specific conversion of acetone into propan-2-ol
at 473 K.

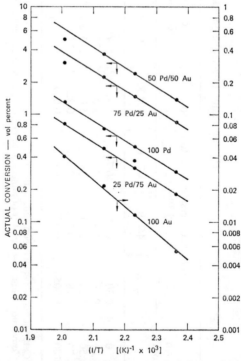

FIGURE 3   Conversion of acetone into propan-2-ol for various SiO_2-supported
Pd/Au catalyst.

886

of acetone and 545 torr of hydrogen was introduced into the reactor. The flow rate of hydrogen was kept at $5 \cdot 0 \pm 0 \cdot 1$ cm^3/min. The reaction products were analysed with a gas chromatograph containing a 10-foot column of bismethoxyethyl adipate supported on Chromosorb W. The reaction was carried out for 40 min over a range of temperatures, and conversion of acetone into propan-2-ol was measured at 10-min intervals, and was found to reach steady-state conditions well before the first sample was taken.

EXPERIMENTAL RESULTS

Over a range of catalyst composition and temperature the rate of acetone formation exhibited the variations shown in Table 3. At each temperature studied the conversion into propan-2-ol goes through a maximum with increasing Au addition. This effect is not associated with changes in surface area of the catalysts, as evidenced by the data plotted in Figure 2 in which the specific conversion is examined as a function of metal composition of the crystallites.

The experimental results lend themselves to the determination of the activation energy for propan-2-ol formation; a parameter independent of the surface area of each catalyst system (Fig. 3). The slopes of the resulting lines in the Arrhenius plots indicate little change in activation energy as the composition of the catalyst is varied from 100 Pd to 25 Pd/75 Au. Only in the case of pure Au/SiO$_2$ a higher activation energy is demonstrable (Table 3).

DISCUSSION

The kinetic aspects of the Pd/Au system for the hydrogenation of acetone have several features in common with the data reported by Couper and Eley in their study of the ortho–para hydrogen conversion on Pd/Au alloy wires.[16] In the latter case little variation in activation energy was observed in the range from 0 to 60 atom % Au, with a sharp rise beyond this value. In addition a maximum in reaction rate occurred near the 60 Pd/40 Au (atom %) composition. Both studies point to the fact that the activation energy is insensitive to the electronic structure of the metal alloy over a wide range of Pd/Au ratios. However, the variation in reaction rate with metal composition (Figure 2) at a specified temperature definitely points to the existence of an electronic effect in which the chemical nature of the interaction between a surface metal atom with its neighbours affects the bond formed with the adsorbate.

In the composition region of constant activation energy the observed variations in reaction rate for the Pd/Au system must be associated with changes of the entropy of activation. The magnitude of this change, as deduced from the data in Figure 2, amounts to about 4 E.U. per mole in going from 100 Pd to 50 Pd/50 Au (atom percent). The increase in activation entropy indicates a larger probability of formation of the activated state on the alloy surface than on Pd, possibly due to an increase in rotational or translational freedom of a reaction intermediate in the adsorbed state. It has been reported[17] that the rate of hydrogen-atom recombination on Pd/Au surfaces exhibits a maximum at alloy compositions near those encountered in our study. This result strongly points to an increase in translational freedom of hydrogen atoms as a major cause for the change in reaction kinetics. Since the hydrogenation of acetone involves stepwise addition of hydrogen adatoms to the acetone admolecule we may conclude that the higher mobility of hydrogen atoms over the metal surface of the alloy is responsible for the rate increase in the range from 100 Pd to 50 Pd/50 Au (atom percent).

The decline in activity at Au > 50 atom percent is undoubtedly the result of d-band filling. Molecular orbital theory indicates that for a fcc metal of the transition group (VIII$_3$) the five d-orbitals are split into e_g and t_{2g} orbitals. The collective t_{2g} electrons occupy a broad band containing six electrons per atom; the e_g electrons are split into two sub-bands, each with two electrons per metal atom.[18] Studies of the De Haas–Van Alphen effect in Pd[19] indicate a distribution of $0 \cdot 55$ d-band holes in the ratio of

887

nearly 2:1 between the t_{2g} and e_g bands. Thus filling of the t_{2g}-band holes with s-electrons from Au is completed at an alloy composition of about 40 Pd/60 Au, a value close to the point of maximum activity in the hydrogenation of acetone. It may be surmised that further dilution with gold causes modification of the d-band character of the alloy crystallite to the extent that hydrogen chemisorption becomes less rapid.

If the formation of the adatom involves electron interaction with a localized band of the solid the magnitude and direction of the entropy change per electron may be estimated from the gradient of the density of state curve at the Fermi level.[20] To a first approximation this parameter is given by the variation of the thermoelectric power (TEP) of the alloy with composition. Although TEP data for Pd/Au are not available, the Pd/Ag system exhibits changes in TEP in qualitative agreement with these conclusions.[21] At low concentrations of IB-metal the addition of Ag causes an increase in the absolute value of TEP, corresponding to a net gain in entropy change. Near 40 atom percent IB-metal the TEP goes through a maximum, in close correlation with catalytic hydrogenation activity.

The adsorption bond at a given metal site is affected by the neighbouring foreign metal atoms on the surface, their presence causing changes in the entropy of activation and modification in the surface orientation and mobility of the adspecies. As a result one would expect the heteroatomic metal ensemble on the surface of the crystallite to influence structure-sensitive reactions in terms of selectivity as well as activity. The variations of hydrogenation activity observed in our experiments cannot be interpreted in terms of the behaviour of the individual metal atoms making up the heteroatomic crystallite; rather they point to the existence of solid-state electronic interactions between Pd- and Au-atoms in the crystallite.

REFERENCES

[1] E. J. Allison and G. C. Bond, *Catalysis Rev.*, 1972, **7**, 233.
[2] O. J. Kleppa, *J. Physique Radium*, 1962, **23**, 763.
[3] C.T.H. Stoddart and C. Kemball, *J. Colloid Sci.*, 1956, **11**, 532.
[4] E. Takai and T. Yamanaka, *Sci. Papers I.P.C.R.*, 1961, **55**, 194.
[5] P. B. Babkova, A. K. Avetisov, G. D. Lyubarskii, and A. I. Gel'bshtein, *Kinetics and Catalysis*, 1972, **13**, 915.
[6] Y. Orito, T. Imai, and S. Niwa, *Chem. Abs.*, 1972, **76**, 7169.
[7] A. D. Semenova, N. V. Kropotova, and G. D. Vovchenko, *Zhur. fiz. Khim.*, 1973, **47**, 2909; *Chem. Abs.*, 1974, **80**, 314.
[8] Z. Csuros, J. Petro, and J. Heiszman, *Magyar Kem. Folyoirat*, 1967, **73**, 181; *Chem. Abs.*, 1968, **68**, 3277.
[9] M. Nakamura, P. Wentrcek, E. Farley, and H. Wise, to be published.
[10] B. J. Wood and H. Wise, *Surface Sci.*, 1975, **52**, 151.
[11] P. Wentrcek, K. Kimoto, and H. Wise, *J. Catalysis*, 1974, **33**, 279.
[12] Y. Soma-Noto and W. Sachtler, *J. Catalysis*, 1974, **32**, 315.
[13] Y. Soma-Noto and W. Sachtler, *J. Catalysis*, 1974, **34**, 162.
[14] H. Wise, *J. Catalysis*, in the press.
[15] P. Wentrcek and H. Wise, to be published.
[16] A. Couper and D. D. Eley, *Discuss. Faraday Soc.*, 1950, **8**, 172.
[17] P. G. Dickens, J. W. Linnett, and W. Palczenska, *J. Catalysis*, 1965, **4**, 140.
[18] J. B. Goodenough, *Magnetism and the Chemical Bond*, Interscience, New York, 1963.
[19] J. J. Villemin and M. G. Priestley, *Phys. Rev. Letters*, 1965, **14**, 307.
[20] H. Wise, *J. Catalysis*, 1968, **10**, 69.
[21] W. Geibel, *Z. anorg. Chem.* 1911, **70**, 240.

DISCUSSION

V. Ponec (*Univ. of Leiden*) said: It surprises me that the authors assume that, by alloying, the chemisorption bond strength of the reaction components varies as well as the entropy of activation, but they find no corresponding variation in the activation energy of the reaction! Therefore my first question is: could not the results be explained by the assumption that, with increasing palladium content, (i) the number of active sites decreases, which has a different effect on the reaction followed and on the side reactions; (ii) the self-poisoning by side reactions (dissolution of hydrogen, carbonaceous residues) decreases as well? Simultaneous operation of (i) and (ii) can lead to a maximum in activity as a function of composition, but without changes in activation energy of the reaction. The decrease in activity can still be related to the changes in the density-of-states variation (as you mentioned in your oral presentation).

Secondly, Auger spectra supply us with information about the composition, more or less a weighted average over several of the upmost layers. Have you any idea how many layers we have to consider in the case of palladium–gold alloys?

Lastly, the so-called 'bridged' carbon monoxide is most likely to be a molecule sitting among three or four palladium atoms. Would this change the results of your calculations?

H. Wise (*Stanford Res. Inst., Menlo Park*) replied: In our interpretation we consider the adsorption process to cause changes in activation entropy with alloy composition. By considering electron transfer at the gas–solid interface to be involved in adsorption we can relate the activation entropy to a solid-state parameter, such as the density-of-state curve of the solid. This correlation is demonstrated in the Figure (p. 890) in which catalytic activity and thermoelectric power are plotted as a function of alloy composition. To a first approximation the thermoelectric power of a metal is proportional to the density-of-state gradient at the Fermi level.[1]

A priori, one cannot exclude the interplay of two opposing processes as you mention. However, it would be rather fortuitous that these same processes would have exactly the same effect in all the other palladium–gold catalysed hydrogenations, each of which shows a maximum of 40 atom % Au. Included in this series is the hydrogenation of benzene, olefins, and the *ortho-para* hydrogen conversion.

As for the escape depths of Auger electrons (73 eV for gold the 331 eV of palladium) we estimate two to three atomic layers. As a result we are averaging over this depth below the surface of the solid. Theoretical calculations of depth profile[2] indicate rapid convergence to bulk composition at three to four atomic layers even for two-component systems with large difference in surface free energy between the components. Thus our results are an adequate semi-quantitative indication of surface composition.

A. Sárkány (*Inst. of Isotopes, Budapest*) asked: Would you provide some additional information on the surface area titration by carbon monoxide? What were the partial pressures of carbon monoxide and oxygen? Did you investigate the reaction orders for acetone hydrogenation?

H. Wise (*Stanford Res. Inst., Menlo Park*) replied: Pulse gas chromatography was employed for measurement of surface area by carbon monoxide titration. A sample of the catalyst, previously reduced in hydrogen as described in the text of the paper, was placed in a reactor, heated at 373 K in a stream of helium for 20 min, and then exposed to ten pulses of carbon monoxide. Subsequently, it was exposed to flowing oxygen at atmospheric pressure for 5 min, and flashed with helium for 5 min to remove the residual oxygen from the system. After this pre-treatment, pulses of carbon monoxide were admitted to the reactor at 373 K, and the quantities of carbon dioxide formed and residual carbon monoxide were measured.

The order of reaction with respect to acetone was measured at 473 K using a constant hydrogen partial pressure of 545 Torr, and partial pressures of acetone in the range 50–215 Torr. Helium was used to raise the total pressure to one atmosphere. The order

[1] H. Wise, *J. Catalysis*, 1968, **10**, 69.
[2] F. R. Williams and D. Mason, *Surface Sci.*, 1974, **45**, 377.

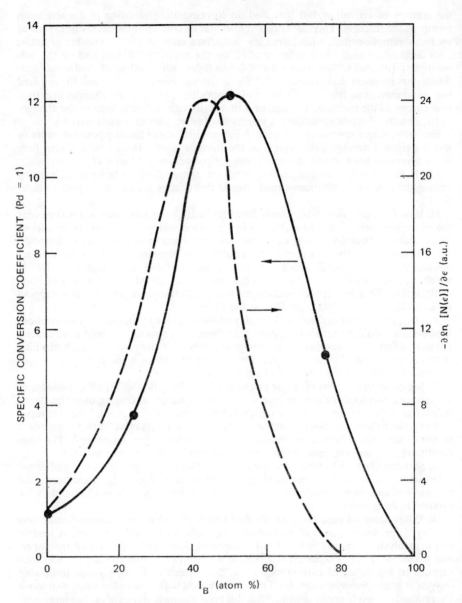

FIGURE Correlation of density-of-state gradient at Fermi level and catalytic activity for Group VIII–IB alloy catalyst.

in acetone was zero for Pd:Au = 0:100 atom %, 0·3 for Pd:Au = 27:75, and 0·4 for Pd:Au = 50:50, 75:25, and 100:0. These values indicate fairly strong chemisorption of acetone on the surfaces of pure palladium and of palladium–gold alloy catalysts.

G. Parravano (*Univ. of Michigan, Ann Arbor*) asked: In the determination of the specific surface area of palladium in mixed-metal catalysts, did you consider the possibility that oxygen may be adsorbed on gold also? Cha and I have reported[3] that gold supported on γ-alumina chemisorbs oxygen and is active for ^{14}C-exchange between carbon monoxide and carbon dioxide. The latter reaction involves adsorbed oxygen as an intermediate.

H. Wise (*Stanford Res. Inst., Menlo Park*) replied: A number of studies indicate that chemisorption of oxygen on gold is an activated process and does not occur to any degree at room temperature. Cha and Parravano, in the reference quoted, state that onset of catalytic activity for oxygen transfer in the $CO-CO_2$ reaction was detectable at about 623 K. In experiments carried out under ultra-high vacuum conditions[4] (background pressure about 10^{-9} Torr) the mass of oxygen adsorbed on gold at room temperature amounted to less than 1% monolayer coverage. Thus in the surface-area determination the error introduced from this source was very small.

W. M. H. Sachtler (*Shell, Amsterdam*) said: I am pleased to notice that you confirm for supported palladium–gold alloys, a high degree of surface enrichment with the Group IB metal, which is analogous to that which we reported for palladium–silver films.[5]

Re-evaluation of our i.r. results for carbon monoxide adsorption, and recent measurements in collaboration with Primet and Matthieu have shown that there is definitely a strong electronic ligand effect of alloying involved in this system. This statement is based on two observations, (i) there is a small but significant shift in i.r. band position due to alloying palladium with silver, and (ii) there is a drastic increase in the ratio of band intensities due to linear and multi-centre adsorbed carbon monoxide respectively which cannot be rationalized by the geometric model according to which only those palladium atoms which are completely surrounded with silver atoms adsorb carbon monoxide in the linear mode.

In your paper you 'conclude that the higher mobility of hydrogen atoms over the metal surface of the alloy is responsible for the rate increase in the range from 100% Pd to 50% Pd/50% Au'. This surprises me. It can only be correct if hydrogen mobility over the surface is one of the slow steps in the hydrogenation of acetone over palladium. Observations with metal films and with FEM tips show that hydrogen surface mobility is very fast at 423 K on transition metals. The activation energy for hydrogen surface diffusion is much lower than 3 kcal mol^{-1}, the value which you report for the activation energy of acetone hydrogenation. How is this discrepancy to be explained?

H. Wise (*Stanford Res. Inst., Menlo Park*) replied: In the geometric adsorption model of the palladium–silver system the dimensions of the adsorbate and especially the van der Waals dimensions may have some relevance. For carbon monoxide the van der Waals shape approximates to a cylinder ($\sim1\cdot4$ Å in radius) with hemispherical caps at both ends (overall length $\sim4\cdot1$ Å). Thus in the linear bonding mode on the surface of palladium the van der Waals dimensions of carbon monoxide will affect more than one surface atom, so that linear-mode adsorption may not be limited to one palladium atom surrounded by four foreign atoms.

While hydrogen adatoms form a mobile surface layer on transition metals at the temperatures of interest to our study, the surface diffusion mechanism on an alloy surface is not considered to govern the activation energy. The reaction orders (see my reply to Sárkány above) indicate the formation of an intermediate from acetone adsorption as the slow step in the reaction, a process common to all the alloy compositions studied.

[3] D. Y. Cha and G. Parravano *J. Catalysis*, 1970, **18**, 200.
[4] N. Endow, B. J. Wood, and H. Wise, *J. Catalysis* 1969, **15**, 316. W. R. MacDonald and K. W. Hayes, *J. Catalysis* 1970, **18**, 115.
[5] R. Bouwman, G. J. M. Lippits, and W. M. H. Sachtler, *J. Catalysis*, 1972, **25**, 350.

J. Basset (*CNRS, Villeurbanne*) said: You have assumed in your lecture that there are three types of carbon monoxide adsorption on transition metals, namely linear, bridged, and another form which would be parallel to the surface. Do you have any reason to ascribe the low-frequency band usually obtained below 1900 cm^{-1} with some transition metals to this rather unexpected species? Why would you reject the now well-accepted attribution to carbon monoxide interacting with three or four transition-metal atoms *via* the carbon atoms, as occurs in cluster compounds.[6]

H. Wise (*Stanford Res. Inst., Menlo Park*) replied: Bonding concepts as applied to metal complexes involving isolated metal atoms have been useful in the interpretation of adsorption on solid surfaces. In line with a molecular orbital approach and the concept that optimal bonding matches maximum overlap of surface and adsorbate orbitals, three possible configurations for carbon monoxide adsorption on a f.c.c.-lattice can occur: (i) by a carbon-to-metal bond (as postulated for metal carbonyls); (ii) by formation of a bond between a carbon atom and two next-nearest metal atoms (by interaction of ground state-sp_z orbitals or sp^2 hybridization of carbon), and (iii) by formation of carbon-to-metal and oxygen-to-metal bonds between carbon monoxide and the two next-nearest metal atoms (a 'prone' configuration involving interactions of metal orbitals with carbon sp^2 and oxygen p or sp^2 orbitals). Identification of configuration (iii) in terms of a specific i.r. frequency has not been made as yet. However, as you point out, the low-frequency bands (<1900 cm^{-1}) observed in the case of palladium may be responsible for this type of adspecies.[7]

M. Boudart (*Stanford Univ.*) said: The finding by the authors of a maximum in the activity for hydrogenation of acetone as gold is dissolved in palladium is most interesting. Many explanations are possible. A likely one should be compatible with our recent findings of a similar but much more pronounced maximum in the activity of similar alloys in the reaction between hydrogen and oxygen performed in excess oxygen.[8]

H. Wise (*Stanford Res. Inst., Menlo Park*) said: Our paper interprets the catalytic behaviour of the palladium–gold system for hydrogenation in terms of surface bonding of adspecies involving electron transfer for covalent bond formation. This process is considered to be governed by electronic solid-state properties as manifested in the density-of-state curve at the Fermi level of the alloy mixtures. To obtain this correlation between activity and catalyst composition the surface composition of the metal crystallite needs to be established. While Auger electron spectroscopy shows that, in the presence of molecular hydrogen the palladium–gold surface composition is the same as that of the bulk, an entirely different behaviour is exhibited in the presence of oxygen. For palladium-rich bulk mixtures considerable enhancement in surface palladium content was noted.[9] The opposite effect was found in gold-rich mixtures. We wish to suggest that in your studies of the oxidation of hydrogen in the presence of excess oxygen the changes in surface composition may accentuate the activity pattern observed.

R. Maatman (*Dordt College, Sioux Center*) asked: You speak of acetone admolecules reacting with hydrogen adatoms. We have shown recently that this type of Langmuir–Hinshelwood process, one in which both reactants adsorb appreciably, can be analysed by means of transition-state theory.[10] (Until now, the usual Langmuir–Hinshelwood analysis has been made for systems in which not more than one reactant adsorbs appreciably.) We have examined the results you report for four catalysts (excepting only that for the pure gold catalyst) and in all cases our calculations indicate that a

[6] R. Ugo, *Proc. 5th Internat. Congr. Catalysis*, Miami Beach, 1972, Vol. 1, p. 19.
[7] Y. Soma-Noto and W. M. H. Sachtler, *J. Catalysis*, 1974, **32**, 315.
[8] M. Boudart and Y. L. Lam, unpublished work.
[9] B. J. Wood and H. Wise, *Surface Sci.*, 1975, **52**, 151.
[10] R. Maatman, *J. Catalysis*, in press.

possible slow step is one in which hydrogen adatoms and acetone admolecules, both appreciably adsorbed, interact. Is this consistent with what you know? That is, do you think that both acetone and hydrogen are appreciably adsorbed before they react?

H. Wise (*Stanford Res. Inst., Menlo Park*) replied: Indeed, the order with respect to acetone (see my reply to Sárkány above) suggests that fairly strong chemisorption of acetone (or of a surface intermediate derived from it) occurs on palladium and the palladium–gold alloys studied. Similar considerations apply to hydrogen chemisorption. Consequently, a Langmuir–Hinshelwood mechanism would be a good representation for this reaction.

D. D. Eley (*Univ. of Nottingham*) said: The maximum rate for reactions of hydrogen occurs at about 20–40 atomic % Au in palladium–gold alloys, but this maximum tends to move towards 0% Au for larger hydrocarbon molecules.[11] It has been reported for hydrogen solubility by Sieverts and by some (not all) other workers. This has led me to suggest that the intermetal atom force constant may decrease as palladium is alloyed with gold. This might lower the activation energy for reactions involving two or more metal sites. I think this possibility, which I would call a second order geometric effect, should be examined. LEED measurements of surface Debye temperatures should help here.[11] Also chemisorptions involving only one site should not show a maximum on this particular hypothesis.

W. K. Hall (*Univ. of Wisconsin, Milwaukee*) (communicated): Several years ago we studied the oxidation of ethylene over palladium–gold alloys.[12] At the same time we studied the method of preparation and found, in agreement with the present work, that hydrazine reduction was the preferred method. We also studied the chemisorption of carbon monoxide and of oxygen. The former gave a nearly linear plot passing through the origin for 0% Pd, but the chemisorption of oxygen was not simple and continued to be greater than monolayer capacity. This raises some questions concerning the actual surface areas and hence the specific rates given in the present paper, since they depend on the titration technique. We interpreted our carbon monoxide chemisorption results as indicating that the surface composition was equal to that in the bulk; this was confirmed by ESCA measurements made at Gulf Research on these same catalysts in 1971; this result was expected, based on the exothermic heat of alloying for this system. The changes in activity with composition were also attributable to the frequency factor, not the activation energy.

[11] D. D. Eley, *Chem. and Ind*, 1976, 12.
[12] H. R. Gerberich, N. W. Cant, and W. K. Hall, *J. Catalysis*, 1970, **16**, 204.

Binary Palladium Alloys as Selective Membrane Catalysts

Vladimir M. Gryaznov,* Viktor S. Smirnov,† and Mikhail G. Slin'ko‡

* P. Lumumba Peoples' Friendship University, Moscow, U.S.S.R.
† A. V. Topchiev Institute of Petrochemial Synthesis, Moscow, U.S.S.R.
‡ Institute of Catalysis, Novosibirsk, U.S.S.R.

ABSTRACT Catalytic properties of metallurgically prepared binary palladium-based alloys have been studied as functions of the nature and content of the second component. Pd–Ni and Pd–Ru alloys proved to be more active for cyclohexane dehydrogenation than other alloys investigated. An activity maximum was observed in Pd–Ni, Pd–Ru, and Pd–Pt alloy composition ranges. On the other hand, increasing the silver content decreased the catalytic activity of Pd–Ag systems, including the intermetallic compounds Pd_3Ag_2 and PdAg. Oxidation by air at 750°C and reduction by hydrogen at 350°C of Pd–Ru and Pd–Rh alloys enhanced the activities and the roughness of the surfaces of these foils. X-Ray photoelectron spectoscopic study of Pd–Rh foils showed that the alloy surfuces were enriched in rhodium after use as catalysts for cyclohexane dehydrogenation. The highest selectivity in the hydrogenation of cyclopentadiene to cyclopentene was shown by Pd–Ru and Pd–Rh alloys which had been used as hydrogen-permeable membrane catalysts.

Introduction

Palladium-based alloys are permeable for hydrogen only, and possess catalytic activity towards some reactions. That is why these alloys may be used as membrane catalysts for dehydrogenation with hydrogen discharge through the catalyst, or for hydrogenation with hydrogen supply through the catalyst. The influence of reagent transfer through the membrane catalyst on the rate and selectivity of catalytic reactions has been described previously.[1]

The numerous data concerning the catalytic activity of palladium alloys were obtained mainly in respect of the *para–ortho*-hydrogen conversion.[2-5] The very important effect of hydrogen dissolved in Pd and Pd–Au alloys on the catalytic activity was observed and explained.[2]

The purpose of the present study was to obtain information on the dependence of the catalytic properties of palladium-based binary alloys, with respect to hydrocarbon dehydrogenation and hydrogenation reactions, upon the nature and the content of the second metal; upon oxidation–reduction treatments, and, finally, upon the content of the surface layer of the alloys.

Experimental

PdX alloys (X = Ni,Rh,Ru,Ag,Pt) were prepared in a vacuum arc furnace with a non-consumable tungsten electrode. The smelting of the initial powders was repeated several times in order to achieve homogenisation of the alloys. The content of tungsten in the alloys was less than 0·1 %. In this paper all concentrations are given in weights percent. The alloys were annealed at 1000°C *in vacuo* and rolled into foils of 0·1 mm thickness. The local X-ray spectroscopic analysis did not show any distinction between bulk and surface concentrations of the second component of the alloys.

The catalytic activity experiment was carried out in a pulse system and in a flow system. Each catalyst was treated *in situ* with hydrogen at 350°C before each catalytic run and with air at 450°C after each run. The products of dehydrogenation of chromatographically pure cyclohexane were benzene and hydrogen. For hydrogenation of

894

cyclopentadiene, cyclohexa-1,4-diene, and cyclo-octa-1,5-diene, hydrogen was introduced through the alloy foil into the stream of hydrocarbon vapour and carrier gas.

RESULTS AND DISCUSSION

Dehydrogenation of cyclohexane proved to be a first order reaction on all alloys investigated. Pd–Ni 5% and some other alloys are more active than pure palladium. The temperature dependence of the degree of conversion, α, of cyclohexane on Pd–Ni 5%, Pd and Pd–Ni 10% is shown in Figure 1a. It is evident that the former alloy surpasses Pd. The temperature of α = 0·3 was chosen as a quantitative measure of the

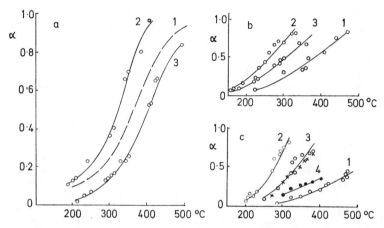

FIGURE 1   Degree of cyclohexane dehydrogenation against temperature for foils:
(a) Pd, Pd–Ni 5%, and Pd–Ni 10% alloys (curves 1, 2, and 3, respectively);
(b) Pd–Rh 10% alloy air-treated at 350° and hydrogen-treated at 350°C (curve 1), air-treated at 750°C and hydrogen treated at 350° (curve 2), air-treated at 750°C and hydrogen-treated at 750°C (curve 3);
(c) Pd–Rh 15% alloy treated as Pd–Rh 10% alloy (curves 1 and 2), argon-treated at 750°C (curve 4) and hydrogen-treated at 350° (crosses along curve 3).

catalytic activity. The corresponding data for Pd–Ni, Pd–Ag, and Pd–Pt alloys are presented in Table 1. This evaluation of catalytic activity shows that the activity of Pd–Ni and Pd–Pt alloys passes through a maximum as a result of increasing second component concentration. The same result is found for Pd–Ru alloys, the activity of which is maximal at 8·4% Ru, decreasing at higher Ru content. The catalytic activity of palladium–silver alloys and the intermetallic compounds Pd_3Ag_2 and PdAg is lower than the activity of pure Pd.

Pretreatment of Pd–Ru 6% and Pd–Ru 7·5% alloys with air at 350°C for 2 h and then with hydrogen at 350°C for 0·5 h does not change the catalytic activity. The apparent activation energy of cyclohexane dehydrogenation on non-treated Pd–Ru 6% alloy is 9·78 ± 1·50 kcal/mol, and after the aforementioned treatment is 9·35 ± 1·35 kcal/mol. The same oxidation–reduction treatment enhances the activity of Pd–Ru 8·4% alloy. An increase in the temperature of air treatment to 750°C without changing the temperature of hydrogen treatment, enhances the catalytic activity of all Pd–Ru alloys studied (4·4; 6·0; 7·5; 8·4; and 9·2% Ru). Electron micrographs of the treated catalysts show the increase in roughness.

The catalytic activity of Pd–Rh alloys treated with air at 750°C is higher than after the same treatment at 350°C. It is interesting that hydrogen treatment at 750°C after oxygen treatment at 750°C depresses the catalytic activity of Pd–Rh alloys. Figure 1b shows these effects for the Pd–Rh 10% alloy. The increase in temperature of hydrogen treatment may cause a decrease in the amount of hydrogen sorbed in the alloy. In fact, the catalytic activity of Pd–Rh 15% alloy was decreased drastically after argon treatment

Table 1

Temperatures corresponding to degree of conversion = 0·3 for cyclohexane on palladium and palladium alloys

Catalyst	t°C	Catalyst	t°C	Catalyst	t°C
Pd	320	Pd–Ag (20%)	357	Pd–Pt (0·1%)	357
Pd–Ni (5%)	255	Pd–Ag (40%)	422	Pd–Pt (0·5%)	272
Pd–Ni (10%)	345	Pd–Ag (50%)	437	Pd–Pt (1·5%)	282

at 750°C during 4 h and hydrogen treatment at 750°C during 0·5 h (compare curves 3 and 4 of Figure 1c). But the next hydrogen treatment at 350°C restores the catalytic activity completely (see crosses along curve 3). It does mean that a definite amount of sorbed hydrogen is essential for the dehydrogenative activity of Pd–Rh alloys. This had previously been found for Pd foil[6] and Pd–Ag alloy.[7]

Palladium–rhodium alloy foil surfaces were analysed by means of X-ray photoelectron spectroscopy before and after cyclohexane dehydrogenation in the flow system. The X-ray photoelectron spectra (XPS) were obtained on an AEI ES-100 spectrometer. The method of handling the spectra has been described elsewhere.[8] The ratio of the intensities of the Rh $d_{5/2}$ and Pd $d_{5/2}$ peaks for non-treated Pd–Rh 10% foil, heated *in vacuo* and in hydrogen at 400°C for 2 h, are the same (0·95 ± 0·05) and very close to that for the bulk rhodium content. The composition of the surface layer does not change after annealing the foil in air as well. But the use of this foil as a catalyst for cyclohexane dehydrogenation at 350°C results in an increase in the rhodium concentration in the surface layer. In each catalytic experiment the mixture of cyclohexane vapour (3·2 Torr) and helium (756·8 Torr) was passed for 1·5 h over the foil (30 cm²) with a velocity 20 ml/min. The rhodium enrichment at the foil surface is shown in Figure 2 as a function of the catalytic experiment number. This enrichment is accompanied by an increase in catalytic activity, which is shown by the lower curve in Figure 2. The XP spectra of the foil in its initial state and after 12 catalytic experiments are shown in Figure 3.

Similar results were obtained for Pd–Rh 5% foil. The initial Rh/Pd ratio was 0·05 ± 0·003 and rose to 0·29 after 14 catalytic experiments. In both cases the surface rhodium concentration increases faster during early experiments than in the course of subsequent experiments. But the catalytic activity still rises despite the slower increase in rhodium surface concentration. These results can be understood if the enrichment of the surface layer in rhodium is followed by the formation of active spots as a result of the lateral mobility of metal atoms with little, if any, rhodium transfer from the bulk of the alloy.

Surface free energies for Rh and Pd may be estimated from the data.[9] Comparison of these values leads to the prediction of preferential adsorption by Pd, in contrast to the XP spectra. The surface composition may be changed in the presence of a substance

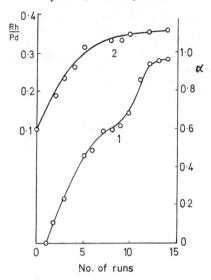

FIGURE 2   Degree of cyclohexane dehydrogenation (curve 1) and Rh/Pd ratio in surface layer of Pd–Rh 10% alloy (curve 2) against number of catalytic runs.

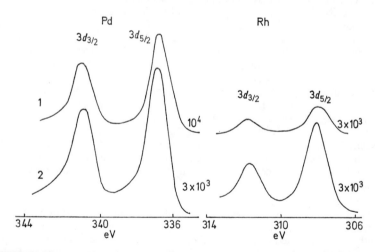

FIGURE 3   Intensities of Pd and Rh peaks in XP spectra of Pd–Rh 10% alloy non-treated (curve 1) and used as a catalyst for cyclohexane dehydrogenation (curve 2), against binding energy (eV).

which forms stronger bonds with one metal than with the other. In our case this substance may be carbon, formed by the partial decomposition of hydrocarbons. The other cause of surface enrichment may be envelopment of the grains of the Pd–Rh alloy by rhodium. This explanation was suggested[10] for the results of a benzene hydrogenation study on Pd–Rh catalysts which were obtained by the reduction of solutions of salts of these metals.

897

The observed dependence of surface layer composition and of catalytic activity of Pd–Rh alloys upon the duration of their use as catalysts is a new example of active surface formation under the influence of a catalytic reaction. Another phenomenon which strongly influences the catalytic activity of palladium alloys is hydrogen sorption. Some evidence of the importance of hydrogen content for the catalytic activity of Pd–Ru and Pd–Rh alloys was given above. In such cases the mathematical simulation[1] of the reactions occurring on the membrane catalysts should take into account what happens at the molecular level, and also the length of the membrane.

The optimal value of the hydrogen transfer coefficient β for hydrogenation with introduction of the hydrogen through the membrane catalyst was found by means of mathematical simulation. The hydrogenation rate is maximal at complete usage of diffused hydrogen if $\beta = 0.5\ kb_1P_1/(1 + b_1P_1 + b_2P_2')$, where k is the hydrogenation rate constant, b_1 and b_2 are the adsorption coefficients of the initial hydrocarbon and its hydrogenation product, and P_1 and P_2 are the partial pressures of these substances.

The experimental study of cyclopentadiene (CPD) hydrogenation with hydrogen transfer through the palladium alloy foils shows that Pd–Ru 9·8% alloy is more active and selective in respect of cyclopentene (CPE) formation than Pd–Rh and Pd–Ni alloys. The yield of CPE is 90% and unconverted CPD is not observed on Pd–Ru 9·8% alloy. The selectivity, *i.e.*, the cyclopentene to (cyclopentene + cyclopentadiene) ratio, is 0·9. Such selectivity was observed on the usual palladium catalyst only at CPD conversions lower than 90%. A high selectivity for transformation of CPD into CPE on a Ni catalyst was reported[11] but there was no data for CPD conversions above 84%. The complete conversion of CPD is very important, as this substance is a heavy poison for the CPE polymerisation catalyst. The products of CPD hydrogenation on the membrane catalyst may be used directly for polymerisation into polypentenomer, a valuable synthetic rubber. The special selectivity of the membrane catalyst is caused by the independent control of the surface concentrations of CPD and hydrogen. The hydrogen diffusion through the palladium alloy produces a surface hydrogen concentration which is low and uniform on the whole catalyst.

Hydrogenation on palladium alloys of cyclodienes with more than six carbon atoms in the ring also gives the corresponding cyclo-olefins. For instance, cyclo-octa-1,5-diene on Pd–Ru 9·8% at 60°C is transformed into cyclo-octene with a selectivity of 0·91. But cyclohexa-1,4-diene hydrogenation on Pd–Rh and Pd–Ru alloys is accompanied by benzene formation and the yield of benzene increases with increasing temperature. This is in accordance with the data[12] concerning hydrogen redistribution in cyclo-olefins and cyclodienes accompanying dehydrogenation and hydrogenation of the initial substance.

ACKNOWLEDGEMENTS. These results were obtained in collaboration with M. M. Ermilova, E. V. Khrapova, A. P. Mishchenko, N. V. Orekhova, and T. S. Ustinova. The authors are indebted to Kh. M. Minachev, G. V. Antoshin, and E. S. Shpiro for the XPS study of Pd–Rh alloys, to E. M. Savitskii, V. P. Polyakova, and N. R. Roshan for the preparation of palladium alloys, and to B. N. Luk'yanov for participation in the mathematical simulation.

REFERENCES

1 V. M. Gryaznov, V. S. Smirnov, and M. G. Slin'ko, *Proc. 5th Internat. Congr. Catalysis*, Miami Beach, 1972, Vol. 2, p. 1139.
2 A. Couper and D. D. Eley, *Discuss. Faraday Soc.*, 1950, **8**, 172.
3 G. Rienäker and G. Vormum, *Z. anorg. Chem.*, 1956, **283**, 287.
4 G. Rienäker and S. Engels, *Z. anorg. Chem.*, 1965, **336**, 259.
5 S. Engels, *Z. anorg. Chem.*, 1966, **348**, 211.

[6] V. M. Gryaznov, L. F. Pavlova, P. Rivera, A. Rossen, and H. Juares, *Kinetika i Kataliz*, 1971, **12**, 1197.

[7] B. Wood, *J. Catalysis*, 1968, **11**, 30.

[8] Kh. M. Minachev, G. V. Antoshin, E. S. Shpiro, and T. A. Navruzov, *Izvest. Akad. Nauk S.S.S.R., Otd. Khim. Nauk.*, 1973, 2134.

[9] B. C. Allen, *Trans. AIME*, 1963, **227**, 1175.

[10] S. Engels, G. Hager, and G. Hille, *Z. anorg. Chem.*, 1974, **406**, 45.

[11] H. Sakamoto, K. Takasaki, J. Harano, and T. Imoto, *J. Appl. Chem. Biotechnol.*, 1974, **24**, 759.

[12] W. M. Gryaznov and V. I. Shimulis, *Doklady Akad. Nauk S.S.S.R.*, 1961, **139**, 870.

DISCUSSION

I. V. Kalechits (*State Committee for Science and Technology, Moscow*) said: The use of membrane catalysts is only one possibility for overcoming thermodynamic limitation by conjugation of reactions; another is to catalyse the conjugated reactions by metal complexes. In this case the energy expended carrying out a principal endoergic reaction is compensated by gain of energy from an additional exoergic reaction. In this way the overall process can be realized under mild conditions, for example at much lower temperatures than usual. This is valuable in particular for manufacture of acetic anhydride.

Compounds of Pd^{II} are reduced by carbon monoxide to metal or complexes of Pd^0 in aqueous or non-aqueous media.[1-4] There are no publications on the interaction of Pd^{II} with CO in solutions of carbonic acids. We found that the anhydrous salts of Pd^{II} are reduced quantitatively to metal by carbon monoxide in solutions of aliphatic carbonic acids at temperatures of 20–150 °C. In the case of the nitrate, the reaction proceeds *via* equation (1). The reduction of palladium acetate leads to the formation of acetic anhydride [equation (2)]. The yield of acetic anhydride is as high as 95–100% with respect to Pd.

$$Pd(NO_3)_2 + 2CO \rightarrow Pd + 2CO_2 + 2NO_2 \tag{1}$$

$$Pd(OCOCH_3)_2 + CO \rightarrow Pd + CO_2 + (CH_3CO)_2O \tag{2}$$

Palladium chloride and other salts of Pd^{II} react similarly to the acetate in the presence of alkali metal salts. If we add to reaction (2) a stage of regeneration of Pd^{II} by means of any oxidizing substance (as an example oxygen is very effective) [equation (3)], the reaction (2) is transformed into a catalytic process where the palladium complexes are the catalyst [equation (4)]. Formally speaking the reaction (4) is a sum of an endoergic process (ΔG^0, positive) [equation (5)] which proceeds spontaneously only at temperatures above 900 K, and an exoergic process (ΔG^0, negative) [equation (6)].

$$Pd + \tfrac{1}{2}O_2 + 2CH_3COOH \rightarrow Pd(OCOCH_3)_2 + H_2O \tag{3}$$

$$2CH_3COOH + CO + \tfrac{1}{2}O_2 \rightarrow (CH_3CO)_2O + H_2O + CO_2 \tag{4}$$

$$2CH_3COOH \rightarrow (CH_3CO)_2O + H_2O \tag{5}$$

$$CO + \tfrac{1}{2}O_2 \rightarrow CO_2 \tag{6}$$

The energy loss of the principal reaction (5) is compensated by energy gain of step (6). Due to this fact the summary process could be realized even at very low temperatures.

[1] F. C. Phillips, *J. Amer. Chem. Soc.*, 1894, **16**, 255.

[2] A. B. Fasman, B. A. Gologov, and D. V. Sokolsky, *Doklady Akad. Nauk, S.S.S.R.*, 1974, **155**, 434.

[3] T. V. Kingston and G. R. Scollary, *Chem. Comm.*, 1969, 455.

[4] P. M. Maitlis, *The Organic Chemistry of Palladium* (Academic Press, New York, 1971), Vol. 2.

The principle of free energy compensation used here by means of the accomplishment of the conjugated process is analogous to the conjugation principles realized in biological systems, for instance, by respiration. But unlike biological systems, the conjugated process in this case does not require the use of several catalysts, one for each stage, but proceeds *via* a single metal complex catalyst which facilitates both the general and complementary reactions in a single process.

This work was carried out in collaboration with Professors I. I. Moiseev and M. N. Wargaftik and is to be published.

V. M. Gryaznov (*Peoples' Friendship Univ., Moscow*) replied: Prof. Kalechits has demonstrated a very interesting example of conjugation of reactions by metal complexes. He has pointed out the possibilities of carrying out an endoergic reaction on one side of the membrane catalyst by gain of free energy from an exoergic reaction occurring on the other side of the same catalyst. Recently Mikhalenko at the Peoples' Friendship University in Moscow has investigated the coupling of an alcohol dehydrogenation to aldehyde with cyclopentadiene hydrogenation to cyclopentene. The coupling of isoamylene dehydrogenation to isoprene with combustion of the evolved hydrogen on the other surface of a membrane catalyst has been studied by Smirnov, Ermilova, Orekhova, and Mischenko at the Institute of Petrochemical Synthesis in Moscow.

R. L. Moss (*Warren Spring Laboratory, Stevenage*) said: The authors find an activity maximum for cyclohexane dehydrogenation over palladium–nickel foils at 5% Ni. We have found an enhancement using palladium–nickel alloy films as catalysts for ethylene hydrogenation.[5] Our maximum rate at 45 atom % Pd, is about twelve times faster than observed over pure nickel and three times faster than that over pure palladium. Examination of palladium–nickel films prepared *in situ* by Auger spectroscopy showed that the outermost surface of the palladium-rich alloy films (with some certainty, beyond 65 atom% Pd) was a complete palladium monolayer.[6] We ascribed the lower activity of palladium and palladium-rich alloys to hydrogen sorption in the surface layers and maximum activity to the occurrence of a completed or a nearly-completed palladium monolayer which exhibited 'true' palladium activity. I should like to ask the authors, how far they ascribed the occurrence of the activity maxima in their systems to the decreased activity of palladium by hydrogen solution in the catalyst surface? Also, do they now have X-ray photoelectron spectra of their alloy of maximum activity which apparently has only a small surface Ni-content (5%)?

V. M. Gryaznov (*Peoples' Friendship Univ., Moscow*) replied: Your explanation of the activity maximum of palladium–nickel alloys as a function of nickel content probably applies more widely than to ethylene hydrogenation. Wood[7] has proposed a mechanism according to which sorbed hydrogen atoms are required for cyclohexane dehydrogenation on a hydrogen-porous membrane. In isobutene dehydrogenation over a palladium–nickel membrane catalyst containing 5·9 wt % Ni we have observed a maximum in the rate as the hydrogen:hydrocarbon pressure ratio is increased.[8] Increase in hydrogen pressure enhances the rate when $p_{H2}/p_{hc} \leqslant 1$, but diminishes the rate at higher concentrations. However, hydrogen solubility in the catalyst is not the only factor that influences catalytic activity; the composition of the surface layer is also very important. An XPS investigation of the palladium–nickel alloy having 5·9% Ni is in progress.

W. M. H. Sachtler (*Shell, Amsterdam*) said: You have discussed the surface composition in particular with respect to palladium–rhodium alloys. From our results[9] I

[5] R. L. Moss, D. Pope, and H. R. Gibbens, *J. Catalysis* (submitted for publication).
[6] C. T. H. Stoddart, R. L. Moss, and D. Pope, *Surface Sci.*, 1975, **53**, 241.
[7] B. J. Wood, *J. Catalysis*, 1968, **2**, 30.
[8] V. S. Smirnov, V. M. Gryaznov, N. V. Orekhova, M. M. Ermilova, and A. P. Mischenko, *Doklady Akad. Nauk. S.S.S.R.*, 1975, **224**, 391.
[9] R. Bouwman, G. J. M. Lippits, and W. M. H. Sachtler, *J. Catalysis*, 1972, **25**, 350.

would expect even more pronounced effects for the palladium–silver alloys which you investigated. We found that, after annealing *in vacuo*, palladium–silver films have a pronounced enrichment of silver in the surface, while adsorption of carbon monoxide causes enrichment with palladium due to chemisorption-induced surface aggregation. This was reversible; after desorption of carbon monoxide the surface returned to the silver-rich state. Have you studied the surface composition of your palladium–silver alloys and the change induced by the catalytic reaction?

Did you study reactions of n-alkanes on these alloys?

V. M. Gryaznov (*Peoples' Friendship Univ., Moscow*) replied: We are familiar with your studies of surface transformations in palladium–silver alloys, and that is why we have not studied the surface composition of our palladium–silver alloys in detail. However, we have found enrichment of silver in the surface of palladium–silver foils by XPS in collaboration with Minachev, Antoshin, and Spiro of the Zelinsky Institute of Organic Chemistry, Moscow.

We have studied n-alkane dehydrogenation to alkenes and alkadiene as well as dehydrocyclization to aromatic hydrocarbons on palladium–nickel, palladium–rhodium, and other membrane catalysts.

R. Gonzalez (*Univ. of Rhode Island, Kingston*) said: We have attempted, using X-ray methods, to characterize the surface structure of platinum–ruthenium clusters supported on silica and have found that redispersion occurs following treatment in air at about 673 K. We think this is due to the formation of RuO_4 which is quite volatile. Although this may not be the case on an unsupported alloy, it is an effect which could lead to significant surface reconstruction.

We find that, on exposing supported ruthenium to oxygen at 673 K for 15 s, a layer of strongly adsorbed oxygen (possibly an oxide) is formed. Ruthenium treated in this manner becomes very difficult to reduce; in fact, a 6 h reduction in flowing hydrogen at 623 K is necessary. Since you have reduced your catalyst for 0·5 h at 623 K I wonder if the catalytic enhancement which you observe might be due to an electronic perturbation of the palladium arising from oxygen adsorbed on ruthenium.

V. M. Gryaznov (*Peoples' Friendship Univ., Moscow*) replied: Data from X-ray photoelectron spectra of palladium–ruthenium foils give no indication of RuO_4 or other oxide formation during our experiments. Nevertheless your idea about the participation of a volatile ruthenium compound in the surface reconstruction is an interesting one.

We have not found any evidence of an electronic perturbation of palladium due to oxygen adsorbed on ruthenium.

H. Wise (*Stanford Res. Inst., Menlo Park*) said: The rate of dehydrogenation of alkanes and of cycloalkanes on palladium is controlled by the rate of diffusion of hydrogen through the membrane, and that of hydrogenation by the amount of hydrogen dissolved in the membrane. Thus in comparing different metal and bimetallic systems one should take these effects into account.

V. M. Gryaznov (*Peoples' Friendship Univ., Moscow*) replied: Our rates of cyclohexane dehydrogenation on palladium alloys were measured without hydrogen removal through the alloy. That is why the catalytic activity of the alloys was not greatly influenced by the effects that you mention.

A. Sárkány (*Inst. of Isotopes, Budapest*) said: I have two questions concerning cyclopentadiene hydrogenation. First, what is the influence of the addition of nickel, ruthenium or platinum to palladium on the rate of diffusion of hydrogen through the membrane? Second, did you find evidence for the surface enrichment of rhodium in cyclopentadiene hydrogenation too?

V. M. Gryaznov (*Peoples' Friendship Univ., Moscow*) replied: The addition to palladium of nickel, ruthenium or platinum at concentrations of a few atom% increases the rate of hydrogen diffusion through the membrane. Further increase in the second component concentration depresses the rate of hydrogen diffusion dramatically.

The surface layer composition of the membrane catalysts in cyclopentadiene hydrogenation is being investigated.

M. Shelef (*Ford Motor Co., Dearborn*) asked: Would you comment on the useful lifetime of the membrane catalyst?

V. M. Gryaznov (*Peoples' Friendship Univ., Moscow*) replied: The lifetime of the membrane catalysts in cyclopentadiene hydrogenation is hundreds of hours without regeneration.

A. Farkas (*Media, Philadelphia*) said: A few years ago some interesting selectivities were reported for the hydrogenation of unsaturated compounds by hydrogen diffused through palladium. Are you familiar with that work?

V. M. Gryaznov (*Peoples' Friendship Univ., Moscow*) replied: I am familiar with several investigations of the hydrogenation of unsaturated compounds by hydrogen diffused through palladium and palladium alloys, but in most cases the products of the selective hydrogenation were not mentioned. In 1973 we reported the selective formation of cyclohexene during benzene hydrogenation on a palladium–nickel membrane catalyst.[10]

J. Turkevich (*Princeton Univ.*) said: As part of the Princeton programme of preparing catalysts as colloidal metals in solutions and then mounting them on alumina supports, and as part of the collaborative U.S.S.R.–U.S.A. programme on membrane catalysis, Namba, Miner, and I have prepared palladium–platinum alloys as monodisperse particles in aqueous solution with particle size in the range 25 to 35 Å. The optical absorption spectra are continuous and are associated with transitions of metallic electrons. The absorption spectra of alloys are an additive function of those of the components indicating that there is no distortion of one electronic structure by the other; this contrasts with the behaviour of platinum–gold alloys where there is a drastic change in individual spectra on alloying. The catalytic activity of platinum–palladium alloys for ethylene hydrogenation shows a peak at 75% Pt with a constant activation energy of 6 kcal mol^{-1} until this peak is reached, and then an increase of 8 kcal mol^{-1}. The number of catalytic sites increases by 50% until 50% mixture is reached, then drops to a minimum at 85% Pt, and finally rises at 100% Pt. The turnover number for this ethylene hydrogenation is constant as platinum is added to palladium, reaches a maximum at 75% Pt and then drops. This complex relationship is not reflected in the electronic behaviour of metal electrons which is additive in the two components. On the other hand, the complex changes in the optical spectra of the platinum–gold alloys find no counterpart in the catalytic activity; gold serves merely as a dilutant to the activity of the platinum. We offer this as evidence that 'metallic electrons' have no direct relationship to catalytic activity in the hydrogenation of ethylene.

[10] V. M. Gryaznov, V. S. Smirnov, and M. G. Slin'ko, in *Mechanisms of hydrogenation reactions* (Akademiai Kiado, Budapest, 1975), p. 107.

Magnetic Study of Ethane and Benzene Adsorption on Ni–Cu/SiO$_2$: Correlation with the Catalytic Activity

J. A. DALMON, J. P. CANDY, and G. A. MARTIN*

Institut de Recherches sur la Catalyse, C.N.R.S., 39, boulevard du 11 Novembre 1918, 69626 Villeurbanne, France

ABSTRACT The chemisorption of ethane and benzene at various temperatures on homogeneous Ni–Cu alloys has been studied by magnetic methods (high field techniques). For benzene, two species are observed: the first corresponding to a high magnetic bond number ($n = 8$) is observed on pure nickel, and the second, corresponding to $n = 3$, is observed either on Ni–Cu or on hydrogen-precovered nickel surfaces. The two species have similar chemical reactivities. It is assumed that in both cases C$_6$H$_6$ is adsorbed as benzyne ligand.

On pure nickel, ethane is chemisorbed at room temperature with a high bond number ($n = 6$) indicating a strong dehydrogenation. At higher temperatures, complete cracking occurs, leading to superficial carbon atoms. Addition of Cu (alloying) or H (chemisorption) to Ni inhibits the chemisorption of ethane and increases the temperature of complete cracking, presumably for geometric reasons (dilution of the active phase).

From this work it appears that Cu addition does not alter greatly the nature of adsorbed benzene species, but inhibits the C–C bond rupture. Hence the sharp decrease in the hydrogenolysis activity and the small variation of the hydrogenation activity which are generally observed as Cu is added, are fully accounted for.

INTRODUCTION

It is well known that reactions of C–C bonds (hydrogenolysis) of organic compounds adsorbed on Ni are inhibited by alloying Ni with Cu, and that the activity towards C–H bonds (hydrogenation, dehydrogenation) is less altered.[1,2] Various reasons are invoked to account for these observations: the decrease of the d character upon alloying, the decrease of the chemisorption heats of hydrocarbons, phase segregation, and also geometric effects[2] (dilution of the active phase by Cu).

A systematic study of the nature of adsorbed hydrocarbon species seemed to us of interest and was expected to shed some light on the phenomenon and its interpretation. In this paper the results of the study of ethane and benzene adsorption on Ni–Cu alloys at various temperatures, and of their reactivities, are discussed. The magnetic high field method was used to determine the "magnetic bond number".[3] High field techniques were preferred to low field methods which give in some cases only a rough estimation of the bond number. It was previously shown that they give coherent and meaningful information in the case of CO and hydrocarbon adsorption on nickel.[4,5] The catalysts studied here were Ni–Cu alloys supported on silica, with a surface composition nearly equal to that of the bulk[6] (no phase segregation).

EXPERIMENTAL

Ni–Cu alloys supported on silica (Aerosil) were obtained by adding the support to a solution of hexa-ammine-nickel and -copper nitrates. A detailed description of the method of preparation and of the morphological characteristics was given in refs. 4 and 6. The total metallic amount was about 15%. Samples were called NiCu z, where z was the Cu percentage. After compression (2 t/cm^2), the reduction was performed in a stream of very pure hydrogen for 15 h at 650°C. This temperature was high enough to ensure a complete reduction and a good homogeneity of the metallic phase. (Curie points and saturation magnetization at 4 K of the metallic phase were very close to the corresponding bulk values). The mean diameters of metallic particles estimated from X-ray line broadening and magnetic properties were about 70Å. After outgassing at 400°C during 3 h, the samples were cooled. Then, volumetric determination of the quantity of adsorbed gas and magnetic measurements were performed.

Magnetization M was measured in moderate fields (20 kOe) at 300, 77, and 4 K. The variations of saturation magnetization (ΔM_s) upon chemisorption were obtained on plotting ΔM against $1/H$ (H, field strength). The magnetic bond number per adsorbed molecule, n, was calculated assuming that nickel atoms in interaction with the adsorbed species cease to participate in the collective ferromagnetism[3,6] *i.e.* $n = \alpha/\mu$, where μ is the magnetic moment in Bohr magnetons (B.M.) of the nickel atom in the alloys, α the decrease of saturation magnetization in Bohr magnetons per adsorbed molecule (B.M. mol^{-1}). It was shown that α values measured at 77 or 4 K were almost identical, so in this work most of the magnetic measurements were performed at 77 K.

Physisorbed benzene was removed by pumping (10^{-7} torr for 3 h) and the desorbed quantity measured by volumetric techniques. In benzene exchange ^{14}C labelled benzene[7,8] was adsorbed. A relatively large amount of unlabelled benzene was subsequently introduced. Some radioactive gas appeared in the gaseous phase and gas chromatography showed that it was C_6H_6. Gaseous and physically adsorbed benzene was trapped at 77K and its radioactivity measured.

RESULTS

The plots of saturation magnetization against the quantity of chemisorbed benzene on Ni and Ni–Cu at 20°C are shown in Figure 1. On pure nickel the decrease is linear, indicating a unique state of chemisorption ($\alpha = 4\cdot8$ B.M. mol^{-1}). On Ni–Cu alloys the curves are not linear and α values are smaller. Figure 2 shows the variations of the bond number as a function of the Cu content: n decreases as the Cu content increases.

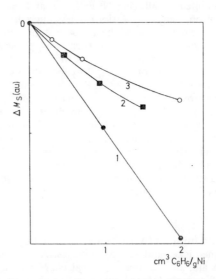

FIGURE 1 Variations of saturation magnetization ΔM_s, against the quantity of chemisorbed benzene.
1:Ni, 2:NiCu 8, 3:NiCu 14.

904

The total amount of chemisorbed benzene at 20°C compared to the number of surface nickel atoms estimated from H_2 chemisorption at 20°C under 100 torr shows that one chemisorbed benzene corresponds to 24 and 27 nickel atoms, respectively, for pure Ni and NiCu 14.

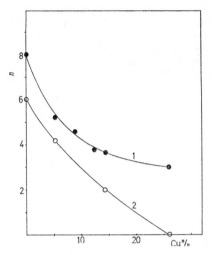

FIGURE 2   Bond number *n* at 20°C *versus* Cu content
1:C_6H_6, 2:C_2H_6

In another set of experiments, the adsorption was performed at —78°C and the system heated stepwise up to 500°C. Results are shown in Figure 3. Below 0°C C_6H_6 is not chemisorbed ($\alpha = 0$, the gas is physically adsorbed). In the 20—80°C range, α is

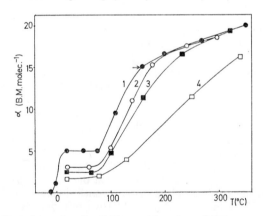

FIGURE 3   Variations of α for C_6H_6 as a function of the temperature.
1:Ni, 2:NiCu 8, 3:NiCu 14, 4:NiCu 26

temperature independent and equal to the slope observed in Figure 1 (direct adsorption at 20°C), suggesting that the corresponding species is the same. For pure Ni the curve shows a peculiarity at about 160°C, with $\alpha = 15.0$ B.M. molecule^{-1}, indicated by an arrow. As Cu is added this becomes less clear.

905

When the nickel surface is precovered by H_2 (irreversibly adsorbed at 20°C), $\alpha_{C_6H_6} =$ 2·1 B.M. mol^{-1} ($n = 3$·5). This value is very close to that observed on Ni–Cu alloys (Figure 1) and on Raney nickel catalysts precovered by hydrogen[9] and suggests that the corresponding species are identical.

These results can be interpreted assuming two chemisorbed species: species I corresponding to $\alpha = 4$·7 B.M. mol^{-1} ($n = 7$·9) observed on the untreated Ni surface and species II ($\alpha = 2$, $n = 3$) observed on Ni–Cu surfaces and Ni surfaces precovered with adsorbed hydrogen. They are similar to those observed on Raney nickel catalysts (species I is observed on outgassed samples and species II on H_2 precovered surfaces). The hydrogenation of adsorbed species is performed in the following way: after C_6H_6

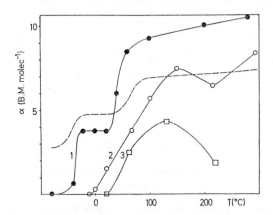

FIGURE 4   Variations of α for C_2H_6 as a function of the temperature.
1:Ni, 2:NiCu 14, 3:Ni partially covered with hydrogen variations of α for C_2H_4 on Ni:
broken line.

chemisorption on Ni (species I) or Ni–Cu alloys (species II), H_2 is subsequently introduced at 20°C under 10 torr. Condensable gases are trapped at 77 K; 15 h later, the chemisorbed benzene is completely hydrogenated into cyclohexane, suggesting no C–C bond rupture in the two adsorbed species. Moreover, both species I and II are exchanged by gas-phase benzene, indicating that the adsorbed species are not dramatically altered.

The α –T curves for ethane are shown in Figure 4. Results obtained for ethylene on pure nickel are also reported. The curve relative to ethane on nickel has a horizontal section between —20 and +20°C with $\alpha = 3$·8 B.M. mol^{-1}. At 60°C a singularity is noticed ($\alpha = 8$·0 B.M. mol^{-1}). The hydrogenation of adsorbed species at 20°C gives C_2H_6 and of adsorbed species at 70 and 120°C CH_4, indicating that the C–C bond fission occurs between 20 and 70°C, presumably at the temperature corresponding to the singularity. Similar experiments lead us to the conclusion that the C–C bond rupture in ethylene occurs at 75°C ($\alpha = 6$·12 B.M. mol^{-1}), where a similar singularity is noticed. On an NiCu 14 sample, ethane is not chemisorbed below —15°C. Moreover, no adsorbed thermostable species are detected at higher temperature. The hydrogenation of the species present on the surface at 20 and 70°C gives C_2H_6, and of the species at 120°C, CH_4, indicating that the C–C bond rupture occurs between 70 and 120°C. This temperature is higher than that corresponding to pure nickel, showing that the temperature of complete cracking increases with Cu content. The plot of bond number against Cu content is shown in Figure 2: it decreases from 6 to about 0 for Cu content higher than

26%, indicating that Cu inhibits the ethane chemisorption; this observation is in good agreement with the increase of the temperature of complete cracking as Cu content increases.

When the surface is partially precovered with hydrogen (irreversibly adsorbed at 20°C), no ethane chemisorption is observed at 20°C, as in the case of Ni–Cu alloys, the corresponding $\alpha - T$ curve (Figure 4) is similar to that of Ni–Cu.

DISCUSSION

Nature of chemisorbed species

For benzene, two species are observed with 7·9 and 3 as bond numbers. The problem of the nature of these species has been partially discussed in a paper on Raney nickel.[9] In spite of the fact the magnetic characteristics of the two species are very different, the chemical properties remain quite similar, (hydrogenation, exchange, temperature range of stability), suggesting a unique model. This apparent contradiction has been tentatively solved by assuming that benzene is adsorbed as benzyne ligand (C_6H_4) as in cluster complexes of osmium[10,11] on three metallic atoms.

This hypothesis accounts for the observed bond number on Ni–Cu and Ni–H surface ($n = 3$). The benzyne radical in triosmium complexes has a fluxional behaviour on the three atoms. On bare nickel it is postulated that benzyne is adsorbed as a benzyne ligand and also that fluxionality occurs on a larger number of surface atoms, thus explaining the larger magnetic bond number.

The magnetic bond number of C_2H_6 on pure nickel is approximately 6. Hydrogenation of the corresponding species does not give CH_4, so the C–C rupture is not achieved. Possible models for C_2H_6 adsorption are:

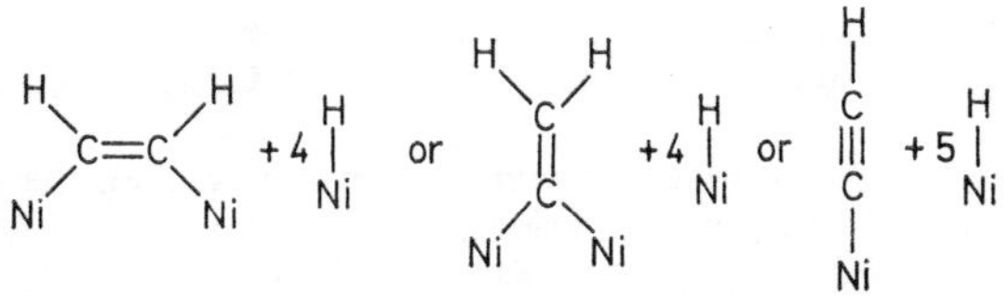

They are in agreement with models proposed to account for observed bond number for C_2H_2 ($n = 2$) and C_2H_4 ($n = 4$) on pure nickel.[5] Whatever the model, the main fact is that ethane is strongly dehydrogenated.

It has been shown that the singularities observed in the $\alpha - T$ curves at $\alpha = 6·12$, 7·5, and 15·0 B.M. mol^{-1}, respectively, for C_2H_4, C_2H_6, and C_6H_6 on pure nickel are likely to be due to C–C bond fission. An interesting problem is to determine the number of H atoms which are bonded to the carbon residue. The reaction may be written in the following way:

$$C_nH_m = n\,CH_x + (m - nx)\,H, \text{ then}$$
$$\alpha_{C_nH_m} = n\,\alpha_{CH_x} + (m - nx)\,\alpha_H$$

The calculation of x, α_H and α_{CH} by optimization gives $x = 0$

$$\alpha_H = 0·68 \text{ B.M. mol}^{-1}$$
$$\alpha_C = 1·75 \text{ B.M. mol}^{-1}$$

The value $\alpha_H = 0·68$ B.M. mol^{-1} is in good agreement with experimental data for H_2 chemisorption ($\alpha_{H_2} = 2\alpha_H = 1·4$ B.M. mol^{-1} [6]). The zero value for x indicates that the carbon residue is completely dehydrogenated.

The observed α_C can be explained assuming the formation of a superficial carbide Ni_3C, (one C for 3 nickel atoms); this carbide would correspond to $\alpha_C = 3 \times 0.6 = 1.8$ B.M. atm^{-1}, a figure which is very close to the observed value. Another possibility is the formation of superficial adsorbed carbon. In this case it is very difficult to predict the α value. No experimental data are available. The only point of comparison is α_C for interstitial carbon in nickel: $\alpha_C = 3.2$ B.M. mol^{-1}.[12] The further increase of α with T at higher temperatures may be explained by a migration of C from the surface ($\alpha_C = 1.75$ B.M. mol^{-1}) into the bulk ($\alpha_C = 3.2$ B.M. mol^{-1}).

On Ni–Cu alloys, the singularity in the α–T curve vanishes, leading to a loss of accuracy in the cracking temperature determination. However, the α values for alloys containing more than 14% Cu are markedly smaller than that corresponding to pure nickel, indicating a strong inhibition of the cracking by Cu addition; this is in good agreement with the results of the hydrogenation of adsorbed species at various temperatures.

Geometric effects

Addition of Cu to Ni leads to a decrease of the bond number of both C_2H_4 and C_6H_6 (Figure 2) probably for geometric reasons. Ni multiplet ensembles corresponding to $n = 8$ and 6 for C_6H_6 and C_2H_6 are required for the adsorption. As Cu is added, the probability of the existence of such ensembles decreases rapidly due to the dilution of the active metal. In the case of C_6H_6, another species corresponding to $n = 3$ may be formed on a small Ni ensemble, so that the bond number decreases from 8 to 3 as the Cu content increases. For C_2H_6, no species other than that corresponding to $n = 6$ seems possible so that no chemisorption occurs.

Catalytic selectivity

From this work, it is clear that Cu addition does not alter greatly the nature of adsorbed benzene species. If it is admitted that this species is the intermediate in C_6H_6 hydrogenation, then the small variations of activity with Cu content are fully accounted for.

In the case of ethane, it has been shown that on pure nickel, the adsorbed species undergoes a strong dehydrogenation prior to C–C bond fission. Addition of small quantities of Cu prevents the formation of this species presumably for some geometric reasons and inhibits the C–C bond rupture. Then the sharp decrease in the hydrogenolysis activity as Cu content increases is also fully accounted for. It is noticeable that purely geometric considerations are used in this work.

Another point of interest is the similar effect of H_2 preadsorption and of Cu addition: C_6H_6 species II is observed either on Ni–Cu or NiH surfaces; in the same way, the addition of Cu or H prevents the formation of the species corresponding to $n = 6$ for ethane. These results suggest that H is not mobile on the surface, and plays the same geometric role of dilution of the active Ni atoms as does Cu. Some recent data on hydrogen mobility at the same temperature on Ni surfaces obtained by neutron scattering (quasi-elastic diffusion) are in agreement with this view.[13] The negative order with respect to H_2 pressure generally observed in ethane hydrogenolysis kinetics at relatively low temperature (198°C) could be explained in this way.

REFERENCES

[1] V. Ponec, *Catalysis Rev.*, 1975, **11**, 41.
[2] J. H. Sinfelt, *AIChE Journal*, 1973, **19**, 673.
[3] P. W. Selwood, *Adsorption and Collective Paramagnetism* (Academic Press, New York, London 1962).
[4] J. A. Dalmon, M. Primet, G. A. Martin, and B. Imelik, *Surface Sci.*, 1975, **50**, 95.
[5] G. A. Martin, and B. Imelik, *Surface Sci.*, 1974, **42**, 157.

[6] J. A. Dalmon, G. A. Martin and B. Imelik, *Proc. 2nd Internat. Conf. on Solid Surfaces*, 1974, *Jap. J. Appl. Phys.*, *Suppl.* 2, *pt* 2, 1974, 261.
[7] P. Tetenyi and L. Babernics, *J. Catalysis*, 1967, **8**, 215.
[8] J. P. Candy and P. Fouilloux, *J. Catalysis*, 1975.
[9] J. P. Candy, J. A. Dalmon, P. Fouilloux, and G. A. Martin, *J. Chim. phys.*, 1975, in the press.
[10] A. J. Deeming and M. Underhill, *J. Organometallic Chem.*, 1972, **42**, C 60.
[11] A. J. Deeming, R. S. Nyholm and M. Underhill, *Chem. Comm.*, 1972, 224.
[12] M. C. Cadeville and J. Deporte, *J. Phys. Letters*, 1972, **41A**, 237.
[13] R. Stockmeyer, H. M. Conrad, A. J. Renouprez, and P. Fouilloux, *Surface Sci.*, 1975, **49**, 549.

DISCUSSION

R. B. Moyes (*Univ. of Hull*) said: We have some results (see Table) which may support the tentative assignment of a benzyne species (p. 907). At the Fifth Congress[1] we interpreted some of these results as arising from the formation of α,β-diadsorbed-C_6H_4 intermediates but they may also be consistent with a benzyne intermediate. They

Table

Reaction	M-values obtained for reactions over metal films							
	Ni	Co	Fe	Mn	Cr	V	Ti	Sc
$C_6H_6 + C_6D_6$	1·6	1·9	1·7	1·6	1·5	1·7	3·5	—
$C_6H_6 + D_6$	2·3	1·9	1·9	1·2	1·6	1·9	1·8	1·5

demonstrate that the intermediate formed by loss of two hydrogens is of wide relevance to considerations of the chemisorption of benzene at metal surfaces.

However, the properties of nickel films differ from those of nickel–silica. We find[2] that chemisorbed benzene occupies four surface atom sites, as estimated from deuterium chemisorption, unlike the 24–27 found by these authors. Our value has been confirmed by other workers.[3]

Further, the amount of labelled benzene chemisorbed on a metal film which can exchange with a subsequently admitted charge of unlabelled benzene is relatively small at 273 K. The values are 15% when the label is ^{14}C, and 25% when the label is D. We found three types of benzene adsorbed on films at 273 K. (a) Physically adsorbed benzene which can be removed by pumping and which is normally less than 5% of the total adsorbed layer. (b) Benzene which is able to exchange with gas-phase benzene (15–20% of the adsorbed layer). (c) Irreversibly adsorbed material, unable to exchange carbon, but able to exchange deuterium with gas-phase benzene. Our results seem to present a picture closer to that the authors draw for adsorption above 423 K on nickel–silica. Can Dr. Martin please indicate how much of the chemisorbed labelled benzene exchanges with unlabelled benzene at 273 K?

G. A. Martin (*CNRS, Villeurbanne*) replied: At 273 K and at room temperature the exchange of chemisorbed labelled benzene with unlabelled benzene is much more complete on nickel–silica than on nickel films, about 80% on pure nickel and 70% on nickel–copper in about 10 min. This indicates clearly that the situation on film is not

[1] K. Baron, R. B. Moyes, and R. C. Squire, *Proc. 5th Internat. Congr. Catalysis*, Palm Beach, 1972, Vol. 1, p. 123.
[2] R. G. James and R. B. Moyes, unpublished results.
[3] D. Brennan, private communication.

exactly the same as on nickel powders. We agree that the irreversibly adsorbed material unable to exchange carbon on nickel films is possibly similar to species observed on nickel–silica at 423 K. It suggests that experiments similar to those that you have performed on nickel films would be of great interest in the case of our samples. However, in spite of these differences, it is stimulating to see that your results support our tentative assignment.

V. Ponec (*Univ. of Leiden*) said: It is very important that the earlier explanation of catalytic data obtained with nickel–copper alloys,[4-6] based on an assumed role of dehydrogenation of surface adsorbed complexes, is further supported by results in this and other papers. In our laboratory Dr. Franken has performed measurements of work function changes upon chemisorption of ethylene at different temperatures.[7] These work function measurements can be considered as complementary to the magnetic measurements in this paper and they lead to the same conclusions.

However, there is one point where the conclusions differ. Franken concluded that the surface composition (or, more exactly, the surface dipole) was the same for alloys equilibrated below (438 K) and above (up to 693 K) the critical temperature of phase separation (about 443 K) and revealed a substantial surface enrichment in Cu. A recent analysis[8] of Auger spectra in the low energy region (<100 eV) has confirmed this conclusion which is, however, at variance with assumptions made by the present authors. If there is an enrichment it would probably change the 'bond-order *versus* % Cu' curves, would it not?

G. A. Martin (*CNRS, Villeurbanne*) replied: Certainly, if a substantial surface enrichment in copper was to occur, some of our calculations would require re-examination. However, it seems more reasonable to admit of no significant enrichment for several reasons. (i) The volume of hydrogen adsorbed per square metre of metal decreases more or less linearly with the fractional copper content, x, from $x = 0$ to $x = 1$ (to be published). This indicates that the surface concentration of nickel is nearly the same as that of the bulk. (ii) The decrease of saturation magnetization per adsorbed molecule for hydrogen, α, is given by $\alpha = 2\mu_{N1}$ (ref. 6). This can be considered as an additional indication that the surface moment of nickel is equal to the average one, suggesting no surface enrichment. (iii) In the case of oxygen chemisorption (low coverage, low temperature) α is related to x by the equation $\alpha = 1\cdot2 - 2x$ (to be published). If it is admitted that the surface stoicheiometry is NiO and CuO (as in the case of pure nickel and copper) this observation can be considered as a proof that no enrichment in copper occurs.

We think that our conclusions differ on this point because the nickel–copper systems are different (film in your case, supported powder in ours), and that the phenomenon of surface enrichment depends also upon the morphology of the system. A similar conclusion was previously reached by Sachtler and co-workers who have shown that the 'cherry model' was not applicable to finely divided nickel–copper alloys.

P. Tétényi (*Inst. of Isotopes, Budapest*) said: First, I would like to support the results and conclusions made by the authors of this paper. Their results are in close agreement with our experiments on the adsorption of benzene, cyclohexane, and ethane. Nevertheless, these conclusions cannot be general because of the great differences between the various metals. As we have found, nickel (and cobalt) adsorb benzene highly dissociatively, but on some other metals (*e.g.* platinum and copper) the

[4] J. H. Sinfelt, J. L. Carter, and D. J. C. Yates, *J. Catalysis.*, 1972, **24**, 283.
[5] V. Ponec and W. M. H. Sachtler, *J. Catalysis*, 1972, **24**, 250.
[6] V. Ponec and W. M. H. Sachtler, *Proc. 5th Internat. Congr. Catalysis*, Palm Beach, 1972, Vol. 1, p. 645.
[7] P. E. C. Franken and V. Ponec, *J. Catalysis*, 1972, **42**, 398.
[8] C. R. Helius, *J. Catalysis*, 1975, **36**, 114.

proportion of dissociative chemisorption was much smaller, amounting to only a few per cent of the adsorption.

Regarding the chemisorption of ethane, in our experiments (reported partly in this Congress) highly dissociative adsorption was again observed. Therefore I would ask: why do the authors consider that the surface species produced by irreversible adsorption have π-bonds? Why not take as a possible alternative for ethane adsorption the model:

$$HC\!\!-\!\!CH_2 + 3H$$
$$\overset{**}{}\ \overset{*}{}\ \ \ \ \ \overset{*}{}$$

in which each asterisk represents a surface nickel atom and each nickel–carbon bond has simple σ-character?

G. A. Martin (*CNRS, Villeurbanne*) replied: Your model can be accepted, since it leads to a bond number of six. This illustrates one of the limitations of the magnetic method: various models may each provide an interpretation for a given bond number. It is for this reason that we are now combining magnetic methods with other techniques, such as i.r. spectroscopy (*e.g.* see ref. 4 in our paper). Exchange experiments with deuterium, or with perdeuterohydrocarbons, as reported in your paper and by Moyes, Garnett, and others, would be of great interest if combined with magnetic measurements.

Another model compatible with $n = 6$ which was not discussed in our paper is (1). It corresponds also to a C_2H_3 radical, σ- and π-bonded to three nickel atoms, and

(1)

exhibiting fluxional behaviour (such a species has been recognized in triosmium complexes).

R. Montarnal (*French Inst. of Petroleum, Rueil Malmaison*) said: Your direct characterization of the chemisorbed state is interesting. One important point of your conclusion is that you consider, in a parallel way, H and Cu as additives exerting an essentially rigid geometric influence.

My comment and question concern the ethane reaction. Concerning the influence of hydrogen, the geometric influence is allowed for by the classical Langmuir expressions if the chemisorption of hydrocarbon and hydrogen is competitive. The site occupied by a hydrogen atom cannot be occupied by hydrocarbon. One has also to invoke, for the dissociative chemisorption of hydrocarbon, the occupation of several sites. But in your case the presence of gaseous hydrogen governs the number of abstracted H-atoms and thence of occupied sites. This can be achieved assuming reversible chemisorption at temperatures where catalysis occurs whereas you invoke rather irreversible chemisorption. By use of such classical expressions one can explain not only negative orders with respect to hydrogen but also eventual positive order (with Re or Fe) with a maximum as observed for hydrocarbons higher than ethane. I am not sure that your rigid model can interpret this. My question is this: considering that catalysis occurs in presence of molecular hydrogen and at a temperature above 373 K, would it be possible to characterize by the magnetic method, the bond number for chemisorption under these conditions and its variation with hydrogen pressure? Moreover would it be possible to obtain the eventual distribution curve for the percentages of the various chemisorbed states, each having a different number of abstracted hydrogen atoms?

G. A. Martin (*CNRS, Villeurbanne*) replied: The ultimate aim of our studies is to perform magnetic measurements under the exact conditions of the chemical reaction. However, we encountered difficulties when working with the higher pressures because the volume of chemisorbed gas has to be known accurately. (As it can be seen from our paper, there are no major problems associated with use of higher temperatures.) It is for this reason that, as a first step, we have studied the system using low hydrogen pressures under which conditions hydrogen is irreversibly chemisorbed. You are right to point out this limitation. We plan to do similar studies using higher hydrogen pressures; preliminary experiments suggest that the behaviour of benzene on nickel–silica precovered with both reversibly and irreversibly adsorbed hydrogen is not the same as that reported here for nickel–silica precovered with irreversibly adsorbed hydrogen.

R. Z. C. van Meerten (*Univ. of Nijmegen*) said: You show that, at 293 K, benzene adsorbs on bare nickel, or on nickel precovered with hydrogen, the number of chemisorption bonds being different in the two cases. However, in your opinion, the chemical properties are similar.

We have studied benzene adsorption and hydrogenation on nickel–silica by gravimetric and low field magnetic methods. The gravimetric experiments showed that, at 293 K, the amount of benzene adsorbed was independent of the hydrogen pressure, and covered about one-third of the metal surface. Of this adsorbed quantity two-thirds reacted rapidly with hydrogen, whereas the remainder reacted very slowly but was hydrogenated after 15 h. You note only that all of the benzene reacted after one night. The difference in reactivity of benzene with hydrogen points to different forms of adsorption. Possibly a π-adsorbed form exists in company with a σ-bonded form, the π-form (with a low bond number) being highly reactive *via* the mechanism proposed by Rooney in 1963, and the σ-form (with a high bond number) being of low reactivity.

G. A. Martin (*CNRS, Villeurbanne*) replied: The change of reactivity which you observed in your experiments, and we in ours, may reflect the strength of the bond between the metal and benzene, the less reactive species being the more strongly adsorbed. This change does not imply different forms of adsorbed benzene. We have no calorimetric data on benzene adsorption, but can illustrate this assertion by reference to hydrogen chemisorption.[9] Calorimetric data shows a decrease in the heat of chemisorption from 18 to 4 kcal mol^{-1}, whereas magnetic isotherms indicates a magnetic bond number constant over the whole range of coverage. This shows clearly that a given species may have various reactivities. We think that these σ-, π-adsorbed species are possibly intermediates in exchange reactions with molecular hydrogen. They may be converted more or less rapidly into π-adsorbed species which participate in hydrogenation.

M. V. Matthieu (*CNRS, Villeurbanne*) said: Dr. Primet and I wish to comment on geometric and electronic effects in these nickel–copper alloys which the magnetic measurements show to be homogeneous solids. First we have shown clearly that electron transfer from copper to nickel occurs (magnetic measurements and i.r. spectrum of chemisorbed carbon monoxide). Band frequencies (ν_{CO}) on nickel decrease when those on copper increase.[10] Clearly there is an electronic effect: the same has been observed for palladium–silver alloys.[11] Second, we followed the change in the ratio of bridged-CO to the total amount of CO chemisorbed as a function of the increasing percentage of copper in the alloy. This decrease could not be correlated with the

[9] J. A. Dalmon, G. A. Martin, and B. Imelik, *Colloques Internationaux du C.N.R.S.*, 1972, No. 201, p. 593.
[10] J. A. Dalmon, M. Primet, G. A. Martin, and B. Imelik, *Surface Sci.*, 1975, **50**, 95.
[11] M. Primet, W. M. H. Sachtler, and M. V. Mathieu, *J. Catalysis*, in press.

probability of finding pairs of nickel atoms on the surface as composition was varied. But by use of Kobosev's theory of ensembles, as developed recently by Dowden,[12] we could interpret our results. We considered a pair of nickel atoms in an ensemble of 20 atoms and calculated, using the binomial probability law, the likelihood of finding copper atoms in this ensemble. To fit one curve representing the variation of concentration of bridged-CO with alloy composition we found, for copper concentrations below 10%, that one copper atom inhibits nineteen nickel atoms. This efficiency of copper decreases with increasing copper content; the limit is reached when one copper atom is near a nickel atom of a pair. When the copper concentration is small, the action of copper is a relatively long range interaction, due to an electronic effect. As the concentration is increased, the electronic effect is coupled with a geometric effect.

This type of calculation, with a greater number of nickel neighbours (perhaps six for ethylene) might explain the sharp decrease of hydrogenolysis observed for alloys containing 10% copper.

G. A. Martin (*CNRS, Villeurbanne*) replied: This interesting type of statistical calculation is based on the assumption of a random distribution of nickel and copper atoms. However, in bulk alloys it is known that nickel atoms have as tendency to be surrounded by other nickel atoms; this short range order depends upon the copper concentration and has been measured by neutron techniques. Obviously, this phenomenon should also occur at the nickel–copper surface, but as far as we know, no determinations of short range order for this surface are available.

J. L. Garnett (*Univ. of New South Wales*) said: The present paper and the preceding comment by Moyes accentuate two important features involving the catalytic chemistry of benzene under homogeneous and heterogeneous conditions; (i) the role of dissociatively as distinct from associatively complexed benzene species in these reactions especially in isotope exchange;[13] (ii) the significance of the M-value[14] in the understanding of reaction mechanism. It is essential to realize that in order to obtain data for the M-value calculation in isotope exchange, deuterium must be used as isotope. In

recent homogeneous metal-catalysed exchange with benzene and benzyl alcohol, Gold and co-workers[15,16] have used tritium as source of isotope. Because they chose tritium instead of deuterium, they have been unable to obtain from their studies M-value calculations, n.m.r. orientation of incorporated isotope in labelled aromatic, or a satisfactory comparison of their homogeneous results with our own homogeneous data.[13] Radiolytic effects from tritium may further complicate Gold's system. We therefore believe that the mechanisms and intermediates proposed by Gold and co-workers are untenable. In particular, the proposed reversible formation of a cyclohexadiene

[12] D. A. Dowden, *Proc. 5th Internat. Congr. Catalysis*, Palm Beach, 1972, Vol. 1, p. 621.
[13] J. L. Garnett, *Catalysis Rev.*, 1971, **5**, 229.
[14] J. R. Anderson and C. Kemball, *Adv. Catalysis*, 1957, **9**, 51.
[15] L. Blackett, V. Gold, and D. M. E. Reuben, *JCS Perkin II*, 1974, 1869.
[16] V. Gold, S. E. Gould, and D. M. E. Reuben, *JCS Perkin II*, 1974, 1873.

species,[15] (2), has been already rejected by a number of workers on energetic considerations.[13] The role of the associative intermediate (3) in both homogeneous and heterogeneous exchange is also considered to be minimal, π-dissociative processes appearing to predominate.[13] There are a number of other controversial points raised by Gold *et al.* concerning our original deuterium work[17] to which we shall reply in a paper to be published elsewhere.[18] Finally, we note that data from other techniques such as isotope exchange are consistent with the results of the present paper, for benzene exchange.

G. A. Martin (*CNRS, Villeurbanne*) replied: We thank Prof. Garnett for his comment, and would add that the non-dissociative π-adsorbed form was not detected in our experiments. Naturally, this does not mean that it does not exist, and experiments are in preparation in which we shall attempt to observe it particularly under conditions of high hydrogen coverage.

[17] J. L Garnett and R. J. Hodges, *J. Amer. Chem. Soc.*, 1967, **89**, 4546; K. Davis, J. L. Garnett, K. Hoa, R. S. Kenyon, and M. A. Long, *Proc. 5th Internat. Congr. Catalysis*, ed. J. W. Hightower, North-Holland, Amsterdam, 1973, Vol. 1, p. 491.
[18] J. L. Garnett, R. J. Hodges, R. S. Kenyon, and M. A. Long, unpublished work.

Selectivity in Reactions of n-Hexane on Supported Pt–Cu Alloys

HERMAN C. DE JONGSTE,* FRED J. KUIJERS, and VLADIMIR PONEC

Gorlaeus Laboratoria, Rijksuniversiteit Leiden, P.O. Box 75, Leiden,
The Netherlands

ABSTRACT   Reforming reactions of n-hexane have been studied on platinum and platinum–
copper alloys. The alloys were prepared in a powder form, with silica as a carrier. The total
activity and the selectivity for isomerisation, dehydrocyclisation, and hydrogenolysis (hydrocracking) have been determined. With increasing copper content the total activity and the
selectivity for isomerisation both decrease, while selectivity for dehydrocyclisation increases.
The results are rationalised by a picture which assumes that the atomic properties of alloy
components are retained also in alloys. The difference between platinum–copper and platinum–
gold alloys is discussed briefly.

Introduction

Alloy catalysts in general and Pt–Cu catalysts in particular are no new subjects of fundamental research in catalysis. However, the objectives of investigations vary substantially with time. The earliest experiments tried to detect a possible role of alloy ordering in catalysts[1,3] and later these alloys seemed to be ideal material to check the predictions concerning the role of the d-holes in catalysis.[2,3] In spite of the fact that the very early papers on alloy catalysts already mention alloying as a means of increasing the selectivity in a desired sense,[4] it was only recently that alloying was recognised to be a convenient way to change the *selectivity* of catalysts in reactions like the reforming reactions of hydrocarbons.[5-9] Results obtained in this direction up to now substantiate a hope that as a result of a systematic study on catalytic behaviour of alloys the general principles governing the selective operation of catalysts would finally be formulated.

Experimental

All chemicals used were described in ref. 5 as were the apparatus used and the procedure adopted for data evaluation.

Catalysts were prepared by coimpregnation, as already described.[19] Alloying was tested by X-ray diffraction (Philips Goniometer PW 1050/25, X-ray tube PWCu2103/00). Particle size was checked by electron microscopy (electron microscope Philips EM200), samples were prepared in the form of a suspension (finely ground catalyst suspended in ethanol by an ultrasonic cleaning apparatus).

Results

(1) Alloy Characterisation

Results of the X-ray diffraction studies are summarised in Fig. 1. No signs of ordering were detected in spite of the temperatures used being below the critical temperature, most likely because the time of the equilibration (and of the use of catalyst) was too short for it.[10] However, this fact does not exclude a short-range ordering which can lead to a higher than random concentration of Pt in clusters.[10]

Results of the electron microscope investigation and application of the simple formula for X-ray line broadening[11] are summarised in Table 1. (Because the results are not corrected for a possible heterogeneity in the lattice constant, the real size could be larger.)

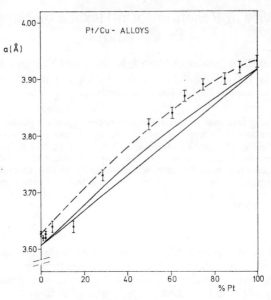

FIGURE 1 Lattice constant a of Pt–Cu alloys used in this work (dots). Drawn full line: data according to Ref. 27. Straight line: Vegard law.

Table 1

Particle Size of Pt–Cu Alloys

Composition % Pt	X-ray, broaden-ing[a] $2r$[b], particle diam. Å	Electron microscopy	
		Range diam. Å	Most particles, diam. Å
0	250 ± 40	200–900	300–500
1·8	200 ± 30	200–800	300
5·2	130 ± 10		
15	60 ± 5		
29	50 ± 5		
50	45 ± 5	50–200	50–100
61	60 ± 5		
75	50 ± 5	40–120	
85	50 ± 5		
92	40 ± 5		
100	50 ± 5	15–90	50

[a] Average from (111) and (200) X-ray diffraction.
[b] Specific surface of alloys $S = 3/rg$; g = density.

(2) Catalytic Activity of Alloys

Activity of the alloys was determined as α, the total conversion of hexane (proportional to the total rate, at low α) per gram or per cm² of catalyst. The surface of alloys—S—was calculated using the values r (Table 1) and the formula $S = 3/rg$. Results are summarised in Fig. 2.

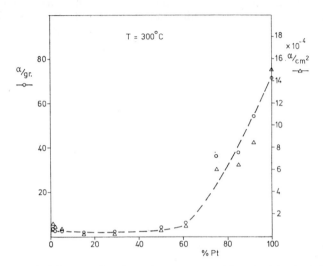

FIGURE 2   Activity as a function of alloy composition. Activity is defined[5] as the overall conversion α of hexane at 300°C and constant flow per gram catalyst (left) or cm² alloy surface (right).

Conversions of hexane, α, were measured at increasing and decreasing temperature; the steady-state was usually achieved readily. Activity of the catalysts varied reversibly and reproducibly if the measurements were performed at T < 300°C. At higher temperatures self-poisoning occurred.

By plotting log α (for α < 10%) against $1/T$, the apparent activation energy of the total reaction was also determined (see below).

(3) Selectivity of Alloys

In contrast to the behaviour of various alloys studied previously,[5,20] these alloys revealed a temperature-dependent selectivity. Some typical examples for the selectivity against T plots are in Fig. 3. All variations with temperature are reversible.

By plotting log α_1 against T^{-1}, where α_1 represents the total conversion in either all isomerisation or cracking and dehydrocyclisation products, apparent activation energies of these particular processes were also determined. Results are in Fig. 4. Fig. 5 shows the particular selectivities as a function of alloy composition.

DISCUSSION

Several authors[12-18,21-23] have contributed much to a better understanding of the reforming reactions on metals. It appears that various mechanisms suggested by different authors most likely operate simultaneously but to a different extent at different temperatures and on different catalysts. Alloying, as well as particle-size changes may lead to variations in the contribution of various mechanisms.[21,22] Mechanisms which have to be considered are the following.

For isomerisation: (1) McKervey, Rooney, and Samman,[12] a "one-site" mechanism for methyl shift; (2) Anderson and Avery,[13] a "two-site" bond-shift mechanism; (3) Barron, Maire, Cornet, and Gault,[14] isomerisation *via* a cyclic intermediate. For dehydrocyclisation: (4) a five-ring closure and eventually ring enlargement;[15] (5) a direct six-ring closure and dehydrogenation;[16,17] (6) dehydrogenation up to tri-enes and ring closure (proved to be operating on metallic as well as oxidic catalysts[18]).

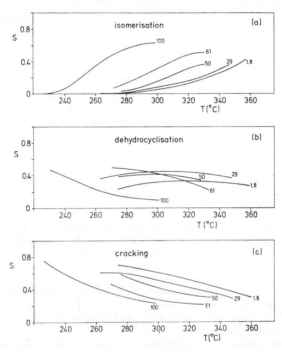

FIGURE 3   Selectivity of alloys as a function of temperature. Composition of alloys (% Pt) is indicated.

Hydrogenolysis (cracking) is usually occurring parallel to these reactions; it has a quite different character on Pt on the one hand and Ni or Cu on the other. Ni and Cu show a distinct preference for the so-called terminal splitting[5,19,23] while Pt breaks the bond in the middle most easily (a direct propane formation).[23] In this sense Ni diluted in Cu behaves more like Pt than like pure Ni.[5] All reactions which must be considered are summarised in the scheme in Fig. 6.

Further, it is known[22] that smaller particles favour formation of cyclic products[14,21,22,24] and isomerisation *via* a cyclic intermediate. Massive metals and large particles favour a bond-shift mechanism of isomerisation[21,22,24] and isomerisation with respect to dehydrocyclisation. Low co-ordinated atoms on stepped surfaces (atoms near lattice defects) favour dehydrocyclisation.[25] These are the conclusions well established by the literature data on this subject.

Let us assume now that isomerisation *via* a cyclic intermediate is the main mechanism with the catalysts used here. The particle size of our catalysts is favourable for it.[22,24] Now, we have to consider the following. Cyclic intermediates are formed by a five-ring closure on one or more (two, most likely) active sites and after they are formed they must be opened again, but on a position in the adsorbed intermediate different from the place of closure. It is highly probable that the intermediate must be adsorbed during this process on active sites different from those used for closure, if isomerisation is to proceed fast.

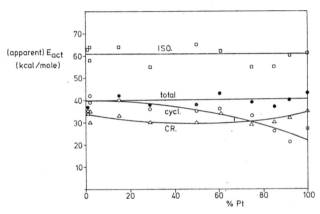

FIGURE 4   Apparent activation energy as a function of alloy composition. Activation energy of the overall conversion of hexane and of the three partial processes: isomerisation, dehydrocyclisation and hydrogenolysis (cracking).

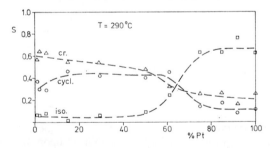

FIGURE 5   Selectivity of alloys at 290°C as a function of alloy composition.

When Pt is diluted in Cu or poisoned by carbon deposits (see Fig. 3 in ref. 19) we may expect that a smaller number of intermediates will react right through to the isomerisation products and more of them will desorb as a cyclic molecule, methylcyclopentane; such behaviour can be seen in Fig. 7. On the other hand, other factors being constant at higher temperatures, more intermediates react right through to the isomerisation products (see Fig. 3, this paper).

In terms of the scheme in Fig. 6 we conclude that dilution of Pt in Cu leads to a decrease in rate of the process (9) and results in a distribution of products like that shown in Fig. 7. Dilution of Pt in Au leads, as our previous work has shown, to a quite different product distribution. This can be seen in Fig. 8. We believe this is because the diluted Pt–Au alloys do not form any cyclic intermediate at all and the main route of isomerisation here is *via* reactions such as 1 and 2 in Fig. 6. This conclusion is confirmed by our results with methylcyclopentane. All diluted Pt–Au alloys produce only benzene but they are inactive in ring opening, *i.e.* in the reaction which is a reverse reaction to the ring closure (reaction 3). Gault *et al.* found by their isotopic method that the highly diluted Pt–Au alloy indeed did not isomerise *via* a cyclic intermediate (we thank Professor Dr. F. G. Gault, Strasbourg, for kindly communicating to us his unpublished results). Summarising, Cu is mainly responsible for blocking reaction (9);

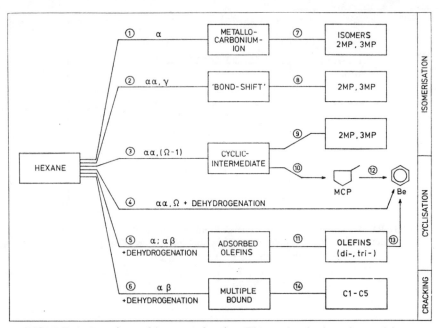

FIGURE 6   Reactions of hexane reforming. The mode of adsorption and intermediates are indicated.

and Au, at very high dilution of Pt, for blocking reaction (3) of the scheme. For dehydrocyclisation several sites are naturally necessary: some of them to bear the intermediate and some others to bind hydrogen split off upon formation of the intermediate. It is conceivable that some of these sites could be the Cu sites, but not the Au sites.

Another interesting point is demonstrated by Figs. 5 and 6. Strong dilution of Pt in Cu increases the relative contribution of hydrogenolysis (cracking) to the overall reaction, in contradiction to all previous experiences[5-9] with Group VIII and Ib metal alloys. However, we must not forget that the intrinsic cracking activity of Pt is lower than the activity of other metals studied previously so that a possible contribution of the "inactive" alloy component can be more easily detected. If the contribution of Cu is really responsible for the increased selectivity for cracking at highly diluted Pt alloys then it is very interesting to note that the cracking activity has a typical Pt character. We observed similar behaviour with Ni–Cu alloys, too: both Ni and Cu (as pure metals) favour terminal cracking but the character of cracking changes when a Ni–Cu alloy is formed; it becomes Pt-like. This might be an indication of "mixed" ensembles of active sites such as Pt–Cu, or Ni–Cu, in cracking and dehydrocyclisation.

The activity of very diluted Pt–Au and Pt–Cu alloys deserves more attention. According to the theories based on the postulated role of *d*-holes (see *e.g.* refs. 2 and 3 for a review), these alloys should be completely inactive. Some authors reported that

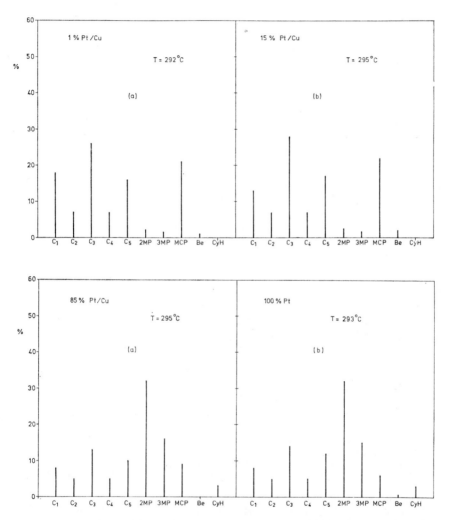

FIGURE 7 A comparison of four Pt–Cu catalysts in the same temperature range. Displacement of 2MP and 3MP by MCP in the product distribution can easily be seen.

they found experimental support for this statement.[2,26] According to the ideas mentioned there should not be a substantial difference in catalytic behaviour of Pt alloys when the *d*-holes are filled by electrons of either Cu or Au. However, our findings are at variance with these predictions. Diluted alloys have a low but not negligible activity and Fig. 8 shows that there is a difference between Pt–Cu and Pt–Au alloys. We have

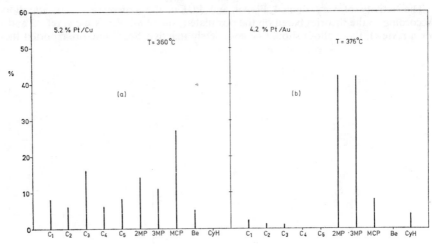

FIGURE 8   A comparison of diluted Pt–Au and Pt–Cu alloys.

seen that an explanation can be suggested in terms of the ensembles of active sites—which are different in size for different reactions—and properties of individual metal atoms, *i.e.* Pt, Au, and Cu, respectively.

REFERENCES

1 G. Rienäcker and H. Hildebrandt, *Z. anorg. Chem.*, 1941, **248**, 52.
2 G. Rienäcker, *Abhandlungen der Deutschen Akademie der Wissenschaften zu Berlin* (Akademie Verlag Berlin—DDR No. 3, 1956), pp. 1–53.
3 G. C. Bond, *Catalysis by Metals* (Academic Press, London, 1962).
4 W. Ipatieff, *Ber.*, 1910, **43**, 3387.
5 V. Ponec and W. M. H. Sachtler, *Proc. 5th Internat. Congr. Catalysis*, Miami Beach, 1972, Vol. 1, paper 43, p. 645.
6 J. M. Beelen, V. Ponec, and W. M. H. Sachtler, *J. Catalysis*, 1973, **28**, 376; V. Ponec and W. M. H. Sachtler, *ibid.*, 1972, **25**, 250.
7 J. H. Sinfelt, J. L. Carter, and D. J. C. Yates, *J. Catalysis*, 1972, **24**, 283; J. H. Sinfelt, *ibid.*, 1973, **29**, 308.
8 T. J. Gray, N. G. Masse, and H. G. Oswin, *Act. 2-ème Congr. Int. Catal.*, (Technip, Paris, 1961), Vol. 2, p. 1697.
9 W. G. Reman, A. H. Ali, and G. C. A. Schuit, *J. Catalysis*, 1971, **20**, 374.
10 E. Torfs, L. Stals, J. van Landuyt, P. Delavignette, and S. Amelinckx, *Phys. Stat. Sol.*, 1974, **22**, 45.
11 M. H. Jellinek and I. Fankuchen, *Adv. Catalysis*, 1948, **1**, 257.
12 A. A. McKervey, J. J. Rooney, and N. G. Samman, *J. Catalysis*, 1973, **30**, 330.
13 J. R. Anderson and N. R. Avery, *J. Catalysis*, 1966, **5**, 446; 1967, **7**, 315.
14 Y. Barron, G. Maire, D. Cornet, and F. G. Gault, *J. Catalysis*, 1963, **2**, 152; Y. Barron, G. Maire, J. M. Müller, and F. G. Gault, *ibid.*, 1966, **5**, 428.
15 V. Haensel, G. R. Donaldson, and F. J. Riedl, *Proc. 3rd Internat. Congr. Catalysis*, (Amsterdam, 1964), Vol. 1, paper I, 9, p. 295.
16 B. H. Davis, *J. Catalysis*, 1973, **29**, 395; 398.
17 F. M. Dautzenberg and J. G. Platteeuw, *J. Catalysis*, 1970, **19**, 42.
18 V. A. Kazanskii, G. V. Isagulyants, H. I. Rozengart, Yu. G. Dubinskii, and L. I. Kovalenko, *Proc. 5th Internat. Congr. Catalysis*, Miami Beach, 1972, Vol. 2, paper 92, p. 1277; Z. Paal and P. Tetenyi, *Acta chim. Acad. Sci. Hung.*, 1968, **58**, 105; *J. Catalysis*, 1973, **30**, 350.

[19] H. C. de Jongste, F. J. Kuijers, and V. Ponec, *Proc. Symposium Scientific Bases for the Preparation of Heterogeneous Catalysts*, Brussels, 1975.

[20] J. R. H. van Schaik, R. P. Dessing, and V. Ponec, *J. Catalysis*, 1975, **38**, 273.

[21] A. O'Cinneide and F. G. Gault, *J. Catalysis*, 1975, **37**, 311.

[22] C. Corolleur, S. Corolleur and F. G. Gault,, *J. Catalysis*, 1972, **24**, 385; C. Corolleur, D. Tomanova and F. G. Gault. *ibid.*, p. 401.

[23] J. E. Germain, *Catalytic Conversion of Hydrocarbons* (Academic Press, London, 1969); H. Matsumoto, Y. Saito, and Y. Yoneda, *J. Catalysis*, 1971, **22**, 182.

[24] H. J. Maat and L. Moscou, *Proc. 3rd Internat. Congr. Catalysis* (Amsterdam, 1964), Vol. 2, paper II, 5, p. 1277.

[25] G. A. Somorjai, R. W. Joyner, and B. Lang, *Proc. Roy. Soc.*, 1972, **A331**, 335; R. W. Joyner, B. Lang, and G. A. Somorjai, *J. Catalysis*, 1972, **27**, 405.

[26] A. A. Alchudzhyan and M. A. Indzhikyan, *Zhur. fiz. Khim.*, 1959, **33**, 983.

[27] A. Schneider and U. Esch, *Z. Elektrochem.*, 1944, **50**, 290.

DISCUSSION

R. L. Moss (*Warren Spring Laboratory, Stevenage*) said: I refer to published work on platinum–gold alloys quoted by Ponec and co-workers, and by O'Cinneide and Gault, and to unpublished work by the latter on 'highly-diluted' Pt–Au alloys, mentioned in the paper. At concentrations of platinum in gold of around 10–15%, isomerization is an important reaction and tracer studies support a cyclic rather than a bond-shift mechanism. It seems to be suggested now that 'diluted' platinum–gold alloys do not form any cyclic intermediate and that an alternative mechanism, metallo-carbonium ion or a bond-shift operates. Can you quantify these terms 'highly-diluted' and 'diluted'? At what dilution does isomerization not involve the cyclic mechanism?

H. C. de Jongste (*Univ. of Leiden*) replied: We have studied silica-supported alloy catalysts with about 16 wt% of the metallic component.[1,2] Alloys were prepared by hydrogen reduction in solution. The alloy selectivity depends on the composition; as an example we quote the following results: alloy containing 12·5% Pt (temperature, 643 K), isomerization 50%, cyclization 40%, hydrogenolysis (cracking) 10%; alloy containing 3·7% Pt (temperature, 622 K), isomerization 92%, cyclization ∼0, hydrogenolysis 8%. The latter alloy has been tested in Prof. Gault's laboratory and it has been found that isomerization was not occurring *via* the cyclic mechanism, in agreement with the observation that the production of cyclic molecules was also depressed by alloying.

Prof. Gault and his co-workers have studied alumina-supported catalysts. These were prepared by impregnation, and alloying was achieved by repeated and prolonged oxidation (*i.e.*, volatilizing of platinum) and a reduction procedure at high temperature. Alloys with 10–15% of platinum in gold revealed isomerization proceedings *via* a cyclic mechanism. Alloys with a lower platinum content were not studied.

So much for the observations. At the moment we can only speculate on the reasons for the different behaviour of alloys prepared on the two carriers and by different methods. We are inclined to believe that platinum forms larger clusters in gold in catalysts prepared on alumina by impregnation and sintering than in the catalysts prepared by hydrazine/carbon monoxide reduction.

F. G. Gault (*Univ. of Strasbourg*) said: My first comment concerns the mechanism of isomerization. The paper reports that, on highly diluted platinum–gold catalysts, the cyclic mechanism of isomerization does not take place. That is true for the alloys prepared by co-reduction of metal halides by hydrazine but not for the platinum–gold

[1] J. R. H. van Schaik, R. P. Dessing, and V. Ponec, *J. Catalysis*, 1975, **38**, 273.

[2] H. C. de Jongste, F. J. Kuijers, and V. Ponec, in *Preparation of Catalysts*, ed. B. Delmon, P.A. Jacobs, and G. Ponecelet (Elsevier, Amsterdam, 1976), p. 207.

catalysts prepared by co-impregnation, exactly like your platinum–copper catalyst. For these platinum–gold catalysts, on the contrary, the cyclic mechanism predominates although the particle size is large, and this we explain by the fact that the segregation of gold towards the faces and edges of the crystallites leaves the platinum atoms completely isolated. The reason why one platinum–gold catalyst behaved in an entirely different manner is not clear.

My second comment concerns the cyclic mechanism. We believe that the non-selective cyclic mechanism, which provides an equal chance of breaking any carbon–carbon bond in the cyclopentane intermediate, requires one single atom. Any change in position for the attachment of the intermediate to the metal may proceed by interconversion between σ-alkyl and π-olefinic species which are likely to involve the very same platinum atom, and proceed very rapidly at the reaction temperature. The reason for the increasing amounts of cyclic product when platinum is alloyed with copper could be due to a change in concentration of active adsorbed hydrogen. If one assumes that copper segregates more than gold to the surface, one would expect that the concentration of hydrogen atoms would be lower on platinum–copper than on platinum–gold alloys and, according to the literature, that the desorption of cyclic molecules would be favoured over the desorption of acyclic molecules.

H. C. de Jongste (*Univ. of Leiden*) replied: Our views on the problem mentioned in your first comment are contained in our reply to Dr. Moss.

Regarding your second comment, we do not believe that our results can be explained in terms of a concentration of hydrogen on platinum–copper alloys that is lower than that on the platinum–gold alloys, for the following reasons. First, copper is more active in chemisorption than gold (Pritchard *et al.* found hydrogen adsorption at high pressures on copper, but not on gold). Secondly, there is no indication of a substantial difference in Cu- and Au-segregation on the surface of these alloys. Thirdly, the relatively high activity of platinum–copper alloys in hydrogenolysis (cracking) suggests that the concentration of hydrogen should also be sufficient for isomerization (cracking consumes hydrogen, isomerization makes no nett use of hydrogen). Lastly, an increase in temperature must certainly lead to a decrease in the hydrogen surface coverage and that should lead to an increase of the dehydrocyclization selectivity. However, Figure 3b in the paper shows little variation of the dehydrocyclization with an increase in temperature and for some alloys there is even a decrease. These reasons convince us that the selectivity changes are not controlled solely by changes in the surface coverage of hydrogen.

V. Haensel (*Universal Oil Products, Des Plaines*) said: Since the product distributions and conversions in catalytic reforming depend a great deal on the experimental conditions (temperature, pressure, space velocity and hydrogen/hydrocarbon ratio), I suggest that the authors discuss the effect of these variables, particularly from the standpoint of a variation in space velocity. The results may throw some light on the reaction path. Also, since methylcyclopentane is one of the more interesting products, have the authors done any work on the conversion of this compound using the catalysts described in the paper?

H. C. de Jongste (*Univ. of Leiden*) replied: Because all our experiments were performed at low conversions in a differential reactor, the influence of the space velocity on the product distribution was limited. Nevertheless, we observed a marginal increase in selectivity for dehydrocyclization when the space velocity was increased. This indicates that some consecutive reactions were not excluded completely under our conditions and that, for example, gaseous MCP can be an intermediate in isomerization. Therefore, the existence of a cyclic adsorbed intermediate having an MCP-like structure is quite possible.

Also, Dorgelo performed some experiments in our laboratory with MCP as a feed and platinum–copper alloys as catalysts. While the earlier experiments showed that the dilute platinum–gold alloys do not catalyse ring-opening in MCP, platinum–copper alloys *are* active in ring-opening. The curve of activity for ring opening as a function of composition was similar in shape to that of the total activity in hexane reforming. This agrees with the conclusions of our paper. Highly diluted platinum–gold alloys are inactive in ring formation and ring-opening and do not isomerize *via* a cyclic intermediate either. In contrast, platinum–copper alloys are active in ring-opening and formation, and isomerization *via* a cyclic intermediate is in principle possible. Whether it actually occurs must be tested either by the methods used in Prof. Gault's laboratory, or by comparison of the isomerization activity of alloys towards pentane (which cannot isomerize *via* a cyclic intermediate) and hexane. We hope to do these experiments in the near future.

P. Ratnasamy (*Indian Inst. of Petroleum*) said: In a recent e.s.r. investigation of platinum–alumina and platinum–iridium–alumina catalysts we have obtained evidence for the presence of an unreduced species of platinum on the surface of the former. We have tentatively assigned the e.s.r. spectrum to $[PtCl_2]^-$. Its concentration is very low, corresponding to about 5–10% of the total platinum present, this amount being within the limits of accuracy of the methods used for determining metal dispersion values. It can, however, be sufficient to influence catalytic activity and selectivity. Can you comment on the role, if any, of such unreduced platinum species in the reforming of hydrocarbons?

H. C. Jongste (*Univ. of Leiden*) replied: My reply can only be speculative since no data are available. However, because no homogeneous catalysts are known for isomerization, dehydrocyclization, or hydrogenolysis, I would suggest that such species behave more or less as inactive surface-blocking additives. By analogy with other systems, I would expect hydrogenolysis to be suppressed most by such additives.

H. Charcosset (*CNRS, Villeurbanne*) said: Several participants have commented on the difficulty of measuring the composition of the first layer in alloy particles by Auger electron spectroscopy. This should be especially true in the case of supported catalysts having a low metal loading. Should it be possible to measure the concentration of exposed platinum atoms in supported platinum–gold and platinum–copper alloys by selective chemisorption of hydrogen on the platinum?

H. C. de Jongste (*Univ. of Leiden*) replied: From experience with nickel–copper alloys we have learned that the conclusion from chemisorption measurements (hydrogen chemisorption, not carbon monoxide adsorption!) was correct; platinum–gold and platinum–copper should not differ greatly. However, we must not forget that the final picture of the surface layer of nickel–copper alloys was obtained only when the results of several methods (work function measurements, chemisorption measurements, Auger spectra obtained in the two energy regions around 0·1 and 0·9 eV) were compared. Only then was the difference in behaviour of hydrogen and of carbon monoxide on chemisorption understood.

Z. Paál (*Inst. of Isotopes, Budapest*) said: I wish to comment on the reactions (3), (9), and (10) in your Figure 6. We have developed, with Prof. Tétényi, a dual site mechanism for catalytic C_5-cyclization and C_5-ring-opening. It involves the binding of the hydrocarbon in a half-hydrogenated state (through a tertiary carbon atom if such is present in the molecule) whereas an adjacent metal atom will act in ring opening and closure as illustrated in the Figure. Such a mechanism is in agreement with your suggestion about active ensembles, and may explain why C_5-ring-opening is preferred in β- and γ-positions with respect to the substituent, and why C_5-cyclization of methylpentanes is enhanced in comparison with that of n-hexane.

Two conditions should be fulfilled for this mechanism to operate. First, it has been found experimentally that sufficient hydrogen should be present, presumably to prevent

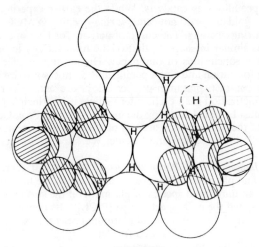

FIGURE

multiple hydrocarbon–metal interactions. Second, the diameter of metal atoms should be in a rather narrow range. Only platinum, palladium, iridium and rhodium were found to fulfil this condition. With decreasing atomic diameter cyclization and isomerization rates decrease. It should be stressed that geometric correspondence must be related to the periodicity of atoms and not to specific planes. The (111) plane is shown for convenience; one of the sites may also be at an edge or at a step.

In this context one of the effects of alloying may be to change the surface geometry. (The atomic diameter of gold is larger than that of platinum, whereas that of copper is smaller.) Another effect may be an influence on hydrogen adsorption. Both effects may alter the abundance of active ensembles that are able to behave as dual catalytic sites.

The Origin of Synergy between MoS_2 and Co_9S_8 in Hydrotreating Catalysts

P. Canesson, B. Delmon,* G. Delvaux, P. Grange, and J. M. Zabala

Groupe de physico-chimie minérale et de catalyse, Université Catholique de Louvain, Place Croix du Sud 1, B-1348 Louvain-la-Neuve, Belgium

ABSTRACT Catalytic activity measurements and physicochemical studies (X-rays, R.E.D., E.P.R., X.P.S., magnetic susceptibility) are reported on cobalt–molybdenum unsupported catalysts prepared in different ways (co-maceration or mechanical mixing of oxides or sulphides). Two composition ranges were investigated with particular attention: (i) to very low $Co/(Co + Mo)$ ratios, Co substituting for Mo in the MoS_2 lattice, where the catalytic activity is depressed; (ii) for concentrations where Co and Mo exhibit a synergistic effect on catalytic phenomena $[Co/(Co + Mo) \sim 0.2—0.6]$ no mixed phase, solid solution, or doped phase shows up. These results apparently support our model of synergy by contact or "junction effect".

Introduction

Three hypotheses[1-3] have been put forward to explain the action of the promoter (*i.e.* cobalt) on the assumed active phase or phases (molybdenum compounds) in cobalt–molybdenum catalysts, namely (i) the "monolayer model",[1,4,5] (ii) the "intercalation or pseudo-intercalation model",[2] and (iii) the "synergy by contact" or "junction effect" hypothesis.[3,6]

This paper summarizes results providing new arguments for a discussion of the mechanism of the synergy between cobalt and molybdenum. These results rest on catalytic activity measurements and physicochemical investigations on unsupported catalysts, prepared at low or moderately high temperatures, either by mechanically mixing oxides or sulphides or by thorough treatment of oxides in aqueous solutions of ammonium sulphide (co-maceration method). These results support and complement those obtained previously on crystallized samples prepared at 1,000°C.[7]

Experimental Details

1. Preparation of the Active Phases.

The co-maceration method has been described in detail in an earlier work.[7] We have prepared catalysts having an atomic ratio $Co/(Co + Mo)$ ranging from 0 to 1. The series of catalysts prepared by the co-maceration method will be called M_6 (macerated samples treated at 600°C).

The mechanical mixtures of oxides or sulphides are obtained by grinding in an agate mortar the required proportions of molybdenum and cobalt compounds. Series O_6 was prepared from cobalt and molybdenum oxides previously calcinated for 4 hours at 600°C. O_6S_4 and O_6S_8 are obtained from the O_6 by sulphiding the latter for 4 h in a mixture of $Ar–H_2S$ (25% vol. H_2S) respectively at 400°C and 800°C. S_4 and S_8 series were obtained by mixing mechanically molybdenum and cobalt sulphides which had been prepared by sulphiding the corresponding oxides under an $Ar–H_2S$ mixture at 400°C and 800°C respectively. Compositions in each series are expressed by the atomic ratio $r = Co/(Co + Mo)$.

2. Catalytic Activity

Catalytic activity measurements were made according to a previously described method.[7] The reactant was a mixture of cyclohexene (30% wt.) and cyclohexane

(70% wt.) containing 5,000 ppm of thiophen. For all catalysts the catalytic activity was measured at 305°C under a pressure of 30 bars [H_2 (gas, NTP)/hydrocarbon (liquid) ratio was 600].

3. *Physicochemical measurements*

Most of the physicochemical investigations were made on catalysts of the M_6 series.

(a) *X*-ray analysis.

Two types of *X*-ray analysis have been made, one is the classical powder *X*-ray diffraction and the other one is the radial distribution of electron density (R.E.D.).

Details on the R.E.D. method have been given previously.[8-10] The experimental *X*-ray intensities were recorded by measuring the time required to accumulate 10^5 counts at intervals of half a degree of 2θ.

(b) Magnetochemical measurements.

Two series of magnetic measurements were made, one at room temperature, the other at the temperature of liquid nitrogen. Magnetic susceptibility measurements were made using a Setaram MTB 10–8 balance according to Faraday's method.

(c) E.P.R. measurements.

They were made with a Bruker ER 400 A apparatus.

(d) X.P.S. measurements.

The measurements were made with a Vacuum Generators ESCA 2 system, working with an aluminium anode ($hv = 1486\cdot6$ eV). All the binding energies were determined using, as reference, the $4f_{7/2}$ signal ($E_B = 82\cdot8$ eV) of an evaporated gold film.

RESULTS

1. *Catalytic Activity*

Results obtained with the M_6 series are shown in Fig. 1. It can be seen that the addition of a small amount of cobalt brings about a drop in the thiophen hydrogenolysis activity. The hydrogenation activity is also decreased and the minimum for

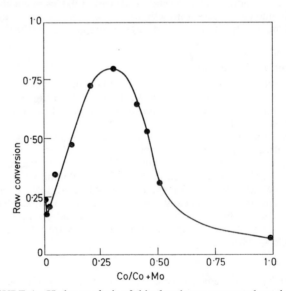

FIGURE 1   Hydrogenolysis of thiophen by co-macerated catalysts.

these two reactions occurs for a ratio r near 0·02. For higher r ratios, the curves corresponding to hydrogenolysis and hydrogenation exhibit a similar shape with a maximum near $r = 0·3$.

With mechanical mixtures of oxides (series O_6) it can be seen, in Fig. 2 (curve 1), that, in spite of the presumably bad contact between phases, there is some synergistic effect. The maximum activity is in the composition range around $r = 0·65$. If the catalyst was only a mixture of oxides, one would expect a linear decrease in activity from MoO_3 to Co_3O_4. Here also, the curves representing thiophen hydrogenolysis and cyclohexene hydrogenation have the same general shape.

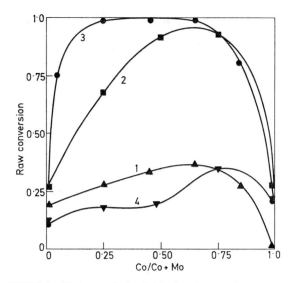

FIGURE 2   Hydrogenolysis of thiophen by mixed compounds:
1. mechanical mixture of oxides (O_6)
2. mechanical mixture of oxides sulphided at 400°C (O_6S_4)
3. mechanical mixture of oxides sulphided at 800°C (O_6S_8)
4. mechanical mixture of sulphides (S_8)

In the case of the O_6S_4 series, the results are plotted in Fig. 2 (curve 2). Here also, one observes an important synergistic effect, for a composition of the active phase near $r = 0·65$. The mixture of sulphides (S_4) only exhibit weak effects, as compared to the O_6S_4 preparation.

For O_6S_8, the maximum activity is observed for catalysts with $r = 0·45$ (Fig. 2, curve 3). This maximum is found at lower concentrations in cobalt as compared to the maximim observed with the O_6S_4 series. For S_8 series, (Fig. 2, curve 4), the effect is more prominent than with S_4 catalysts. This fact may be due to the higher preparation temperature of the sulphides and a more intimate contact between the two phases.

2. *Physicochemical studies.*

(a) *X-ray analysis.*
All the catalysts have been analysed by *X*-ray diffraction after the catalytic test.

In the case of M_6 catalysts, the only phases which can be detected are hexagonal MoS_2 and Co_9S_8. Nevertheless, in the range of low cobalt concentration ($r \leqslant 0·05$), it is possible to detect some reflexions which may be attributed to rhombohedral MoS_2. In all the samples of the O_6 series, the only molybdenum compound which gives a diffraction pattern is MoO_2. The cobalt compounds one can detect, are Co_9S_8 for compositions $r = 0.25, 0.45, 0.65, 0.85$ and metallic cobalt for catalysts with cobalt contents higher than $r = 0.65$. X-ray diagrams of the O_6S_4 series show that the phases are MoO_2, MoS_2, Co_9S_8, whereas MoS_2 and Co_9S_8 are the only forms detected in the O_6S_8 series.

The R.E.D. density has been determined on catalysts of the M_6 series. By substracting the theoretical contribution of MoS_2 and Co_9S_8 from the experimental curves, no new peak is detected, and there is no noticeable variation in the interatomic distances Mo–Mo, Mo–S, and S–S. Accordingly, it is probable that no mixed compounds exist in the concentration region where the synergistic effect is observed.

(b) E.P.R.

For catalysts of the M_6 series, two kinds of signals can be observed. The first kind can be attributed to Mo^{5+} ions, the other could be assigned to $(S-S)^{2-}$. For catalysts having a low concentration in cobalt ($r \leqslant 0.05$), one can observe a substantial signal due to polysulphide species and traces of Mo^{5+} ions. For the composition range $0.25 \leqslant r \leqslant 0.6$, the signal of Mo^{5+} rises to very high values and only traces of polysulphide ions can be observed.

(c) X.P.S. results.

In catalysts with a cobalt content corresponding to the composition exhibiting a maximum synergistic effect, the molybdenum is generally in the form of MoS_2, its 3 $d_{5/2}$ and 3 $d_{3/2}$ energy states being respectively equal to 227.6 eV and 230.8 eV ($\sigma = 0.087$). The binding energy of the 2 p level of sulphur equals 260.7 eV ($\sigma = 0.086$). This indicates that the oxidation state of sulphur is principally -2. However, one can identify, in certain cases, a second state of sulphur, situated 1.65 eV in the direction of higher binding energy. This state could be attributed to elemental sulphur in the 0 state. Cobalt exhibits a simple spectrum, only in cases corresponding to the composition near $r = 0.3$ and characterized by a mean binding energy of 777.9 eV ($\sigma = 0.13$) corresponding to the 2 $p_{3/2}$ level. In the case of other compositions, the 2 $p_{3/2}$ peak of cobalt broadens towards higher binding energies, without any simple deconvolution being possible.

For catalysts with a low cobalt content ($0 < r < 0.05$), a second molybdenum species can be detected. The signal can be attributed to Mo^{5+} ($+ 1.45$ eV). The sulphur peak is not changed. For compositions up to $r = 0.05$, no cobalt signal can be detected, even after an accumulation of 24 h.

(d) Magnetochemical measurements.

The magnetic moments are computed from magnetic susceptibility data.[11] This allows us to identify the nature of magnetic species. The magnetic moment, per mole of Co, of all samples is approximately constant and equal to 3 μ_B, except in the range corresponding to the anomalies already mentioned. In the low concentration and in the medium concentration ranges the values of the moment attain, respectively, 5.9 and 4.17 μ_B. At the temperature of liquid nitrogen, all the samples are ferromagnetic. Assuming that the values of the magnetization at 77 K are due exclusively to CoS_2, it is possible to calculate the concentration of this product in the catalysts. According to this hypothesis, the concentration of CoS_2 per gram of cobalt in the catalyst can be calculated.[11] The value of the magnetic moment (5.9 μ_B) implies the presence of a compound having 5 single electrons, since a compound of this type has a theoretical moment of 5.91 μ_B. Of all the known compounds of cobalt, only one has 5 single electrons namely Co^{4+}. From measurements at low temperature, one can conclude that only 29% of cobalt ions are in the CoS_2 form; consequently the remaining 71% of the cobalt is in other forms e.g. Co^{4+} ions, possibly substituted for Mo^{4+} in MoS_2.

Canesson, Delmon, Delvaux, Grange, Zabala

DISCUSSION

The low ($r \leqslant 0.05$) and medium ($0.25 \leqslant r \leqslant 0.75$) concentration range will be discussed separately.

1. Low Cobalt Concentration.

If the cobalt concentration on the surface is equal to the concentration in the bulk, and if one takes into account the size of the particles, one should expect a cobalt signal to be observed in X.P.S. No signal is actually detected for $r \leqslant 0.05$. This suggests that the concentration in the bulk is higher than that on the surface. This confirms our supposition, from magnetic measurements, that 71% of Co is incorporated inside the MoS_2 lattice. This incorporation could explain the substantial decrease of the c parameter of crystalline MoS_2 in the presence of small amounts of cobalt.[12] Co^{4+} ions have indeed a smaller ionic radius than Mo^{4+}. This hypothesis implies the formation of a MoS_2–CoS_2 solid solution. The solubility limits should be very narrow. On the other hand, the presence of segregated CoS_2 (29% of the cobalt) could possibly explain the formation of rhombohedral MoS_2; CoS_2 somehow inducing the formation of this phase.

We have no explanation for the existence of $(S–S)^-$ and Mo^{5+} ions, as detected by E.P.R., except, perhaps, that these ions could be stabilized by the special phases present in the low concentration range.

Thus we have strong arguments that a solid solution is present in the low concentration range. But cobalt, in these intercalated compounds, instead of acting as an activator[2] actually depresses the catalytic activity (Fig. 1).

2. Medium concentration range.

Catalytic activity measurements show that the maximum of the synergistic effect occurs at different concentrations, according to the catalyst series: $r = 3$ for M_6, 0.45 for O_6S_8, 0.60 for O_6S_4, 0.65 for O_6, and 0.75 for S_8. No solid solution is detected. These observations can easily be explained assuming that the synergistic effect arises because of the mere contact between distinct MoS_2 and Co_9S_8 particles or domains. When the catalysts are prepared by different methods, it can be expected that the sizes and relative sizes of the MoS_2 and Co_9S_8 domains, as well as the number and intimacy of the contacts between them are different. It follows that the maximum of the synergistic effect occurs for different concentrations in cobalt.

A modification of the monolayer model could perhaps explain our results, assuming that Co_9S_8 could play part of the role assigned to alumina in the original monolayer model.[1] In that case, however, it would be difficult to explain why the maximum of the synergistic effect occurs for similar r ratios in supported and unsupported catalysts. And it would be almost impossible to explain the noticeable (however relatively small) synergistic effect observed when the pure sulphides are simply mixed (S_8 series).

The simplest hypothesis, for explaining the synergistic effect is thus, at the moment, that the mere contact between the MoS_2 and Co_9S_8 phases (or domains) promotes the activity of, at least, one of these phases (contact synergy). We have still no explanation for this effect. One possible explanation could be some electronic transfer across the junction between the phases or domains.

Whichever is the exact origin of the synergy, the phenomenon involves, in addition to the change of catalytic activity, a change of other physicochemical properties of the solids (e.g. E.P.R., magnetic measurements).

ACKNOWLEDGEMENTS. Part of this work was supported by the "Centre d'Information du Cobalt". Professeur E. Derouane (Université Namur) and Dr. V. Perrichon (I.R.C.

931

Villeurbanne) are gratefully thanked, respectively, for the E. P. R. and magnetic measurements.

REFERENCES

[1] G. C. A. Schuit and B. C. Gates, *A.I. Ch. E. Journal*, 1973, **19**, 417.
[2] A. L. Farragher and P. Cossee, *Proc. 5th Internat. Congr. Catalysis*, Miami Beach, 1972, (North Holland, Amsterdam, 1973), p. 1301.
[3] P. Grange and B. Delmon, *J. Less Common Metals*, 1974, **36**, 353.
[4] J. M. J. G. Lipsch and G. C. A. Schuit, *J. Catalysis*, 1969, **15**, 163, 174, 179.
[5] V. H. J. De Beer, T. H. M. Van Sint Fiet, J. F. Engelen, A. C. Van Haandel, M. V. J. Wolfs, C. H. Amberg, and G. C. A. Schuit, *J. Catalysis*, 1972, **27**, 357.
[6] G. Hagenbach, Ph. Courty, and B. Delmon, *J. Catalysis*, 1973, **31**, 264.
[7] G. Hagenbach, Ph. Courty, and B. Delmon, *J. Catalysis*, 1971, **23**, 295.
[8] P. Ratnasamy and A. J. Léonard, *Catalysis, Rev.*, 1972, **6**, 293.
[9] P. Ratnasamy and A. J. Léonard, *J. Catalysis*, 1972, **26**, 353.
[10] P. Ratnasamy, L. Rodrique, and A. J. Léonard, *J. Phys. Chem.*, 1973, **77**, 2242.
[11] V. Perrichon, J. Vialle, P. Turlier, G. Delvaux, P. Grange, and B. Delmon, *Compt. rend.*, 1976, **282**, C85.
[12] G. Hagenbach, Ph. Courty, B. Delmon, *Compt. rend.*, 1971, **273**, C1489.

DISCUSSION

M. Lo Jacono (*Univ. of Rome*) and **D. C. Koningsberger** (*Tech. Univ. of Eindhoven*) said: It appears that all your results may be explained by an intercalation effect of cobalt. The maximum activity for sample O_6S_4 is reached at $Co/(Co + Mo) = 0.75$, while for sample O_6S_8 the maximum activity is obtained at a low ratio of about 0.25. Owing to the lower temperature of sulphidation of sample O_6S_4 more cobalt is needed for effective diffusion into the MoS_2 crystallites in order to obtain effective intercalation. The fact that the samples O_6S_4 and O_6S_8 show a larger activity than sample M_6 can be understood in terms of a more effective intercalation process during the reorganization of the structure going from the oxidic to the sulphidic state.

Recently, we have obtained e.p.r. evidence for the existence of Mo^{3+} ions in HDS catalysts. Moreover, preliminary attempts to relate the e.p.r. signal with catalytic activity seem to have been successful. The paramagnetism of these ions contributes to the magnetic susceptibility of your samples. Therefore, the valence state of cobalt (Co^{4+}), which you derived from your susceptibility measurements without taking into account the paramagnetism of these Mo^{3+} ions, cannot be correct. Moreover, it is difficult to accept that, under reducing conditions of treatment with hydrogen sulphide, Co^{2+} becomes Co^{4+}. Intercalation of one ion Co^{2+} produces two ions Mo^{3+}, giving five unpaired electrons (assuming a high-spin cobalt ion and a low-spin molybdenum ion). The g-value of the e.p.r. signal which you attributed to Mo^{5+} is $g = 2.000$. Your identification is based upon cited literature. The g-value of the Mo^{5+} ion cited in this and all other literature is always lower than $g = 1.96$. Therefore, also based upon our e.s.r. results, we do not agree with the proposed valence state of $5+$ for the molybdenum ion.

B. Delmon (*Catholic Univ. of Louvain*) replied: Our observation of Co^{4+} was made using fresh co-macerated catalysts and does not correspond to the steady-state (*i.e.* after the catalytic test). The attribution of the magnetic moment may be questioned for this reason or because of contamination by oxygen.

Secondly, it is true that, in an earlier publication, we had attributed a g-value of 2.000 to Mo^{5+}. In the present work, the generally accepted value of $g = 1.96$ has been adopted.

Concerning the explanation of the observed effects, either by intercalation or by contact synergy, it is clear that neither side has overwhelming arguments. But, if the intercalation model holds true, why would such a large amount of cobalt be necessary for the observation of synergy, and why should minute amounts, amply sufficient for surface contamination or intercalation, bring about only negligible changes in activity, or sometimes even a decrease in the activity per unit surface area (ref. 6, my paper)? Why, if the contact synergy hypothesis is unacceptable, do all systems near the synergic maximum exhibit two phases (MoS_2 and Co_9S_8) rather than a modified MoS_2 phase? Why, if it were wrong, would mechanical admixture of particles containing only cobalt with particles containing only molybdenum almost invariably exhibit synergy?

G. T. Pott (*Shell, Amsterdam*) said: I agree with Dr. Koningsberger. Since the intercalation of cobalt in your catalysts is an inefficient process, because of the way you prepared them, one could envisage that maximum activity occurs at high cobalt concentrations. Could the lowering in activity at very low cobalt concentrations be explained assuming a different morphology of molybdenum disulphide crystallites in the presence of such small amounts of cobalt?

B. Delmon (*Catholic Univ. of Louvain*) replied: In answer to your question I refer you to results obtained by Furimsky and Amberg.[1] These authors prepared unsupported catalysts by carefully impregnating cobalt on molybdenum disulphide. This method certainly maximizes superficial intercalation of cobalt. However, maximum catalytic activity occurs only for relatively high cobalt contents, *i.e.* Co/(Co + Mo) = 0·25. How could the MoS_2 lattice accommodate such a large amount of cobalt? Why would such a large amount of cobalt be necessary for mere contamination? On the other hand, the contact synergy hypothesis accounts easily for the result, because, in Furimsky and Amberg's system, synergy would require an amount of the Co_9S_8 phase comparable to the amount required in other unsupported systems.

The lowering of activity at very low cobalt concentration affects *activity per unit surface area* (ref. 6, my paper). The total exposed area is therefore not at stake. In principle, differences in activity might certainly be related to differences in morphology [*e.g.* ratio of (001) to (100) or (010) planes]. However, scanning electron microscopy and BET measurements do not indicate noticeable anomalies.

J. P. Franck (*Inst. Franç. du Petrole, Rueil Malmaison*) said: From the results presented in Figure 2 (particularly curves 2 and 3) it is difficult to distinguish the optimum cobalt to molybdenum ratio because of the high conversion levels obtained, but it seems to approach the optimum ratio obtained for co-macerated catalysts (Figure 1). Could we not assume that this parallelism is due to species migrating during high temperature sulphidation ($\sim$1073 K) and sublimation of molybdenum compounds?

B. Delmon (*Catholic Univ. of Louvain*) replied: We agree that it is difficult to recognize the optimum cobalt to molybdenum ratio in the curves you mention. Our remark concerning the variability of the optimum refers to what can be concluded from all published results.

Migration of species during preparation cannot be ruled out. Sublimation is less likely because, in our experiments, temperature has been progressively raised between 673 and 1073 K while the sample was in the reducing-sulphiding atmosphere, in order to reduce molybdenum trioxide. Nevertheless, transport phenomena involving other species cannot be excluded and we have no decisive argument against contamination.

B. C. Gates (*Univ. of Delaware*) said: Catalysts such as you describe are used commercially with metal-containing petroleum feedstocks, and they accumulate massive amounts of deposited sulphide of vanadium and nickel; yet they retain a remarkably large fraction of their initial hydrodesulphurization activity. One might expect that the deposits would alter and block the region of contact between the Co_9S_8 and MoS_2

---

[1] E. Furimsky and C. H. Amberg, *Canad. J. Chem.*, 1975, **53**, 2542.

phases. Can you reconcile your explanation of 'contact' synergy with the occurrence of this deposition? Do you agree with the speculation that accessible surface catalytic sites might be continuously formed during the ageing (deposition) processes?

B. Delmon (*Catholic Univ. of Louvain*) replied: We agree with your suggestion that the deposition of nickel might continuously add new contact zones, which could form new sites by a synergic effect. This would support the contact synergy hypothesis. We have no opinion concerning vanadium.

E. J. Newson (*Haldor Topsøe, Lyngby*) said: Referring to curves 2 and 3 of your Figure 2, it does not necessarily follow that 'the maximum in the synergic effect occurs for different concentrations in cobalt'. The catalysts involved were sulphided for 4 h at two different temperatures, 673 K and 1073 K. If we now look at Figure 1 of the next paper by Hargreaves and Ross we see that sulphidation took about 20 h at 573 K to complete. My question is: could the maxima of curves 2 and 3 in your Figure 2 be simply the result of incomplete sulphidation and not the effect of the atomic ratio $Co/(Co + Mo)$? Do you have any other data on the kinetics of sulphidation?

G. C. Stevens (*BP, Sunbury-on-Thames*) said: We have found, using ESCA, a correlation between catalytic activity and the intensities of $Mo(3d)$ signals attributable to Mo^{4+}-S and oxide in CoMo-alumina. Reduced oxide and sulphide states of molybdenum are both important. Your Figure 2 shows that the sulphided oxides (curve 2) are far more active than the mixture of sulphides (curve 4). As I understand it, the oxides were sulphided at 673 K, whereas the sulphide mixture was heated to 578 K during the test. Could you comment on the possible involvement of reduced molybdenum oxides as well as molybdenum disulphide in the active catalyst. Do you think the difference in contact of Co_9S_8 and MoS_2 caused by treating at 673 K instead of 573 K is sufficient to account for the very large activity increase?

B. Delmon (*Catholic Univ. of Louvain*) in reply to the two previous speakers, said: The statement that Dr. Newson quotes is supported by the results we present in Figures 1 and 2 by and all other results published by other authors and ourselves. It is clear that the position of the maximum depends on the method of preparation.

Concerning the question raised by Dr. Newson and Dr. Stevens we may remark that our activity measurements are made after the steady-state has been attained. For the very early stages of sulphidation, the position of the maximum depends on the period of time over which a mixture of oxides has been contacted with the reacting feed under catalytic conditions.[2] But a maximum is always observed. As synergy is observed for fully sulphided samples, there is no doubt that there is an effect of the $Co/(Co + Mo)$ ratio. We believe that the nature and extent of contact between phases determines the position of the maximum. There is probably little effect of cobalt on the reduction-sulphidation of molybdenum trioxide[3,4] so it is extremely unlikely that synergy could be attributed to modifications during the genesis of the catalytic system.

G. C. Stevens (*BP, Sunbury-on-Thames*) said: In our investigations of MoS_3 we have found that $S(2p)$ gave two sets of peaks in the XPS, one assignable to S^{2-} and the other 1·2 eV higher. Furthermore, e.p.r. spectra of molybdenum trisulphide formed by decomposition of ammonium thiomolybdate gave this signal, one attributable to Mo^{5+} at $g = 1·93$ and a complex sulphur signal. Since these spectra are similar to your own, do you think that molybdenum trisulphide could have been formed in your sulphided molybdenum catalysts?

B. Delmon (*Catholic Univ. of Louvain*) replied: We do not mention MoS_3 among the species detected by XPS or e.p.r. MoS_3 is reported to be formed only in the presence

[2] J. M. Zabala, M. Mainil, P. Grange, B. Delmon, *Compt. rend.*, 1975, **280**, *C*, 80.
[3] J. M. Zabala, P. Grange, B. Delmon, *Compt. rend.* 1974, **279**, *C*, 725.
[4] J. M. Zabala, P. Grange, B. Delmon, *IV Iberoamericano Simposio de Catalisis*, Mexico, 1974.

of excess sulphur in non-reducing atmosphere. It is certainly not stable under the conditions of preparation of the catalyst samples, and of the catalytic test.

M. Schiavello (*Univ. of Rome*) said: It is difficult to accept that the oxidation state of cobalt is 4+ in a reducing atmosphere such as that used during the reaction. To accept this, it would be necessary to carry out a crystal field analysis to show whether this cobalt species is in a weak or in a high field, and then to calculate its theoretical magnetic moment. The following process provides an alternative explanation. $Co^{2+} + S^{2-} + (2Mo^{4+})/MoS_2 \rightarrow \dot{S} + (2Mo^{3+} - Co^{2+})/MoS_2$. In this way the presence of two unpaired electrons on two Mo^{3+} species, together with the three unpaired electrons of Co^{2+} interpret the high μ-value obtained. Moreover, by this scheme, the presence of sulphur is also justified.

My second remark concerns the determination of catalytic activity which appears to have been carried out at one temperature. There are many cases in the literature where activity patterns or comparisons of activity have been made at a single temperature. Often conclusions reached for one temperature are found to be invalid when the measurements were extended to a range of temperature.

B. Delmon (*Catholic Univ. of Louvain*) replied: My answer to Lo Jacono and Koningsberger is relevant to your first point. Thank you for suggesting an alternative process, which will be of much benefit to us in our current magneto-chemical investigations.

Concerning your second remark, we agree that the activity pattern changes with temperature. The position of the synergetic maximum may change, but there is ample evidence that it always exists.

P. Mars (*Twente Univ. of Technology, Enschede*) (communicated): In our laboratory Prasad has made similar observations on the synergic effect of cobalt on molybdenum catalysts. He has used mixtures of molybdenum oxide on alumina and cobalt oxide on alumina. After sulphidation and reduction at 783 K he observed a synergic effect which depended on the degree of dispersion. The effect was the largest (50%) for the finest dispersion (4 μm).

The increase in catalytic activity caused by the presence of cobalt is generally different for the various reactions (hydrogenation, isomerization, hydrodesulphurization). However, when the various reactions are carried out in the same gas stream the relative increase in activity is practically the same, as observed by Ahuja *et al.*,[5] Voorhoeve *et al.*,[6] and Hagenbach *et al.*[7] We have observed the same phenomenon in the hydrogenolysis of pyridine on oxidic CoMo–Al$_2$O$_3$.[8] Hydrogenation of pyridine to piperidine, cracking of carbon–nitrogen bonds, and the transalkylation reactions of amines are all equally promoted. My questions are: do you still agree with these observations of 'non-selective promotion' when the reactions are carried out in the same gas stream and do you have any explanation for this fact? Could the phenomenon be related to coke formation on the catalyst?

B. Delmon (*Catholic Univ. of Louvain*) replied: Dr. Gajardo, in our group, has also observed that mixtures of $MoO_3–Al_2O_3$ and $CoO_x–Al_2O_3$ exhibit synergy. Your observation that the effect depends on the degree of dispersion might support the contact synergy hypothesis.

We believe that, in principle, the variations of the various reactions with cobalt content should be different, even if the various reactions occur simultaneously. In our experiments, we quite often observe conspicuously different trends for the various

[5] S. P. Ahuja, M. O. Derrien, and J. F. le Page, *Ind. and Eng. Chem.* (*Product Res. and Development*), 1972, **9**, 272 (1972).

[6] R. J. Voorhoeve and J. C. M. Stuiver, *J. Catalysis*, 1971, **32**, 228.

[7] G. Hagenbach, Ph. Courty, and B. Delmon, *J. Catalysis*, 1971, **23**, 295.

[8] J. Sonnemans and P. Mars. *J. Catalysis*, 1973, **31**, 209.

categories of reactions. Non-selective promotion could result from diffusion limitations or, as you suggest, from some fouling of the catalysts. The possibility of controlling differently the various categories of reactions (by dispersion, activation temperature, support) opens promising avenues for the development of catalysts with better selectivities.

H. Heinemann (*Mobil, Princeton*) (communicated): At what pressure was sulphidation of your catalysts carried out? Can the oxidation states observed be extrapolated to what would prevail in hydrodesulphurization at 400–2000 p.s.i.g. in the presence of hydrogen?

B. Delmon (*Catholic Univ. of Louvain*) replied: Sulphidation was carried out at atmospheric pressure. If some non-stoicheiometry occurs in the sulphides, it will certainly depend on the sulphiding conditions. Your remark implies also that the non-stoicheiometric steady-state catalysts might differ depending on the processes used. We would agree. Our measurements of catalytic activity were made after the steady-state was attained. So we believe that the stoicheiometry should correspond to the catalytic conditions (*i.e.* 30 bars) rather than to the preparation conditions.

An Investigation of the Mechanism of the Hydrodesulphurisation of Thiophen over Sulphided Co–Mo–Al$_2$O$_3$ Catalysts. Part I. The Activation of the Catalysts and the Effect of Co–Mo Ratio on the Activity and Selectivity

Anthony E. Hargreaves and Julian R. H. Ross*

School of Chemistry, University of Bradford, Bradford, BD7 1DP

ABSTRACT   The activation procedure for a Co–Mo–Al$_2$O$_3$ catalyst has been examined and a comparison of the activities and selectivities of a series of related activated catalysts has been made. It has been shown that the catalytically active material is a partially reduced sulphide produced by the interaction of the oxide material with hydrogen sulphide, followed by exposure to hydrogen. Both the hydrodesulphurisation of thiophen and the hydrogenation of but-1-ene are promoted by the inclusion in the catalyst of cobalt, but the latter is affected to a lesser extent. From this, it is implied that two types of active site are present in the catalyst surface, one for C–S bond cleavage, the other for hydrogenation.

INTRODUCTION

The most commonly encountered catalyst for the hydrodesulphurisation (hds) reaction may be represented, in its oxide form, as CoO–MoO$_3$–Al$_2$O$_3$; evidence has been presented for the presence of compounds such as CoAl$_2$O$_4$ and CoMoO$_4$, as well as for solid solutions of MoO$_3$ in Al$_2$O$_3$,[1–5] the degree of interaction depending on the method of preparation. Various methods of catalyst activation have been reported including treatment in hydrogen[6] and in H$_2$/H$_2$S mixtures.[7,8] There have been a number of investigations of the hds of thiophen over such catalysts. It is generally accepted that two types of activity reside in the catalysts, one for hydrogenation–dehydrogenation, the other for carbon–sulphur bond cleavage, the latter occurring on Co promoted Mo sites. Amberg and co-workers favour a mechanism for the reaction involving C–S bond cleavage to give adsorbed butadiene species, followed by hydrogenation to give butenes and, subsequently, butane;[9–11] Kieran and Kemball (using unsupported MoS$_2$) claimed[12] that ring hydrogenation occurred prior to splitting out H$_2$S (dehydrosulphurisation, dhs).

The work discussed above was generally based on either single, or a limited number of unrelated ill-defined, catalyst samples. This paper presents results for a series of related catalysts; catalyst activation and the effect of Co–Mo ratio on activity and selectivity are described. Results for the effect of Co:Mo ratio on the reactions of possible intermediates will be described elsewhere.

EXPERIMENTAL

Catalyst preparation. Table 1 shows the compositions of the catalysts used in this work. They were prepared by sequential impregnation of freshly precipitated pseudo-boehmite [Al(OH)$_3$] of industrial origin using cobalt formate and ammoniacal MoO$_3$ solutions. After filtration and drying at 383 K for 48 h, the material was ground and sieved and particles in the size range 250—350 μm were retained and calcined in air at 823 K for 17 h when required for use. The analyses given in the Table were carried out

Table 1

Details of the composition of the catalysts used in this work; their activities for thiophen hds and for butane formation during hds, as well as for but-1-ene hydrogenation over selected catalysts, are also given.

Catalyst Number	% CoO (by weight)	% MoO_3 (by weight)	Co–Mo Atomic ratio	Initial rate of thiophen hds /10^9 mol g^{-1}s^{-1}	Initial rate of butane formation during hds /10^9 mol g^{-1}s^{-1}	Initial rate of but-1-ene hydrogenation /10^9 mol g^{-1}s^{-1}
1·1	0·0	11·5	0·0	9·26	0·18	2·66
3·1	1·58	10·4	0·29	21·9	0·29	—
3·2	2·93	11·6	0·48	37·7	0·46	5·80
3·3	3·56	10·9	0·62	54·8	0·47	—
3·4	4·10	10·7	0·73	65·7	0·49	—
3·5	4·93	11·6	0·81	93·0	0·81	7·60
4·1	3·74	9·1	0·78	102	0·56	—
4·2	3·74	10·8	0·67	79·2	0·57	—
4·3	3·76	11·9	0·60	69·7	0·44	—
4·4	3·78	12·7	0·57	63·2	0·69	—
4·5	3·57	13·1	0·52	51·3	0·50	—

using normal wet analytical methods and the results are quoted with respect to the freshly calcined material. The areas of the samples were all within the range 270—330 m^2g^{-1}, the pore volumes from 0·47—0·55 cm^3g^{-1}, and the average pore diameters from 6·1—7·0 nm; all these were determined using N_2 physisorption techniques.

Reaction system. The high-vacuum system used (reaction volume = 300 cm³) included a piston-type glass pump for recirculation and a facility for sampling for g.l.c. analysis; the column used comprised 13 % 2-methoxyethyl adipate plus 6 % di-2-ethyl-hexyl sebacate on Chromosorb P. The catalyst sample was supported on a sintered glass disc contained in a glass reaction vessel whose temperature could be controlled to ± 1 K.

Procedure. Samples (0·1 g) of each catalyst were sulphided before use by exposure to H_2S at pressures from 13·3 to 39·9 kNm^{-2}, at temperatures from 473—673 K, and for times ranging from 1—65 h. The activities of catalysts sulphided under various combinations of conditions were measured for the hds of thiophen using the following standard conditions:

$$P_{C_4H_4S} = 420 \text{ N m}^{-2}; P_{H_2} = 2950 \text{ N m}^{-2}; T = 521 \text{ K or } 545 \text{ K}$$

For comparison of the activities and selectivities of catalysts of various compositions, sulphiding was carried out at 573 K using a pressure of H_2S of 13·3 kN m^{-2} for 17 h, and the experiments were carried out using the same partial pressures as above with the catalyst temperature at 521 K; in experiments where the rates of hydrogenation of but-1-ene were measured, its partial pressure was 177 N m^{-2}. Activities were measured as the initial rate of appearance of product molecules (but-1-ene, *cis-* and *trans-* but-2-ene, butane) from the tangent to a plot of the total pressure of C_4 products against time at $t = 0$; the initial rate of production of butane alone was also measured.

Materials. The thiophen used was Aldrich Gold Label (99%+) and was further purified by freezing–pumping cycles in the gas-handling line; the hydrogen was supplied by the British Oxygen Co. (Grade X, 99·9%+) and the H_2S and but-1-ene were supplied by Matheson Inc. (C.P. Grade, 99%+).

RESULTS

Catalyst activation. After outgassing overnight at 573 K, a freshly calcined (oxidic) sample of catalyst 3·5 had negligible activity for the hds of thiophen at temperatures up to 573 K. The same negative result was obtained for a sample which had been reduced in hydrogen at a pressure of 13·3 k Nm^{-2} at 573 K for a period of 17 h followed by outgassing for 2 h at the same temperature; on the other hand, this catalyst had a high activity for the hydrogenation of but-1-ene to butane, giving a constant initial rate of 400 × 10^{-9} mol g^{-1}s^{-1} at a temperature of 521 K (compare the Table and below).

When a freshly calcined sample (3·5) was exposed to an excess of H_2S (with respect to the Co–Mo content of the catalyst) at temperatures between 473 K and 673 K, it was found to have developed an activity for the hds of the thiophen.

Initially, but-1-ene was formed in quantities above the equilibrium proportion with respect to *cis-* and *trans-*but-2-ene, but after a short period, the butenes were at equilibrium; the butenes were then slowly hydrogenated to butane; small traces (~1%) of butadiene were also observed among the products. The level of activity for hds attained seemed to be relatively independent of the pressure of H_2S but depended markedly on the time and temperature of sulphidation; Figure 1 shows the optimum catalyst activity developed after exposure to several doses of the thiophen + hydrogen reaction mixture (see below) plotted as a function of temperature of sulphidation (time = 17 h in each case) and of time of sulphidation (temperature = 573 K in each case). It will be noted that the activity reaches a maximum after ~17 h sulphidation at 573 K. The standard sulphidation conditions of 17 h exposure to H_2S (13·3 k N m^{-2}) at a temperature of 573 K were chosen for subsequent work mainly as a matter of convenience, the latter being slightly above the chosen hds reaction temperature. Under these conditions, the optimum activity for catalyst 3·5 was 93×10^{-9} mol $g^{-1}s^{-1}$; it is interesting to note that when the reduced sample discussed above was exposed to H_2S under the same conditions, it gradually developed maximum hds activity of 139×10^{-9} mol $g^{-1}s^{-1}$, which is slightly higher than, but comparable with, that of the freshly sulphided oxide.

It was found that the hds activity of the freshly sulphided oxide increased markedly with exposure to the thiophen–hydrogen reaction mixture. Figure 2 shows the initial

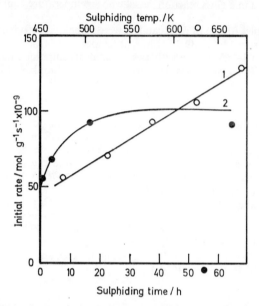

FIGURE 1 The activity of catalyst 3·5 for the hds of thiophen at 521 K as a function of temperature (curve 1, time = 17 h) and time of sulphiding in H_2S (curve 2, temperature = 573 K); see text for hds reaction conditions.

rates of reaction for a series of short experiments carried out over sample 3·5, each allowed to proceed for 15 min (~15% conversion). It can be seen that the activity rises to a maximum after three or four such experiments and then drops off slowly reaching a steady state activity after about six experiments as long as standard reactions are carried out; however, dose 11 of Figure 2 was allowed to stay in contact with the catalyst for 100 min, when the activity during subsequent standard experiments was found to rise above that of the previous maximum. Results similar to those for doses 1—5 in Figure 2 were found for all the catalysts investigated and also for the sample of catalyst 3·5 which was sulphided after reduction; the results shown in Figure 1 and for the comparison of the various catalysts all refer to activities at the maximum. The reason for the attainment of the maximum was further investigated: the maximum activity could be achieved by exposure of the sulphided catalyst to one dose of reaction mixture for 60 min, but no change in activity resulted on evacuation of the reaction system or an exposure to helium, showing that the activation was due either to the hydrogen or to the thiophen; exposure of the freshly sulphided catalyst to hydrogen either for four 15 min periods or for one period of 60 min resulted in an identical activity to that attained in thiophen–hydrogen mixtures, indicating that the activation was due to hydrogen. The results of Figure 2 show the typical reproducibility of results reported in this paper; the points lie within about ±6% of the smooth curve drawn through them. A number of repeat experiments using fresh samples of catalyst showed that the activity at the maximum was reproducible to ±10% for any one catalyst; this is negligible compared with the changes in activity with Co–Mo ratio described below.

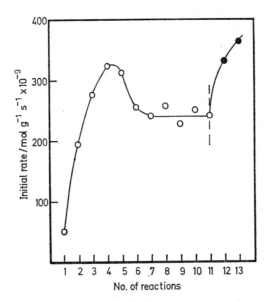

FIGURE 2 The hds activity at 545 K for successive reactions over a freshly sulphided sample of catalyst 3·5; the experimental conditions are given in the text, and all reactions were allowed to proceed for 15 min, apart from reaction 11 which was of 100 min duration.

Catalytic activity as a function of Co and Mo content. The maximum specific activity for the hds of thiophen (computed from the rate of production of butenes and butane) is shown in the fifth column of the Table for each catalyst of the series; the activity increases for the series 3 catalysts as the Co content is increased and decreases for the series 4 catalysts with increasing Mo content. The specific activities are plotted as a function of Co–Mo ratio in Figure 3. It can be seen quite clearly from this Figure that the optimum Co–Mo ratio occurs above 0·8 for the particular method of catalyst preparation used in this work; a range of activities of more than ten is found for the hds of thiophen, and clearly this cannot be accounted for solely by an increase due to the presence of additional cobalt sites but must be due to promotion of the molybdenum sites by cobalt.

Selectivity of the catalyst as a function of Co and Mo content. The initial rate of production of butane for each of the catalysts is shown in the penultimate column of the Table. It can be seen that, whereas the rate of hds varies by a factor of more than ten, the rate of butene hydrogenation only varies by a factor of about four, *i.e.* the selectivity for butane formation decreases markedly (see Figure 3). As the results were obtained by measuring the partial pressures of the reaction products at low conversions and as the butenes are the predominant products, the rates of butane formation can be considered to correspond to equivalent reaction conditions in each case. Several experiments were carried out in which but-1-ene was added to the reaction; rapid isomerisation of but-1-ene occurred and the rates of production of butane showed the same range of activities as for the production of butane in the experiments described above, but the results were less quantitative and are not given in detail. The final column of the Table gives, for catalysts 1·1, 3·2, and 3·5, the rates of hydrogenation of but-1-ene in the absence of thiophen over the activated materials; the rates vary by a factor of three, which is close to the variation in the rates of production of butane during the hds reaction. In these

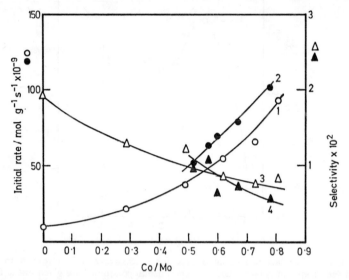

FIGURE 3   The hds activity (curves 1 and 2) and selectivity for butane formation curves 3 and 4) as a function of Co–Mo atomic ratio; curves 1 and 3 refer to the series of catalysts with constant Mo content, and curves 2 and 4 to that with constant Co content.

experiments, isomerisation of but-1-ene to *cis*- and *trans*-but-2-ene occurred rapidly. The reaction appeared to be acceleratory; when a series of experiments was carried out over an activated catalyst, the activity continued to increase rather than dropping to a steady value. It is worth noting the fact that the activities of the activated (sulphided) catalysts for butene hydrogenation are considerably lower than that quoted above for a prereduced catalyst.

DISCUSSION

In the present work, the catalysts were prepared by impregnation of Al(OH)$_3$ by Co and Mo salts, followed by calcination at 823 K. It is probable that some sort of interaction occurs between the adsorbed Co and Mo species and the Al, possibly *via* oxygen bridges, prior to calcination, but interaction with the formation of compounds such as CoAl$_2$O$_4$ is even more likely during calcination at 823 K.[2-4] Catalysts of this formulation are more frequently prepared by impregnation of a precalcined alumina, and it is quite probable that less interaction between the components occurs in such cases. We believe that this accounts for the fact that the optimum catalytic activity in this work occurs at a Co–Mo ratio of > 0·8, whereas other workers[4] have reported optima at ratios as low as ~ 0.2; higher Co–Mo ratios will be required when Co is incorporated to any great extent in the alumina. The freshly calcined material probably therefore consists of a monolayer of Co and Mo species, roughly corresponding to CoO and MoO_3 supported on an alumina base which may include some CoO and/or MoO_3; the surface oxide species may be partially hydroxylated. When this is exposed to H_2S, displacement of oxygen and hydroxyl groups by sulphur will occur, and it has been shown that this reaction is kinetically controlled,[13] which is consistent with the dependence of the activity for thiophen hds (see Figure 1) as a function of time and temperature of sulphiding. If the oxidic material is prereduced, the sulphidation process should occur more readily as it is not now controlled by displacement of oxygen, and this results in a higher hds activity for similar sulphidation conditions.

Under the sulphiding conditions used here, the stable sulphides of Co and Mo are respectively Co_9S_8 (or some higher sulphide) and MoS_3,[14] and the surface phases are therefore expected to be related to these compositions, with the limitation, due to kinetic factors, that incomplete sulphidation may have occurred. The former is stable with respect to Co metal at the reaction temperatures of 521 and 545 K when the ratio P_{H_2S}/P_{H_2} is greater than 10^{-8} whereas the latter requires a ratio of 10^4 to prevent reduction to MoS_2; when hydrogen is introduced after the sulphidation treatment, either in the presence or absence of thiophen, the Co_9S_8 will therefore remain stable as long as P_{H_2S}/P_{H_2} remains greater than 10^{-7}. If the activity of surface phases related to MoS_2 are greater than those related to MoS_3, this explanation accounts for the increase in activity on exposure of the sulphided catalyst to hydrogen shown in Figure 2, and for the similar results for the prereduced catalyst.

The second rise in activity shown in Figure 2 must be explained in a slightly different way. The catalyst was sulphided at 573 K, which did not give the maximum degree of sulphiding (see Figure 1). Hence, further sulphiding is possible, especially when hydrogen is present, and this occurs when a reaction mixture is allowed to remain in contact with the catalyst for longer periods with the production of higher partial pressures of H_2S; a higher activity is subsequently found which is again achieved gradually, probably due to a gradual re-reduction of the surface species.

These conclusions are consistent with the catalyst pretreatment procedures outlined in the literature,[3,9–11,14,15] with the notable exception of the work of Lipsch and Schuit,[6] obtained with a pulse microcatalytic reactor using hydrogen as carrier gas. It is probable that they were in fact working with a predominantly reduced catalyst and that it is maintained in the reduced state by the flow of hydrogen. In more recent work, Schuit and co-workers[7,8] have sulphided their catalysts with H_2S–H_2 mixtures and have obtained data for the sulphur content at various stages of the pretreatment of the catalyst which are consistent with the work reported here.

The activity and selectivity of the reaction varies considerably with the Co–Mo ratio; the catalysts are more active and selective for butene formation at high Co contents. Clearly two types of activity reside in the catalyst, one for hds and the other for hydrogenation. As the former depends much more on the promotion by Co than the latter, we conclude that whereas the hds reaction occurs on a site which involves a complex interaction of Co and Mo sulphides, the hydrogenation reaction occurs predominantly on unpromoted sites; the second type of site is unlikely to be oxidic in form (*e.g.*, $CoAl_2O_4$) as the selectivity of the reaction remains the same during the increase in hds activity as a function of the degree of sulphiding shown in Figure 1. Additional results, which will be discussed elsewhere, for the reactivities of other compounds such as tetrahydrothiophen, over the same series of catalysts, and the relative rates of these reactions as a function of Co–Mo ratio encourage us to believe that one of these sites gives C–S bond cleavage and the other gives hydrogenation–dehydrogenation. Further, the reaction sequence probably involves hydrogenation of the adsorbed thiophen molecule on one site, followed by the transfer of the saturated species to the other site where it undergoes dehydrosulphurisation (dhs).

This sequence is similar to that suggested by Kieran and Kemball,[12] and also resembles to that given by Griffiths *et al.*, [18] but it is considered by Amberg[11] to be of secondary importance.

ACKNOWLEDGEMENTS   We thank Laporte Industries Ltd. for a studentship for A.E.H., and for providing facilities for catalyst preparation and analysis.

REFERENCES

[1] J. H. Ashley and P. C. H. Mitchell, *J. Chem. Soc.(A)*, 1968, 2821.
[2] J. M. J. G. Lipsch and G. C. A. Schuit, *J. Catalysis*, 1969, **15**, 174.
[3] P. C. H. Mitchell, *The Chemistry of Some Hydrodesulphurisation Catalysts Containing Molybdenum* (Climax Molybdenum Co., London, 1967).
[4] J. T. Richardson, *Ind. Eng. Chem. (Fundamentals)*, 1964, **3**, 154.
[5] P. L. Villa, F. Trifiro, and I. Pasquon, *Reaction Kinetics and Catalysis Letters*, 1974, **1**, 341.
[6] J. M. J. G. Lipsch and G. C. A. Schuit, *J. Catalysis*, 1969, **15**, 179.
[7] M. Lo Jacono, J. L. Verbeek, and G. C. A. Schuit, *J. Catalysis*, 1973, **29**, 463.
[8] V. H. J. de Beer, *Some Structural Aspects of the CoO–MoO₃ γ-Al₂O₃ Hydrodesulphurisation Catalyst* (Thesis, Technical University of Eindhoven, 1975).
[9] P. Dekisan and C. H. Amberg, *Canad. J. Chem.*, 1963, **41**, 1966.
[10] S. Kolboe and C. H. Amberg, *Canad. J. Chem.*, 1966, **44**, 2623.
[11] P. J. Owens and C. H. Amberg, *Adv. Chem. Ser. (Amer. Chem. Soc.)*, 1961, **33**, 182.
[12] P. Kieran and C. Kemball, *J. Catalysis*, 1965, **4**, 394.
[13] J. R. H. Ross and D. Snaith, unpublished results.
[14] J. B. McKinley, *Catalysis* (ed. P. H. Emmett, Reinhold, New York, 1957) Vol. 5, p. 405.
[15] See review by S. C. Schuman and H. Shalit, *Catalysis Rev.*, 1970, **4**, 245.
[16] G. H. Hagenbach, Ph. Courty, and B. Delmon, *J. Catalysis*, 1973, **31**, 264.
[17] A. E. Hargreaves and J. R. H. Ross, unpublished results.
[18] R. H. Griffiths, J. D. F. Marsh and W. B. S. Newling, *Proc. Roy. Soc.*, 1949, **A197**, 194.

DISCUSSION

P. C. H. Mitchell (*Univ. of Reading*) said: I wish to make two comments concerning hydrodesulphurization catalysts.

(a) The catalysts should be regarded as hydrogenation catalysts (*i.e.*, cobalt-promoted molybdenum disulphide). You report that the initial rate of butene hydrogenation is much less (1–10%) than that of hydrodesulphurization. Presumably hydrocarbon hydrogenation is rate-limiting under steady-state conditions. Also, Parsons and Ternan (this Congress) find that the activity patterns for promoted hydrodesulphurization catalysts are similar to those for hydrogenation catalysts. Previously we suggested that one function of cobalt was to promote hydrogenation.[1]

(b) With regard to activity-structure correlations, I wish to make the obvious but neglected point that correlations should be sought with the metal species at the catalyst surface. Whereas all the molybdenum is at the surface, only part of the cobalt is at the surface and this is not a constant fraction of the total cobalt.[2] We have determined the surface concentration of cobalt by following to equilibrium the desorption into water of Co^{2+} ions from the catalyst surface. We then find that the activities for butene

[1] P. C. H. Mitchell, *The Chemistry of Some Hydrodesulphurisation Catalysts Containing Molybdenum* (Climax Molybdenum, London, 1967), p. 6.
[2] N. P. Martinez, P. C. H. Mitchell, and P. Chiplunker, paper presented at the *Climax Second International Conference on the Chemistry and Uses of Molybdenum*, September 1976.

hydrogenation for Ross' catalysts correlate strikingly with the surface concentration of cobalt (see Figure 1).[2] We believe that the surface cobalt, as determined in our

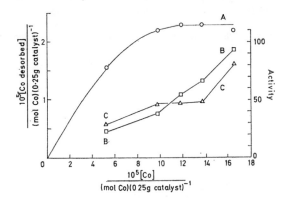

FIGURE 1 Relation between the surface concentration of cobalt for a series of Co–Mo–Al$_2$O$_3$ catalysts, measured by the desorption of cobalt into water (curve A), and the activities of the catalysts in thiophen hydrodesulphurization [initial rate/10^{-9} (mol thiophen) (g catalyst)$^{-1}$ s^{-1}] (curve B) and in butane formation [initial rate/10^{-11} (mol butane) (g catalyst)$^{-1}$ s^{-1}] (curve C). Activities are taken from the paper by Hargreaves and Ross (this Congress).

desorption experiments, is the cobalt which is capable of edge intercalation in MoS$_2$ crystallites in the active catalysts, *i.e.* the cobalt which promotes rate-limiting hydrogenation reactions by creating more active sites.

J. R. H. Ross (*Univ. of Bradford*) replied: The hydrodesulphurization and hydrogenation functions of our catalysts are relatively independent, apart from the fact that thiophen and butenes compete for the hydrogenation sites. Hence, the reaction conditions will determine whether or not hydrogenation is rate-determining; it is dangerous, from results such as ours, to extrapolate to the higher temperatures and pressures of commercial operation without a full knowledge of the kinetic parameters.

The results of Martinez *et al.* are most interesting, but it is odd that the results for catalyst 3·5 do not fit their correlation; the results for 3·5 in our table are fully corroborated by our unpublished results for dehydrogenation of tetrahydrothiophen and hydrogenation of butadiene.[3] One must also remember that catalyst 1·1 (without cobalt) has substantial butene hydrogenation activity.

M. Ternan (*Energy Res. Laboratories, Ottawa*) said: Dr. Mitchell suggests a relationship between hydrogenation and hydrodesulphurization. Some values for the hydrogen:carbon ratio in products obtained in our studies are shown in Figure 2. The hydrogen and carbon analyses were performed using the standard ASTM methods for ultimate analysis of coal. It is seen that the hydrogen content increases substantially with the wt % MoO$_3$ in the catalyst. The twin peak phenomenon observed for the other reactions was also observed for hydrogenation (see the lower part of the figure). It is apparent that both the promoter and the molybdenum have a major role in hydrogenation.

Z. Dudzik (*A. Mickiewicz Univ., Poznan*) said: Are you sure that, after exposure of your catalysts to hydrogen sulphide at 573 K for 17 h, only the stable sulphides

[3] A. E. Hargreaves and J. R. H. Ross, unpublished work.

CoS_8 and MoS_3 and incomplete sulphidation products are present on the surface? In my opinion some paramagnetic sulphur species stabilized on the catalyst matrix may be formed under these conditions. These species are also active for hydrodesulphurization. We have found that these reactions occur on crystalline silica–alumina (sodalite structures) and porous silica containing these sulphur species.

J. R. H. Ross (*Univ. of Bradford*) replied: We have no direct evidence for the nature of the species present on the surface of the sulphided catalyst, and it is possible that

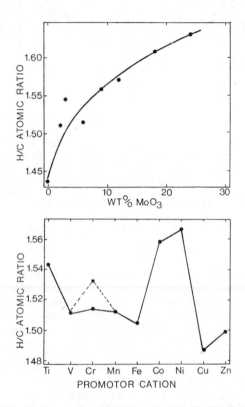

FIGURE 2, TOP: Hydrogen to carbon atomic ratio in the liquid hydrocarbon reaction product against the concentration of MoO_3 in the unpromoted molybdenum–alumina catalysts. BOTTOM: Hydrogen to carbon atomic ratio in the liquid hydrocarbon reaction product against the catalyst promoter cation in the molybdenum–alumina catalysts. (●) catalysts calcined at 500 °C; (■) catalyst calcined at 900 °C. The atomic metal: molybdenum ratio was 1:1 in all the promoted catalysts. All data points obtained at 400 °C.

species such as those you postulate are present. However, our alumina support alone was inactive.

M. Lo Jacono (*Univ. of Rome*) said: Recent publications have shown that, at a fairly low molybdenum content (12% MoO_3), no separate phase MoO_3 can be identified. The strong interaction of molybdenum ions with support, shown for example by the different ease of reduction, indicates that Mo^{6+} ions are accommodated on the surface

946

of the alumina forming a molybdate ion. One model is that of molybdate ions attached on alumina surface, in registry with the structure.[4]

On what grounds do the authors assert that cobalt can be present as CoO, at 600 °C in air? It is known, by experiments and from thermodynamic considerations, that the phase stable in these conditions is Co_3O_4.

J. R. H. Ross (*Univ. of Bradford*) replied: We do not intend to imply that MoO_3 phases are present, but that the thermodynamics of whatever surface compounds are present are related to those of MoO_3; *i.e.* the ease of reduction from $Mo_{surface}^{6+}$ to $Mo_{surface}^{4+}$ is related to that for solid MoO_3 to solid MoO_2. Obviously the kinetics of reduction of each type of species may differ markedly and our conclusions are based entirely on thermodynamic arguments.

In referring to the untreated catalyst as $CoO–MoO_3$, we adopted common usage.[1,5,6] We followed this practice because we were not concerned with the untreated material.

C. J. M. Brew (*Texaco, Ghent*) said: Dr. Ross observes no initial activity with a catalyst reduced in hydrogen. We believe that such catalysts are initially active, but that the products or precursors are adsorbed by the catalyst to form coke; this is not the case with sulphided catalysts. We have carried out experiments with both hydrogen-reduced and sulphided catalysts using a pulse microreactor. The product balance for the reduced catalyst is poor during the initial pulses, and only when hydrogen sulphide is given off by the catalyst does the balance reach the correct level (see Figure 3); this is not observed with sulphided catalysts.

Although we disagree with Dr. Ross concerning the initial activity of hydrogen-reduced catalysts, we do agree that such catalysts are very active for butene hydrogenation. The role of sulphidation is to prevent this initial fouling of the catalyst.

J. R. H. Ross (*Univ. of Bradford*) replied: The difficulty is one of semantics. We define reaction as the production of C_4-hydrocarbons. We did not analyse for thiophens in these experiments, and it is probable that thiophen adsorption occurs rapidly on the reduced catalyst. The rate of C_4-production was ∼6% of the rate on the sulphided material and we considered this to be negligible for the purposes of our presentation; however, even this rate of reaction would cause the production of sulphur, giving gradual sulphidation of the catalyst, and this would be accentuated at higher pressures.

Our butene hydrogenation results show that the sulphided catalyst is completely different to the reduced material, sulphidation decreasing the activity markedly. We therefore believe that the role of sulphur is more complicated that that suggested by Dr. Brew, at least for our catalysts.

H. Heinemann (*Mobil, Princeton*) said: In your paper you conclude that 'the reaction sequence probably involves hydrogenation of the adsorbed thiophen molecule on one site, followed by transfer of the saturated species to the other site, where it undergoes dehydrodesulphurization'. How do you reconcile this with our work[7] which shows that thiophen hydrodesulphurization with deuterium results in little, if any, D_2S?

Dr. J. R. H. Ross (*Univ. of Bradford*) replied: We recognize that there is an apparent disagreement between your work and ours, and we shall endeavour to resolve this in a further paper in which we discuss our mechanism in more detail. We must also take into account that Moyes and co-workers[8] have reacted thiophen with deuterium over unsupported molybdenum disulphide at 609 and 673 K and observe that the hydrogen sulphide formed contained 90% D_2S at 5% conversion. These workers suggest that the absence of D_2S in your work may have resulted from isotope exchange between

[4] M. Lo Jacono, A. Cimino, and G. C. A. Schuit, *Gazzetta*, 1973, **103**, 1281.
[5] See paper by Parsons and Ternan (p. 965).
[6] M. Lo Jacono, J. L. Verbeek, and G. C. A. Schuit, *J. Catalysis*, 1973, **29**, 463.
[7] R. J. Mikovsky, A. J. Silvestri, and H. Heinemann, *J. Catalysis*, 1974, **34**, 324.
[8] M. Eyre, R. B. Moyes, C. T. Screen, and P. B. Wells, private communication.

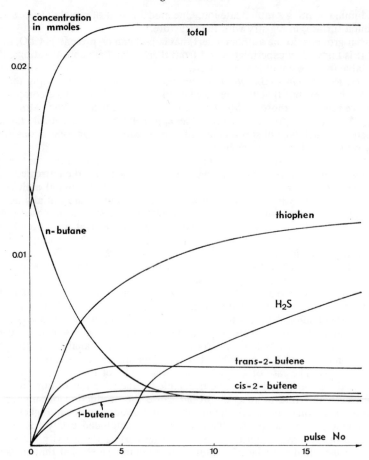

FIGURE 3   Product distribution from pulses of thiophen over a Co–Mo–Al$_2$O$_3$
catalyst reduced in hydrogen at 673 K.

deuterium adsorbed on the catalytically active sites with protium from the alumina
support. This suggestion is supported by their observation that hydrogen exchange
between molecular deuterium and protium atoms of the support in 9%–MoS$_2$–alumina
is fast at 480 K, which is a lower temperature than you used (561–616 K). Furthermore,
they indicate that mass-spectrometric analysis of H$_2$S–HDS–D$_2$S mixtures presents
difficulties, due to exchange of deuterium for protium in the instrument, and this may
lead to an underestimation of the concentration of D$_2$S in such a sample.

T. Edmunds (*British Petroleum, Sunbury-on-Thames*) said: I suggest that your initial
rate of thiophen hydrodesulphurization solely indicates the extent of formation of MoS$_2$
on your catalyst, and that you could prove this by use of ESCA. It is interesting that
this initial rate increases with Co:Mo ratio. The fact that no maximum is achieved
below a Co:Mo ratio of 0·8 is not in conflict with other workers because they measure
an equilibrium rate of thiophen conversion (in particular, the surface state of the catalyst
is in equilibrium) whereas you measure an initial rate following sulphidation in the
absence of hydrogen.

You state that the activity of molybdenum disulphide for thiophen hydrodesulphurization is greater than that of the trisulphide. I believe Amberg has reported the reverse, but, of course, MoS_3 is reduced to MoS_2 in hydrogen above 473 K.

J. R. H. Ross (*Univ. of Bradford*) replied: We recognize that our results are obtained under conditions very different from those of other workers. However, we believe that the ageing of our catalysts by prolonged exposure to a reaction mixture is likely to cause similar changes in activity on all the catalysts, and that our comparisons and the explanation based on the different methods of preparation are both valid. We do not believe that sequential treatment in hydrogen sulphide and molecular hydrogen will cause marked differences from pre-treatment in H_2S–H_2 mixtures or other sulphiding atmospheres.

We postulate only that MoS_2 surface species may be the more active. Amberg has not, to our knowledge, investigated MoS_3; Kolboe and Amberg[9] report the activity of samples of molybdenum disulphide prepared by reduction of the trisulphide in hydrogen, but the higher specific activities of these samples (compared with as received MoS_2) could be explained in many ways without suggesting that they contained unreduced trisulphide. Grange and Delmon[10] state that MoS_2 is the only molybdenum sulphide which has been investigated to date, and we have found nothing which contradicts this statement.

G. C. Stevens (*British Petroleum, Sunbury-on-Thames*) said: Firstly, I was confused by the results in your paper which indicate that the selectivity for butane formation decreased as the Co:Mo ratio rose, and the comment that the rate of butane formation rose as the cobalt content increased. How can the selectivity for formation of butane fall if the rate of formation rises? Secondly, the activities of the basal and edge planes of molybdenum disulphide differ. The basal plane gives hydrogenolysis but the edge plane is chiefly responsible for hydrogenation, so the results you report might indicate that the proportion of basal and edge planes of molybdenum disulphide is influenced by Co:Mo ratio. Finally hydrogen sulphide is a well-known poison. When you did your experiment on hydrogen activation (after sulphiding) did you observe the formation of hydrogen sulphide?

J. R. H. Ross (*Univ. of Bradford*) replied: The selectivity of the reaction is defined as the rate of formation of butane relative to that of the butenes; when the rate of butene formation rises more than that of butane formation, the selectivity drops.

The kinetics of sulphidation[11] lead us to believe that our catalysts consist of monolayers of the active components, and the basal and edge planes of crystalline molybdenum disulphide are unlikely to be present. Hydrogen sulphide is an irreversibly adsorbed poison on catalysts such as nickel–alumina but we believe that it is reversibly adsorbed on our materials, competing only weakly for the active sites. This is in accord with the slight decrease in rate of reaction at higher conversions which we observed. However, we did not analyse for hydrogen sulphide, or do experiments in which hydrogen sulphide was added to the reaction mixture.

P. Grange (*Catholic Univ. of Louvain*) said: You obtain optimum catalytic activity at a Co:Mo ratio of 0·8 whereas other workers have reported optima at ratios as low as 0·2, and you assume the formation of $CoAl_2O_4$. This hypothesis is true; our ESCA measurements on industrial catalysts confirm it. For sulphided catalysts we obtained two cobalt signals one due to $CoAl_2O_4$ and the other due to sulphided cobalt. There exists a relationship between the latter signal and catalytical activity, whereas for cobalt in $CoAl_2O_4$ there is no such effect. This could explain why you need more cobalt when you are promoting $CoAl_2O_4$ formation by impregnation, as on $Al(OH)_3$.

[9] S. Kolboe and C. H. Amberg, *Canad. J. Chem.*, 1966, **44**, 2623.
[10] P. Grange and B. Delmon, *J. Less-Common Metals*, 1974, **36**, 353.
[11] J. R. H. Ross and D. Snaith, unpublished work.

J. R. H. Ross (*Univ. of Bradford*) replied: We imply in the paper that $CoAl_2O_4$ had little activity; this suggestion dates from the work of Richardson (see our ref. 4) and there have since been a number of papers which also indicate the existence of $CoAl_2O_4$ phases. Friedman *et al.*[12] have recently reported ESCA results similar to those which you describe.

J. P. Franck (*Inst. Franç. du Petrole, Rueil Malmaison*) (communicated): Concerning the effect on activity of exposure of hydrosulphurization catalysts to thiophen–hydrogen mixtures, do you think that this effect can be associated with catalyst sulphidation that has not reached its steady-state under hydrogen sulphide alone? Have the authors detected hydrogen sulphide in the gas evolved during exposure to hydrogen alone?

J. R. H. Ross (*Univ. of Bradford*) replied: The apparatus used did not have facilities for the analysis of hydrogen sulphide. Although your explanation is feasible, we believe that the fact that the steady-state activity could be reached by treatment in hydrogen alone as well as in thiophen–hydrogen mixtures favours our interpretation. Also, although we use data for stoicheiometric compounds only loosely related to those on the surface of the catalyst (see my reply to Lo Jacono), the thermodynamic arguments favour our explanation, which is consistent with results advanced in the literature for the active components.

H. Heinemann (*Mobil, Princeton*) (communicated): What is the significance of your results for catalysis at pressures of 600–2000 p.s.i.g.? The oxidation state and surface composition is likely to be quite different from what you and the previous speaker described.

J. R. H. Ross (*Univ. of Bradford*) replied: Our pressure was considerably below atmospheric pressure. The equilibrium constant K_p for the reaction: $MoS_2(s) + 2H_2(g) \rightleftharpoons Mo(s) + 2H_2S(g)$ is given by the usual expression and is independent of the total pressure. The same applies regardless of the nature of the surface phase, and hence the arguments of this and the previous paper will apply at 600–2000 p.s.i.g. ($\sim$4–14 $\times$ 10^6 N m^{-2}).

[12] R. M. Friedman, R. I. Declerk-Grimée, and J. J. Fripiat, *J. Electron Spectroscopy and Related Phenomena*, 1974, **5**, 437.

Testing of hds Catalysts in Small Trickle Phase Reactors

ARIE DE BRUIJN*

Akzo Chemie Nederland bv., Research Centre Amsterdam, P.O. Box 15,
Amsterdam, Netherlands

ABSTRACT The apparent activity of an hds catalyst in a small trickle phase reactor can be influenced very much by channelling and backmixing effects. Dilution of the catalyst bed with small inert particles reduces the influence of both phenomena. The experimental results show that the activity of 50 ml diluted catalyst was equal to that of a 1 l undiluted catalyst bed. Moreover the relation between catalyst particle size and activity was identical in both reactors, and could be explained satisfactorily by pore diffusion limitation. The comparison of the catalyst activity in different reactors demonstrates that the geometry and flow pattern can strongly influence the observed activity. These anomalies, typical for a small trickle phase reactor, can even disturb the activity ranking of catalysts.

The applicability of different reaction orders combined with different reaction time parameters is discussed.

INTRODUCTION

In the literature[1-3] several models for the hydrodesulphurization (hds) of oil fractions in trickle phase reactors have been reported. Apparent reaction orders of one and two were used, and in many cases a relation between mass velocity and hold-up was involved.

First order kinetics were used[2,3] with a correction for the hold-up in the relation between LHSV and reaction time parameter.

Second order kinetics were used, when the spare time τ, defined as 1/LHSV, was taken as time parameter.

Recently Paraskos[3] presented a short review of the models.

The efficiency of the hds process is strongly influenced by the geometry of the reactor. From the literature[4] it appeared that a contacting efficiency of 90% for a pilot plant was observed. This value was based on the measurement of the residence time distribution. In commercial reactors the efficiency can be lower. Schwartz and Roberts[5] compared several mathematical models to describe the data of commercial units, and concluded that the plug flow model may often describe the data. When backmixing was present, the dispersion model gave good results at not too high conversion levels. Mears[6] showed that the amount of backmixing can be related to the ratio of catalyst bed length and particle size (L/dp). This ratio should exceed about 350 for less than 5% deviation from plug flow. The optimal situation seems to exist in a large pilot plant with a long catalyst bed. The influence of the intraparticle diffusion on the residence time distribution was not accounted for in the reported contacting efficiency of 90% in a pilot plant. The hydrodynamic behaviour with respect to plug flow will be better, because the internal hold-up plays a role in the variance of the time distribution curve.

In small bench scale reactors the short catalyst bed causes backmixing. In addition, wall effects occur, resulting in a large degree of channelling. The flow pattern in small 50 ml catalyst beds can seriously influence the results, when two catalysts are compared in activity.

Extrudates with an average length of 3–5 mm and a diameter of 1·6 or 3·0 mm are large compared to the reactor diameter (21 mm) and therefore show a badly reproducible packing. This results in poor repeatability of the activity measurements. Not only the diameter, but also the extrudate length influences the apparent activity very much.

Both parameters have a large impact on the degree of channelling and dispersion. When two catalysts are compared, it is therefore essential that the particles have the same length distribution as the length effects can be very large and may disturb the activity ranking. The backmixing can be decreased[6] by dilution of the catalyst bed with small inert particles. Moreover a good oil distribution will be obtained, thus suppressing the channelling.

In the following sections of this paper the results of activity measurements on hds catalysts will be reported. Experiments in small (50 ml) diluted and undiluted catalyst beds will be compared with data obtained in a large 1 l pilot plant with undiluted catalyst beds. Also the effects of extrudate length and diameter on activity, measured in the different catalyst beds, will be discussed, as these relations are very important for the comparison of catalysts.

Experimental

Apparatus

The experiments were carried out in an isothermal bench scale reactor with an inner diameter of 21 mm. Some experiments were also done in a 1 l pilot plant with a reactor of 38 mm i.d. Both reactors were provided with an axial thermowell of 6 mm O.D. The experiments were performed in once-through operations.

Catalyst loading

The catalyst beds were placed in the midsection of the reactor tube. Below and above the bed, inert alundum beads of 2—4 mm diameter created a good isothermicity, while the beads above the catalyst were also necesasry for the preheating of the reactants and for a good oil distribution. When dilution was applied, the catalyst was loaded in 5 portions. Each portion was mixed up with dilution material outside the reactor in a volume ratio of 1 to 1. The dilution materials used were quartz grains with a sieve size of 20—60 mesh (0·4—0·9 mm) and carborundum with an average particle size of 46 mesh (0·5 mm). The size of the dilution material was chosen after trial and error in a glass model of the reactor. With the above mentioned particle size, a reproducible and stable bed packing was obtained.

Pretreatment of the catalysts

Prior to the loading the catalysts were dried at 550°C for 1 h. Before the measurements, the catalysts were presulphided with the feedstock mixed with 3 wt % of butanethiol. After a presoaking period of 3 h at 204°C without hydrogen, the presulphiding was carried out for 16 h at a temperature of 316°C, a pressure of 21×10^5 N/m^2, LHSV of 1, and with 200 Nm^3 hydrogen per m^3 oil.

Experiments

The experiments were carried out at a fixed pressure of 41×10^5 N/m^2 and with a fixed hydrogen to oil ratio of 540 Nm^3/m^3. Temperatures and LHSV's were varied. The feedstock was a vacuum gas oil from Kuwait with a boiling range of 302—520°C and a sulphur content of 2·45 wt %.

The catalysts used in this investigation were Ketjenfine catalysts containing 12 wt % of molybdenum trioxide and 4 wt % of cobalt oxide on a γ-alumina base. All catalysts had the same particle density as the pore size distributions were equal.

The activities with respect to the sulphur removal have been expressed as rate constants according to a certain order. The reaction time factor for the calculation has been taken as $\tau = \text{LHSV}^{-1}$.

RESULTS

In Figure 1 the results of measurements with crushed (0·4—0·9 mm) catalysts are given. The activity data show a good fit with second order kinetics. The results of the experiments in which the L/dp ratio was 440 or more fit a straight line plot. Different

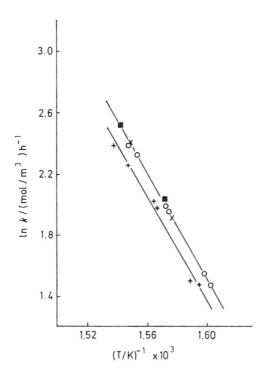

FIGURE 1   Arrhenius plot for crushed (0·4—0·9 mm) Co/Mo catalyst.
Second order kinetics.
Experiments.

O	diluted catalyst: dilution ratio =	1	L/dp = 1000
■	undiluted catalyst		L/dp = 440
×	diluted catalyst: dilution ratio =	2, 3	L/dp = 1000
+	undiluted catalyst		L/dp = 340

LHSV's and linear velocities have been used. Experiments in which the L/dp ratio was 340 showed a lower conversion level. In the literature[6] a minimum value of L/dp for less than 5% deviation of plug flow of 350 is mentioned at $Re_L = 8$.

This criterion holds very well for the experimental data.

On testing 50 ml of $\frac{1}{16}$ inch extruded catalyst in the same reactor with undiluted catalyst bed, the L/dp ratio is found to be 70. So in addition to channelling, a rather strong effect due to backmixing can also be expected. In Figure 2 the effect of dilution of $\frac{1}{16}$ inch extruded catalyst is shown. The activity expressed as a 1·65th order rate constant gave a good fit for the results with respect to different LHSV's. The apparent weight activity of the catalyst increased by about 50% at 360°C on dilution of the catalyst bed.

The repeatability of activity measurements improved. Expressed in terms of variation this value was 10% in undiluted catalyst beds and 3·5% in diluted catalyst beds.

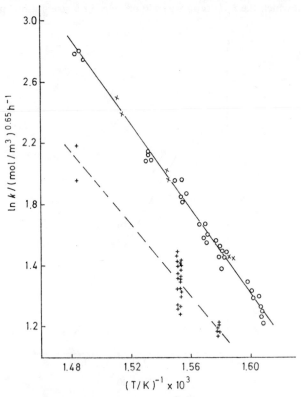

FIGURE 2 Arrhenius plot for extruded Co/Mo catalyst with diameter of 1·6 mm.
Rate constant according to 1·65th order.
LHSV between 1 and 4 for bench scale reactor.
1 and 2 for 1 l pilot plant.
○ Experiments in 50 ml diluted catalyst bed.
× Experiments in 1000 ml undiluted catalyst bed.
+ Experiments in 50 ml undiluted catalyst bed (catalyst was broken to an average length of 3·0 mm).

Particle size effects

The particle size of hds catalysts influences the apparent activity with vacuum gas oil as feedstock. However, in small trickle systems the effects of particle size were very large.

In Figure 3 the results of activity measurements of one catalyst are given. The catalyst was broken in order to create different average lengths of the extrudates at equal diameter. In the undiluted catalyst bed the average length of the catalyst affects the apparent activity very strongly. In the diluted catalyst bed an activity difference of only 4% has been measured for extrudates with an average length of 2 and 3·6 mm.

In Figure 4 the relation between apparent activity and extrudate diameter is shown. In the undiluted catalyst bed the relation is more pronounced than in the diluted catalyst

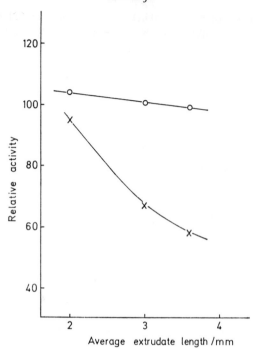

FIGURE 3   Relation between activity and average extrudate length.
Extrudate diameter: 1·6 mm.
Activity expressed in rate constant according to 1·65th
order and compared to the value at an average length
of 3·0 mm.
O   Relation in a 50 ml diluted catalyst bed.
×   Relation in a 50 ml undiluted catalyst bed.

bed. The extrudate length and diameter have a large impact on the flow pattern in the reactor. With particles which are large compared to the reactor diameter, channelling and backmixing will occur and lower the conversion. On comparing two catalysts in a small undiluted catalyst bed not only the diameter, but also the extrudate length, influences the measured activity difference.

Comparison with a 1 l pilot plant

Some experiments, carried out in the 50 ml diluted catalyst bed, were also performed in a 1 l pilot plant with undiluted catalyst bed. The method of presulphiding, the catalyst, the type of oil, and the conditions were the same. The L/dp ratio in the reactor was 380. The ratio of reactor diameter (38 mm) and particle size exceeds the value of 10. For the same conditions the mass velocity is about 5 times higher compared with the bench scale reactor. The higher mass velocity may change the Pé number in a favourable direction. The L/dp ratio meets the criterion of Mears for avoiding dispersion, and agrees with the estimates in our first experiments.

In Figure 2 the results of the activity measurements in the 1 l pilot plant are compared with the measurements in the 50 ml diluted catalyst bed. In Figure 4 the same is done with respect to the relation between activity and extrudate diameter.

The apparent activities of one catalyst in both apparatus agree well, while the influence of the extrudate diameter appears to be the same, within the limits of experimental accuracy.

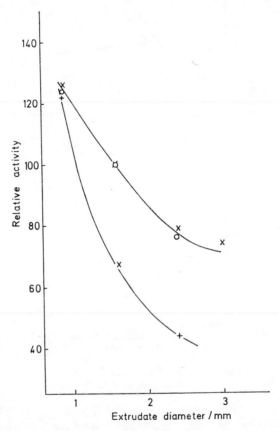

FIGURE 4   Relation between activity and extrudate diameter.
Activity expressed in rate constant according to 1·65th order and compared to the value of a diameter of 1·6 mm.
○   Relation in a 50 ml diluted catalyst bed.
×   Relation in a 1000 ml undiluted catalyst bed.
+   Relation in a 50 ml undiluted catalyst bed (catalyst was broken to an average length of 3·0 mm).

DISCUSSION

Particle size effects

The relation between activity and particle size for diluted catalyst beds as shown in Figures 3 and 4, points to pore diffusion limitation. The reaction rate constants in the experiments were calculated using the space time τ defined as 1/LHSV. For crushed catalyst (0·4—0·9 mm) second order kinetics give a good fit, while for extrudates with a diameter of 1·6 mm a 1·65th order describes the data satisfactorily. The apparent activation energies are 33 and 28 kcal/mole, respectively. The changes in apparent order and activation energy are also in line with diffusion limitation.[9]

The influence of diffusion limitation in terms of the effectiveness factor[9] was estimated by the triangle method. The ratios of the reaction rates were compared with the ratios of particle sizes. In Figure 5 the results of the estimations are shown. The only variable

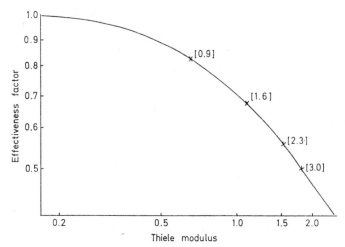

FIGURE 5   Effectiveness factor as function of the Thiele modulus [Extrudate diameter].

in this investigation is the particle size, defined as the ratio of volume and outer surface area of the particles. Other properties of the catalysts are equal. Based on the straight line relationship between Thiele modulus and particle size, the influence of the diameter of the extrudates were recalculated with the known particle size as variable. The largest difference between the measured and calculated activities was 7%.

The particle size effects depend on the type of oil, the conditions, and the type of catalyst. The general pore diffusion model is too simple to allow extrapolation outside the range of measurements or to other types of feedstocks and certainly to catalysts with different porosities. However, the model explains well the large influence of the particle diameter and the small influence of the length of the particles for these experiments. The model was applied successfully for interpolation of particle size effects.

Kinetics in test reactors

In the literature[1,10-12] several authors describe the successful use of a pseudo second order relation for desulphurization. The space time is then defined as 1/LHSV. Also, first order kinetics, in combination with the mean residence time, can result in a good description of the process. The relation between the mean residence time and LHSV is given by the external hold-up in the system or measured contacting efficiencies.[4,13] However, an order of reaction between 1 and 2 is also acceptable.

As pointed out in the literature[1,10-12] the addition of several first order reactions results in a higher apparent overall order with respect to space time variation and in a first order behaviour concerning initial concentration of the reactant.
In the equation

$$r_{hds} = \sum_{i=1}^{P} k_i S_i = \sum_{i=1}^{P} k_i S_{0i} (1 - C) \sim k S_0 (1 - C)^n \qquad (1)$$

957

For an ideal plug flow reactor, integration of the mass balance results in

$$k\tau = \frac{1}{n-1}\left[\frac{1}{(1-C)^{n-1}} - 1\right] \tag{2}$$

or

$$k\tau = \frac{S_0^{n-1}}{n-1}\left[\frac{1}{S^{n-1}} - \frac{1}{S_0^{n-1}}\right] \tag{3}$$

Equation 2 is the well known formula for pseudo second order with $n = 2$. Although eqn 1 can only be solved in a numerical way, some simple calculations can illustrate the effects. In Table 1 the oil fraction for a theoretical gas phase process was divided into 10 fractions with different first order rate constants. The overall reaction order was estimated based on the best fit.

It can be seen from Table 1 that the apparent order with respect to space time depends on the ratio of the gastest and slowest reactions. The initial concentration does not influence the rate constant, as would be expected for first order kinetics.

Structural effects of the sulphur containing molecules alone can cause a ratio[7] of 5, and combined with the influence of the molecular weight in a wide boiling range, rather high ratios of fastest and slowest rate constants can be expected.[11] As illustrated in Table 1 the data can also be described rather well in a limited range of conditions with an overall first order reaction combined with the reaction $\tau = \text{LHSV}^{-0.67}$.

This exponent has no physical background in this example because in a gas phase

Table 1

The fitting of hydrodesulphurization data in a theoretical gas phase
process with different orders

Assumptions: constant values of pressure, hydrogen to oil ratio, and temperature; the feedstock consists of 10 compounds with a concentration of 0.1 S_0 (S_0 is total wt % of sulphur); Each compound has a different rate constant and follows first order kinetics with respect to sulphur removal.

The following formula has been applied: $k = \dfrac{S_0^{n-1}}{n-1}\left[\dfrac{1}{S^{n-1}} - \dfrac{1}{S_0^{n-1}}\right]\text{LHSV}$

Case A: First order rate constants of the 10 compounds are:
1, 2, 3, 4, 5, 6, 7, 8, 9, and 10.
ratio of fastest and slowest reactions = 10.

LHSV	S total in prod.	% hds	k 1st order	k 2nd order	k 1.45th order	k 1st order and $\tau = \text{LHSV}^{-0.67}$
0.5	$0.015\,S_0$	98.5	2.1	32.8	6.2	2.6
1	$0.058\,S_0$	94.2	2.9	16.2	5.6	2.9
2	$0.153\,S_0$	84.7	3.8	11.1	5.9	3.0
3	$0.244\,S_0$	75.6	4.2	9.3	5.9	3.0
4	$0.324\,S_0$	68.6	4.5	8.4	5.9	2.8
5	$0.391\,S_0$	60.9	4.7	7.8	5.8	2.8
6	$0.447\,S_0$	55.3	4.8	7.4	5.8	2.7

Case B: First order constants of the 10 compounds are:

1, 3, 5, 7, 9, 11, 13, 15, 17, 19;

ratio of fastest and slowest reactions = 19

LHSV	S total in prod.	% hds	k 1st order	k 2nd order	k 1·7th order	k 1st order and $\tau = \text{LHSV}^{-0.67}$
0·5	$0.014\,S_0$	98·6	2·1	35·2	13·4	2·6
1	$0.042\,S_0$	95·8	3·2	22·6	11·6	3·2
2	$0.096\,S_0$	90·4	4·7	18·7	11·9	3·7
3	$0.147\,S_0$	85·3	5·8	17·4	12·1	4·0
4	$0.197\,S_0$	80·3	6·5	16·3	12·1	4·1
5	$0.246\,S_0$	75·4	7·0	15·4	11·9	4·1
6	$0.288\,S_0$	71·2	7·5	14·0	11·9	4·0
8	$0.366\,S_0$	63·4	8·0	13·8	11·7	4·0
10	$0.432\,S_0$	56·8	8·4	13·1	11·4	3·9

Case C: First order rate constants of the 10 compounds are;

1/5, 6/5, 11/5, 16/5, 21/5, 26/5, 31/5, 36/5, 41/5, 46/5;

ratio of fastest and slowest reactions = 46

LHSV	S total in prod.	% hds	k 1st order	k 2nd order	k 1st order and $\tau = \text{LHSV}^{-0.67}$
0·1	$0.014\,S_0$	98·6	0·4	7·2	0·9
0·2	$0.036\,S_0$	96·3	0·7	5·3	1·1
0·5	$0.077\,S_0$	92·2	1·3	6·0	1·7
1	$0.129\,S_0$	87·1	2·1	6·8	2·1
2	$0.228\,S_0$	77·2	3·0	6·8	2·4
3	$0.319\,S_0$	68·1	3·4	6·4	2·4
4	$0.395\,S_0$	60·5	3·7	6·1	2·4
5	$0.458\,S_0$	54·2	3·9	5·9	2·3

process $\tau = 1/\text{LHSV}$. In a trickle phase process, however the exponent has the same value,[2] when hold-up correlations are used. It cannot be concluded, therefore, solely on the basis of the results of the best fit and without support of other physical measurements, that the hold-up relations are valid.

The high reaction order has been confirmed in experiments with straight run gas oils with end boiling points below 360°. During this process the oil is mainly in the gas phase (90%) and hold-up variations cannot have a large influence. Apparent reaction orders of 1·25—1·7 were measured. Based on these data, and theoretical considerations also, high orders for the desulphurization of vacuum gas oil can be expected, certainly due to the larger boiling range.

In an investigation,[14] desulphurized vacuum gas oil was mixed with the original feedstock in order to create comparable feedstocks with different sulphur concentrations. The first order behaviour in sulphur concentration and the higher order in reaction time parameter were confirmed and the rate eqn 1 described the data satisfactorily.

For the diluted catalyst bed, separate experiments with 50, 75, and 100 ml of catalyst at equal LHSVs were performed. The activity was the same within the limits of experimental error. As the hold-up is correlated with the mass velocity,[8,13] there were no hold-up variations, and the plug flow assumption was justified.

The observed reaction order of 1·65 depends on the type of oil and catalyst and is independent of the reactor geometry. With the same order the time parameter in the 1 l pilot plant was also $\tau = 1/\text{LHSV}$, while in the undiluted 50 ml catalyst beds the relation $\tau = \text{LHSV}^{-0·8}$ was the best combination.

The relation between LHSV and the time parameter is influenced by the reactor geometry, but will be the same in a reactor for the different processes which occur at the same time as desulphurization, denitrogenation, *etc.*

When different relations between performance and LHSV exist for the different processes, as reported by Paraskos,[3] it is more likely on physical grounds that the apparent orders are different than that the time parameters are not equal. Besides, the different behaviour of catalysts with the same particle size can be attributed to different apparent orders, as the time parameters are equal. These different reaction orders can be a useful tool in catalyst screening with respect to diffusion limitation.

By application of catalyst bed dilution the relation between LHSV and the time parameter is fixed and is less sensitive for random deviations. In this way the catalyst performance is not influenced by anomalies in hydrodynamic behaviour.

CONCLUSIONS

In small trickle phase reactors with 50 ml catalyst the apparent activity of an hds catalyst is influenced very much by channelling and backmixing in the reactor. These phenomena lower the activity level and are responsible for bad reproducibility.

Apart from the diameter of the catalyst extrudates, the length of the particles also has a strong influence upon the activity. This can change the activity ranking in catalyst screening. The anomalies, typical for a small catalyst bed are avoided by dilution of the catalyst.

The activity level and particle size effects are the same in a pilot plant with 1 l of undiluted catalyst and in a 50 ml diluted catalyst bed. The effect of the particle diameter is less pronounced in a diluted catalyst bed. The influence of the extrudate length is of minor importance. The particle size effects can be explained completely by pore diffusion.

In the diluted catalyst bed the space time ($= 1/\text{LHSV}$) is a realistic reaction time parameter. A high apparent reaction order (1·65) is found when applying this parameter.

The applied kinetics hold for a 50 ml diluted catalyst bed as well as for a 1 l pilot plant.

NOMENCLATURE

C	conversion wt %
dp	equivalent spherical diameter m
D	diffusion coefficient m²/s
Di	dispersion coefficient m²/d
k	reaction rate constant h⁻¹
k_i	reaction rate constant of one compound h⁻¹
L	catalyst bed length m
Lp	volume/outer surface of particle m
LHSV	liquid hourly space velocity m³/m³ h.
n	order of reaction
P	pressure N/m²
$Pé$	Péclet number $= v \cdot dp/Di$
Re_L	Reynolds number for the liquid $= v \cdot dp/\nu$
S	sulphur content in product wt %

S_t sulphur content in product for one compound wt %
S_o sulphur content of feedstock wt %
S_{oi} sulphur content of feedstock for one compound wt %
T temperature K
η effectiveness factor

ϕ Thiele modulus $= Lp \sqrt{\dfrac{k}{D}}$

ν viscosity m^2/s
ρ density kg/m^3
τ space time $= 1/\text{LHSV}$ h.

REFERENCES

[1] H. Beuther and B. K. Schmid, *Proc. 6th World Petr. Congress Sect. III*, p. 297 (Verein für Forderung des 6. Welt-Erdöl Kongresses, Hamburg, 1964).

[2] H. H. Clarke and J. B. Gilbert, *Ind. Eng. Process Res. Dev.*, 1973, **12**, no. 3, 328.

[3] J. A. Paraskos, J. A. Frayer, *Ind. Eng. Process. Res. Dev.*, 1975, **14**, no. 3, p. 315.

[4] E. V. Murphree, A. Voorhies, Jr., and F. X. Mayer, *Ind. Eng. Process. Res. Dev.*, 1964, **3**, no. 14, p. 381.

[5] J. G. Schwartz and G. W. George, *Ind. Eng. Process Res. Rev.*, 1973, **12**, no. 3, p. 262.

[6] D. E. Mears, *Chem. Eng. Sci.*, 1971, **26**, 1361.

[7] A. G. Frye and J. E. Mosby, *Chem. Eng. Progr.*, 1967, **63**, no. 9, 66.

[8] P. Way, *The Performance of Trickle Bed Reactors*, Thesis, Massachusetts Institute of Technology, 1971, University Microfilms, Ann Arbor, Michigan.

[9] O. Levenspiel, *Chemical Reaction Engineering*, (Wiley, London, 1972), 2nd edn., ch. 14.

[10] H. Nakamura, H. Satori, K. Tabuchi, S. Nishizaki, S. Kurita, T. Watanabe, and J. Sakuma, *Internat. Chem. Eng.*, 1970, **10**, no. 3, 506.

[11] H. Ozaki, Y. Satomi, and T. Hisamitsu, *Proc. 9th World Petr. Congress*, Tokyo, 1975. Panel discussion 18 (4).

[12] R. M. Massagutov, G. A. Berg, G. M. Kulinick, and T. S. Kirillov, *Proc. 7th World Petr. Congress*, **4**, p. 177.

[13] C. N. Satterfield *et al.*, *A. I. Ch. E. J.*, 1969, **15**, 226.

[14] A. de Bruijn, unpublished results.

DISCUSSION

B. C. Gates (*Univ. of Delaware, Newark*) said: In experiments using a laboratory microreactor operated under conditions similar to yours (we used a commercial cobalt molybdate catalyst presulphided with a mixture of hydrogen and hydrogen sulphide for hydrodesulphurization of dibenzthiophen in a paraffinic carrier oil[1]) it was observed that a break-in period of about eighty hours was required before the catalyst achieved a constant activity. Failure to appreciate that the catalyst requires such a long break-in period can make a determination of the kinetics difficult. Have you also observed such long transients in catalyst operation?

Your Shiele plot showing the effect of pore diffusion on rate would be quantitatively valuable if you would provide additional results, especially the reaction temperature and the catalyst pore size distribution; would you please give this information?

A. de Bruijn (*Akzo Chemie, Amsterdam*) replied: The instability of the catalyst activity, or break-in period, can be a function of the method of presulphidation. We performed presulphidation in the liquid phase with a sulphur compound added to the feedstock. We measured the activity of the catalyst in several runs over a period of 300–600 h under various conditions, but no break-in period was observed.

The additional information that you request is as follows. The temperature was 638 K, and the pore size distribution of the used catalyst is given below.

[1] M. Honalla, V. H. J. de Beer, B. C. Gates, and G. C. A. Schuit, unpublished work.

Pore radius/Å	Pore volume/ml g^{-1} (nitrogen desorption)
22	0·0
25	0·1
30	0·3
40	0·44
50	0·45
60	0·46
70	0·46
80	0·46^5
100	0·47
300	0·47
Total pore volume	0·48

M. Ternan (*Energy Research Laboratories, Ottawa*) said: Several laboratory-scale investigations have been reported[2-4] in which the liquid–vapour mixture enters the bottom of the reactor and leaves from the top. This technique has been applied in cases where a portion of the reaction mixture is in the liquid phase under reaction conditions. Have you carried out any bottom-feed experiments and can you comment on the usefulness of this technique in decreasing channelling and backmixing?

A. de Bruijn. (*Akzo Chemie, Amsterdam*) replied: We carried out several upflow experiments for comparison with downflow experiments. For $\frac{1}{32}$ inch extrudates the reaction rate constant was the same in a diluted catalyst bed in downflow operation as it was in an undiluted catalyst bed in upflow operation. For $\frac{1}{16}$ inch extrudates the results were 15% higher in the diluted catalyst bed in downflow operation compared to the upflow experiment with an undiluted catalyst bed. With a good oil distribution hardly any channelling can be expected in upflow experiments, but with too short a catalyst bed, backmixing still remains an important factor and affects the rate constant.

Apparent particle size effects were indeed smaller in the upflow experiments with undiluted catalyst beds compared to downflow operation with undiluted catalyst beds.

E. J. M. Verehijen (*Dutch State Mines, Geleen*) said: You state that the ratio of catalyst bed length to particle size (L/dp) should exceed a value of about 350 in order that plug flow and high activity may be maintained. In Figure 1 you demonstrate that this value is valid for your experiments. In fact, the condition of a minimum value for L/dp must be a function of the linear velocity or space velocity. You used different space velocities and linear velocities in situations where your L/dp ratio is rather critical; nevertheless, you do not mention the effect of low space velocity on the activity in this case. Can you explain this?

A. de Bruijn (*Akzo Chemie, Amsterdam*) replied: The relation between the Péclet number, Pé, and the Reynolds number, Re, is given by the relation[5]: Pé $\sim$ Re$^{0·5}$ for a large range of Reynolds numbers. However, for the low Reynolds number in our reactor (0·1) there are hardly any relations available. In the experiment with $L/dp = 340$, the linear velocity was varied by a factor 1·5. Extrapolation of the above relation results

[2] E. C. McColgan and B. I. Parsons, *The Hydrocracking of Residual Oils and Tars, Part 2: The Catalytic Hydrocracking of Athabasca Bitumen* (Mines Branch Research Report R253, Department of Energy, Mines & Resources, Ottawa 1972).

[3] J. W. Snider and J. J. Perona, *AIChE J.*, 1974, **20**, 1172.

[4] T. Takematsu, K. Shimada, Y. Kuriki, S. Oshima, M. Suzuke, and J. Kato, *Nippon Kagakukishi*, 1974, **12**, 2384.

[5] R. W. Michell and I. A. Furzen, *Trans. Inst. Chem. Eng.*, 1972, **50**, 334.

in a change of about 20% for the volume of L/dp. Experiments showed that deviation by backmixing occurs when L/dp lies between 340 and 440. If the relation is valid, the effects are not large but noticeable. Apparently the exponent of the Reynolds number is lower in our situation.

R. W. Coughlin (*Lehigh Univ., Bethlehem, Pa.*) said: It is not clear to me whether you ascribe the influence of dilution by inert particles (or even influence of particle size at constant effectiveness factor) to backmixing, or to channelling, or to both. Were any experiments performed with tracer added to a single phase in order to determine how residence-time distribution was influenced by changes in particle size and particle dilution?

A. de Bruijn (*Akzo Chemie, Amsterdam*) replied: The dilution of small catalyst beds effects both channelling and backmixing. The channelling could be observed in a glass model of the reactor. A part of the catalyst bed was not wetted properly. The preferential flow patterns changed with time. In the diluted catalyst bed no channelling could be observed. Tracer studies were made using undiluted catalyst beds. The residence time distributions were very wide, with a rather sharp peak at short times, showing channelling. The position of this peak and the variance of the distribution were heavily influenced by the particle size. The calculated average residence time could not explain the activity differences. We did not carry out tracer studies in diluted catalyst beds. Recently equal residence time distributions for small diluted catalyst beds and a large catalyst bed were found.[6] These results agree with our investigations.

C. McGreavy (*Univ. of Leeds*) said: The paper suggests that significant intraparticle diffusion effects are influencing the reaction rate, based on observations of the variation of activity (presumably the observed effective rate constant) with catalyst particle size. It is also stated that the apparent activation energy is approximately 30 kcal mol^{-1} which implies that the actual value is appreciably higher and consequently that the reaction rate is very sensitive to temperature changes. Yet no mention is made of heat generation effects, which are often quite large in such reactions.

By taking such phenomena into account, some of the rather anomalous results observed with the diluted beds might be explained. Any heat generated by the reaction would be more easily dissipated in the diluted beds while, in the undiluted bed, partial evaporation of the liquid could interfere with the distribution and hence the contact area. This means a lower activity would be inferred, consistent with the experimental results. Furthermore, it offers at least a partial explanation of the dependence of activity on particle size, since heat generation effects would be more severe in larger particles, perhaps even causing evaporation of liquid from the pores. Was any attempt made to establish that these effects were absent, either by calculation or by temperature measurements in the catalyst bed?

A. de Bruijn (*Akzo Chemie, Amsterdam*) replied: The temperature in the catalyst bed was measured in the centre of the catalyst bed in the axial direction. No significantly different temperature profiles were noticed for different particle sizes. As far as can be estimated by calculation, the heat generation has no important influence on diffusion limitation and on particle size effects. Besides, heat effects are not very large, because the concentration of reactants is low and the liquid phase has a high heat capacity. The relatively high activation energy for $\frac{1}{16}$ inch extrudates can be explained by the rather high effectiveness factor. For diffusion-limited reactions the apparent activation energy can be written as

$$-\frac{E_{obs}}{R} = -\frac{E_{intrinsic}}{R} + \frac{d \ln \eta}{d \ln \phi}\left(-\frac{E_{in}}{2R}+\frac{E_D}{2R}\right)$$

[6] Plenary lecture by Dr. v.d. Vusse, *4th ISCRE* (Heidelberg, April, 1976), to be published in proceedings.

As the measured $E_{\text{intrinsic}}$ was 33 kcal mol^{-1} and with $\eta = 0.67$, the term $d(\ln \eta)/d(\ln \phi)$ had a value of about 0.6. E_D is about 7 kcal mol^{-1}. E_{obs} will be 25–26 kcal mol^{-1}, which is close to the measured value of 28 kcal mol^{-1}. On the other hand, if partial evaporation by the heat generation is important, the conversion level and temperature should influence the differences between the behaviour of diluted and undiluted catalyst bed. This was not observed.

J. W. Hightower (*Rice Univ., Houston*) asked: How did you determine the second-order kinetics? If it was found by fitting conversion reciprocal space velocity results to a power law equation, it is possible that the conclusion could be misleading. Results that follow Langmuir–Hinshelwood kinetics coupled with product poisoning can sometimes be plotted successfully according to higher order power law equations. The best way to determine reaction order is to vary the initial concentration and note the effect on the rate in a differential reactor, but I realize that this is impossible in a trickle-bed reactor. Could you comment about the possibility of product poisoning in your system?

A. de Bruijn (*Akzo Chemie, Amsterdam*) replied: The apparent order was determined by varying the linear hour space velocity. The additional effect of several first-order reactions results in a higher order behaviour with respect to space time. With respect to concentrations the order remains unity. In the literature, first-order kinetics have been reported for pure compounds (ref. 7 of the paper) and for narrow oil fractions.[7] This phenomenon has also been described in the present paper. The possibility of product poisoning was present in our experiments. Apparently reversible poisoning by hydrogen sulphide and poisoning by nitrogen compounds does not affect the reaction order.

D. E. Mears (*Union Oil, Brea*) said: Could you describe how liquid was initially distributed over the alundum beads at the top of the reactor, and give the depth of the alundum beads? It seems unlikely that the alundum beads could redistribute a poor initial distribution. Regarding the Mears axial dispersion criterion, the minimum L/dp ratio of 350 cited here applies to 90% conversion with first-order reactions. The minimum L/dp ratio increases both with conversion and with reaction order.

A. de Bruijn (*Akzo Chemie, Amsterdam*) replied: The liquid was distributed by a small pipe, ending just above the alundum beads in the centre of the reactor. Gas and oil streams were combined in the piping just before the entrance to the reactor. With this system, we obtained apparently reproducible distributions of liquids. The length of the alundum bed was about 25 cm. A layer of about 1 cm of 14 mesh carborundum was placed just above the diluted catalyst bed. I agree that alundum beads cannot redistribute the oil effectively, if initial distribution is poor.

I agree with your remark regarding the minimum L/dp ratio. However, for our situation L/dp was calculated with $n = 2$, for 90% conversion, and with a value for the Péclet number of 0.25, which is quite realistic as an average value from the literature. The calculated value is then about 370, which is close to the cited value. A more accurate value is difficult to estimate, and is not very important for the avoidance of backmixing effects. A safe margin should be taken.

[7] H. Hoog, *J. Inst. Petroleum*, 1950, **38**, 738, and ref. 14 of the paper.

The Hydrodesulphurization and Hydrocracking Activity of Some Supported Binary Metal Oxide Catalysts

BASIL I. PARSONS and MARTEN TERNAN*

Energy Research Laboratories, Department of Energy, Mines and Resources, Ottawa, Ontario, K1A OG1 Canada

ABSTRACT   The hydrocracking of heavy residual oils causes rapid catalyst deactivation. Catalysts containing low concentrations of metal oxides on γ-alumina were evaluated in an attempt to reduce the resulting high catalyst replacement costs. Hydrocracking, hydrodesulphurization and hydrodenitrogenation catalytic activity measurements were made at a pressure of $1\cdot39 \times 10^7 \, \mathrm{N\,m^{-2}}$ using a 345—525°C gas oil. Studies with unpromoted catalysts containing 2·2—30%, MoO_3 showed that the lower concentrations (2·2—3·0% MoO_3) produced significant hydrodesulphurization activity. Catalysts promoted by Ti, V, Cr, Mn, Fe, Co, Ni, Cu and Zn oxides containing 2·2% MoO_3 and having metal:molybdenum atomic ratios of 1:1 were also studied. The results are discussed in terms of catalyst surface acidity, the role of hydrogen in the catalytic process and the applicability of the intercalation model and the monolayer model. Some of the promoted catalysts containing 2·2% MoO_3 were found to compare favourably with a commercial catalyst containing 12% MoO_3 and 3% CoO.

INTRODUCTION

An investigation of the hydrodesulphurization and hydrocracking of heavy residual oils and bitumens has been in progress at our laboratories for several years.[1] When these heavy feedstocks are processed, rapid catalyst deactivation occurs due to the deposition of coke and of nickel, vanadium, and iron metals. At high conversions of the high boiling hydrocarbons to lower boiling distillates, catalyst fouling caused by coke formation is several times greater than that caused by metals deposition.[1f] Regeneration of the catalyst by burning off the coke restores much of the original activity However, since this process does not remove the metals, repeated use increases the extent of permanent deactivation. The rapidity and severity of catalyst deactivation cause catalyst replacement costs to be much greater for residual oil hydrocracking processes than for gas-oil or naphtha hydrorefining processes. One method of reducing catalyst cost is to decrease the concentrations of the active metal oxides impregnated in the alumina support. In our laboratory the hydrocracking and hydrodesulphurization activities of catalysts initially containing 2·2 wt % MoO_3 and 1·1 wt % CoO were found to be almost as good as that of a commercial catalyst containing 12 wt % MoO_3 and 3 wt % CoO.[1e] Trifiro et al.[2] have also studied catalysts have lower than normal concentrations of CoO. Rovesti and Wolk[3] found that acceptable hydrodesulphurization results could be obtained using a sequence of two different catalysts. They recommended that a catalyst containing 2 wt % Mo be used for demetallization in a first stage; sulphur removal was to occur in a second stage using a catalyst containing 15 wt % MoO_3 and 3 wt % CoO.

The purpose of this work was to establish some of the characteristics of catalysts having low concentrations of active metal oxides; specifically, the effect of molybdenum concentration and the promotional effect produced by a number of fourth period transition metal oxides were investigated. The catalyst evaluation was performed using a high boiling gas–oil (345–525°C) containing 3·64% sulphur. Previous studies showed that the same relative activities were obtained when catalysts were evaluated with this gas–oil[4] and with Athabasca bitumen.[1e]

EXPERIMENTAL

The catalysts used in this study were prepared by spraying aqueous solutions of the appropriate metal salts on to 5 kg of α-alumina monohydrate powder (Continental Oil Company, Catapal SB) which were being mulled in a mix-muller. Initially a solution consisting of 14 cm^3 of concentrated HNO_3 (70 wt %) and 1,000 cm^3 of distilled water was added to the powder. The first salt solution added contained one of the cations being tested as a catalyst promoter. The ammonium paramolybdate solution was added last. The following compounds, containing the promoter cations, were used: $Zn(NO_3)_2 \cdot 6H_2O$, $CuSO_4 \cdot 5H_2O$, $Ni(NO_3)_2 \cdot 6H_2O$, $Co(NO_3)_2 \cdot 6H_2O$, $Fe(NO_3)_3 \cdot 9H_2O$, $MnSO_4 \cdot H_2O$, $Cr(NO_3)_3 \cdot 9H_2O$, $VOSO_4 \cdot 2H_2O$, and $TiCl_4$. All the promoted catalysts had a metal promoter: molybdenum atomic ratio of 1:1 and contained 2·2 wt % MoO_3. The impregnated mixtures were dried at 120°C for 3 h and calcined at 500°C for 3 h. The catalyst promoted with chromium was also calcined at 900°C. The calcined powder was mixed with 2 wt % stearic acid and pressed into cylindrical pellets ($L = D = 3·18$ mm) in a continuous pelleting press. The stearic acid acted as a binder and lubricant during pelleting. The pellets were recalcined at 500°C for 4 h to remove the stearic acid.

Surface area and surface acidity measurements were made on the catalysts in their oxide form. A gravimetric quartz spring balance was used to perform the nitrogen adsorption measurements; the surface areas were calculated using the BET method. The method of Clark, Ballou, and Barth[5] was used to determine surface acidities. 1 g samples of catalyst powder, smaller than 74 μm (200 mesh U.S. Standard Sieve Series), were calcined at 500°C for 3 h and slurried in acetonitrile. After a stable initial reading was obtained, the potentiometric titrations were performed by adding 5 cm^3 increments of a 0·006 M solution of n-butylamine in acetonitrile and waiting 5 min before taking the reading. The potentiometric reading obtained after a total of 25 cm^3 of n-butylamine solution had been added was used as an indication of catalyst acidity.

The feedstock used to evaluate catalyst activity was prepared in our pilot plant by thermally hydrocracking Athabasca bitumen obtained from Great Canadian Oil Sands at Fort McMurray, Alberta. The 345–525°C portion distilled from the hydrocracked product was used as the feed material. Typical compounds in this gas-oil contained from three to six condensed rings. Properties of the feedstock are listed in Table 1.

Table 1
Properties of the Heavy Gas–oil Feedstock

Property	Value
Boiling range	345—525°C
Specific gravity 16/16°C	0·992
Conradson carbon	0·97 wt %
Sulphur	3·64 wt %
Nitrogen	3800 ppm
Kinematic viscosity at 38°C	108·8 mm^2 s^{-1}

The catalyst pellets were evaluated in a bench-scale fixed bed reactor having a volume of 155 cm^3 and a length:diameter ratio of 12. The reactor was filled sequentially from the bottom with 42 cm^3 berl saddles, 100 cm^3 catalyst pellets and 13 cm^3 berl saddles. The heavy gas-oil, mixed with hydrogen (purity = 99·9 wt %), flowed continuously into the bottom of the reactor and up through the catalyst bed. The product leaving the top of the reactor flowed to receiver vessels where the liquid and vapour were separated. Each experiment was performed at a pressure of $1 \cdot 39 \times 10^7$ N m^{-2}, a liquid volumetric space velocity of $2 \cdot 0$ h^{-1} based on the reactor volume occupied by the catalyst pellets, and a hydrogen flow rate of $71 \cdot 8$ cm^3 s^{-1} at STP. The catalyst was presulphided and stabilized using the mixture of gas–oil and hydrogen at 400°C. Subsequent experiments were performed sequentially at temperatures of 420, 400, and 380°C. Gas–oil and

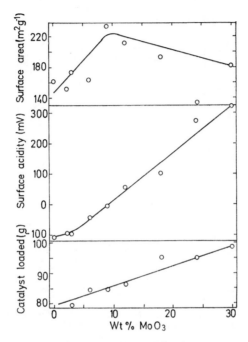

FIGURE 1 Surface area (m^{2}g^{-1}), surface acidity (mV), and weight of catalyst loaded into the reactor (g) against the concentration of MoO$_3$ in the unpromoted molybdenum–alumina catalysts.

hydrogen were in contact with the catalyst for 7 h prior to the 420°C experiment. The reaction system was maintained at steady state conditions for 1 h prior to, and for 2 h during, the period in which each sample of liquid product was collected.

The liquid product was analysed for sulphur, nitrogen, and the weight fraction boiling above 420°C. The sulphur content was determined using an X-ray fluorescence technique.[6] The nitrogen analyses were performed using a hydrogenation-microcoulometric apparatus.[7] A modified version of the U.S. Bureau of Mines Hempel distillation analysis[8] was used to distill the liquid product into nine fractions. The change in the amount of the heaviest fraction (+420°C) was taken as an indication of the extent of hydrocracking.

RESULTS AND DISCUSSION

Measurements of surface area and surface acidity for the oxide form of the unpromoted and promoted catalysts are shown in Fig. 1 and 2, respectively. The weight of the catalyst charge is also shown at the bottom of each figure. The surface areas of the unpromoted catalysts are comparable to those reported by Giordano *et al.*[9] and by de Beer[10] for catalysts prepared by impregnating aqueous solutions of ammonium paramolybdate in γ-alumina powders and subsequently drying and calcining. The surface areas of the promoted catalysts, calcined at 500°C, did not vary by much. However, the Cr promoted catalyst calcined at 900°C had a much lower surface area than the Cr promoted catalyst calcined at 500°C. Similar results (decreasing alumina surface areas with increasing calcining temperature) have been reported by Gitzen.[11]

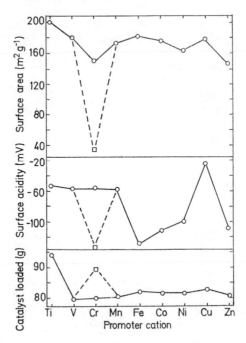

FIGURE 2   Surface area (m^2g^{-1}), surface acidity (mV), and weight of catalyst loaded into the reactor (g) against the catalyst promoter cation in the molybdenum–alumina catalysts. ($\bigcirc$) catalysts calcined at 500°C; ($\square$) catalyst calcined at 900°C.

The Cr promoted catalyst calcined at 500°C was yellow in colour indicating the presence of Cr^{6+} ions.[12] In contrast the Cr promoted catalyst which was calcined at 900°C had a light green colour suggesting the formation of a Cr_2O_3 compound.[13]

All the potentiometric titration curves obtained during the surface acidity measurements had the same general shape as that reported by Clark, Ballou and Barth[5] for F-10 alumina. The surface acidities of the unpromoted catalysts were found to increase fairly uniformly with MoO_3 content. This is in agreement with the finding of Kiviat and Petrakis[14] that Bronsted acidity increased with increasing MoO_3 concentrations on η-Al_2O_3. For the promoted catalysts in their oxide form the Co, Ni and Zn promoted catalysts had almost the same acidity as the unpromoted catalyst containing 2.2 wt % MoO_3, the Fe and Cr (calcined at 900°C) promoted catalysts were slightly more basic, and the Ti, V, Cr (calcined at 500°C), Mn and Cu promoted catalysts were more acidic. Our results for the Co promoted catalysts are different than those obtained by Ratnasamy, Sharma, and Sharma[15] who found that both the distribution and the strength of acid sites on a catalyst prepared by the simultaneous impregnation of Co and Mo salts were different from those of a comparable catalyst containing Mo only.

As shown in Fig. 3, the extents of the three reactions, hydrocracking, hydrodesulphurization, and hydrodenitrogenation, do not change in exactly the same manner as the molybdenum content of the unpromoted catalyst increases. The reaction mixture converts the oxide form of the catalyst to a sulphide form. However for the sake of simplicity, all catalyst concentrations are referred to in their oxide form. The increases

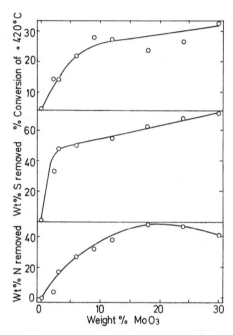

FIGURE 3   Conversion of hydrocarbons boiling above 420°C to lower boiling compounds, weight percent sulphur removed from the liquid hydrocarbon, and weight per cent nitrogen removed from the liquid hydrocarbon against the concentration of MoO_3 in the unpromoted molybdenum–alumina catalysts. All data points obtained at 400°C.

in hydrocracking and hydrodesulphurization with MoO_3 concentration were quite large up to 9% and 3% respectively, but more gradual at higher concentrations. The activity pattern for hydrodenitrogenation was considerably different; the activity increased up to 18% MoO_3 and then decreased.

The change in the hydrocracking activity pattern at 9% MoO_3 may be related to the formation of an epitaxial layer of molybdenum on the alumina surface. Dufaux, Che, and Naccache[16] reported that molybdenum was bonded to γ-alumina forming a complete monolayer at concentrations of 10% MoO_3. At high concentrations of molybdenum in the catalyst, free MoO_3 was formed. Changes in activity patterns with MoO_3-Al_2O_3 catalysts have also been observed at 9–10% MoO_3 concentrations for reactions completely different from the ones considered here.[17,18]

To some degree, hydrodesulphurization and hydrocracking are related. When a sulphur atom is removed from a polycyclic thiophen type compound the number of rings is reduced by at least one and a lower boiling compound is formed. On this basis the hydrocracking activity would be expected to follow the hydrodesulphurization activity pattern. The fact that from 3 to 9 % MoO_3 the extent of hydrocracking increases markedly, while hydrodesulphurization activity does not, suggests that considerable hydrocracking not related to sulphur removal is occurring.

The hydrodenitrogenation reaction was the only one to show a decrease in activity at high catalyst concentrations of MoO_3. Giordano *et al.*[9] found that an $Al_2(MoO_4)_3$ species began to form at MoO_3 concentrations of 20 % and higher. The decrease in hydrodenitrogenation activity shown in Fig. 3 may be related to formation of this compound.

The surface acidity results in Fig. 1 are not directly proportional to any of the reaction results in Fig. 3. This suggests that the reaction rates are not limited by the concentration of acidic sites for all the unpromoted catalysis. Presumably when a sufficient number of acidic sites are present the reaction rate is limited by some step in the reaction process which is not controlled by the number of acidic sites.

The activities of the promoted catalysts are shown in Fig. 4 as a function of increasing atomic number of the promoter cation. The Co and Ni promoters produced the highest activities for all the reactions. The Cr promoted catalyst (calcined at 900 °C) also exhibited a relatively large hydrodenitrogenation activity. These results are somewhat different to those reported by de Beer *et al.*[19] They found that Zn and Co promoted catalysts had the highest hydrodesulphurization activities and that Mn and Ni promoted catalysts had lower but similar activities. Their experiments were performed for shorter periods than ours and their promoted catalysts contained 12 wt % MoO_3 whereas ours contained only 2·2 %. These differences in experimental conditions were probably responsible for the different hydrodesulphurization results. The greater activities we obtained with the Co and Ni promoted catalysts are consistent with the fact that these materials are used extensively as promoters in commercial hydrodesulphurization catalysts.

The twin peak pattern in the hydrodenitrogenation activity shown in Fig. 4 is similar to activity patterns reported for other reactions involving hydrogen. Twin activity peaks have been reported by Dowden, MacKenzie, and Trapnell[20] for hydrogen–deuterium exchange, by Dixon, Nicholls, and Steiner[21] for cyclohexane disproportionation and dehydrogenation, and by Harrison, Nicholls, and Steiner[22] for ethylene hydrogenation. The existence of the twin peak pattern for hydrodenitrogenation activity suggests that the role of the catalyst promoter involves hydrogenation which tends to support the suggestion of Smith.[23]

According to the intercalation model, the catalyst promoter is located between layers of sulphur atoms. For example, Voorhoeve and Stuiver[24] suggested that Ni atoms were inserted between adjacent sulphur layers of WS_2, and that the role of the alumina support was only to disperse the catalyst. Farragher and Cossee[25] found that the increase in benzene hydrogenation activity using tungsten catalysts was proportional to the "goodness of fit" of the promoter in the octahedral holes between the WS_2 layers. Plausible cationic radii of the catalyst promoters (Ti^{4+} = 68 pm, V^{5+} = 59 pm, Cr^{3+} = 63 pm, Cr^{6+} = 52 pm, Mn^{2+} = 80 pm, Fe^{2+} = 74 pm, Fe^{3+} = 64 pm, Co^{2+} = 72 pm, Ni^{2+} = 69 pm, Cu^{2+} = 72 pm, and Zn^{2+} = 74 pm) used in this study do not correlate with the activity patterns shown in Fig. 4. The intercalation model may provide an explanation for systems in which the catalyst does not interact with the support. If the evidence[16] for the interaction of molybdenum with γ-alumina is correct, only one sulphide layer will exist making intercalation impossible. A monolayer model incorporating some of the features proposed by Schuit and Gates,[26] Massoth,[27] or Ueda and Todo[28] might be more appropriate, especially when low concentrations of catalytic materials are used, as was the case in this study.

Finally, the catalytic activities of the Co promoted catalyst shown in Fig. 4 can be compared with a Co promoted commercial catalyst. A pelleted commercial catalyst (Harshaw 0603T) containing 12 wt % MoO_3 and 3 wt % CoO was used to process the same feedstock at the same operating conditions.[4] The percentage conversion of +420°C material, the weight per cent sulphur removed, and the weight percent nitrogen removed were 31, 86, 50, respectively. While the activities of the low-metals catalysts in Fig. 4 are less than those of the commercial catalyst, they do approach it quite

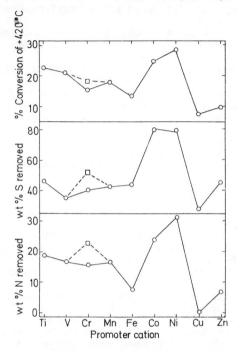

FIGURE 4   Conversion of hydrocarbon boiling above 420°C to lower boiling compounds, weight percent sulphur removed from the liquid hydrocarbon, and weight percent nitrogen removed from the liquid hydrocarbon against the catalyst promoter cation in the molybdenum-alumina catalysts. (○) catalysts calcined at 500°C; (□) catalyst calcined at 900°C. The atomic metal : molybdenum ratio was 1:1 in all the promoted catalysts. All data points obtained at 400°C.

closely. The activities of low-metals catalysts compare just as favourably with that of the commercial catalyst when a heavier feedstock such as Athabasca bitumen is used.[1e] The low-metals catalysts have the advantage that the cost for catalyst ingredients will be significantly less than for the high-metals commercial catalyst. This feature may be of interest when processing heavy residual feedstocks which cause rapid catalyst fouling.

ACKNOWLEDGEMENTS. The authors wish to thank E. C. McColgan, P. S. Soutar, R. W. Taylor, M. J. Whalley, R. J. Williams and G. R. Lett for technical assistance.

REFERENCES

[1]   E. C. McColgan, P. S. Soutar, and B. I. Parsons, *The Hydrocracking of Residual Tars and Oils*, Mines Branch Research Report (*a*) R-246, (*b*) R-253, (*c*) R-256, (*d*) R-261, (*e*) R-263, (*f*) R-273 (Department of Energy, Mines and Resources, Ottawa, 1971–74).

[2] F. Trifiro, P. L. Villa, I. Pasquon, A. Iannibello, and V. Berti, *Chimica e Industria*, 1975, **57**, 173.

[3] W. C. Rovesti and R. H. Work, *Demetallization of Heavy Residual Oils*, EPA-650/2-73-041 (Office of Research and Development, U.S. Environmental Protection Agency, Washington, D.C., 1973).

[4] R. J. Williams, R. G. Draper, and B. I. Parsons, *Catalysts for Hydrocracking and Refining Heavy Oils and Tars, The Effect of Cobalt to Molybdenum Ratio on Desulphurization and Denitrogenation*, Mines Branch Technical Bulletin TB-187 (Department of Energy, Mines and Resources, Ottawa, 1974.)

[5] R. O. Clark, E. V. Ballou, and R. T. Barth, *Analyt. Chim. Acta*, 1960, **23**, 189.

[6] G. Frechette, J. C. Hebert, T. P. Thinh, and Y. A. Miron. *Hydrocarbon Processing*, 1975, **54**, 109.

[7] L. A. Fabbro, L. A. Filachek, R. L. Iannacone, R. T. Moore, R. J. Joyce, Y. Takahashi, and M. E. Riddle, *Analyt. Chem.*, 1971, **43**, 1671.

[8] N. A. C. Smith, H. M. Smith, O. C. Blade, and E. L. Garton, *The Bureau of Mines Routine Method for the Analysis of Crude Petroleum* (U.S. Bureau of Mines, Bulletin No. 490, 1951).

[9] N. Giordano, J. C. Bart, A. Vaghi, A. Castellan, and G. Martinotti, *J. Catalysis*, 1975, **36**, 81.

[10] V. H. J. de Beer, *Some Structural Aspects of the* $CoO–MoO_3–Al_2O_3$ *Hydrodesulphurization Catalyst* (Thesis, Eindhoven University of Technology, 1975).

[11] W. H. Gitzen, *Alumina as a Ceramic Material* (The American Ceramic Society, Columbus, Ohio, 1970), p. 36.

[12] P. Cossee and L. L. Van Reijen, *Actes 2me Congr. Internat. Catalyse* (Editions Technip., Paris, 1961), Vol. 2, p. 1679.

[13] F. A. Cotton and G. Wilkinson, *Advanced Inorganic Chemistry* (Interscience, New York, 3rd edn., 1962), p. 832.

[14] F. E. Kiviat and L. Petrakis, *J. Phys. Chem.*, 1973, **77**, 1232.

[15] P. Ratnasamy, D. K. Sharma, and L. D. Sharma, *J. Phys. Chem.*, 1974, **78**, 2069.

[16] M. Dufaux, M. Che, and C. Naccache, *J. Chim. phys.*, 1970, **67**, 527.

[17] N. Giordano, M. Padovan, A. Vaghi, J. C. J. Bart, and A. Castellan, *J. Catalysis*, 1975, **38**, 1.

[18] K. S. Seshadri and L. Petrakis, *J. Phys. Chem.*, 1970, **74**, 4102.

[19] V. J. H. de Beer, T. H. M. Van Sint Fiet, J. F. Engelen, A. C. Van Haandel, M. W. J. Wolfs, C. H. Amberg, and G. C. A. Schuit, *J. Catalysis*, 1972, **27**, 357.

[20] D. A. Dowden, N. Mackenzie, and B. M. W. Trapnell, *Proc. Roy. Soc. A*, 1956, **237**, 245.

[21] G. M. Dixon, D. Nicholls, and H. Steiner, *Proc. 3rd Internat. Congr. Catalysis*, (North Holland, Amsterdam, 1965), p. 815.

[22] D. L. Harrison, D. Nicholls, and H. Steiner, *J. Catalysis*, 1967, **7**, 359.

[23] G. V. Smith, *J. Catalysis*, 1973, **30**, 218.

[24] R. J. H. Voorhoeve and J. C. M. Stuiver, *J. Catalysis*, 1971, **23**, 243.

[25] A. L. Farragher and P. Cossee, *Proc. 5th Internat. Congr. Catalysis*, (North Holland, Amsterdam, 1973), p. 1301.

[26] G. C. A. Schuit and B. C. Gates, *AIChE J.*, 1973, **19**, 417.

[27] F. E. Massoth, *J. Catalysis*, 1975, **36**, 164.

[28] H. Ueda and N. Todo, *Bull. Chem. Soc. Japan*, 1970, **43**, 3698.

DISCUSSION

G. C. Stevens (*British Petroleum, Sunbury-on-Thames*) said: I have been comparing the conditions you used with those used by de Bruijn. From your paper I have tried to estimate the quantity (bed length/particle diameter) and make it 64; this is very much lower than the values reported by de Bruijn. Would you comment on the possibility of diffusion limitation in your experiments, particularly looking at Figure 3 and the dependence of wt% sulphur removed on wt% MoO_3. What was the particle size distribution of your samples?

M. Ternan (*Energy Res. Laboratories, Ottawa*) replied: To check for possible diffusion limitations, experiments were performed with catalyst particles having different sizes. A relatively active catalyst formulation containing 15 wt% MoO_3 and 3 wt% CoO was used. Experiments were performed with the same feedstock and at the same conditions described in our paper. The same sulphur removal, 95·3%, was obtained with catalyst particles having diameters of 3·18 and 1·59 mm. The fact that identical results were obtained with decreasing catalyst particle size indicates that neither pore diffusion nor diffusion from the bulk to the catalyst surface was the rate-controlling step. The fact that diffusion was not controlling for the more active catalysts described above indicates that it was also not controlling for the less active catalysts described in Figure 3.

De Bruijn, in his paper emphasized that channelling and backmixing can influence results obtained in bench-scale fixed-bed trickle-flow reactors. In our studies a bottom-feed reactor was used. This eliminated the channelling problems which sometimes occur in top-feed trickle-flow reactors.

De Bruijn also noted that Mears has developed a criterion for the absence of dispersion effects (backmixing) in trickle-bed reactors.[1] Mears states that significant backmixing will be absent if the ratio of catalyst bed-length to catalyst particle diameter exceeded 350 for a liquid phase Reynolds number of 8. Mears also noted that his criterion was 20 times more conservative than an alternative criterion developed by Petersen.[2] Therefore, significant backmixing effects in our reactor would be absent according to one criterion[2] but present according to the other criterion.[1] If backmixing were present in our reactor it would only involve mixing in the bulk liquid. This phenomenon would have been exactly the same in each experiment. Therefore, it would have no effect on the relative catalyst activities.

De Bruijn has reported pore diffusion limitations whereas we have not. It is possible that this difference can be explained in terms of the experimental conditions employed. Different catalysts were used. Our reactor pressure ($13·9 \times 10^6$ N m^{-2}) was higher than de Bruijn's ($4·1 \times 10^6$ N m^{-2}). We used a bottom-feed reactor configuration; de Bruijn used top-feed. This could have caused the extent and effectiveness of catalyst wetting to be different in the two studies. One could speculate that some of the effects attributed to pore diffusion by de Bruijn could have been related to changes in catalyst wetting. For example, Satterfield and Ozel[3] report non-uniform catalyst wetting at Reynolds numbers as high as 55. In general, studies in bench-scale trickle-bed reactors are performed at much lower Reynolds numbers. In contrast, in bottom-feed reactors the catalyst particles are constantly surrounded by liquid, regardless of the Reynolds number.

We did not make pore size distribution measurements on all the catalysts described in our paper. However, the catalysts on which mercury porosimetry measurements were made all had similar pore size distributions. The penetration curves indicated that the micropores had typical diameters of approximately 7 nm.

J. Bousquet (*Elf-Erap, Solaize*) asked: Are you sure that the up-flow reactor that you use for this study is the best one considering that you have tested cracked products? The selectivity of the reaction could depend critically on residence time distribution.

Is your industrial catalyst the best for your study? It seems to me that the industrial process for hydrotreating this gas oil might be more efficient with a Ni–Mo catalyst than with a Co–Mo catalyst.

The observation that a catalyst having a low metal content gives the same result as an industrial catalyst is very interesting. The influence of the support, of the impregna-

[1] D. E. Mears, *Chem. Eng. Sci.*, 1971, **26**, 1361.
[2] E. E. Petersen, *Chemical Reaction Analysis* (Prentice-Hall, Engelwood Cliffs, 1965), pp. 198 and 223.
[3] C. N. Satterfield and F. Ozel, *A.I.Ch.E. Journal*, 1973, **19**, 1259.

tion method, and of the activation procedure, are also very important. Thus, by screening industrial catalysts one finds that different catalysts having the same metal (high) content can have activity levels which differ by a factor of three.

M. Ternan (*Energy Res. Laboratories, Ottawa*) replied: In making catalyst activity measurements it is essential that the type of the reactor used does not cause mass transfer limitations. Our measurements indicate that the results presented in this study were not controlled by mass transfer. (See my response to Dr. Stevens.) While we do not pretend to have optimized commercial-scale reactor configuration, it has been shown elsewhere[4] that, for laboratory studies where both liquid and vapour are present under reaction conditions, higher conversions are obtained in up-flow reactors than in down-flow reactors. I agree that residence time distribution affects selectivity; however, it is reasonable to compare catalyst activities if the experiments are performed at the same residence time.

We have performed catalyst activity measurements with several industrial catalysts including $Co-Mo-Al_2O_3$, $Ni-Mo-Al_2O_3$ and nickel–tungsten formulations. We do not claim that any of them are optimum for our feedstocks. The industrial catalyst described in our papers does not differ greatly from others we have studied.

C. J. M. Brew (*Texaco, Ghent*) (communicated): The catalysts were prepared by impregnating the base with the promoting ion first and the molybdenum second. This could account for the differences between Dr. Ternan's results and those reported by de Beer *et al.*[5] Impregnation of the promoter ion first encourages the formation of a bulk aluminate, which is inactive as a promoter, and this also means that the molybdenum will interact more weakly with the base. We consider that the Al–O–Mo linkage is essential for an active catalyst.

The importance of the order of impregnation is shown in the Figure. Two cobalt molybdate catalysts were prepared in the same manner except that the order of impregnation was reversed. The order is given by the order of the symbols. When the two catalysts were reduced with hydrogen, they had essentially identical activity patterns: high initial activity falling off rapidly due to coking. However, with the sulphided catalysts, Mo–Co was much more active than Co–Mo and this difference was maintained. Thus the discrepancy between Dr. Ternan's results and those of de Beer *et al.*,[5] concerning the effect of various promoter ions on activity, are possibly due to the different order of impregnation.

M. Ternan (*Energy Res. Laboratories, Ottawa*) replied: Experiments in which the order of impregnation with molybdenum and cobalt was varied, have been performed in our laboratory.[6] After the catalysts had been in the presence of the reaction mixture for more than 20 h any differences in the activities were within experimental error. It would be interesting to know whether or not the large difference in thiophen conversion obtained by Dr. Brew would have persisted if the sulphided catalysts had been exposed to the reaction mixture for longer periods of time.

In my opinion, the differences between our results and those reported by de Beer *et al.*[7] are not due to the order of impregnation. For example, they initially reported

[4] T. Takematsu and B. I. Parsons, *A Comparison of Bottom Feed and Top Feed Reaction Systems for Hydrodesulphurization*, Mines Branch Technical Bulletin TB 161 (Department of Energy, Mines & Resources, Ottawa, 1972).

[5] V. H. J. de Beer, T. H. M. van Sint Fiet, J. F. Engelen, A. C. van Haandel, M. W. J. Wolfs, C. H. Amberg, and G. C. A. Schuit, *J. Catalysis*, 1972, **27**, 357.

[6] E. C. McColgan, P. S. Soutar, M. A. Rethier, and B. I. Parsons, *The Hydrocracking of Residual Oils and Tars, Part 5: Surface Coated Cobalt Molybdate Catalysts for Hydrotreating*, Mines Branch Research Report R263 (Dept. of Energy, Mines and Resources, Ottawa 1973).

[7] V. H. J. de Beer, T. H. M. van Sint Fiet, J. F. Engelen, A. C. van Haandel, M. W. J. Wolfs, C. H. Amberg, and G. C. A. Schuit, *J. Catalysis*, 1972, **27**, 357.

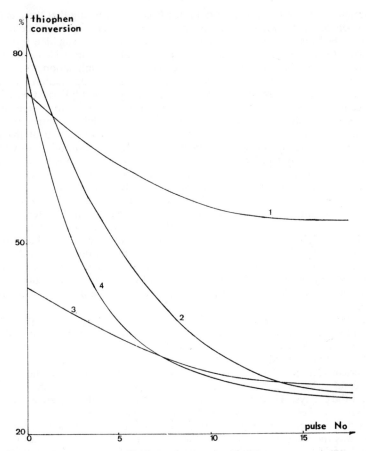

FIGURE Activity of sulphided and non-sulphided laboratory catalyst *versus* pulse number. 1, Mo–Co sulphided; 2, Mo–Co non-sulphided; 3, Co–Mo sulphided; 4, Co–Mo non-sulphided.

that zinc-promoted Mo–Al$_2$O$_3$ catalysts were more active than either cobalt- or nickel-promoted Mo–Al$_2$O$_3$. A later study by de Beer *et al.*[8] showed that the activity of zinc-promoted Mo–Al$_2$O$_3$ decreased rapidly. After eight hours exposure to the reaction mixture their zinc-promoted Mo–Al$_2$O$_3$ was only slightly more active than the corresponding unpromoted Mo–Al$_2$O$_3$ catalyst. The data from the second study by de Beer *et al.*[8] are consistent with our results.[9] The catalysts used in their initial study[7] were apparently exposed to the reaction mixture for only a relatively short period of time, and as a result, they may not have been in a steady state condition. I suspect that the differences between the results we have reported here and those initially reported by de Beer *et al.* are related to the steady-state working conditions of the catalysts rather than the order of impregnation.

[8] V. H. J. de Beer, T. H. M. van Sint Fiet, G. H. A. M. van der Steen, A. C. Zwaga, and G. C. A. Schuit, *J. Catalysis*, 1974, **35**, 297.
[9] M. Ternan, R. J. Williams, and B. I. Parsons, *J. Catalysis*, 1976, **42**, 177.

P. Grange (*Catholic Univ. of Louvain*) said: It is not surprising that there is no relationship between activity and surface acidity of the precursor in the oxidic form. It is well established that, in the steady-state, the catalyst is in the sulphide form and we have shown (*e.g.* by ESCA) that the monolayer of molybdenum oxide on the alumina carrier is segregated into free molybdenum disulphide by sulphidation.

M. Ternon (*Energy Res. Laboratories, Ottawa*) replied: Sulphidation studies on a Ni–Mo–Al$_2$O$_3$ catalyst used for hydrocracking and hydrodesulphurization of gas oil were recently performed in our laboratory,[10,11] It was essential to prevent air from contacting the catalyst if accurate analyses of catalyst sulphur content were to be obtained. The catalyst sulphur content was found to be a function of the method employed to remove hydrocarbon oil (a portion of the reaction mixture) which was adhering to the catalyst pellets at the conclusion of the reaction experiments. When hydrogen distillation was used[11] the catalyst sulphur content corresponded closely to the stoicheiometric amount required for MoS$_2$ and Ni$_3$S$_2$. When the adhering oil was removed by benzene extraction[10] sulphur contents considerably in excess of the stoicheiometric amount were found. In contrast some extremely interesting X-ray photoelectron spectroscopy by Patterson *et al.*[12] shows that, under certain conditions, much of the molybdenum on the surface is attached to only one sulphur atom. One explanation which might reconcile the above observations would be that the environment of the catalyst (hydrogen, benzene, or vacuum) could affect its sulphur content. The same reasoning suggests that truly meaningful surface acidity measurements must be performed at reaction conditions with the catalyst in the presence of the reaction mixture.

J. P. Franck (*Inst. Franç. du Petrole, Rueil Malmaison*) asked: What happens to catalyst life when the content of the active species, for example, cobalt and molybdenum oxides, is decreased from about 15 wt% to about 3 wt%? Have you measured carbon deposition on used catalysts as a function of the concentration of active species?

M. Ternan (*Energy Res. Laboratories, Ottawa*) replied: We have not performed enough catalyst life studies to determine the effect caused by concentration of active species in the catalyst. However, from a limited study the following statements can be made. When the heavy gas oil was used as feedstock, the rate of catalyst deactivation was very slow.[13] In contrast, extremely rapid catalyst fouling occurred[14] when Athabasca bitumen was the feedstock, even when the concentration of active species was 15%.

Coke deposition on our catalysts has been measured as a function of active species content with gas oil used as feedstock. The coke decreased from 8 to 3 wt% as the concentration of MoO$_3$ increased from 0 to 3 wt%. Used catalysts having higher concentrations of MoO$_3$ also contained 3 wt% coke. When the promoter cation was varied, the Ni-, Co- and Cr(3+)-promoted catalysts contained the smallest amounts of coke.

H. Topsøe (*Haldor Topsøe, Lyngby*) said: Several questions in this discussion have

[10] M. Ternan and M. J. Whalley, *Catalysts for Hydrocracking and Refining Heavy Oils and Tars, Part 3: The Effect of Presulphiding Conditions on Catalyst Performance*, CANMET Report C-76-9 (Dept. of Energy, Mines & Resources, Ottawa, 1976).

[11] M. Ternan and M. J. Whalley, *Canad. J. Chem. Eng.*, in press.

[12] T. A. Patterson, J. C. Carver, D. E. Leyden, and D. M. Hercules, *J. Phys. Chem.*, in press.

[13] R. J. Williams, M. Ternan, and B. I. Parsons, *Catalysts for Hydrocracking and Refining Heavy Oils and Tars, Part 2: The Effects of Molybdenum Concentration and of Zinc to Molybdenum Ratio on Desulphurization and Denitrogenation*, Energy Research Laboratories Report 75–85 (R) (Dept. of Energy, Mines & Resources, Ottawa, 1975).

[14] E. C. McColgan and B. I. Parsons, *The Hydrocracking of Residual Oils and Tars, Part 6: Catalyst Deactivation by Coke and Metals Deposition*, Mines Branch Research Report R273 (Dept. of Energy, Mines & Resources, Ottawa, 1974).

[15] E. Furimsky, M. Ternan, and B. I. Parsons, *Some Aspects of Coke Formation on Promoted and Unpromoted Molybdenum–Alumina Catalysts*, Energy Research Laboratories Research Report ERP/ERL 76–65 (J) (Dept. of Energy, Mines & Resources, Ottawa, 1976).

dealt with the chemical state of cobalt. The answers to many of these questions can be obtained by use of Mössbauer spectroscopy.[16] Recently we have performed such experiments on several catalysts and model systems.[17] From these experiments information (oxidation state, low spin–high spin, symmetry of surroundings, *etc.*) was obtained concerning cobalt in the calcined state as well as the effects of different reaction conditions such as sulphidation. The technique allows the relative amounts of cobalt in different surroundings to be determined. For example it is possible to determine *in situ* the fraction of cobalt atoms at the surface.

[16] J. A. Dumesic and H. Topsøe, *Adv. Catalysis*, **26**, in press.
[17] H. Topsøe and S. Mørup, *Proc. Internat. Conf. Mössbauer Spectroscopy*, Krakow, 1975, p. 305; B. S. Lausen, H. Topsøe, S. Mørup, and R. Candia, *J. Phys.* (*Paris*), in press.

A Simultaneous Infrared and Kinetic Study of the Reduction of Nitric Oxide by Carbon Monoxide over a Chromia–Silica Catalyst

STUART S. SHIH, DAVID S. SHIHABI, and ROBERT G. SQUIRES*

School of Chemical Engineering, Purdue University, West Lafayette, Indiana 47907, U.S.A.

ABSTRACT   Infrared spectroscopy has been used to observe the surface species on a chromia-silica catalyst during the reduction of NO by carbon monoxide. Spectra taken during the reaction (250—300°C) show that the catalyst surface is predominantly covered with chemisorbed NO species. Stable carbonates are also observed; however, chemisorbed carbon monoxide and nitrous oxide are not observed at reaction temperature.

The catalyst activity can be correlated with the concentration of an active surface species of NO having an infrared absorption band at 1735 cm^{-1}. The rate expression of carbon dioxide formation was found to be:

$$r_{CO_2} = k_2 P_{CO} \log (I_0/I)_{1735\ cm^{-1}}$$

Nitric oxide was reduced to N_2O and small amounts of N_2.

A reaction model is suggested which accurately represents the rate of reaction and selectivity of the products.

INTRODUCTION

The kinetics of the reduction of nitric oxide by carbon monoxide and the simultaneous measurement of the infrared spectra of the adsorbed species were studied in order to gain better understanding of the selectivity of chromia toward N_2O formation. The information obtained from this approach was used in putting together a consistent reaction model, identifying the kinetically active surface species, and elucidating the active site character and poison effects.

EXPERIMENTAL

The infrared spectra of the adsorbed species and the reaction kinetics were simultaneously measured at reaction conditions in a differential plug flow recycle reactor in which the reacting gas (50 cm³/min) was recycled (at 4000 cm³/min) by a stainless-steel bellows pump, assuring differential gradientless conditions in the reactor.

The reactor consisted of two 2·9 cm i.d., 10 cm long stainless-steel infrared cells (with Irtran II windows) connected in series. A catalyst disc was placed in the sample cell and a silica reference disc in the reference cell. The i.r. cells were maintained within 250 ± 1°C by externally wound Nichrome heaters.

Oxygen-free helium and carbon monoxide were obtained by passing the gases over copper turnings at 350—400°C. In addition, carbon monoxide was passed through ascarite and molecular sieve columns to remove carbon dioxide and water vapour. The carbon dioxide and water vapour impurities in the nitric oxide gas were removed by passing the gas through ascarite and drierite desiccants. Before entering the reactor, the mixture of reactants (NO and CO) and helium, flowed through a Dry Ice–acetone cold trap to remove water vapour and higher-molecular-weight oxides of nitrogen. Ultrahigh purity helium, purified in a diffusion cell, was used during catalyst activation runs.

A gas chromatograph, using two 3·3 m columns in series packed with Porapak Q was used to separate the gaseous products. The first column, operated at room temperature, separated CO_2 and N_2O. N_2, CO, and NO were separated in the second column which was immersed in a Dry Ice–acetone bath.

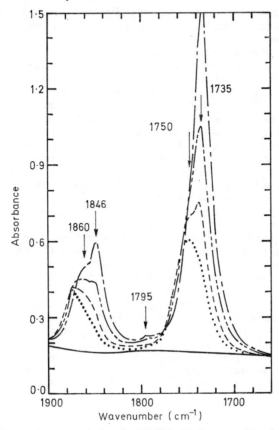

FIGURE 1   Ir spectra during the NO/CO reaction over chromia–silica.
——————   Background in He.
— · —   1·5 h of reaction.
— ·· —   14·5 h of reaction.
- - - - - -   46 h of reaction.
· · · · · ·   100 h of reaction.
All recorded at 250°C, $P_{CO} = 2·46 \times 10^4$ Nm^{-2}, $P_{NO} = 1·22 \times 10^4$ Nm^{-2}.

The catalyst used in the present study was prepared by impregnation of silica powder, Cab-o-Sil M5, with chromic acid in such proportion that the final product contained 10% chromium. This catalyst was dried in an oven at 250°C for 24 h. 0·070 g of catalyst powder was then pressed at $8·27 \times 10^7$ Nm^{-2} to form a 22 mm diameter, 0·1—0·2 mm thick self supporting catalyst disc. The disc was then calcined in air in an oven at 500°C for 12 h. Similarly, a silica reference disc with no chromia was prepared and placed in the reference cell. Prior to the reaction the catalyst was activated in helium at 330°C for 5 h and at 250°C for 3 h. The background spectrum was recorded at 250°C in helium and the gas mixture was then introduced. The partial pressure ranges of the reactants are shown in Figure 3. The kinetics and i.r. spectra were taken after at least 1 h of reaction. The i.r. spectral region from 2500 to 1300 cm^{-1} was recorded at reaction conditions on a Perkin-Elmer Model 180 Spectrometer having a resolution of better than 3 cm^{-1}.

RESULTS

The experimental results are summarized in Figures 1, 2, and 3. The solid lines in Figures 2 and 3 are curves predicted by our suggested reaction model, which is discussed in the next section.

Figure 1 shows the spectra taken during reaction at 250°C. The fresh catalyst, calcined at 500°C, shows a high activity and gives rise to two strong i.r. bands at 1735 and 1846 cm^{-1} under the reaction conditions. The activity decreased during the course of the experiments. At the beginning the deactivation was more rapid, becoming slower after several hours. During this deactivation, changes in i.r. spectra were noted.

A linear correlation between the band intensity of the 1735 cm^{-1} band and the rate of CO_2 formation has been found (see Figure 2). This correlation holds for a wide range of partial pressures of CO and NO. The 1735 cm^{-1} band in Figure 1 includes a shoulder at 1750 cm^{-1}, which becomes more prominent as the intensity of the 1735 cm^{-1} band decreased. Similarly and more clearly, the fresh sample has a shoulder at 1860 cm^{-1} associated with the 1846 cm^{-1} band. As the sample deactivates, the 1846 cm^{-1} band becomes the shoulder of the 1860 cm^{-1} band as shown in Figure 1.

Integration of the band areas (from absorbance spectra) of these two bands, the 1735 and 1846 cm^{-1} bands, yields a relatively constant ratio of 0·35—0·40. During the reaction infrared bands were observed in the region 2000—1200 cm^{-1}: a weak band at 1790 cm^{-1}, a strong band at 1550 cm^{-1} and two weak bands at 1430 cm^{-1} and 1365 cm^{-1}. The intensity of the 1550 cm^{-1} band slowly increased and seemed to increase in width. A slow increase in intensity of both the 1430 and 1365 cm^{-1} bands was found.

In addition to CO_2, the NO/CO reaction produces N_2O and smaller amounts of nitrogen. In Figure 3 the selectivity of N_2O (defined as $r_{N_2O}/ - r_{NO}$) is plotted against the partial pressure of NO at fixed values of CO partial pressure and chemisorbed NO concentration [$\log (I_0/I)$]. The selectivity of N_2O is dependent on the partial pressure of CO as well as NO, and surface concentration of chemisorbed NO as indicated by the intensity of the band at 1735 cm^{-1}. In general the selectivity decreases as either partial pressure of CO or the surface concentration of chemisorbed NO increases. However, a higher partial pressure of NO leads to a higher selectivity of N_2O if other variables are fixed.

DISCUSSION

Catalyst surface behaviour under reaction conditions.

During the reaction the catalyst surface is predominantly covered by nitric oxide. No chemisorbed CO or N_2O was observed. The observed catalyst deactivation is accompanied by a decrease in the intensity of the chemisorbed NO bands and an increase in the intensity of carbonate bands. After several hours of reaction the following bands are observed in the spectra:

1735 cm^{-1}	strong	chemisorbed NO (dinitrosyl ligands)
1846 cm^{-1}	strong	
1750 cm^{-1}	shoulder	chemisorbed NO species catalytically inactive for
1860 cm^{-1}	shoulder	NO/CO reaction
1795 cm^{-1}	weak	chemisorbed NO (mononitrosyl ligand)
1550 cm^{-1}	medium	bidentate carbonate
1365 cm^{-1}	weak	
1430 cm^{-1}	weak	unco-ordinated carbonate
1390 cm^{-1}	very weak	monodentate carbonate
1500 cm^{-1}	very weak	

980

As deactivation continues the band at 1550 cm^{-1} increases in width. A broad band at 1620 cm^{-1} and a weak band at 1510 cm^{-1} are observed. It is difficult to assign these two bands, especially the band at 1620 cm^{-1}. Some species such as adsorbed water, chemisorbed NO_2, and bicarbonate may give rise to a band near 1620 cm^{-1}. Perhaps, the 1620 cm^{-1} is a combination of two or all three of these species (*i.e.* the shape of this band

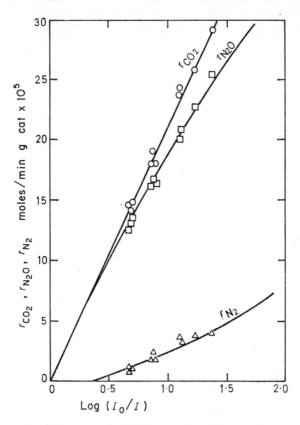

FIGURE 2   Rates of CO_2, N_2O, and N_2 formation against log (I_0/I).
$P_{CO} = 3\cdot26 \times 10^4$ Nm^{-2}, $P_{NO} = 1\cdot2 \times 10^4$ Nm^{-2}.
○   Rate of CO_2 formation,
□   Rate of N_2O formation,
△   Rate of N_2 formation,
—   Calculated.

is not well-defined). However, chemisorbed water is likely to be the candidate, since both the chemisorbed NO_2 and bicarbonate are unstable at 250°C. The 1510 cm^{-1} band can be ascribed to either the monodentate carbonate or the monodentate nitrato-complex.

The chemisorption of NO is tentatively assigned as dinitrosyl complex on the surface. Other interpretations of i.r. bands of chemisorbed NO have been found in the literature. Chemically, chemisorbed NO can be ascribed as follows based upon Roev and Alekseev's classifications:[1]

1846 cm^{-1} double-bond ionic bonding
1795 cm^{-1} co-ordinative bonding
1735 cm^{-1} covalent bonding

Since a constant band area ratio of the 1846 cm^{-1} band to 1735 cm^{-1} band (*ca.* 0·35—0·40) was found in the current study, it indicates that both bands can probably be attributed to the same species. Thus, this assignment seems tenuous. Lunsford and Windhorst[2] observed i.r. bands ascribed to a dinitrosyl complex at 1910 and 1830 cm^{-1} after adsorption of NO on cobalt-exchanged Y zeolites. Chromium dinitrosyl complexes

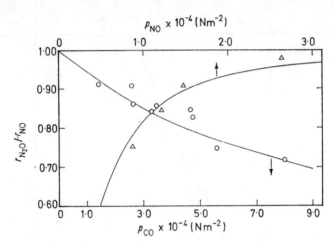

FIGURE 3   Selectivity as a function of P_{CO} and P_{NO}.

○ $r_{-N_2O} - r_{NO}$ against P_{CO}
   $P_{NO} = 1·13 \times 10^4$ Nm^{-2}, log $(I_0/I) = 1·0$, 250°C.

△ $r_{N_2O} - r_{NO}$ against P_{NO}
   $P_{CO} = 3·07 \times 10^4$, log $(I_0/I) = 1·0$, 250°C.

are well known.[3] The vibrational frequencies of dinitrosyl complexes are slightly lower than the frequencies observed in this study. However, the separation of two vibrational frequencies is 100 cm^{-1} in the current study, which is very close to the value for chromium dinitrosyl complexes (about 105 cm^{-1}).

The weak band at 1795 cm^{-1} could be attributed to a single NO molecule co-ordinatively bonded to a surface chromium site. The shoulders at 1750 and 1860 cm^{-1} are attributed to catalytically inactive dinitrosyl species. The sites for the 1735 and 1846 cm^{-1} are energetically different from the sites for the shoulders at 1750 and 1860 cm^{-1}. The difference may be associated with the different oxidation states between the two sites. Eley *et al.*[4] have studied the NO chemisorption on partially reduced chromium oxide catalysts and have observed two bands at 1740 and 1860 cm^{-1}. They also have reported that the 1740 cm^{-1} band is a combination of separate bands at 1743 and 1738 cm^{-1} due to NO covalently bound to Cr^{3+} and Cr^{5+} sites, respectively. The band at 1860 cm^{-1} is similarly ascribed to NO chemisorbed on two types of sites.

Recently Kugler *et al.*[5,6] observed i.r. bands, ascribed to a *cis*-dimer structure, at 1875 and 1745 cm^{-1} following NO adsorption on reduced chromia–silica samples at —78°C. It is difficult to differentiate between a dinitrosyl complex structure or dimer structure on the basis of our infrared evidence alone. Further isotopic studies would be useful.

Reaction model

A reaction model which is consistent with infrared and kinetic analyses can be represented by the following four steps:

$$[*] + NO \xrightarrow{k_1} [*-NO] \tag{1}$$

$$[*-NO] + CO \xrightarrow{k_2} [*-N] + CO_2 \tag{2}$$

$$[*-N] + NO \xrightarrow{k_3} [*] + N_2O \tag{3}$$

$$2[*-N] \xrightarrow{k_4} 2[*] + N_2 \tag{4}$$

where [*] is active site; [*–NO] and [*–N] are chemisorbed NO and N on the surface, respectively. The concentration of chemisorbed NO,[*–NO], the rate-determining active centre, is proportional to the band intensity at 1735 cm^{-1}:

$$[*-NO] = C \log (I_0/I) \tag{5}$$

where C is the proportionality constant. Assuming the rate of formation of [*–N] in step (2) is equal to the rate of consumption of [*–N] in step (3) and step (4); *i.e.*

$$k_2 [*-NO] P_{CO} = k_3 P_{NO} [*-N] + k_4 [*-N]^2 \tag{6}$$

where P_i is the partial pressure of the ith component, the surface concentration of [*–N] can be calculated by Equation (7):

$$[*-N] = \frac{-k_3 P_{NO} + \sqrt{k_3^2 P_{NO}^2 + 4k_2k_4 [*-NO] P_{CO}}}{2k_4} \tag{7}$$

The rate of CO_2, N_2O, and N_2 formation can be written as:

$$r_{CO_2} = k_2' P_{CO} \log (I_0/I) \tag{8}$$

$$r_{N_2O} = \frac{k_3^2 P_{NO}^2}{2k_4} \left(-1 + \sqrt{1 + \frac{4k_2'k_4 P_{CO}}{k_3^2 P_{NO}^2} \log (I_0/I)} \right) \tag{9}$$

$$r_{N_2} = \frac{k_3^2 P_{NO}^2}{4k_4} \left(-1 + \sqrt{1 + \frac{4k_2'k_4 P_{CO}}{k_3^2 P_{NO}^2} \log (I_0/I)} \right)^2 \tag{10}$$

where k_2' is the product of k_2 and C. From Equation (8) and either Equation (9) or Equation (10), k_2' and k_3^2/k_4 can be calculated. Values of k_2' and k_3^2/k_4 at 250°C are shown below:

k_2'	k_3^2/k_4
$1.27 \times 10^{-5} \pm 3\%$	$1.91 \times 10^{-6} \pm 7.8\%$

Based on these values of k_2 and k_3^2/k_4, kinetic data were simulated and plotted with the observed data as shown in Figures 2 and 3. Good agreement was obtained.

Other, more complex reaction sequences were considered which led to models with additional parameters. The added complexity did not seem justified, based on the excellent fit of the two parameter model to the data shown in Figures 2 and 3. In particular, steps such as:

$$\text{Cr} \overset{O-N}{\underset{O-N}{\big\|}} + CO \xrightarrow{k_5} \text{Cr}\diagdown_{ON_2} + CO_2 \quad (11)$$

$$\text{Cr}\diagdown_{ON_2} \xrightarrow{k_6} \text{Cr} + N_2O \quad (12)$$

$$\text{Cr}\diagdown_{ON_2} \xrightarrow{k_7} \text{Cr}\diagdown_{O} + N_2 \quad (13)$$

$$\text{Cr}\diagdown_{O} + CO \xrightarrow{k_8} \text{Cr} + CO_2 \quad (14)$$

which might be postulated if the NO was adsorbed as a dimer, or if there was an interaction between two adsorbed NO groups in an adsorbed dinitrosyl complex, were considered. The distribution of N_2O and N_2 predicted by this sequence of steps is independent of the partial pressure of CO and NO, and thus does not agree with the kinetic data shown in Figure 3.

If, as indicated by the infrared evidence, the predominate adsorbed NO species is a dinitrosyl complex, then both the i.r. and kinetic data can be explained if the surface species represented by [*] in Equations (1)*(4) is —Cr—NO.

Catalyst deactivation

Our data clearly show that the catalyst deactivation is directly related to the reduced surface concentration of the 1735 cm^{-1} species. This effect may be due to some or all of the following: (i) competitive adsorption of CO complexes (*i.e.*, carbonates), water, or oxygen; (ii) change of surface oxidation state; (iii) sintering.

The formation of irreversibly adsorbed CO (carbonates corresponding to i.r. bands at $1550, 1430$, and 1365 cm^{-1}) most probably contribute to the deactivation of the catalyst. Efforts have been made to correlate the catalyst activity with the intensity of 1550 cm^{-1} band. Quantitative correlations are not obtained. Qualitatively, the deactivation is accompanied by the increase in the intensity of the 1550 cm^{-1} band. Treatment of the sample with oxygen at low temperature (at $250°C$) removes the bands associated with carbonates species, but the activity is not fully restored. Water is a strong poison for the NO/CO reaction over chromia/silica catalysts. It is strongly adsorbed on the surface. Kobylinski and Taylor[7] have reported that adsorbed water can be removed only by activation above $500°C$.

Sintering decreases the dispersion of catalysts and decreases the active site density. In the present study the samples have been calcined at $500°C$ for 12 h. The degree of sintering is expected to be small at the reaction temperature of $250°C$.[8]

ACKNOWLEDGEMENT. This research was supported jointly on the ARPA-IDL Program Grant DAHC-0213, and the ARPA-MATS Grant DA1573611.

REFERENCES

[1] L. M. Roev and A. V. Alekseev, *Elementary Photoprocesses in Molecules*, B. S. Neporent Ed. (English Translation, Consultants Bureau, N.Y., 1958) p. 250.
[2] K. A. Windhorst and J. H. Lunsford, *J. Amer. Chem. Soc.*, 1975, **97**, 1407.
[3] W. P. Griffith, *Adv. Organometallic Chem.*, 1968, **7**, 211.
[4] D. D. Eley, C. H. Rochester, and M. S. Scurrell, *JCS Faraday I*, 1973, **69**, 660.
[5] E. L. Kugler, R. J. Kokes, and J. W. Gryder, *J. Catalysis*, 1975, **36**, 142.
[6] E. L. Kugler and J. W. Gryder, *J. Catalysis*, 1975, **36**, 152.
[7] T. P. Kobylinski and B. W. Taylor, *J. Catalysis*, 1973, **31**, 450.
[8] M. Shelef, K. Otto, and H. Gandhi, *J. Catalysis*, 1968, **12**, 361.

DISCUSSION

E. L. Kugler (*Stanford Univ.*) said: Work by Kugler and Gryder (to be published) has shown that the reaction of nitrous oxide with carbon monoxide over chromia is very rapid at room temperature. Therefore, I would expect that this reaction would also be important in your system at 523 K. Do you have evidence that N_2O is not an intermediate in the formation of molecular nitrogen?

Secondly, Kobylinski and Taylor[1] have used a similar chromia catalyst for the reduction of nitric oxide with hydrogen. These authors reported that N_2O and N_2 were the only nitrogen-containing products present and that ammonia was not formed. Chromia was unique among the transition-metal oxides studied in this regard. Is your mechanism consistent with these observations?

R. G. Squires (*Purdue Univ., West Lafayette*) replied: We did not study the reaction of nitrous oxide with carbon monoxide. Our only indirect evidence is that the process you mention does not fit our overall kinetic results.

With regard to your second point, we did not have molecular hydrogen in our reacting system, and hence I cannot be sure what products we would have found. Our model does presume an adsorbed nitrogen species that could possibly be unique to chromia. However, we have no evidence that it is unique.

A. L. Dent (*Carnegie-Mellon Univ., Pittsburgh*) (communicated): My questions concern the nature of the two i.r. bands reported in your paper and their relationship to possible reaction intermediates in the NO–CO reaction system. First, have you any results which demonstrate the transient behaviour of these two bands and the manner in which they behave in the absence of one or both of the reactants? If the bands are assignable to an NO-complex, do they appear in the absence of carbon monoxide? I raise these questions because of several reports (*e.g.* Kokes and Dent) of the transient behaviour of a CH (or CD) band attributable to an ethyl radical species during ethylene hydrogenation over zinc oxide, whereas Sheppard and co-workers have reported that for this same reaction on alumina, similar bands appeared to be too stable to be attributed to a reactive species.

Secondly, would you indicate the method of integration of the two bands; were the shoulders included as part of the band envelope or were peak heights used?

Finally, what is the basis for your speculation that the reduction in intensity of the band at 1735 cm^{-1} might be due to water or oxygen? Deliberate introduction of these poisons might serve to clarify the situation.

R. G. Squire (*Purdue Univ., West Lafayette*) replied: We did not specifically study the transient behaviour of these peaks. They are, however, relatively stable in the absence of carbon monoxide under reaction conditions. The fact that a species may be relatively stable *in vacuo* (or under some other condition) does not preclude its being a catalytically active entity when exposed to a proper reactant under appropriate conditions.

[1] T. P. Kobylinski and B. W. Taylor, *J. Catalysis*, 1973, **31**, 450.

With regard to your second point, we tried to eliminate the contribution of the shoulder. For this reason we focused our attention on the 1735 cm^{-1} band where the effect of the shoulder was smaller. I agree with your final suggestion. Such experiments will be executed in the future.

A. Zecchina (*Univ. of Turin*) said: In our laboratory we have obtained[2] spectra of the NO–O$_2$ interaction which are very similar to those shown in Figure 1 of your paper. Moreover, the same results are obtained when a reduced catalyst is contacted with nitrogen dioxide which acts as an oxidizing agent. So it seems reasonable that the observed deactivation is due to a progressive surface oxidation caused either by oxygen or by NO$_2$, or both.

The bands at 1845 and 1735 cm^{-1} can be definitely assigned to a dinitrosyl complex as our work on this system has shown (see Figure). The small frequency difference between your bands and ours can be explained in terms of different operating conditions such as temperature and concentration. The two NO-groups are strongly coupled and interacting. This is demonstrated by the following results. On chromium ions of sufficiently low co-ordination number NO-groups always enter in pairs and no mono-nitrosyl complexes are observed. The central band at 1795 cm^{-1} is due to a mono-nitrosyl complex involving more highly co-ordinated ions. Furthermore, every perturbation caused by extra ligand insertion affects the two frequencies simultaneously. It is not possible to perturb a single NO-group without affecting the properties of the other one. I wish to ask whether it is possible to reconcile your kinetic mechanisms with our results?

R. G. Squires (*Purdue Univ., West Lafayette*) replied: I see no inconsistency between your results and ours. We conclude that if the predominantly adsorbed NO entity is a dinitrosyl complex, then both i.r. and kinetic data can be explained.

H. Arai (*Univ. of Tokyo*) said: The author says that adsorbed carbon monoxide was not detected on his chromia catalyst during reaction; this must mean that it is adsorbed very weakly or not at all.

We have adsorbed NO–CO mixtures on rhodium–alumina and have observed that adsorbed–N$_2$O is formed when nitric oxide was in excess in the gas phase, and that adsorbed–NCO is formed when carbon monoxide was in excess. Such a test might distinguish slight adsorption from no adsorption in your system.

R. G. Squires (*Purdue Univ., West Lafayette*) replied: We could not observe adsorbed carbon monoxide under reaction conditions. However, a strong adsorption band attributable to adsorbed carbon monoxide appears when the temperature is reduced from reaction conditions (where $\rho_{CO} = 2 \cdot 7 \times 10^4$ N m^{-2}) to room temperature.

T. G. Alkhazov (*Petroleum and Chem. Inst., Baku*) (communicated): Recently we have studied the reduction of NO by carbon monoxide using unsupported chromium oxide as catalyst. We observed reduction to N$_2$O and N$_2$; the N$_2$:N$_2$O ratio increased with temperature. However, activity of the catalyst did not change during a run. Results of our kinetic and adsorption study enable us to conclude the catalyst surface in the steady-state is reduced and covered by adsorbed carbon monoxide. Adsorption of NO is the rate-determining step. To my mind there is a contradiction between our results. This may be explained by a change in the properties of chromium oxide when it interacts with silica, which you used as a support. I should like your opinion.

R. G. Squires (*Purdue Univ., West Lafayette*) replied: I agree with your suggestion.

H. Knözinger (*Univ. of Munich*) said: The temperature of a wafer used in i.r. spectroscopy may differ considerably from that of the cell due to the heating effect of the beam. Were the authors able to measure the temperature of the sample directly? And also, how did they account for the sample emission, which should affect the band intensities at temperatures around 523 K?

[2] E. Garrone, G. Ghiolti, and A. Zecchina, unpublished work.

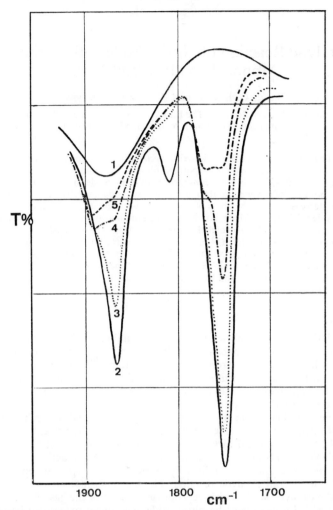

FIGURE Transmission *vs.* wavelength. 1: Background; 2: NO-covered sample; 3–5: after contact with molecular oxygen at room temperature for 10 min, 30 min, and 60 min respectively.

R. G. Squires (*Purdue Univ., West Lafayette*) replied: The temperature was measured under reaction conditions by a thermocouple which touched the sample wafer. The Perkin Elmer 180 spectrometer has a beam chopper located between the radiation source and the sample. Thus the radiation from the sample is steady while the radiation from the source is pulsed. These two signals are differentiated by the instrument and the radiation from the heated sample is filtered out.

Catalytic Reaction of Nitric Oxide with Carbon Monoxide on Iron and Nickel

BRUCE G. BAKER* and RAYMOND F. PETERSON

School of Physical Sciences, The Flinders University of South Australia,
Bedford Park, South Australia, 5042, Australia

ABSTRACT   The decomposition of nitric oxide and the reaction of nitric oxide with carbon monoxide have been investigated on evaporated films of iron and nickel. The initial reaction, on both metals, at 500—550 K is the decomposition of nitric oxide to nitrogen and nitrous oxide and the incorporation of oxygen into the film to form an oxide layer. The rate of the decomposition reaction decreases as the thickness of the oxide layer increases and becomes immeasurably slow at an oxide thickness > 50 nm. After reaction, examination of the film by electron diffraction shows that the oxide is NiO on nickel and Fe_3O_4 on iron. The oxidized film surface catalysed the reduction of NO by CO to form N_2 and CO_2. The activation energies are the same for both catalysts but the reaction is slower on the iron oxide and is inhibited by nitric oxide. In further experiments on a single crystal of nickel it is shown that the breaking of the N–O bond can be detected by Auger spectroscopy and that the rate is equal to the initial decomposition rate for NO on nickel films.

INTRODUCTION

The catalytic reactions of nitric oxide on metals and oxides have been investigated in many laboratories. The decomposition to nitrogen and oxygen and reduction by carbon monoxide or by hydrogen have received most attention.[1,2] Much of this work has been directed toward the development of practical catalytic systems to control nitric oxide emissions. For this reason catalyst materials have generally been in a form likely to survive practical reaction conditions.

This interest in nitric oxide has also stimulated fundamental research on the chemisorption of nitric oxide on well characterized materials. The present experiments on evaporated films of iron and nickel were designed to complement other experiments on metal single crystals in progress in this laboratory.[3] The reactions of an adsorbed monolayer on the crystal, determined by Auger electron spectroscopy, provide an insight into the overall catalytic reaction. The larger surface area of the metal film facilitates the catalytic study which depends on analysis of the gas phase.

Iron and nickel are oxidized by nitric oxide. The metal oxides can be reduced by carbon monoxide. Under reaction conditions the state of oxidation of the catalyst surface will be established. In the present experiments the metal is oxidized by nitric oxide at the temperature of reaction of NO with CO. An objective of the work is to examine the oxide which forms and to compare its reactivity with that of prepared oxide catalysts of known initial composition.

EXPERIMENTAL

The metal films were prepared in a Pyrex ultrahigh vacuum system pumped by a three-stage oil diffusion pump and protected by a baked alumina trap. The pressure after baking at 600 K and outgassing of components was *ca.* 10^{-7} N m^{-2}. Spectroscopically standardized iron and nickel wires (0·5 mm) were subjected to a prolonged outgassing and partial evaporation in a separate vacuum system before mounting in the cleaned reaction vessel. Films were deposited at 500 K and a pressure of 10^{-6} N m^{-2}.

The reactant gases were $> 99\%$ pure. Prepared mixtures and pure gases were introduced, usually to a pressure of *ca.* 650 N m^{-2}. The reaction system, of total volume 1·8 dm^3, was connected *via* a leak to a mass spectrometer (MS10). The cracking patterns at 70 eV and sensitivities for NO, CO, N$_2$O, CO$_2$, and N$_2$ were determined. The mass peaks 14, 16, 28, 30, 44 were recorded during reactions and the composition of the gas in the vessel was calculated with regard to corrections for sensitivity and leak rate. The rates of reaction were calculated from the pressure change with time, the system volume, and by assuming that the film surface area was 2·2 dm^2.

The films, after reaction, were recovered by breaking the vessel and stripping the film from the glass by an adhesion technique. After dissolving the adhesive, the film specimen was examined by transmission electron diffraction in the electron microscope (AEI 802, 100 kV).

RESULTS

Reactions on Metal Films

In a preliminary experiment an equimolar mixture of nitric oxide and carbon monoxide (650 N m^{-2}) was allowed to react on a freshly prepared iron film. It was found that the nitric oxide decomposed to nitrogen but that initially there was very little oxidation of CO to CO$_2$. When 70% of the NO had been converted, $< 5\%$ of the CO had reacted. No oxygen was observed and the mass balance indicated that oxygen had been incorporated into the film to form an oxide. Subsequent experiments were designed to investigate the oxidation of the metal in the absence of carbon monoxide, followed by the reaction of an NO/CO mixture on the resulting oxide.

The decomposition of nitric oxide on iron at 513 K is shown in Figure 1. The products are nitrogen and nitrous oxide. No gas-phase oxygen and no higher oxide of nitrogen was observed. The gaseous products reasonably account for the nitrogen but the loss of oxygen from the gas phase would require the formation of an oxide layer of thickness 50—70 nm. The reaction rate decreases with time and effectively stops after 100 min at 513 K. The reaction on nickel is qualitatively similar but the initial rate is slower and the time taken to reach the limiting oxide thickness is in excess of 20 h. The initial rates for iron and nickel are compared in Table 1.

Table 1

Decomposition on nitric oxide

Catalyst	Temperature (K)	Initial rate molecules m^{-2} s^{-1}
Iron	513	$3\cdot6 \times 10^{18}$
	553	$8\cdot9 \times 10^{18}$
Nickel	513	$8\cdot2 \times 10^{16}$

The partial pressure of nitrous oxide was found to pass through a maximum, suggesting that this product also undergoes decomposition. In a separate experiment on an iron film at 513 K and an initial pressure of 200 N m^{-2} the decomposition rate for N$_2$O was 7×10^{15} molecules m^{-2} s^{-1}.

Reactions on oxidized films

The oxidized films from the nitric oxide decomposition experiments were tested as catalysts for the reduction of nitric oxide by carbon monoxide. For an oxidized iron film the reaction products at 553 K are shown in Figure 2. Only carbon dioxide and nitrogen were detected. Nitrous oxide was not observed. The quantities of carbon dioxide and nitrogen were in accordance with the equation.

$$2NO + 2CO \rightarrow N_2 + 2CO_2$$

The same reaction products were observed on oxidized nickel films over the temperature range 473—553 K.

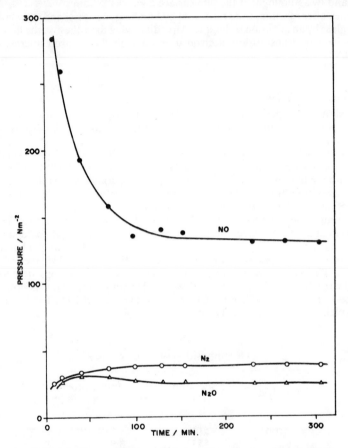

FIGURE 1   Decomposition of nitric oxide on an iron film at 513 K.

The relative pressure dependence of the reaction rate was investigated by reacting 650 N m^{-2} of prepared mixtures having NO/CO ratios of 2:1, 1:1 and 1:2. Within this range the initial rates at 553 K could be represented by equations of the form

$$\text{Rate} = kP_{CO}^a \times P_{NO}^b$$

with values of a and b in Table 2.

The temperature dependence of the reaction on oxidized films of nickel and iron were determined from initial rates of the equimolar mixture. The activation energies are 61 ± 4 kJ mol^{-1} for oxidized nickel and 60 ± 4 kJ mol^{-1} for oxidized iron. These data were obtained in the temperature range 473—593 K. In this range, the reaction on oxidized nickel is 50× faster than on oxidized iron.

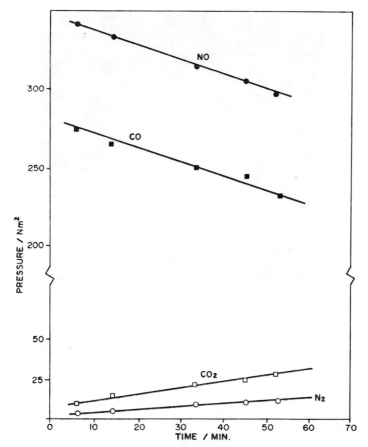

FIGURE 2   Reaction of nitric oxide with carbon monoxide on an oxidized iron film.

Table 2

Partial pressure dependence of the NO/CO reaction at 553 K.
$$\text{Rate} = kP_{CO}^a \times P_{NO}^b$$

Catalyst	a	b
Oxidized Iron	0.24 ± 0.02	-0.14 ± 0.02
Oxidized Nickel	0.58 ± 0.06	0.04 ± 0.01

Examination of films by electron diffraction

At the conclusion of a reaction experiment the film was examined by transmission electron diffraction. For nickel the rings in the diffraction pattern indicated that the film consisted of comparable amounts of metallic nickel and nickel oxide (NiO). For iron, the indexing of the pattern (Plate 1) is of greater importance since various oxides

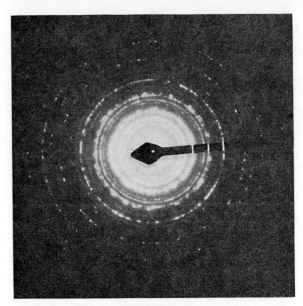

PLATE 1   Transmission electron diffraction pattern from an iron film oxidized by nitric oxide at 513 K.

could have resulted from the reaction. The indexing of the pattern is facilitated by the rings from metallic iron which act as an internal standard. The calculated interplanar spacings from the remaining rings are compared with tabulated values[4,5] for Fe_3O_4 and $\gamma\text{-}Fe_2O_3$ in Table 3. Other iron compounds $\alpha\text{-}Fe_2O_3$, FeO, Fe_2N, Fe_2C were considered and proved absent. Many of the spacings for Fe_3O_4 and $\gamma\text{-}Fe_2O_3$ are indistinguishable but there is sufficient data to identify the oxide as Fe_3O_4. This identification is based on the very close agreement of all of the observed spacings with Fe_3O_4 and on the relative intensities of the rings from the spacings 2·99, 2·55, and 1·33 Å.

Chemisorption of nitric oxide on nickel

During the kinetic studies on films another series of experiments on nickel single crystals was in progress. These involved LEED, Auger spectroscopy, and thermal desorption measurements of nitric oxide adsorbed on the (110) face of nickel. Experimental details are published elsewhere[3] but a result which can be correlated with the reaction on nickel films is reported here.

The decomposition of a monolayer of nitric oxide was studied by measuring the Auger peaks from nitrogen (382 eV) and oxygen (510 eV), relative to the nickel peak at 860 eV. The Auger spectra were recorded at 298 K but between each analysis the crystal was heated at a rate of 10 K s^{-1} to successively higher temperatures. The results are recorded in Figure 3. The calculated peak heights for a monolayer of nitrogen and a monolayer of oxygen are indicated by A_N and A_O, respectively. At low temperatures the observed oxygen peak height agrees with A_O but the nitrogen peak is less than A_N. A ready explanation is that NO is adsorbed *via* the nitrogen so that the oxygen overlays the nitrogen and attenuates emission.

Table 3

Indexing of electron diffraction pattern from an iron film oxidized by nitric oxide

Measured ring radius (on original plate)	Interlayer spacings (Å)			
	α-Fe	$d = \dfrac{17 \cdot 8}{r}$	Fe_3O_4	γ-Fe_2O_3
5·96		2·99	2·97	2·95 (s)
6·99 (s)		2·55	2·53 (s)	2·64
8·42		2·11	2·10	2·21
8·75 (s)	2·03	2·03	1·93	2·01
10·35		1·72	1·71	1·70
11·02		1·62	1·62	1·61
11·99		1·48	1·48	
12·44 (s)	1·43	1·43	1·42	1·43
13·37		1·33	1·33	1·32 (s)
13·87		1·28	1·28	
14·73		1·21	1·21	1·21
15·27 (s)	1·17	1·17	1·17	
16·30		1·09	1·09	
16·98		1·05	1·05	

When the crystal is heated beyond 500 K the spectrum changes. The nitrogen peak height increases; oxygen decreases. A more detailed study of this process showed that the rate at 513 K was 10^{17} molecules $m^{-2} s^{-1}$ in close agreement with the rate of decomposition of nitric oxide on a nickel film. The change in the Auger spectrum is therefore interpreted as due to the breaking of the N–O bond.

A mass spectrometric study of thermal desorption from the NO monolayer on the nickel crystal is also recorded in Figure 3. A small desorption of NO occurs at low temperature but no gas is evolved at 500—550 K, the region of the decomposition. The desorption of nitrogen occurs at about 850 K and is accompanied by the decay of the nitrogen Auger peak. The oxygen Auger peak does not decay further until after the nitrogen is desorbed. No desorbed oxygen is observed, indicating that oxygen must be incorporated into the crystal.

In another experiment the competitive adsorption of CO and NO was studied by Auger electron spectroscopy. At monolayer coverage the surface was covered by NO with CO present only in the gas phase. A similar result is obtained after the crystal has been slightly oxidized by NO.

Discussion

The oxidation of iron by nitric oxide proceeded faster than the reaction on nickel. Iron is known to have the greater affinity for nitric oxide.[1] Both nitrogen and nitrous oxide must initially form since the slow rate of decomposition of N_2O, observed for N_2O alone, shows that gas phase N_2O could not be the sole intermediate for the formation of nitrogen.

The study of Auger spectroscopy was interpreted as indicating that the oxygen is on top of the nitrogen when nitric oxide is chemisorbed on nickel. A survey of infrared spectroscopic evidence[1] has shown general agreement that NO is bound *via* the nitrogen to iron and nickel surfaces. The Auger experiment also shows that the breaking of the N–O bond in adsorbed nitric oxide is the rate limiting step in the onset of reaction on nickel. In this case there is no gas phase of nitric oxide, and thermal desorption of nitrogen does not occur at this temperature. In the reaction on the films the gas-phase nitric oxide apparently displaces nitrogen and so propagates further reaction.

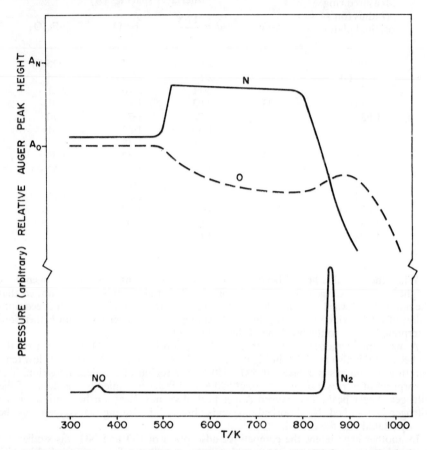

FIGURE 3 Auger peak heights for nitrogen (N) and oxygen (O) from an adsorbed monolayer of nitric oxide on (110) nickel; A_N and A_O are the respective calculated values for a monolayer, (top); corresponding thermal desorption spectrum, (bottom).

The decrease in the rate of nitrogen production as the reaction proceeds can be attributed to the slowness of incorporation of oxygen into the oxide layer. This effectively stops the reaction when the oxide is 50—70 nm thick. Nitric oxide can decompose to nitrogen and gaseous oxygen on NiO and on Fe_2O_3 but at higher temperatures.[6]

The oxidation of iron by oxygen at these temperatures would result in Fe_2O_3.[7] It is of particular interest that oxidation by nitric oxide forms Fe_3O_4. The electron diffraction evidence for this is supported by the kinetics observed when the NO/CO reaction was conducted on the oxide surface. The rate of this reaction was much slower on the iron oxide and was more strongly inhibited by nitric oxide than was the case for nickel oxide. The order of reactivity for the NO/CO reaction has been shown to be $Fe_2O_3 >$ NiO $> Fe_3O_4$[1] and the order of heats of adsorption of nitric oxide to be $Fe_3O_4 >$ NiO $> Fe_2O_3$.[2]

The properties of these semiconducting oxides are related to their non-stoicheiometry.[8] Fe_2O_3 would be expected to have an excess of Fe and be an *n*-type semiconductor. NiO has an excess of oxygen and is a *p*-type. Fe_3O_4 is an intrinsic semiconductor and in addition would be expected to have an excess of oxygen giving a more *p*-type character. The adsorption of nitric oxide is apparently favoured by *p*-type, whereas the reduction of nitric oxide is not. This raises the question of the state of Fe_2O_3 when acting as an effective catalyst for the NO/CO reaction. It has been shown that some reduction of the surface must occur to account for the catalytic activity of Fe_2O_3, and that this oxide has relatively much greater activity for the NO/CO reaction than for the O_2/CO reaction.[1] Carbon monoxide can reduce Fe_2O_3, and in the present work it is shown that NO oxidizes iron only to Fe_3O_4. However if reduction of Fe_2O_3 to Fe_3O_4 occurred under reaction conditions a low catalyst activity would result. It is suggested therefore that the active Fe_2O_3 catalyst surface must be only partially reduced and certainly not to the state of bulk Fe_3O_4. It might also be suggested that an Fe_2O_3 catalyst under some conditions might lose activity due to excessive reduction. Oxidation by oxygen would then be expected to restore catalytic activity.

The activity of an iron oxide catalyst in an oxidation–reduction reaction can be understood only in terms of the actual surface composition existing under reaction conditions. The application of Auger electron spectroscopy to the study of the competitive adsorption of NO and CO at various stages of growth of the oxide layer is feasible and could give the necessary insight into the behaviour of oxide catalysts.

ACKNOWLEDGEMENTS. The Auger spectra were measured by B. A. Sexton. This work was supported by the Australian Research Grants Committee.

REFERENCES

[1] M. Shelef and J. T. Kummer, *Chem. Eng., Prog. Symp. Series*, 1971, **67**, 74.
[2] M. Shelef, *Catalysis Rev.*, 1975, **11**, 1.
[3] G. L. Price, B. A. Sexton, and B. G. Baker, submitted to *Surface Sci.*
[4] K. W. Andrews, D. J. Dyson, and S. R. Keown, *Interpretation of Electron Diffraction Patterns* (Plenum Press, 1971).
[5] G. D. Renshaw, C. Roscoe, and P. L. Walker, *J. Catalysis*, 1970, **18**, 164.
[6] T. M. Yur'eva, V. V. Popovskii, and G. K. Boreskov, *Kinetika i Kataliz*, 1965, **6**, 1041.
[7] U. R. Evans, *The Corrosion and Oxidation of Metals: 1st Suppl.* (Edward Arnold, London, 1968), p. 25.
[8] F. E. Kröger, *J. Phys. Chem. Solids*, 1968, **29**, 1889.

DISCUSSION

M. A. Chesters (*Atomic Energy Res. Est., Harwell*) said: I wish to comment on your use of Auger spectroscopy to follow the decomposition of nitric oxide on nickel (110). You correlate an increase in the nitrogen Auger signal with decomposition of the molecule. In the case of carbon monoxide decomposition on tungsten, studied recently at Southampton University, the ratio of the carbon to oxygen signals, as measured by the area under the respective Auger peaks, remained constant on decomposition. However, if derivative signal peak heights are used, an apparent change in the carbon:oxygen ratio occurs which arises because of a change in the Auger peak shapes. Thus the shielding effect referred to in your paper appeared to be negligible in this case. Do you have any comment?

W. N. Delgass (*Purdue Univ., West Lafayette*) said: I should like to pursue this comment a little further. When the Auger line involves valence electron levels, changes in the chemistry of the atom in question can change the cross-section for Auger emission, since new orbitals are involved in calculating the emission probability. Thus, the total area, as well has the shape of the peak, can change with the chemistry. The effect would not change qualitative interpretation but could be a significant factor in quantitative interpretation.

B. G. Baker (*Flinders Univ.*), in reply to the two previous speakers, said: I agree that our explanation of the attenuating effect of oxygen on the nitrogen Auger signal may be an oversimplification and that more subtle chemical effects may be involved. We did not observe any change in the peak shape when NO dissociated. The i.r. data (refs. 1 and 2) show that the molecule is bound *via* nitrogen. The observed magnitude of the increase in the nitrogen Auger signal is reasonably accounted for by the removal of the overlying oxygen. I should emphasize that the observed correlation of the rate of change in the Auger spectrum with the rate of the decomposition of NO on nickel films is independent of any speculation concerning the Auger mechanism.

P. H. Emmett (*Portland State Univ.*) said: I am puzzled as to the conditions under which you find nitrogen retained by your iron surface. Is it proper to assume that all of the nitrogen retention was the result of the initial impact of NO with the fresh iron surface? Certainly one would not expect any retention of nitrogen on an oxidized surface.

B. G. Baker (*Flinders Univ.*) replied: Nitrogen is not retained on the surfaces of iron or nickel films in the presence of gaseous NO. Adsorbed nitrogen atoms, from the decomposition of NO are retained on the (110) surface of a nickel crystal after adsorbing and dissociating one monolayer of NO.

J. W. Hightower (*Rice Univ., Houston*) said: Several years ago Winter suggested that nitrous and nitric oxides might be adsorbed through oxygen, these atoms fitting into oxide vacancies in the surface. This model interprets the strong adsorption of the released oxygen, but is in obvious disagreement with your Auger results. Would you comment on Winter's model in the light of your result?

B. G. Baker (*Flinders Univ.*) replied: Our Auger results are obtained using metallic nickel. I.r. data, more recent than the work to which you refer, generally supports the view that NO is bound *via* nitrogen. We have favoured this interpretation of our results.

Oxidation of Nitrogen Mono-oxide over Transition-metal Ion-exchanged Zeolites

Hiromichi Arai,* Hiroo Tominaga, and Junichi Tsuchiya

Engineering Research Institute, Faculty of Engineering, University of Tokyo,
Hongo, Bunkyo-ku, Tokyo, 113, Japan

ABSTRACT Oxidation of NO by molecular oxygen was studied over zeolites in which sodium ions were substituted by some transition-metal ions. The trend in activity was found as Cu^{I}, Cu^{II}, $Cr^{III} > Fe^{II}$, Fe^{III}, $Co^{II} > Ni^{II}$, Mn^{II}. With the most active group of catalysts, the kinetics and mechanism of NO oxidation were investigated. Co-operative adsorption of NO and O_2 was suggested as they key step of the reaction over Cu^{I} and Cu^{II} on zeolites. A quasi-Rideal mechanism was suggested for NO oxidation over Cr^{III} on zeolites, which strongly activates oxygen on adsorption. The activity of a Cr^{III} on zeolite catalyst was greatly enhanced by water, and the presence of SO_2 did not result in deactivation of catalyst at high temperatures. Some infrared spectroscopic observations on adsorption and reactions of NO and related substances on the catalysts were performed, and evidence was obtained for the reaction mechanisms suggested.

Introduction

In this study an attempt has been made to develop a new catalytic system for oxidation of NO to NO_2. The catalyst is expected to be applied successfully to the cleaning of combustion flue gases which contain NO as well as oxygen in large amounts. NO is not reactive enough to be adsorbed on any solid adsorbent, or absorbed in any solvent, at a sufficiently high rate for industrial practice. Upon oxidation, however, it gives NO_2 which is much more reactive and less difficult to remove.

The homogeneous gas phase reaction,

$$2NO + O_2 \rightleftharpoons 2NO_2,$$

has been extensively studied and is characterized by 3rd order kinetics with a negative activation energy. As a consequence, the rate of oxidation of NO in such a low concentration as is usual in stack gases is extremely slow. This necessitates, in practice, the employment of the catalyst which is active at low temperatures where the oxidation is favoured thermodynamically.

Recently, dissociative adsorption of oxygen on zeolite catalysts containing transition-metal ions was reported by M. Boudart et al.[1] and also by the present authors.[2,3] These catalysts were found to be highly active for oxidation of carbon monoxide[2] and dehydrogenation of cyclohexanes.[3,4]

In view of this some zeolite catalysts were prepared by replacing the sodium ion with several transition-metal ions, and their activities were evaluated for NO oxidation by molecular oxygen. The reaction mechanism discussed below is based on a kinetic study of oxidation and infrared spectroscopic observation of the adsorbed NO.

Experimental

Catalyst preparation

Linde Molecular Sieves of type NaX, NaY, and Norton Co.'s Zeolon, both in powder form, were used to prepare catalysts by the method of ion-exchange.[2] The zeolites, after being treated with boiling distilled water to remove excess of sodium, were slurried in distilled water. An aqueous solution of transition-metal chloride (except for $FeSO_4$, and a liquid-ammonia solution of CuI) was added dropwise to the slurry being stirred, and the slurry was warmed up to 363 K for several hours and kept standing overnight at room temperature. In the cases of Cu^I, Co^{II}, Fe^{II} and Mn^{II}, the ion-exchange was carried out under a stream of nitrogen in order to prevent oxidation of the ions. After filtration, the slurry was washed with distilled water and dried at 383 K. The extent of ion-exchange was evaluated from an analysis of the residual transition-metal ion in the filtrate and washings.

Apparatus

Oxidation of NO was studied by use of a conventional atmospheric flow reactor, made of quartz, where a given amount of catalyst, usually 2 g, was loaded. A mixture of NO and oxygen, diluted with helium, flowed downwards through the catalyst bed. The reactor effluent was analysed for NO and $NO + NO_2$ by a chemiluminescence method using a Shimazu NO_x Analyzer CLM-201.

Adsorption of NO on the catalyst was studied by an infrared spectroscopic method described elsewhere.[5] The catalyst powder was pressed into a disc 20 mm in diameter and 0·1 mm in thickness, and the disc was placed for i.r. inspection in a cell equipped with the facilities for adsorption and desorption experiments at elevated temperatures. A Hitachi IRA-2 spectrophotometer was employed.

RESULTS AND DISCUSSION

Relative activities of transition-metal ions.

Over X-type zeolites loaded with various transition-metal ions, the percent oxidation of NO was measured as a function of reaction temperature. Several examples of the experiment are shown in Fig. 1. With a rise in temperature, the percent oxidation of NO

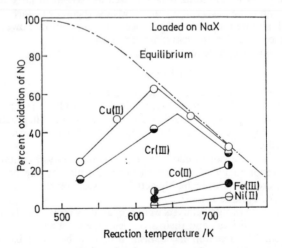

FIGURE 1   Activities of transition-metal ions loaded on zeolite X (S.V. 23,400 cm³/g-cat·h, NO 300 ppm, O_2 6·4 vol%).

generally increases, but up to an equilibrium value which is inversely proportional to the reaction temperature. In view of this, activities of the catalyst were compared at 623 K where the ceiling value of percent oxidation is as high as 75% if the oxygen partial pressure is 1×10^4 Pa.

The results summarized in Table 1 indicate that the faujasite zeolites both NaX and Y are endowed with good catalytic activity for NO oxidation when some transition-metal ions are incorporated by ion-exchange with Na^I. The activity series is: Cu^I, Cu^{II}, $Cr^{III} > Fe^{II}$, Fe^{III}, $Co^{II} > Ni^{II}$, Mn^{II}. The above sequence is rather tentative since the activity is strongly dependent upon the extent of ion-exchange as is shown below. Anyway, it is noteworthy that Cu^{II} and Fe^{III} ions are as active as Cu^I and Fe^{II}, respectively, for the oxidation of NO.

Table 1

Activities of transition-metal ions loaded on zeolites X and Y for NO oxidation. (623 K, S.V. 23,400 cm^3/g-cat·h, O_2 6·4 vol%)

Catalyst (% ion-exchange)		Percent oxidation[a]	
		$[NO]_0 = 100$ ppm	$[NO]_0 = 300$ ppm
$Cr^{III}X$	(60)	50·0	41·8
$Fe^{II}X$	(n.d.)[b]	16·3	—
$Fe^{III}X$	(n.d.)	6·0	4·1
$Co^{II}X$	(74)	16·3	8·3
$Ni^{II}X$	(47)	1·9	1·7
$Cu^{II}X$	(95)	68·4	62·1
Na^IX	(0)	~0	~0
$Cr^{III}Y$	(44)	53·4	37·8
$Mn^{II}Y$	(n.d.)	~0	~0
Fe^{II}	(76)	3·4	—
$Fe^{III}Y$	(100)	22·4	8·4
$Co^{II}Y$	(73)	14·5	13·4
$Ni^{II}Y$	(50)	0	0
Cu^IY	(n.d.)	40·4	29·8
$Cu^{II}Y$	(67)	40·0	32·8
Na^IY	(0)	~0	~0

[a] Measured at two levels of initial concentration of NO, 100 and 300 ppm.
[b] Not precisely determined.

In our previous paper[2] it was shown that, in the oxidation of CO, Cu^I and Fe^{II} loaded on zeolite are far more active than Cu^{II}, and Fe^{III}. The activity difference was ascribed to the fact that the former two ions can adsorb an oxygen molecule in a dissociative way while the latter two can not. Kinetic study, in addition, suggested a Rideal mechanism for CO oxidation where CO molecules in the gas phase attack the oxygen atoms, on the catalyst, in a bridged form which is given by the reaction;[1]

$$2Fe^{II} + \tfrac{1}{2} O_2 \rightarrow Fe^{III}-O^{2-}-Fe^{III}.$$

In the oxidation of NO, accordingly, the reaction mechanism and hence the kinetics cannot be the same as those in the oxidation of CO. For the elucidation of these, further studies were carried out on the catalysis of Cu^{II} and Cr^{III} ions loaded on zeolites.

Catalyst activity as a function of the extent of ion-exchange.

Two series of catalysts $Cu^{II}X$ and $Cr^{III}Y$ were prepared at various levels of Na^I ion-exchange with the respective metal ions, and their catalytic activities were examined, as shown in Fig. 2. In the both cases, the activities increase in an exponential way with the progress of ion-exchange. It should be noted, however, that the activity of Cr^{III} becomes appreciable in the first half-stage of the ion-exchange, but that of $Cu^{II}X$ does in the second half. In the case of Cr^{III} and NaY, the maximum degree of ion-exchange was found to be *ca.* 50%.

Sodium ions in faujasite-type zeolites are located at three different crystallographic sites, namely S_I, S_{II}, and S_{III}. Energetically, S_I is the most stable position but located

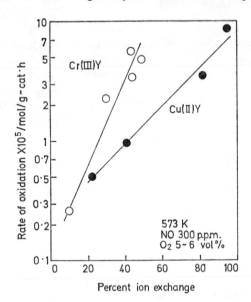

FIGURE 2 Catalyst activities as a function of extent of ion-exchange (573 K, NO 300 ppm, O_2 5–6 vol %).

within a hexagonal prism connecting sodalite cages, and hence accessible only for the smaller species. S_{II} and S_{III} are in a large cage and open to the larger species. Accordingly, the cations, Cu^{II} or Cr^{III}, are not evenly distributed over these sites but located at some specific sites in preference to others, depending on the cation size as well as on the extent and conditions of the ion-exchange.

The results shown in Fig. 2 might be interpreted as follows. Cu^{II} ions occupy S_I preferentially and then S_{II} and S_{III}, and Cr^{III} ions take up S_{II} and S_{III} and not S_I. The cations located at S_I, if at all, are not available for oxidation of NO, and only those at S_{II} and S_{III} are active for the reaction. These hypotheses should be verified by structural analyses of the catalysts, such as determination of cation distribution by sites and the change in pore size and surface area.

Kinetics of NO *oxidation over* $Cu^{II}X$ *and* $Cr^{III}Y$.

The rate of catalytic oxidation of NO can be defined by:

$$\frac{dx}{d(W/F)} = k_1 p_{NO}{}^m p_{O_2}{}^n - k_2 p_{NO_2}{}^r$$

where x = mole fraction of NO oxidized; W = catalyst employed, g; F = feed rate of NO, mol/h; p = partial pressure; k = rate constant. The kinetic parameters were determined under reaction conditions where the reverse reaction, or decomposition of

NO_2, is negligible. The initial rates of oxidation of NO over $Cr^{III}Y$ were plotted against partial pressures of NO and O_2, respectively. The dependence of the initial rate on the reciprocal of the absolute temperature was measured. Similar experiments were performed with $Cu^{II}X$. These results are summarized in Table 2.

The observed difference in the reaction orders with respect to the partial pressures of NO and O_2, m and n, may be assigned a significance in terms of the reaction mechanism. From the experimental fact that n is smaller than m, it is suggested that O_2 adsorbs more strongly than NO does on both of the catalysts in their working states.

Our previous experiments,[2] however, demonstrated that $Cu^{II}Y$ does not chemisorb O_2 while $Cu^{I}Y$ does so dissociatively. On the other hand, it has been reported that Cu^{II}

Table 2

Kinetic parameters of NO oxidation

	$Cu^{II}X$		$Cr^{III}Y$		
% Ion-exchange	41	95		44	
Temperature, K	581	581	533	628	673
m^{a}	—	0·59	0·71	0·79	0·81
n^{a}	—	0·31	0·35	0·24	0·36
E_{app}, kJ/mol	69	30		48	

a $r_0 = k_1 p_{NO}{}^{m} p_{O_2}{}^{n}$.

supported on solids adsorbs NO, while Cu^{I} does not, or does so very weakly if at all.[6-8] In view of these results, adsorption of O_2 on $Cu^{II}X$ catalyst should be preceded by NO adsorption on Cu^{II} to give Cu^{I}.

$$
\begin{array}{ccccc}
& & & NO^+ & \\
& & & | & \\
2\begin{array}{c} Cu^{II} \\ | \\ Z \end{array} & \xrightarrow{2NO} & 2\begin{array}{c} Cu^{I} \\ | \\ Z \end{array} & \xrightarrow{\frac{1}{2}O_2} & \begin{array}{cc} NO^+ & NO^+ \\ | & | \\ Cu^{II}\!-\!O^{2-}\!-\!Cu^{II} \\ | & | \\ Z & Z \end{array}
\end{array}
$$

(Z: anionic site of zeolite)

The last complex is supposed to decompose to give NO_2, $Z\text{–}Cu^{II}$ and $Z\text{–}Cu^{I}\text{–}NO^+$. In the case of $Cu^{I}X$, which is as active for NO oxidation as $Cu^{II}X$, adsorption of O_2 may precede that of NO, *viz*:

$$
\begin{array}{ccccc}
2\begin{array}{c} Cu^{I} \\ | \\ Z \end{array} & \xrightarrow{\frac{1}{2}O_2} & \begin{array}{cc} & \\ & \\ Cu^{II}\!-\!O^{2-}\!-\!Cu^{II} \\ | & | \\ Z & Z \end{array} & \xrightarrow{2NO} & \begin{array}{cc} NO^+ & NO^+ \\ | & | \\ Cu^{I}\!-\!O^{2-}\!-\!Cu^{I} \\ | & | \\ Z & Z \end{array}
\end{array}
$$

The last complex possibly decomposes into NO_2, NO, and $2\,Z\text{–}Cu^{I}$. A similar mechanism, or co-operative adsorption of NO and O_2, might be applicable for the catalysis of NO oxidation over Fe^{II} and Fe^{III} supported on zeolites.

The catalytic action of $Cr^{III}Y$ for oxidation of NO, however, seems almost exclusively due to the activation of oxygen. On an oxidized surface of chromia supported on alumina,

NO adsorption was reported to be very small and slow compared with the corresponding reduced surface.[9] Furthermore, Cr^{III} exchanged zeolites were found to adsorb NO only when reduced so as to contain Cr^{II}, and not in the oxidized state, Cr^{III}, Cr^V or Cr^{VI}.[8] Accordingly, in the oxidative atmosphere employed in this study, NO adsorption on $Cr^{III}Y$ catalyst is supposed to be extremely limited especially at high temperatures. The increase in the experimental m values with the rise in temperature reflects this. Hence, a quasi-Rideal mechanism may operate for NO oxidation over $Cr^{III}Y$ catalyst, especially at high temperatures.

Experimental support for the above suggestions will be presented in the last section dealing with infrared spectroscopic studies.

Effects of the addition of H_2O and SO_2 in the feed on the catalyst activity

Since the ultimate aim of this study is to develop an industrial process for flue gas cleaning, the effects of fossil fuel combustion products on the catalyst activity should be examined. Fig. 3 indicates the changes in percent oxidation of NO in the test run,

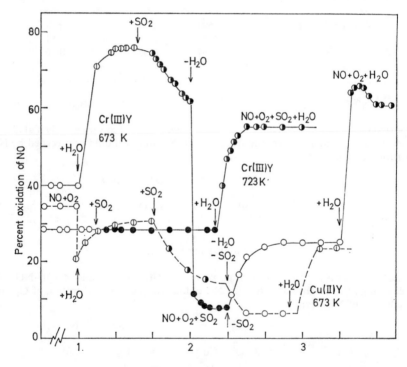

Hours on stream

FIGURE 3 Effect of presence of H_2O and/or SO_2 on NO oxidation. Experimental Conditions.

	Temp. (K)	S.V.	NO(ppm)	O_2(vol %)	SO_2(ppm)	H_2O(vol %)
$Cu^{II}Y$	673	30,000	100	5	200	8·3
$Cr^{III}Y$	673	30,000	100	5	200	8·3
$Cr^{III}Y$	723	30,000	120	5	135	2·4

beginning with the NO + O_2 (diluted with helium) reaction, where H_2O and/or SO_2 were added or stopped, after stated intervals, into the feed.

The action of water is most interesting. It reduced the catalytic activity of $Cu^{II}Y$, but on the other hand promoted that of $Cr^{III}Y$, in that the percent oxidation of NO nearly doubled and exceeded the equilibrium conversion. Based on the findings by spectroscopic study to be mentioned below, it is inferred that H_2O displaces NO_2 fairly strongly adsorbed on $Cr^{III}Y$ catalyst and thus promotes the reaction rate. It also prevents the re-adsorption of NO_2 on the catalyst, which leads to its decomposition;

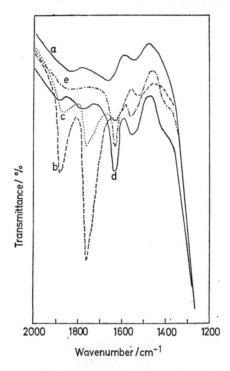

FIGURE 4   I.r. spectra of NO adsorbed on $Cr^{III}Y$ zeolite. a: blank. The sample was evacuated at 673 K for 2 h, b: 50 torr NO introduced at 298 K and evacuated at 298 K, c: 373 K. The sample was evacuated at 673 K for 2 h and oxidized by 10 torr oxygen at 673 K for 2 h and evacuated at 298 K, d: 50 torr NO introduced at 298 K and evacuated at 298 K, e: 373 K.

the reverse reaction. Thermal decomposition of NO_2 in the gas phase is so slow at the reaction temperature that the oxidation of NO to an extent beyond the equilibrium is possible. Another inhibiting action of water was observed with Cr^{III} supported on Zeolon particularly at low temperatures where capillary condensation of water in the small pores occurs.

Inhibition by SO_2 was remarkable with both of the catalysts at 673 K. However, with $Cr^{III}Y$ catalyst at 723 K, inhibition by SO_2 was not appreciable, possibly because NO adsorption is not the key step at this temperature. Promotion of oxidation of NO by H_2O was also observed when SO_2 was present in the feed. The SO_2 was oxidized almost completely to SO_3 under the reaction conditions.

In view of these results, $Cr^{III}Y$ is expected to be employed for the treatment of flue gases containing SO_2 without any decay, in the short term at least, in its catalytic activity.

Infrared spectroscopic study of NO adsorption

A number of papers have been published on the structure and reactivity, studied by i.r. and e.s.r. techniques, of NO adsorbed on catalysts, including transition-metal ions loaded on zeolites. Several findings in this laboratory are mentioned briefly, particularly as regards NO adsorption on $Cr^{III}Y$ zeolite, and are presented in Figs. 4 and 5.

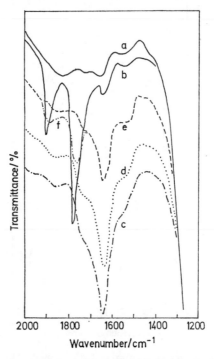

FIGURE 5   Effect of water on NO adsorbed on $Cr^{III}Y$ zeolite. a: blank, b:50 torr NO introduced at 298 K and evacuated at 298 K, c: 5 torr water introduced at 298 K and evacuated at 298 K, d: 373 K, e: 473 K, f: 50 torr NO introduced again at 298 K and evacuated at 298 K.

$Cr^{III}Y$ turned from light blue-green to black when evacuated (10^{-5} torr) at 673 K for 2 h in the i.r. cell. Admission of NO on to this reduced catalyst at room temperature, gave two strong absorption bands at 1895 and 1770 cm^{-1}. Both of these bands disappeared completely on evacuation at 423 K, and also on introduction of O_2 at temperatures up to 423 K, giving rise to the absorptions at 1625 and 1540 cm^{-1} which were found when NO_2 was introduced on to the catalyst. On the other hand, $Cr^{III}Y$ pre-oxidized in the i.r. cell by 100 torr O_2 at 673 K for 2 h did not give appreciable absorption in 1700–1900 cm^{-1} region upon admission of NO at room temperature. However, absorption bands appeared corresponding to NO_2, which did not desorb on evacuation at temperatures below 473 K. These facts suggest that oxidized $Cr^{III}Y$ catalysts do not adsorb NO, but readily give NO_2 which, once formed, is apt to remain strongly adsorbed on the catalyst.

NO and NO_2 adsorbed on the reduced $Cr^{III}Y$ were found to be easily displaced by introduction of water (5 torr). Preadsorption of water prevented the adsorption of NO and NO_2. Addition of SO_2 on the adsorbed NO, and *vice versa*, had little effect on the the respective bands.

These infrared observations are seemingly in harmony with the kinetics and mechanism suggested in the preceding section for NO oxidation over $Cr^{III}Y$ catalysts.

REFERENCES

[1] R. L. Garten and M. Boudart, *A.C.S. Preprints*, D7, 1973.
[2] T. Kubo, H. Tominaga, and T. Kunugi, *Bull. Chem. Soc. Japan*, 1973, **46**, 3549.
[3] T. Kubo, H. Tominaga, and T. Kunugi, *Nippon Kagaku Kaishi*, 1972, 196.
[4] T. Kubo, T. Hino, H. Tominaga, and T. Kunugi, *Nippon Kagaku Kaishi*, 1973, 2257.
[5] J. B. Peri and R. B. Hannan, *J. Phys. Chem.*, 1960, **64**, 1526.
[6] H. S. Gandhi and M. Shelef, *J. Catalysis*, 1973, **28**, 1.
[7] C. C. Chao and J. H. Lunsford, *J. Phys. Chem.*, 1972, **76**, 1546.
[8] C. Naccache and Y. Ben Taarit, *JCS Faraday I*, 1973, **69**, 1475.
[9] K. Otto and M. Shelef, *J. Catalysis*, 1969, **14**, 226.

DISCUSSION

L. J. R.Vand amme (*Catholic Univ. of Louvain*) said: In your paper you mentioned that Cu^{II}-zeolites do not adsorb oxygen whereas Cu^{I}-zeolites adsorb oxygen dissociatively. We agree as far as Cu^{II}-Y-zeolites are concerned. However, for the $Cu^{I}Y$-zeolite, we find that oxygen adsorbs but not dissociatively in the oxidation of carbon monoxide (see Beyer, Jacobs, Uytterhoeven and Vandamme, this Congress). You introduced the Cu^{I} directly, whereas we reduced a Cu^{II}-zeolite. Do you think the difference is due to the method of preparation of the zeolite?

You appear to compare X- and Y-zeolites having different metal ions, without making a distinction between the X or Y character, but attributing differences to the ion. You show that the adsorption of oxygen is stronger than that of nitric oxide in NO oxidation, and stronger than that of carbon monoxide in CO oxidation. So, I assume we can compare both reactions. We have found a completely different behaviour, kinetics, and mechanism for CO oxidation catalysed by Y- and X-zeolites (unpublished work). Would you comment please?

Lastly, how did you make a CuY-zeolute with such a high degree of ion exchange (95% is represented in Figure 2)? Are you sure all of the copper is stoicheiometrically exchanged?

H. Tominaga (*Univ. of Tokyo*), in reply, said: We prepared $Cu^{I}Y$-Zeolite by ion-exchange using Cu^{I} in liquid ammonia in the presence of inert gas. The catalytic behaviour of this $Cu^{I}Y$ could differ considerably from yours which is probably prepared by reduction of $Cu^{II}Y$ with carbon monoxide. Please refer to our ref. 2 for further details.

D. Cornet (*Univ. of Caen*) said: You state that chromium-containing zeolites adsorb NO only after the chromic ions have been reduced to the bivalent state. We have been able to observe an appreciable adsorption on a chromium-exchanged A-zeolite after oxidative treatment. Analysis for chromium shows that it is mostly in the form of Cr^{IV}. The extent of adsorption of NO on this material amounts to 2-3 molecules per chromium atom at room temperature. The resulting complex is fairly stable; it decomposes upon heating above 423 K, and the gas evolved is still nitric oxide. In addition, we have some experience with Cr-exchanged Y-zeolites, and have found them very susceptible to oxidation. Have you observed the presence of oxidized states beyond Cr^{III} in your work?

H. Arai (*Univ. of Tokyo*) replied: Our e.p.r. results show that Cr^{III} on zeolite Y was reduced to Cr^{II} after evacuation at temperatures above 400 °C, and was oxidized to Cr^{V} in the presence of oxygen at 400 °C. Nitric oxide was not adsorbed as such but was adsorbed on the Cr^{V} to give a Cr^{III} NO_2 species which was identified by i.r. spectroscopy. This species decomposed to NO and NO_2 on heating.

G. Munura (*Univ. of Seville*) said: You obtain reduction of Cr^{III} to Cr^{II} simply by outgassing your samples at about 673 K. Can you determine the percentages of each ion in your evacuated samples?

H. Arai (*Univ. of Tokyo*) replied: The intensity of the e.p.r. signal of the Cr^{III} species indicated that almost all ions were reduced to Cr^{II} which has no signal.

J. C. Conesa (*Inst. of Catalysis and Petrochem., Madrid*) asked: You describe a structure Z—Cu^{II}—O—Cu^{II}—Z as appearing on oxidation of Cu^{I} with molecular oxygen in a $Cu^{I}Y$-zeolite. Such copper pairs have been described in the literature: several authors including ourselves have recorded their e.p.r. spectra. Did you detect the e.p.r. spectrum of this species?

H. Arai (*Univ. of Tokyo*) replied: To our regret, our e.p.r. spectra were not sufficiently well resolved to show clearly the presence of Z—Cu^{II}—O^{2-}—Cu^{II}—Z, so that structure was assumed by analogy with Z—Fe^{III}—O^{2-}—Fe^{III}—Z which has been proposed by several workers.

H. Bremer (*Tech. Hochschule "Carl Schorlemme", Merseburg*) said: It is well known that zeolites exchanged with transition elements may, on oxidation and reduction, undergo lattice breakdown. This is especially true with CrY-zeolites. By what means have you proved the stability of your zeolite lattices? What is the special effect or advantage of zeolites, with respect to your reactions, as compared with transition element oxides?

H. Arai (*Univ. of Tokyo*) replied: X-ray analysis of our CrY-zeolite, to demonstrate structure retention, has not been carried out yet. Zeolites have the advantage of a well-defined crystallographic structure and a large surface area. They provide a support into which ions can easily be introduced by the ion-exchange method. Moreover, active sites and species can be identified easily by spectroscopic methods.

Oxidation of Ammonia to Nitrogen Mono-oxide in a Fluidized Bed of a Cobalt Oxide Catalyst

TODORKA F. POPOVA, D. GEORGIEV-KLISSURSKI,* and IVAILO A. ZRUNCHEV

Higher Institute of Chemical Technology, Sofia 56, and Institute of General and Inorganic Chemistry, Bulgarian Academy of Sciences, Sofia 1113, Bulgaria

ABSTRACT   A systematic study has been carried out on the oxidation of ammonia in a fluidized bed of a Co_3O_4 catalyst. The dependence of the degree of oxidation of ammonia on the temperature in the catalyst zone was determined for ammonia concentrations between 7 and 12% and 720—725°C was found to be the optimum temperature. At a production rate of ca. 4.5×10^4 kg of oxidized ammonia per 24 h and for 1 m^2 of the cross section of the reactor a 97% degree of oxidation of NH_3 to NO was achieved. The optimum contact time was $0.95–1.10 \times 10^{-2}$ s. It was established that a 97% degree of oxidation remains constant at fluidization numbers below 5. Pilot-plant experments carried out over a period of three months showed that within such a period the degree of oxidation of ammonia to nitrogen mono-oxide remains constant and close to 97%. The method offers important advantages in comparison with the known fixed bed and two-stage processes. It is very interesting from the industrial point of view and could compete with the present industrial process using platinum catalysts.

INTRODUCTION

The current industrial production of nitric acid is based on the selective oxidation of ammonia to nitrogen mono-oxide over platinum catalysts:

$$4\ NH_3 + 5\ O_2 \rightarrow 4\ NO + 6\ H_2O + 907\ kJ$$

Considerable reduction of the capital investment involved can be achieved by replacement of the platinum catalysts with non-platinum, predominantly oxide-type catalysts.

The catalytic activity of oxides of transition metals with respect to this reaction has been subjected to a systematic study.[1-13] In previous papers of ours,[1,4] devoted to the mechanism of the reaction, a distinct correlation was established between the specific catalytic activity of the oxides and the binding energy of oxygen in their surface layer. The highest activity and selectivity were observed[1,2] with Co_3O_4.

Up till now, non-platinum catalysts have had very limited application in industrial nitric acid production. It has been established that the reaction, on account of its high rate, is limited by diffusion phenomena.[2,8] As a result the degree of utilization of the catalyst surface is very low.[8,14] The carrying out of the reaction in the external diffusion region also creates conditions for local overheating. Besides, when a nonplatinum catalyst is used in a stationary layer, there occurs a relatively fast deactivation of the upper part of the catalyst layer, where the bulk of the ammonia is oxidized.

With a view to eliminating these disadvantages, intensifying the process, and making it applicable on an industrial scale, we have studied the reaction using a fluidized bed Co_3O_4 catalyst.

EXPERIMENTAL

Materials

Cobaltous–cobaltic oxide was prepared by thermal decomposition of CoC_2O_4, $2H_2O$ at 400°C in air and subjected to additional thermal pretreatment and granulation.[13]

By a programmed increase of temperature in the range 400—920°C over a period of 10 h, specimens with different densities were prepared. An optimum apparent density of $3 \cdot 6 \times 10^3$ kg m^{-3} and optimum mechanical strength were observed with specimens calcined at 900°C. The specific surface area of the catalyst measured and calculated by low temperature krypton adsorption according to the BET method was in the range 0·4—0·9 m^2 g^{-1} for the various specimens. Homogeneous catalyst fractions of 0·5 to 2 mm in diameter were used for different kinetic measurements.

The ammonia and air were purified by methods described previously.[8]

Apparatus and procedure

The activity and selectivity of the thermally pretreated cobalt oxide catalyst were studied by a flow-type installation described elsewhere.[14] The stainless-steel reactors were 49 and 150 mm in diameter[14] and enabled experiments with fluidized bed or fixed bed oxide catalysts to be carried out. The analytical methods have been described previously.[8]

The flow rate at which the fluidization starts ("first critical rate") was calculated according to the equation of Todess,[5]

$$Re_f = Ar/(1400 + 5 \cdot 22 \ Ar)$$

where $Re_f = w_f \ d/\nu$ stands for the Reynolds' number; w_f is the flow rate at which the fluidization starts; d is the average diameter of the particles, and ν the kinematic viscosity. The criterion of Archimedes, Ar, is defined by the expression: $Ar = (g \ d^3/2) (\rho_s - \rho_g)/\rho_g$ where ρ_g and ρ_s are the densities of the gaseous and the solid phases. The porosity of the catalyst layer was calculated by the expression:

$$\zeta = (18 \ Re + 0 \cdot 36 \ Re^2)^{\ 0 \cdot 21}/Ar$$

and varied under the experimental conditions from 0·40 to 0·72. The contact time τ was defined by the expression $\tau = v_e/V$; where V denotes the volume of the reaction mixture passing the reactor for unit of time, and v_e the volume of the catalyst layer exluding the catalyst itself. In this formula v_e was replaced by the porosity of the layer. Other symbols used are: H_0 the depth of the catalyst layer and D_a the diameter of the apparatus.

RESULTS AND DISCUSSION

1. *Influence of various parameters on the oxidation process*

1.1 *Temperature and ammonia concentration*

The dependence of the degree of oxidation of ammonia to NO on the temperature in the catalyst zone is shown in Fig. 1. The data correspond to different concentrations of ammonia in the ammonia–air mixture within the range 7—11% by vol. It was established that the optimum temperature for Co_3O_4 catalyst used in a fluidized bed is *ca.* 725°C. This value, which is lower than the optimum temperature for the same catalyst in a stationary layer, can be explained by the limitation of the role of the temperature and diffusion gradients in the catalyst layer. As was shown previously[2,8] the reaction on a fixed bed Co_3O_4 catalyst is limited by diffusion phenomena. The diffusion of reagents to the external catalyst surface is a rate limiting step.[2] Applying a fluidized catalyst, a transient and even an external kinetic region can be attained.[14] The heat transfer and mass transfer coefficients are from 80 to 100 times higher than in the case of a fixed bed catalyst.

1.2 *Contact time*

As is well known, when the process is carried out in a stationary catalyst layer, the degree of oxidation of ammonia to NO obtained is strongly affected by the contact time. For a process performed with a fluidized bed catalyst this parameter is also important and reflects other basic parameters of the process. In Fig. 2 the dependence of α, the degree of oxidation of ammonia to NO on the contact time is shown; 0.95—1.0×10^{-2} s can be regarded as an optimum value. The increase of the reaction rate due to the increase in temperature leads to a reduction of the optimum contact time.

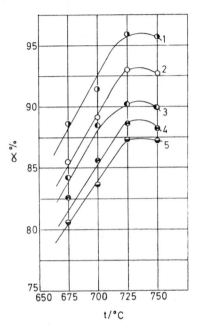

FIGURE 1   Temperature dependence of α, the degree of oxidation of ammonia to NO: $H_o/D_a = 0.34$; $W = 5 \times 10^4$ h^{-1}; $d = 0.7$ mm (1) 7% by vol. NH_3; (2) 8% NH_3; (3) 9% NH_3; (4) 10% NH_3; and (5) 10.5% NH_3.

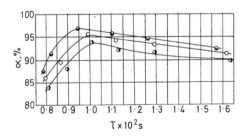

FIGURE 2   Dependence of α, the degree of oxidation of ammonia to NO, on the contact time τ: $H_o/D_a = 1.07$; $d = 1.30$ mm.

1.3 *Flow rate and mixing*

The influence of the flow rate is shown in Fig. 3. At a twofold increase of the space velocity in the range 2×10^4—4×10^4 h^{-1} (calculations for an equivalent stationary layer), the degree of oxidation of ammonia to NO remains constant and even a tendency towards an increase is observed. This can be explained by an acceleration of the transport phenomena leading to a better admission of the reacting gases to the catalyst surface and to its more complete utilization. The intensive mixing leads to a decrease of the effective thickness of the diffusion layer, δ.

The intensity of the axial mixing can be varied by variation of the fluidization numbers, defined as the ratio between the actual flow rate and the flow at which the fluidization starts: $K = w/w_f$. The axial mixing is accompanied by an undesirable

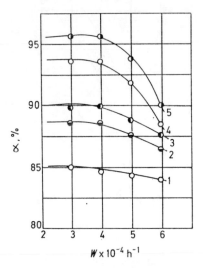

FIGURE 3 Relationship between the space velocity W, and α the degree of oxidation of ammonia to NO: $d = 0.7$ mm.

phenomenon, namely, the mixing of the unreacted ammonia with NO. This could lead to a loss of ammonia, due to secondary reactions:

$$4\,NH_3 + 6\,NO = 5\,N_2 + 6\,H_2O + 1800\ kJ$$

The graphs in Fig. 4 show that for fluidization numbers up to 5, the degree of ammonia oxidation to NO remains constant and in the range 700—750°C it even increases. This leads to the conclusion that in this range the mixing does not play its negative role. Consequently, an intensification of the process could be achieved at a high degree of selective oxidation of ammonia to NO.

2 Catalyst productivity and stability

The dependence of the degree of oxidation of ammonia to NO on the productivity of the catalyst with the smallest grain size used ($d = 0.7$ mm) is shown in Fig. 5. The productivity can be further increased by an increase of the catalyst grain size. This permits higher flow rates to be applied. The experiments showed that at a productivity of 5—6 $\times$ 10³ kg NH_3 for 24 h and 1 m² of the cross section of the reactor, the degree of selective oxidation can be kept at 97%. Such a high degree of oxidation can be attributed to a maximum limitation of the secondary reactions and a result of homogeneous distribution of the catalyst particles in the fluidized bed.

The Co_3O_4 catalyst used in a fluidized bed possesses a high stability. The pilot plant

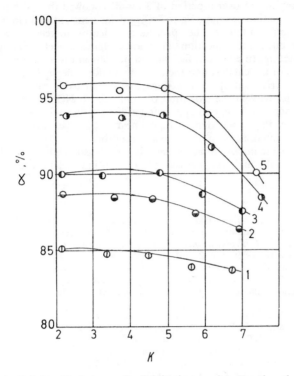

FIGURE 4   Relationship between the fluidization number K and α, the degree of oxidation of ammonia to NO: $d = 0.7$ mm.

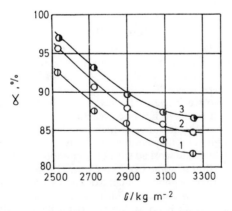

FIGURE 5   Relationship between the catalyst productivity and α, the degree of oxidation of ammonia to NO: $H_o/D_a = 0.64$; $d = 0.7$ mm; (1) 675°C; (2) 700°C; (3) 725°C.

experiments carried out over a period of 3 months showed that within such a period the degree of oxidation of ammonia to NO remains constant and close to 97%.

The application of a fluidized bed process considerably increases the stability of the Co_3O_4 catalyst, due to intensification of the heat exchange and elimination of the danger of local overheating. In addition, the catalyst poisons in this case are distributed over the whole mass of the catalyst and their effect should be negligible.

The studies carried out up to now show that cobalt oxide catalysts possess a high activity and selectivity in the oxidation of ammonia to NO, which could satisfy the practical demands. The increased apparent density, together with improved mechanical properties, allow the application of high linear and space velocities, and offer the possibility of an intensification of the oxidation process.

The method offers important advantages in comparison with the fixed bed and two-stage processes and could compete with the present industrial process using platinum catalysts.

REFERENCES

[1] S. Kunev, D. Georgiev-Klissurski, and E. Vateva, *Izvestya na Fizicheskia Institute na BAN* (Reports of the Institute of Physics, Bulgarian Acad, Sci.), 1960, **9**, 53.
[2] M. I. Temkin *et al.*, *Problemy fizicheskoi Khimii* (Physical Chemistry Problems) (Goskhimizdat, Moscow, 1959), vol. 2, p. 14.
[3] N. G. Kurin and M. S. Sakharov, *Khimiya i khimicheskaya Tekhnologiya*, 1960, **3**, 141.
[4] S. Kunev, D. Georgiev, and E. Vateva, *Compt. rend. Acad. bulg. Sci.*, 1962, **15**, 61.
[5] I. P. Mukhlenov, *Kataliz v kipyashtem Sloe* (Fluidized bed catalysis) (Khimiya, Leningrad, 1971) p. 15.
[6] M. Novak and J. Vosolsobe, *Chem. Průmysl*, 1970, **20**, 11.
[7] R. Griesbach and P. D. Reinhardt. *Z. anorg. Chem.*, 1955, **281**, 241.
[8] D. G. Georgiev, *Ph.D. Thesis*, Bulgarian Acad. Sci., 1963, p. 17.
[9] L. E. Apel'baum and M. I. Temkin, *Zhur. fiz. Khim.*, 1948, **22**, 179.
[10] D. A. Epstein, *Doklady Akad. Nauk. S.S.S.R.*, 1950, **74**, 1101.
[11] C. Predoiu, *Rev. Chim.* (*Bucharest*), 1967, **18**, 3.
[12] I. A. Zrunchev and T. F. Popova, *Compt. rend. Acad. bulg. Sci.*, 1975, **28**, 11.
[13] P. F. Popova, D. Georgiev, and I. A. Zrunchev, *Bulgarian Patent* 21825.
[14] T. F. Popova, *Ph. D. Thesis*, Higher Institute of Chemical Technology, Sofia, 1975.

DISCUSSION

G. Rienäcker (*Zentralinst. fur Phys. Chem. der AdW., Berlin*) said: Your results are very impressive. You carried out experiments over a period of three months with constant activity of the catalyst, yielding 97% NO. How high was the correlation of ammonia in these experiments?

Twenty-five years ago we investigated technical cobalt oxide catalysts which had been used in production plants in fixed-bed reactors, and analysed the oxidation state of the cobalt. Fresh catalysts and those which had worked with 7% ammonia had a constant and high activity and they contained cobalt in a high oxidation state. On the other hand, catalysts which had worked with 11% ammonia showed decreased activities and a remarkable reduction of some Co^{3+} to Co^{2+}. Did you investigate the oxidation state of your catalysts? In our opinion, a high oxidation state of cobalt is very important, and therefore a relatively low concentration of ammonia is favourable. A mixture of 10·5 or 11% ammonia with air is not sufficiently oxidizing to maintain the high oxidation state of the catalysts and their high and constant activity in the normal fixed-bed reactors.

D. Georgiev-Klissurski (*Inst. of Gen. and Inorg. Chem., Sofia*) replied: Experiments with various concentrations of ammonia in the reaction mixture showed that the optimum concentration was 7%. This concentration ensures a high oxidation state of the catalyst.

X-Ray analysis of the catalyst has been performed after various working periods. Our results are in a good agreement with your ideas.

J. W. Geus (*Univ. of Utrecht*) said: I wish to pursue Prof. Rienacker's remark. We have found that the presence of Co_3O_4 is a prerequisite for an active and selective catalyst. When CoO is formed the activity and selectivity go down. However, it is difficult to ascertain the presence of CoO because it is oxidized to Co_3O_4 on cooling the catalyst in an oxidizing atmosphere.

The reason for changing from platinum to oxidic catalysts is that the platinum losses increase intolerably at higher pressures. Have you any experience with your process at higher pressures?

A. Georgiev-Klissurski (*Inst. of Gen. and Inorg. Chem., Sofia*) replied: High-pressure experiments are now in progress. It is expected that an improvement in the process will be achieved.

N. Pernicone (*Montedison, Novara*) said: Concerning the choice between fixed-bed and fluidized-bed reactors, you have found that the latter has some definite advantages. However, some disadvantages should be mentioned too. For instance, it may be very difficult to avoid formation of fines during fluidized-bed operation of this catalyst. Moreover, the separation of these fines from a gaseous phase containing water vapour and nitrogen oxides without any contamination of the nitric acid produced will require the solution of some technological problems. Have you any comments?

D. Georgiev-Klissurski (*Inst. of Gen. and Inorg. Chem., Sofia*) replied: The catalyst possesses a high mechanical strength which is registered in our patent. The catalyst fines remain in the condenser after the first cooler in the 5% nitric acid formed. This diluted acid can be utilized in the cycle.

T. P. Wilson (*Union Carbide, South Charleston*) said: Following up Dr. Pernicone's comments, I would like to emphasize that the use of a fluid-bed for a chemical process introduces problems of its own. Dr. Pernicone refers to the generation of fines and loss of catalyst in the flowing gas stream. Perhaps an even more serious problem is the design of a fluid bed to obtain the desired contact time. Dr. Georgiev-Klissurski refers to contact times of the order of 0·01 s. The maximum gas velocity in a fluidized-bed can hardly be greater than 10 ft s^{-1} and is more likely to be 1·0—1·5 ft s^{-1}. The bed depth required to achieve the required contact time is, then, 0·01—0·1 ft, probably less than one inch. Design of a large fluid-bed reactor to operate at this bed depth would be virtually impossible.

D. Georgiev-Klissurski (*Inst. of Gen. and Inorg. Chem., Sofia*) replied: The bed depth is a function of the desired contact time and linear velocity. The theoretical depth is smaller than one inch, but when taking into account the wall effect, the eventual non-homogeneous distribution of the gas stream, and the different sizes of the particles, the practical bed depth will be 30—40 mm. This depth is favourable for a large variety of reactors for oxidation of ammonia on a catalyst in the fluidized state.

I would like also to point out that in this work the contact time is defined in a different manner.

N. N. Bakhshi (*Univ. of Saskatchewan, Saskatoon*) asked: Can the catalyst be regenerated? If so, what is its activity, and have you used regenerated catalyst in your work?

D. Georgiev-Klissurski (*Inst. of Gen. and Inorg. Chem., Sofia*) replied: The catalyst can be regenerated, but we have not used regenerated catalyst.

Mechanism of Enantioselective Hydrogenation over Modified Nickel Catalysts

JOOP A. GROENEWEGEN and WOLFGANG M. H. SACHTLER*

Gorlaeus Laboratoria, Rijksuniversiteit Leiden, P.O.B. 75 Leiden, The Netherlands

ABSTRACT   The present work tries to elucidate the general laws of enantioselectivity, a particular case of stereoselectivity. As an example the enantioselective hydrogenation of methylacetoacetate (MAA) to methyl 3-hydroxybutyrate (MHB) on a modified nickel catalyst has been studied. The structure of the adsorption complexes of the relevant compounds has been established, using infrared spectroscopy as the analysing tool and silica supported nickel as the catalyst.

The results show that MAA is dissociatively adsorbed on nickel, forming an unsaturated chelate ring with one surface nickel atom. α-Amino-acids form saturated chelate rings, but hydroxy-acids tend to form carboxylates rather than chelates.

A mechanism is suggested, which not only rationalizes the opposite enantioselectivities due to an amino-acid and a hydroxy-acid having the same absolute configuration, but is also capable of qualitatively rationalizing a number of empirical observations reported in the literature.

INTRODUCTION

Catalytic reactions yielding molecules with asymmetric carbon atoms are of special interest if the possible enantiomers are formed in quantities deviating significantly from the equilibrium distribution. Such a process is called an enantioselective reaction and can be regarded as a particular case of stereoselectivity. To perform a catalytic enantioselective synthesis a requirement is that either one of the initial compounds or the catalyst already possesses a centre of chirality.

A prominent example of asymmetric synthesis with a heterogeneous catalyst is the Ziegler–Natta polymerisation of olefins which displays a high stereospecificity ($\sim$100%). The mechanism of this reaction has been thoroughly studied[1,2] and reveals some elements of general importance to stereospecific and enantioselective heterogeneous catalysis. Here, the chirality centre is created by the formation of a new carbon–carbon bond. A molecule with a double bond first forms a π-complex with the transition metal and is subsequently inserted in a M–R bond of the same transition metal atom, where M–R stands for Ti–C_nH_{2n+1}.

Stereospecificity results from two causes: (1) Of the four possible configurations in which the adsorbing propylene molecule can form a π-bond with the Ti^{3+} centre, one configuration is energetically most favourable. (2) For the insertion step, where the asymmetric carbon atom is formed, two reaction paths are possible, but one of these is clearly preferred. These two conditions essentially determine the tacticity of the product. The syndiotactic polymer is formed if, after insertion, a new monomer is adsorbed on the vacancy of the polymer chain, while the isotactic polymer results if after each insertion step the polymer chain shifts to its original position.

In the context of the present paper it should be noted that stereospecificity follows from the two basic conditions that for both the adsorption complex *and* the transition state of the insertion step one of a discrete number of possibilities is clearly preferred. Inspection of the detailed geometry then easily shows which configuration of the ad-complex and which reaction path leading to insertion are most favourable. Another example of an enantioselective catalytic reaction is the asymmetric hydrogenation of methyl acetoacetate (MAA) to methyl 3-hydroxybutyrate (MHB), to which the present

1014

study is related. Much work on this reaction has been done by Izumi[3] in Osaka, who usually performed this hydrogenation in an autoclave under high hydrogen pressures, Raney nickel has most frequently been used as the catalyst. Enantioselectivity was observed if the catalyst had previously been immersed in a solution of an optically active compound (the "modifier").

Amino-acids and hydroxy-acids are efficient modifiers.

The most important features of this reaction, as established experimentally by Izumi are:

(a) With (S)-amino-acid as a modifier, (R)-(—)-MHB is preferentially obtained, whereas with (S)-hydroxy-acid the (S)-(+)-MHB enantiomer is predominantly formed, although both modifiers have an identical absolute configuration. Replacement of the modifier by its antipode causes sign reversal of enantioselectivity in either case.

(b) The optical yield passes through a maximum if the pH of the modification solution is varied.

(c) If the hydrogen atom of the asymmetric carbon atom in the modifier is substituted by a bulkier group, the optical yield decreases.

(d) With increasing chain length of the modifier the optical yield decreases if hydroxy-acids are used as modifier, but increases in the case of amino-acids.

(e) A second carboxyl group increases the optical yield, in particular for hydroxy-acids. Monocarboxylic amino-acids are better modifiers than monocarboxylic hydroxy-acids, whereas dicarboxylic amino-acids are less effective than the corresponding hydroxy-acids.

(f) The presence of water during the hydrogenation induces sign reversal of enantioselectivity, but only in the case of amino-acid modifiers.

In the initial phase of this work in our laboratory it was established that for this reaction enantioselectivity is neither confined to Raney nickel nor to the presence of a liquid phase, but it is observed with equal sign if the hydrogenation is performed in the vapour phase, *e.g.* over supported nickel modified by solution[4] or modified by sublimation from the vapour.[5]

We therefore conclude that enantioselectivity is only due to the chemisorption complexes on nickel.

The aim of the present work is to elucidate the general laws of enantioselectivity with heterogeneous catalysts and, more in particular, from a knowledge of the structures of the adsorption complexes involved, to suggest a mechanism for Izumi's reaction.

EXPERIMENTAL

(1) *Catalyst*

Silica-supported nickel in the form of a pressed disc was used as the catalyst. It was prepared according to the precipitation method.[6] After the reduction in purified hydrogen, the ultimate catalyst contained 17% by weight of nickel. Average metal particle diameter, determined by hydrogen adsorption was about 35Å.

(2) *Analysing Technique*

Infrared spectroscopy has been used as the analysing tool since this technique is most useful for the determination of the identity of adsorbed species such as substituted carboxylic acids or β-diketones. Infrared spectra were recorded on a Perkin Elmer model 325 spectrophotometer, using the crossed chopper mode in order to eliminate infrared emission by the sample.

(3) *Modification procedure*

Because of the low volatility of amino-acids and hydroxy-acids, a special sublimation technique had to be developed to adsorb these compounds *in situ* on to a reduced nickel surface. The modifiers could be adsorbed from the vapour by subliming them at temperatures roughly 100°C lower than their melting points.

RESULTS AND DISCUSSION

(1) *Structure of the adsorption complexes.*

(1.1) *Amino-acids.*

From the resemblances of the infrared spectra to those of metal–amino-acid complexes[7] it was concluded that α-amino-acids are dissociatively adsorbed on nickel, forming a chelate ring with one nickel atom (see Figure 1).

It appeared to be possible to distinguish between adsorption on the metal and adsorption on the silica support. Upon heating the modified Ni/SiO$_2$ catalyst, amino-acid was desorbed from silica at a lower temperature than from nickel. Characteristic for α-amino-acids adsorbed on nickel are the strong carboxylate bands at 1400 and 1600 cm^{-1} and the amino stretching frequencies at about 3250 and 3350 cm^{-1}.

Amino-acids adsorbed on silica show a strong carbonyl band at 1740 cm^{-1}.

(1.2) *MAA and Related Compounds.*

MAA is adsorbed on nickel as an *O*-bonded chelate (see Figure 1) characterized by strong bands at 1600 and 1530 cm^{-1} (due to C$=$O and C$=$C respectively). This result, again, has been obtained by comparison of the spectra with the large amount of infrared data of metal complexes of β-diketones.[7]

The saturated product of MAA, MHB, is also readily adsorbed on an evacuated nickel surface, its spectrum being identical with that obtained if MAA is admitted. Obviously, dehydrogenation occurs leading to the resonance-stabilised six-membered ring, well known in β-diketone co-ordination chemistry.

Upon further investigation of this interesting phenomenon it appeared that it might be possible to follow spectroscopically the interconversion of the saturated and the unsaturated forms by admitting hydrogen or by evacuation at 75°C, respectively. Which of the two forms prevails on the surface depends on the hydrogen pressure rather than on the compound initially admitted.

Heating the catalyst *in vacuo* with the adsorbed hydrogenated species thus causes an *increase* in intensity of the bands due to MAA on nickel, as mentioned, while simultaneously the bands due to MAA adsorbed on silica *decrease* in intensity because of desorption of this less strongly bound complex. This feature permits an unambiguous distinction of the bands characteristic of MAA adsorbed on Ni and silica, respectively. It shows in particular that the 1740 cm^{-1} band is definitely due to the form adsorbed on silica.

(1.3) *Hydroxy-acids.*

Hydroxy-acids tend to form carboxylates rather than chelates, so their OH groups remain available for bridging with adjacent molecules. Carboxylates and chelates cannot be expected to display different infrared spectra if adsorbed on nickel.

However, on palladium, known for its strong co-ordination tendency, we were able to observe weak bands at 1675, 1600, and 1400 cm^{-1} upon adsorption of hydroxyacetic acid. The 1675 cm^{-1} band, which is ascribed to a C$=$O stretching frequency of an asymmetric carboxylate group[7] seems to show that, on palladium, chelate is indeed present as well as carboxylate. In view of the inferior co-ordination power of nickel it appears justified to assume that hydroxy-acids are almost exclusively chemisorbed in the carboxylate form on the latter metal.

(1.4) *Coadsorption.*

The spectra of the complexes remain essentially unchanged if MAA and either modifier are coadsorbed, although there are also indications for some interaction between the coadsorbed compounds. Coadsorbed MAA is more easily desorbed after hydrogenation than pure adsorbed MAA. Separate tests in a flow reactor proved that the Ni/SiO_2 catalyst modified by sublimation indeed displays enantioselectivity towards the same MHB enantiomer as reported by Izumi for Raney nickel catalysts modified by means of a solution. Optical yields are, however, lower for the silica-supported samples modified by sublimation.

(2) *Mechanism of the Asymmetric Hydrogenation.*
(2.1) *Postulates.*

The mechanism rationalizing the cause of asymmetric catalysis with modified nickel can be formulated in terms of three postulates, related to the structures of the adsorption complexes.

I   Enantioselectivity results from the exclusive stereochemical interaction of an MAA molecule and a modifier molecule adsorbed on two adjacent nickel atoms.

From this postulate it follows that for the animo-acid-modified catalysts the skeletons of the two chelates are located in planes roughly perpendicular to the surface and parallel to each other, the interplanar distance being equal to the nickel–nickel distance. This identification leaves open four possibilities for the coadsorption complex, with respect to the mutual position of the two chelates. Similarly four possibilities exist for the MAA–hydroxy-acid coadsorption complex.

II   One of the four arrangements, consistent with postulate I, has the lowest free energy.

It is a matter of discussion which arrangement has the lowest free energy. It should be noted, however, that any arrangement which is distinctly preferred will lead to enantio-selectivity. We believe that the preferred arrangements are:

(a) For amino-acid-modified catalysts, MAA and the alkyl group R are at opposite sides of the amino-acid plane to avoid the steric hindrance expected for arrangements having these moieties on the same side of this plane. Also for steric reasons the methoxy-group of MAA is as far as possible from the R-group of the amino-acid.

(b) For hydroxy-acid-modified catalysts the most favourable arrangement is such that a hydrogen bond is formed between the OH group of the hydroxy-acid and the methoxy-oxygen atom of MAA.

III   The two transition states arising from the interaction of hydrogen with adsorbed MAA have different free energies.

In the MAA–Ni chelate the carbon atom of the *C*-bonded methyl group is in the plane of the chelate skeleton. When the complex is hydrogenated a hydrogen atom is attached to the β-carbon atom and the methyl group is pressed out of the plane. There are two possibilities and for obvious reasons we believe that the transition state where the methyl group is bent away from the modifier molecule has the lower free energy, since severe steric hindrance would arise if the CH_3 group were pressed towards the modifier complex.

The most favourable adsorption complexes and transition states according to postulates (*II*) and (*III*) are shown schematically in the Figures 1 and 2.

(2.2) *Discussion of the postulates.*

In our opinion, *I* is the only really new postulate. For a number of reasons we reject alternatives such as the complex suggested by Klabunovskii and Petrov,[8] where MAA and modifier are ligands of one and the same metal atom. Such a coadsorption complex would be impossible for steric reasons on most crystal faces, the formation of two chelates, for example, requiring four vacant ligand sites on one Ni atom, which are not available. If, however, the nickel atom were pulled out of the surface by corrosive chemisorption a planar complex would result with the two ligands.

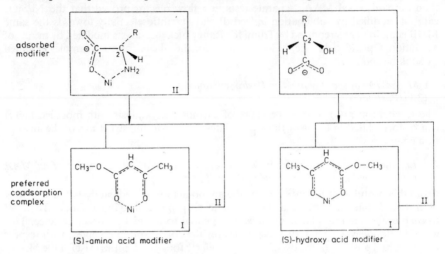

FIGURE 1   Preferred arrangements of the coadsorption complexes.

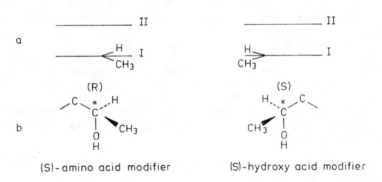

FIGURE 2   Preferred configurations of the new centre of chirality. (a) after formation of the new C*–H bond. (b) after desorption of MHB.

Such complexes are known[9] to have different infrared spectra and can hence be rejected on the basis of our spectroscopic data. Moreover, they fail to explain the phenomenon of enantioselectivity, as in these complexes the centre of chirality of the modifier ligand and the centre of prochirality of the substrate ligand are far apart and not in mutual stereochemical interaction.

The postulates (I) and (II) are basically identical to those responsible for stereospecificity in Ziegler–Natta catalysis as discussed in the introduction. These principles—the term "postulate" may appear overcautious—are apparently of general validity in enantioselectivity with heterogeneous catalysts. The assumption that hydrogen bonding determines the preferred complex with a hydroxy-acid modifier is supported by the observation of Yasumori[10] that the reaction order in MAA is 0·8 with tartaric-acid-modified Raney nickel but 1·0 in the case of amino-acid-modified or unmodified catalysts.

We fully agree with Yasumori's conclusion that this suggests a chemical interaction between adsorbed MAA and tartaric acid. Postulate (*III*) assuming that the reaction path over the lowest transition state prevails, also applies to those homogeneous hydrogenations where hydrogen attack is more favourable from one side than from the other. This obvious statement may, however, not be reversed; a preferred path of attack by a hydrogen atom is only one of a number of possible causes why one of the two transition states has a lower free energy. In the present case the difference in steric hindrance against bending the methyl group out of its original plane is assumed to be responsible for the difference in transition-state energy.

(2.3) *Confrontation of the suggested mechanism and published data.*

In this section it will be shown that the mechanism described not only rationalizes the opposite enantioselectivity induced by an amino-acid and a hydroxy-acid of identical absolute configuration, but also correctly predicts a number of observed phenomena. These include the observations b–f as mentioned in the introductory section.

(b) In the case of a small surface coverage with modifier, only a fraction of the adsorbed MAA is sterically influenced while the majority of the MAA not interfering with adsorbed modifier will, therefore, be hydrogenated statistically to racemic MHB. Also at high coverage the surface is not modified optimally because then an adsorbed MAA molecule will become sandwiched between two modifier molecules, their stereochemical interactions disturbing each other.

Also in this case racemic MHB will be predominantly formed. Accordingly, the optical yield in MHB should pass through a maximum while the surface coverage of the modifier varies from zero to one. This is quite in agreement with the independent observation that the amount of both glutamic acid and tartatic acid in the catalyst varies almost linearly with the pH of the modification solution, whereas the optical yield plotted against the pH of modification in either case shows a maximum. On a surface ideally modified in the sense of postulate (*I*) the distance between modifier molecules should be large enough to ensure that an adsorbed MAA molecule interacts with only one modifier molecule. On the other hand, the distance should be small enough to minimize the number of racemic sites, *i.e.* sites on which an adsorbed MAA molecule does not interact with any modifier molecule.

(c) If the hydrogen atom at the asymmetric atom of the modifier molecule is substituted by a more bulky group, both sides of the modifier will be equally unattractive for MAA adsorption on adjacent Ni atoms. As a consequence the optical yield will decrease.

(d) Since the most favourable arrangement in the sense of postulate (*II*) for amino-acids is due to steric hindrance with the alkyl group R, it follows that the optical yield increases as the size of R increases.

Because of the rotational freedom around the $C(1)$–$C(2)$ bond in the adsorbed hydroxy-acid the bulky group R hinders adsorption at both sides of the hydroxy-acid plane. Increasing size of R decreases adsorption adjacent to an adsorbed hydroxy-acid molecule and, consequently, the optical yield. Since steric hindrance plays a role, the bulkiness of the ester group can also be expected to be important.

This is confirmed by the experimental result that in the sequence ethyl acetoacetate, propyl acetoacetate, isopropyl acetate the optical yield decreases if tartatic acid is used as modifier.

(e) With dicarboxylic acids the second carboxyl group will also be adsorbed on the surface, fixing the molecule tightly. The side of the modifier which was merely hindered in monocarboxylic acids will be completely blocked for MAA adsorption in the case of dicarboxylic acids and consequently the enantioselectivity must be larger.

The presence of a second carboxyl group therefore increases the optical yield as is indeed found. The rotational freedom around the C(1)–C(2) bond of adsorbed hydroxy-acids mentioned in (d) is believed to be responsible for the smaller enantioselectivity of hydroxy-monocarboxylic acids as compared to amino-monocarboxylic acids. The adsorption complex of the latter, being a strong bidentate ligand, cannot have rotational freedom around the C(1)–C(2) bond.

Hence, fixation of the molecule by the presence of an additional carboxyl group also attached to the surface is expected to increase the optical yield much more in the case of hydroxy-acids than in the case of amino acids since adsorption adjacent to the hydroxy-acid plane is no longer hindered by the bulky group R. Indeed, it has been reported that hydroxy-dicarboxylic acids are even better modifiers than amino-dicarboxylic acids.[3] MAA fixation by the hydrogen bond is apparently more effective than by the steric hindrance between alkyl and methoxy-groups.

(f) Water will protonate the amino-group of an adsorbed amino-acid molecule. As a consequence the amino-group loses contact with the surface. The resulting species is very similar to an adsorbed hydroxy-acid as it is adsorbed through the carboxyl group only while the NH_3^+ group is capable of forming a hydrogen bond with the methoxy-oxygen of MAA. The optical selectivity will then be similar to that with a hydroxy-acid as modifier, *i.e.* opposite to that of the initially adsorbed anionic amino-acid. The adsorption complex of a hydroxy-acid is not likely to be modified by water.

Concluding Remarks

The preceding discussion shows that many of the empirical rules established experimentally can be readily understood on the basis of the model suggested. We know of no evidence which is in conflict with the model. There are, however, some observations reported in the literature which seem to defy a simple explanation in terms of this or any other mechanism put forward so far.

For the continuation of research in asymmetric catalysis it might be useful to draw attention to these puzzling phenomena, in particular the following:

Some (S)-amino-acid modifiers (*e.g.* glutamic acid) induce the preferred formation of either (R)- or (S)-MHB depending on the temperature of the modifying solution.[12]

Sometimes the optical yield is sensitive to the base used for adjusting the pH of the modifying solution.[13]

Acetylacetone and α,α-dimethyl-MAA,—the latter being unable to form an unsaturated chelate—have been reported to be hydrogenated with considerable enantioselectivity.[14]

The observations which illustrate that enantioselective catalysis is at least as complicated as it is fascinating, stress the need for further research.

In our opinion this should include:

a quantitative determination of the metal surface coverage with adsorbed modifier, substrate, solvent, and product molecules in the steady state of the reaction;

a determination of the quantitative kinetic parameters of the hydrogenation of MAA and other unsaturated compounds on nickel and other metals in the presence of various modifiers;

a study of enantioselectivity and of adsorption complexes on alloy surfaces, in particular those where modifier and substrate are adsorbed on only one of the alloy constituents.

ACKNOWLEDGEMENTS. The investigations were supported by the Netherlands

Foundation for Chemical Research (S.O.N.) with financial aid from the Netherlands Organization for the Advancement of Pure Research.

REFERENCES

[1] E. I. Arlman and P. Cossee, *J. Catalysis*, 1964, 3, 99.
[2] E. I. Arlman, *J. Catalysis*, 1966, 5, 178.
[3] Y. Izumi, *Angew. Chem.*, 1971, 83, 956.
[4] J. J. Stephan and R. Hijmans, *Master's Thesis*, Leiden, 1970.
[5] J. A. Groenewegen, *Thesis*, Leiden 1974.
[6] J. W. Geus, *Dutch Patent Application*, 1967, 6, 705, 259; 1968, 6, 813, 236.
[7] K. Nakamoto, *Infrared Spectra of Coordination Compounds*, (Academic Press, London, New York, 1968).
[8] E. I. Klabunovskii and Yu. I. Petrov, *Doklady Phys. Chem.*, 1967, 173, 269.
[9] Y. Fujii and T. Ejiri, *Bull. Chem. Soc. Japan*, 1972, 45, 283.
[10] I. Yasumori, Y. Inoue, and K. Okabe, *Catalysis, Heterogeneous and Homogeneous*, eds. B. Delmon and G. James (Elsevier, Amsterdam, 1975), p. 41.
[11] Y. Izumi, M. Imaida, T. Harada, T. Tanabe, S. Yajima, and T. Ninomiya, *Bull. Chem. Soc. Japan*, 1969, 42, 241.
[12] Y. Izumi, Y. Tatsumi, M. Imaida, Y. Fukuda, and S. Akabori, *Bull. Chem. Soc. Japan*, 1965, 38, 1206.
[13] T. Tanabe, K. Okuda, and Y. Izumi, *Bull. Chem. Soc. Japan*, 1973, 46, 514.
[14] Y. Izumi, personal communication.

DISCUSSION

I. Yasumori (*Tokyo Inst. of Technology*) said: The authors refer to our kinetic study in which a nickel powder catalyst modified with tartaric acid (TA) was used. I agree with their view that the decrease in the reaction order with respect to methyl aceto-acetate (MAA) on the modified catalyst suggests the existence of an attractive inter-action between the modifier and the reacting molecules on the surface. However, this change in reaction order is an averaged one in nature: the hydrogenation takes place simultaneously on the uncontrolled sites and on the controlled sites with modifier. A more direct evidence for its existence is obtained from the thermal desorption spectra of adsorbed MAA. The MAA molecules adsorbed on unmodified surface give a desorption peak at around 76 °C which shows no shift with increasing amount of MAA adsorbed. However, molecules adsorbed on the surface modified with TA give a peak shifted to the high temperature side by 15 °C when the amount of adsorbed MAA is below 7% of monolayer capacity and the heating rate is 12·5 °C min^{-1}. This suggests that MAA molecules occupy the site near the modifier because of the attractive interaction between them.

Further, various poisons have different effects on the optical yield when they are introduced onto the modified surface covered with a small amount of MAA molecules. Adsorbed thiophen causes no change in the optical yield, carbon monoxide decreases the yield, whereas pyridine increases the yield (though the total rate of hydrogenation is decreased). This evidently shows that molecules of the poison may preferentially occupy the site controlled by the modifier, depending on their relative bond strength with the surface in comparison with that of adsorbed MAA.

If the orientation of the modifier and reacting molecules on the surface are those postulated in Figures 1 and 2, it may be difficult for surface hydrogen to move up the gap between these molecules and to react with carbon atom of β-carbonyl in the reacting molecules. What do you think about the possibility of such a hydrogen transfer or intra-molecular hydrogen spill-over?

W. M. H. Sachtler (*Shell, Amsterdam*) replied: The results of your temperature programmed desorption and catalyst poisoning experiments are highly relevant. I am pleased to notice that they appear to be in excellent agreement with our proposed model.

As to your question, I agree that it would be of great interest to know in atomic detail, how the three hydrogen atoms become attached to MAA_{ads} to transform it into MHB. We have considered several paths for hydrogenation of the chelate and find that the required elongations in the $Ni \cdots H \cdots C$ transition states can indeed be very small for several of these paths. I do not wish to speculate, however, as to which of these will be preferred, since we have no experimental data on the matter.

A. Tai (*Osaka Univ.*) said: According to your results, the carboxyl group of the modifying reagent is in a dissociated form on the nickel surface. What kind of counter ion do you think is involved; is it a Ni^{2+} ion or does the metal surface accommodate the positive charge? According to our results, the direction of the optical yield changes with temperature. Do you think this phenomenon can be explained by a static stereochemical model? Your results suggest that, in gas phase reactions, MAA and MHB are interconvertible on the catalyst surface. Might this not be one reason for the lower optical yield?

W. M. H. Sachtler (*Shell, Amsterdam*) replied: For a metal surface, partially covered with adsorbed negative ions, the compensating positive charges will largely be concentrated on the metal atoms adjacent to the negative ions. The two alternatives which you mention thus become indistinguishable. Previously we studied the chemisorption of HCOOH on Ni or W and benzaldehyde on V_2O_5 surfaces.[1,2,3] Just as in the present case of multifunctional carboxylic acids adsorbed on Ni, these adsorptions are dissociative, a hydrogen atom being split off, and the i.r. spectra show that the organic fragments are very similar to those in corresponding bulk compounds (formates, benzoates). Our work function measurements of HCOOH on W further showed that the formate is negative with respect to the positive surface atoms (or ions).

With regard to your second question, our model predicts that enantioselectivity depends on the relative surface concentrations of the various adsorbates, and these concentrations should vary with temperature. This would, however, not predict a reversal of the sign of enantioselectivity. Sign reversal has, however, been reported by your group for amino-acid modifiers upon adding water. Our model explains this by admitting that a protonated amino-acid molecule will be adsorbed as a carboxylate rather than a chelate. From your present paper I notice that sign reversal upon lowering the temperature seems to be significant only for amino-acid modifiers, and I wonder whether this is, again, indicative of a preferred presence of this modifier in the carboxylate form at low temperature.

Lastly, interconversion of MHB and MAA would be a possible path for racemization, but in the experiments to which you refer we believe that the high flow rate and H_2:MAA ratio render this path less important than the increase in the number of non-selective nickel sites during hydrogenation.

G. L. Haller (*Yale University*) said: I wish to comment on the predictions of your model with respect to the effects of coverage on selectivity and ask if these predictions are supported by experiments. In your model, low enantioselectivity at low coverage is due to hydrogenation on Ni atoms far removed from adsorbed modifier, whereas at high coverage it is due to interactions when the MAA becomes 'sandwiched between two modifier molecules'. This model would therefore predict (i) a maximum in selec-

[1] W. J. M. Rootsaert and W. M. H. Sachtler, *Z. phys. Chem.* (*Frankfurt*), 1960, **26**, 16.
[2] W. M. H. Sachtler and J. Fahrenfort, *Actes 2me Congr. Internat. Catalyse*, Paris, 1960 (Editions Technip, Paris, 1961), Vol. 1, p. 831.
[3] W. M. H. Sachtler, G. J. H. Dorgelo, J. Fahrenfort, R. J. H. Voorhoeve, *Proc. 4th Internat. Congr. Catalysis*, Moscow, 1968 (Akademai Kiado, Budapest, 1971), Vol. 1, p. 454.

tivity at a modifier coverage below 0·5 and (ii) an increase in selectivity at low coverage using a poison that selectively covered nickel sites far removed from the modifier. Prediction (ii) appears to be supported by the results presented in the comment of Prof. Yasumori, but a preferred poison would be one with a functional group identical to the modifier, *e.g.* acetic acid on the surface modified by hydroxy acid. Have you measured the coverage at which the maximum in the selectivity occurs or performed poisoning experiments to test prediction (ii)?

W. M. H. Sachtler (*Shell, Amsterdam*) replied: I agree that, in the absence of additional complicating factors, the predictions which you have formulated would follow from our model. I would add that certain alloys might offer the conditions for preferential adsorption of substrate near modifier molecules. As regards your question, none of the measurements described have, to my knowledge, been carried out, but experiments where the pH of the modifying solution was varied have been described by the group in Osaka, and in our paper we say that these results appear to agree with the predictions of our model.

Comparative Study of the Rate and Optical Yield on the Enantioface Differentiating Hydrogenation of Acetoacetate with the Raney Nickel Modified with Optically Active Substance (MRNi)

Tadao Harada, Yuji Hiraki, Yoshiharu Izumi, Junzaburo Muraoka, Hiroshi Ozaki, and Akira Tai*

Institute for Protein Research, Osaka University, Yamada Kami, Suita, Osaka, Japan 565

ABSTRACT   Comparative studies of the rate and optical yield on the enantioface differentiating hydrogenation with modified Raney nickel reveal that the role of modifying reagent can be divided into two independent functions. One is to regulate the rate by occupying a part of the active area of the catalyst by a simple competitive inhibitor, and the other is to differentiate the enantioface of the substrate as a reference chiral system. The optical yield was a monotonic function of temperature and its sign changed at different temperatures (T_0) with different modifying reagents. The value of T_0 was found to be a useful parameter for the evaluation of enantioface differentiating reactions. Weak interactions between the substrate and the modifying reagent, rather than obvious factors such as steric bulk, were expected to determine T_0. The rate-determining step of hydrogenation was the addition of hydrogen to the adsorbed substrate; therefore the enantioface of the substrate was ultimately differentiated when the adsorption of the substrate occurred prior to the rate-determining step.

Introduction

Asymmetric hydrogenation of methyl acetoacetate with Raney nickel modified with optically active substances (MRNi) was first reported by one of us in 1963.[1] Since then, systematic studies on this catalyst have been continued by our research group.[2]

Other research groups have also studied on this catalyst or its analogues, and many instructive results have been reported.[3-6] However, the mechanism of reaction with this catalyst is still not clearly determined. Although stereochemical models have been produced in order to elucidate the function of this catalyst,[4,5] these intuitive models can explain only limited parts of our experimental results.

We have taken the view that it is essential to clarify the nature of the catalyst rather than to suggest tentative stereochemical models, and have carried out various investigations.

The reactions achieved by this catalyst are enantioface differentiating reactions according to the classification of asymmetric reactions proposed by one of us,[2] because the formation of optically active product is effected when the catalyst possesses the ability to differentiate between the *si* and *re* faces (enantiofaces) of a substrate.

In the course of studies on MRNi originally presented by our group, it was noticed that the reproduciblity of optical yields was obtained so long as the catalyst is prepared under the same conditions, although the rate of the reaction was not reproducible. These facts suggested that the hydrogenation process and the enantioface differentiation process are independent in the course of reaction.[7]

This report presents a part of our efforts in connection with comparative studies on the hydrogenation activity and enantioface differentiating ability of MRNi.

Experimental

Gas-chromatographic analysis was carried out with a Shimadzu GC-3AH apparatus using 300 × 0·5 cm stainless-steel column packed with PEG 20-M (10%) on Chromo-sorb W. Optical rotations were measured with a Perkin-Elmer 241 polarimeter at 578 nm with a quartz cell (length of light path 1 dm).

All materials except those listed below were obtained from commercial sources and used without further purification. n-Propyl, n-butyl, isopropyl, and t-butyl aceto-acetate,[8] L-2-hydroxy-3-methylbutyric acid,[9] R-3-hydroxybutyric acid,[10] L-*NN*-dimethylalanine,[11] and L-*N*-methylvaline were prepared by published procedures. Optically pure R-methyl 3-hydroxybutyrate was prepared by the esterification of R-3-hydroxybutyric acid with diazomethane. Other esters of R-3-hydroxybutyric acid were prepared from R-methyl 3-hydroxybutyrate by ester exchange with the corresponding alcohols. The optical rotation (rad) of these esters (neat) are —7·48 (methyl), —5·06 (ethyl), —4·62 (n-propyl), —4·58 (isopropyl), and —4·55 (n-butyl), respectively.

Preparation of Catalyst.

In order to obtain reproducible results with respect to the rate and optical yield, the procedure for preparation of the catalyst had to be improved. The reproducibility of rate on four repeated runs with new catalyst are as follows, 13·3, 13·0, 13·1 and 13·2 (mmol/h). The standard procedure for the preparation of the new catalyst is as follows; into an alkaline solution (52 g of sodium hydroxide in 260 ml of deionized water), which had been cooled to 20°C on an ice bath, 19·5 g of well pulverized Raney alloy (Ni 42%, Al 58%, Kawaken Finechemical Co.) was added in portions at such a rate that the temperature of the mixture was maintained at 20 ± 2°C with occasional shaking. The resulting suspension of Raney nickel was kept at 75–78°C for 45 min, then allowed to stand at room temperature for 1–2 h. After removal of the alkaline solution by decantation, the Raney nickel was washed four times with 650 ml portions of deionized water. The modification of the catalyst was carried out at 0°C by a published procedure.[12] The values of modifying pH were adjusted to an isoelectric point for each monoamino-monocarboxylic acid; to 5·1 for D_s-tartaric acid, L-2-hydroxy-3-methyl-butyric acid, and L-glutamic acid, and to 9·6 for L-lysine and L-ornithine.

Hydrogenation.

The liquid-phase hydrogenation under atmospheric pressure was carried out with a specially designed 100 ml flask having a liquid inlet funnel, a gas inlet side-tube, and a heating jacket containing hot water. Into the flask, which had been filled with hydrogen and connected with a graduated hydrogen reservoir through a side tube, the fixed amount of MRNi suspended in 17·5 ml of substrate (or solution of substrate) was introduced. The flask was fixed to a reciprocal shaker and the jacket was connected with a hot-water circulation system. The heating and shaking of the reaction mixture were started simultaneously. The hydrogen uptake was followed by taking measure-ments every 10 min. The rate of hydrogenation was determined from the amount of hydrogen consumed in 1 h after the initial 10 min.

Determination of Optical Yield.

The optical yield of product was determined from the optical purity of the product obtained at less than 40% conversion. After removal of catalyst from the reaction mixture by filtration, the filtrate was distilled under reduced pressure. The optical rotation and the ratio of product and unreacted substrate in the distillate were deter-mined by means of polarimetry at 25°C and gas chromatography at 90°C, respectively. The optical purity of product was determined by means of a calibration chart.

Determination of Modifying Reagent Adsorbed on the Catalyst.

Determination of amino-acids adsorbed on the catalyst was performed by the micro-Kjeldahl method. Determination of hydroxy-acids adsorbed on the catalyst was carried out by a published method.[13]

RESULTS

The Effect of Concentration of Substrate [s] and Amount of Catalyst [c] on the Rate of Hydrogenation (v).

Hydrogenation of methyl acetoacetate (MAA) diluted with methyl 3-hydroxybutyrate (MHB) ranging in concentration 1·86 mol/l to 9·29 mol/l (neat) was carried out at 70°C with 0·6 g of Raney nickel modified with D_s-tartaric acid (D_s-tartaric acid MRNi). The results are listed in Table 1. A plot of log (v) against log $[s]$ gave a straight line and with its slope, the rate of hydrogenation was expressed by $v \propto [s]^{0.2-0.3}$.

Table 1

Effect of Concentration of Substrate on the Rate of Hydrogenation

Concentration of Substrate $[s]$ (mol/l)	Rate of Hydrogenation (mmol/h)		
	D_s-Tartaric Acid-MRNi	L-Valine -MRNi	L-Glutamic Acid-MRNi
1·86	10·06	6·15	3·50
5·58	12·71	8·05	4·83
9·29 (neat)	13·11	10·26	5·11

The rates of hydrogenation of MAA diluted with MeOH (1·33 mol/l) were determined at 40°C in the presence of various amounts of D_s-tartaric acid MRNi. The results are summarized in Table 2. A plot of v against $[c]$ gave a straight line through the origin, and hence the rate of hydrogenation was described as $v \propto [c]$.

Table 2

Effect of the Amount of Catalyst on the Rate of Hydrogenation

Amount of Catalyst (g)	3·6	2·4	1·2	0·6	0·3	0·15
Rate of Hydrogenation (mmol/h)	11·6	7·0	3·6	1·2	0·5	0·22

The Effects of Modifying Reagent on the Rate of Hydrogenation, Amount Adsorbed on the Catalyst, and Optical Yield of Product.

The hydrogenations of MAA (neat) were carried out at 70°C with various sorts of MRNi. The resulting rates and optical yields are summarized in Table 3, together with the amount of modifying reagent adsorbed on the catalyst.

Table 3
The Effect of Modifying Reagent on the Rate, Optical Yield, and
Amount Adsorbed on the Catalyst

Modifying Reagent	Rate of Hydrogenation (mmol/h)	Optical Yield (%)	Amount Adsorbed of Modifying Reagent (mmol/g cat.)
D_s-Tartaric acid	18·1	25·6	
L-Alanine	10·6	1·1	
L-Butyrine	10·3	4·1	
L-Valine	10·7	13·2	0·185
L-Leucine	10·3	4·2	
L-2-Hydroxy-3-methylbutyric acid	14·8	—0·7	0·133
L-NN-Dimethylalanine	13·2	—0·1	
L-N-Methylvaline	13·2	3·8	
L-Gultamic acid	7·2	8·9	0·216
L-Ornithine	3·5	6·7	
L-Lysine	3·9	7·3	0·273

Both the rates and optical yields depend upon the nature of the modifying reagent, but no direct correlations were observed between them. The rate of hydrogenation decreased in the following order with respect to types of modifying reagent, hydroxy-acid and N-alkyl-monoamino-monocarboxylic acid > monoamino-monocarboxylic acid > monoamino-dicarboxylic acid > diamino-monocarboxylic acid. The rate was identical with MRNi where the modifying reagents were of homologous structure, regardless of the difference in optical yield.

A plot of the rate against amount of modifying reagent adsorbed resulted in a straight line. No systematic relationships were found between the optical yield and amount of modifying reagent adsorbed.

Temperature Dependence of Rate and Optical Yield

The effects of temperature upon the optical yield and rate of hydrogenation were investigated in the range 30°–80°C with various sorts of MRNi. The Arrhenius plot with respect to the rate of hydrogenation with each MRNi is shown in Figure 1. All plots were parallel to each other and the plots corresponding to modifying reagents of homologous structure overlapped.

The values of apparent activation energy were 44 ± 2 J/mol in all cases. The rate of hydrogenation with all the kinds of MRNi studied depended solely on the pre-exponential term. The apparent kinetic parameters of hydrogenation with MRNi's of homologous structure were found to be equal regardless of the difference in the optical yield.

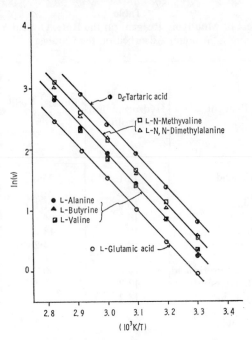

FIGURE 1   Arrhenius Plots of Hydrogenations with Various Sorts of MRNi

Figure 2 shows a plot of the optical yield (the sign is taken as + in the case of an excess of R-isomer) against reaction temperature. The optical yield was found to be a monotonic function of reaction temperature. All the plots crossed the zero value of optical yield at different temperature with respect to the kind of modifying reagent, except for D_s-tartaric acid MRNi.

Hydrogenation of Homologous Esters of Acetoacetic Acid with MRNi.

The hydrogenations of homologous esters of acetoacetic acid were carried out at 70°C with L-valine MRNi. The rates and optical yields are listed in Table 4. The rate of reaction decreased with increases of chain length and side-chain. However, no systematic changes were found in the optical yield of product with the changes of ester group. The Arrhenius plots with respect to the rates of hydrogenation of butyl acetoacetate and MAA are shown in Figure 3. The plots were straight lines and were parallel to each other. The values of apparent activation energy were 44 ± 2 J/mol in both cases.

DISCUSSION

General Features of Hydrogenation with MRNi.

The fact that the rate of hydrogenation is 0·2—0·3 in order with respect to the concentration of substrate, and the linear relationship between the rate and amount of catalyst, indicate that the supplies of substrate and hydrogen to the active area of the catalyst are sufficient under our experimental conditions. By taking the value of apparent activation energy into consideration, the rate-determining step of the hydrogenation is shown to be a surface reaction.

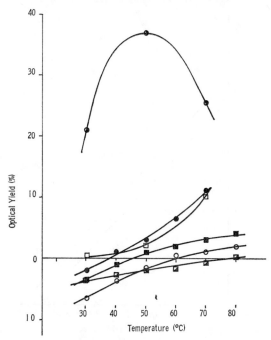

FIGURE 2   Temperature Dependence of Optical Yield

Modifying reagent
◐ D$_s$-Tartaric acid
□ L-Glutamic acid
● L-Valine
▦ L-Butyrine
○ L-Alanine
◪ L-3-Hydroxy-2-methylbutyric acid

Table 4

Effect of Structure of Ester group of Acetoacetate on the Rate and
Optical Yield
$CH_3 \cdot CO \cdot CH_2 \cdot CO_2R$

R	Concentration* Neat (mol/l)	Rate of Hydro-genation (mmol/h)	Optical Yield (%)
Methyl	9·22	10·7	11·7
Ethyl	7·87	9·7	16·4
n-Propyl	6·67	8·7	14·9
n-Butyl	6·18	7·0	11·1
IsoPropyl	6·84	6·2	14·3
t-Butyl	6·11	3·8	—

* The values at 25°C.

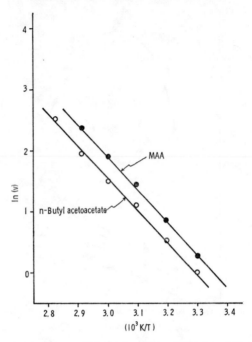

FIGURE 3   Arrhenius Plots of Hydrogenations of MAA and Butyl Acetoacetate

The Role of Modifying Reagent and Ester Moiety of Substrate.

Various investigations on the state of the modifying reagent on the catalyst surface[13-15] have revealed that the modifying reagent is adsorbed on the nickel surface as a chelate. In general, the nickel chelate formed with an amino-acid is more stable than that formed with a hydroxy-acid. Hence, the order of adsorption ability of modifying reagents is expected to be hydroxy-acid < monoamino-monocarboxylic acid < diamino-mono-carboxylic acid. The order corresponds to the reverse of that found for the reaction rates. When the linear relationship between the amount of modifying reagent adsorbed and rate of reaction, and the constancy of activation energy with change of the kind of modifying reagent, are taken into account, it is clear that one of the functions of the modifying reagent is to decrease the active area of the catalyst as a simple inhibitor without changing the mechanism of hydrogenation. Thus, the value of the pre-exponential term in the Arrhenius equation with respect to each MRNi determines the active area remaining in a given weight of catalyst.

Since there are no correlations between the rate and the optical yield, while both of them vary with changes in the nature of the modifying reagent, the role of the modifying reagent in the enantioface differentiating reaction can be divided into two independent functions. One is to regulate the rate of hydrogenation as a simple inhibitor, and the other is to differentiate the enantioface of substrate as a chiral reference system. The kinetic results make it clear that neither of the functions of the modifying reagent changes the mechanism of hydrogenation.

In the rate studies on the hydrogenation of the homologous esters of acetoacetic acid, no quantitative relationship was found between the rate and the structure of the ester moiety. However, it may be expected that two factors, the steric bulk of the ester group and the molar concentration of neat substrate, determine the rate of hydrogenation. As the values of the apparent activation energy for MAA and butyl acetoacetate are equal, and hence the mechanism of hydrogenation is expected to be the same, it is reasonable to suppose that the ester group is only responsible for the equilibrium amount of substrate adsorbed on the catalyst surface. The lack of correlation between the rate and the optical yield suggests that the structure of the ester group in the substrate independently affects the rate of reaction and the enantioface differentiating ability of MRNi.

All the considerations mentioned in this section lead to the notion that "hydrogenation" and "enantioface differentiation" are independent on the surface of catalyst.

The Nature of Enantioface Differentiation.

The temperature dependence of the ratio of enantiomers (C_R/C_S) is often expressed by the equation (1).

$$\ln(C_R/C_S) = (\Delta G_R^{\ddagger} - \Delta G_S^{\ddagger})/RT = \Delta\Delta G^{\ddagger}/RT = \Delta\Delta S^{\ddagger}/R - \Delta\Delta H^{\ddagger}/RT \quad (1)$$

The plot of $\ln(C_R/C_S)$ against $1/T$ should give a straight line so long as the differences of entropies $(\Delta\Delta S^{\ddagger})$ and enthalpies $(\Delta\Delta H^{\ddagger})$ are temperature-independent. However, none of our results gave linear relationships between $\ln(C_R/C_S)$ and $1/T$. In the first place, the equation (1) is an approximate expression of first order, and is not sufficient to evaluate $\Delta\Delta G^{\ddagger}$ properly because the major part of $\Delta G^{\ddagger}$ is already cancelled out in the value of $\Delta\Delta G^{\ddagger}$ and the accumulation of minor energetic differences plays an important role in the value of $\Delta\Delta G^{\ddagger}$. In other words our results indicate that the phenomenon of differentiation must be expressed by the summation of the differences in minor mutual interactions between modifying reagent and substrate.

The continuous change of optical yield from — to + (or *vice versa*) with elevation of temperature is not compatible with any kind of explanation based on the static stereochemical models.

The temperature at which the direction of differentiation changes (T_0) is an important parameter in investigating the mechanism of enantioface differentiating reactions. The empirical rule[2] that the direction of enantioface differentiation with $RCHXCO_2H$ MRNi is reversed when X is changed from NH_2 to OH can be uniformly understood as the change of T_0.

The exceptional behaviour of D_s-tartaric acid MRNi can be explained by the predominant effect of hydrogen bonding at temperatures below 60°C, as has been reported.[14] The above explanation is also supported by the fact that the temperature giving the optimum optical yield corresponds to the initial temperature of the dissociation of hydrogen bonding.

Although the correlation between T_0 and the structure of modifying reagent is still equivocal, we are engaged in studying this problem.

As a result of rate studies, the eqn. (2) is obtained from the eqn. (1) when we presume an Arrhenius intermediate,

$$\ln (C_R/C_S) \propto [(\Delta\Delta G^{\ddagger} - \Delta\Delta G^{A}) + \Delta\Delta G^{A}] \quad (2)$$

where $\Delta\Delta G^{A}$ means the difference of free energies of adsorption. In eqn. 2, the first term and second term are related to the differentiation at the rate-determining step and to that at the adsorption step, respectively. The different optical yields regardless of their identical reaction rates with respect to MRNi of homologous modifying reagents mean that the rates of formation of enantiomers, v_R and v_S, are not independent but restricted to $v = v_R + v_S$. The relation, $v = v_R + v_S$, can be rationalized on the assumption that modifying reagents interact with the substrate on the catalyst and the interaction results in the disproportionate formation of the diastereomeric adsorbed species without changing the total amount of adsorbed substrate.

It can be concluded at the present stage that the ratio of enantiomers depends principally on the term $\Delta\Delta G^A$, and that ultimately enantioface differentiation of the substrate is effected prior to the rate-determining step.

REFERENCES

[1] Y. Izumi, M. Imaida, H. Fukawa, and S. Akabori, *Bull. Chem. Soc. Japan*, 1963, **36**, 21.
[2] Y. Izumi, *Angew. Chem. Internat. Edit.*, 1971, **10**, 871; Y. Izumi, A. Tai, K. Hirota, and T. Harada, *Kagakusosetsu*, 1974, **4**, 85.
[3] I. Yasumori, Y. Inoue, and K. Okabe, *Internat. Symp. on Relation between Heterogeneous and Homogeneous Catalytic Phenomena*, 1974, p. 141.
[4] E. I. Klabunovskii and Yu. I Petrov, *Doklady Akad. Nauk S.S.S.R.*, 1967, **173**, 1125.
[5] J. A. Groenewegen and W. M. H. Sachtler, *J. Catalysis*, 1975, **38**, 501.
[6] L. H. Gross and R. Rys, *J. Org. Chem.*, 1974, **39**, 2429.
[7] H. Ozaki, A. Tai and Y. Izumi, *Chem. Letters*, 1974, 935.
[8] S. O. Lawesson, S. Gronwall, and R. Sanberg, *Org, Synth.*, Coll. Vol. V, 1973, p. 155.
[9] M. Winitz, L. Bloch-Frankenthal, N. Imaizumi, S. M. Birnbaum, G. C. Baker, and J. P. Greenstein, *J. Amer. Chem. Soc.*, 1956, **78**, 2423.
[10] P. A. Levene and H. L. Haller, *J. Biol. Chem.*, 1925, **65**, 49.
[11] R. E. Bowman and H. H. Stroud, *J. Chem. Soc.*, 1950, 1342.
[12] Y. Izumi, T. Harada, T. Tanabe, and K. Okuda, *Bull, Chem. Soc. Japan*, 1971, **44**, 1418.
[13] T. Harada, *Bull. Chem. Soc. Japan*, 1975, **48**, 3236.
[14] I. Yasumori, Y. Inoue, K. Okabe and Y. Izumi, *Japan and U.S.S.R. Seminar on Catalysis* (Alma-Ata, U.S.S.R.), 1974, preprint.
[15] J. A. Groenewegen and W. M. H. Sachtler, *J. Catalysis*, 1974, **33**, 176.

DISCUSSION

W. M. H. Sachtler (*Shell, Amsterdam*) said: As to your new discovery that, for some modifiers, the sign of selectivity is reversed upon lowering the reaction temperature, I fully agree that it will not be easy to explain this in terms of a static stereochemical model. However, your final conclusion, that enantioselectivity results from the existence of two adsorbed states, differing in free energy by $\Delta\Delta G^A$, and two transition states differing by $\Delta\Delta G^{\pm}$, appears to be exactly equivalent to our postulates II and III respectively based on static stereochemical models (p. 1017). Would you care to give us your views as to the causes of the sign reversal of enantioselectivity for some modifiers at a temperature T_0?

A. Tai (*Osaka Univ.*) replied: Equation (2) in the text holds for any kind of competitive reaction in which the Arrhenius intermediate is involved. I do not think that our terms, $\Delta\Delta G^A$ and $\Delta\Delta G^{\pm}$, are exactly equivalent to your postulates II and III, because the entropy term is not taken into account in your stereochemical model. The question is puzzling so long as we persist in an explanation based on stereochemical models. We are seeking to interpret these phenomena in terms of the participation of the entropy term and of the temperature dependent enthalpy term, as mentioned in the text.

J. A. Groenewegon (*Univ. of Leiden*) said: The constancy of the activation energie you have measured seems to contradict our third postulate (see p. 1017), which says that a difference in activation energy causes optical activity in the product. However, a difference in activation energy of 2 J mol^{-1}, which is your experimental error, already causes the two velocity constants to differ by a factor of three, which means an optical yield of 75%. So the optical yield may vary considerably, without there being noticeable departure from linearity in your Arrhenius plots.

A. Tai (*Osaka Univ.*) replied: I agree with your comment with respect to the experimental error, but I would remind you that your postulate will not be confirmed experimentally unless the rate constants, k_S and k_R, are independently determined. As mentioned in the text, our results are explained rationally if we introduce the idea that the enantio-differentiation is independent of the rate-determining step is and determined by $\Delta\Delta G^A$.

J. A. Groenewegen (*Univ. of Leiden*) said: With tartaric acid as a modifier, increasing bulkiness of the ester group in the substrate causes a decrease of the optical yield.[1] According to your Table 4, for L-valine modifier, no relationship is found between the bulkiness of the ester group and the optical yield. These facts support our view that amino-acids and hydroxy-acids act differently in directing the absolute configuration at the asymmetric carbon atom in the product. Do you agree?

A. Tai (*Osaka Univ.*) replied: The difference could be explained by the effect of the functional group of the modifier, as you propose. However, the results are not extensive enough to lead to a general rule.

A. Hoek (*Univ. of Leiden*) said: On the fourth page of your paper you state that 'no systematic relationships were found between the optical yield and the amount of modifier adsorbed'. Does this invalidate your earlier reports which claimed that, upon varying the pH of the modifying solution, the amount of adsorbed modifier barely changes,[2,3] whereas at the same time the optical yield passes through a maximum?[4] My second question concerns your assignment of the rate-determining step. What reasons have you to believe that the rate-determining step is the addition of hydrogen to the adsorbed substrate and not, for instance, desorption. As you know Prof. Yasumori found a reaction order of unity in MAA for the overall rate[5] so the adsorption step can also be rate-determining.

A. Tai (*Osaka Univ.*) replied: The optical yield is expected to be determined by the nature of adsorbed species and of modifiers.[6] Although various factors determine the amount of adsorption, it is not reasonable to correlate the optical yield to the amount of adsorbed modifier without considering the nature of adsorbed species.

As we say in our text, if the rate-determining step is a diffusion controlled process other than a surface reaction, the value of activation energy must be less than 30 kJ mol^{-1}. Although detailed studies of the mechanism of the surface reaction have not been carried out, the rate-determining step should be one of the elementary processes of the surface reaction.

Your last statement is not sound because the order in substrate changes with a change in concentration of the substrate.

A. H. Weiss (*Worcester Polytechnic Inst., Worcester, Mass.*) said: You show that optical yield changes from negative to positive as temperature increases (Figure 2). Does this mean that racemic species can be used as modifiers? Secondly, what kind of physical change could have occurred on the modified surface to permit the negative to positive yield to proceed? For example, could a shift from D to L have occurred in the chemisorbed modifier?

A. Tai (*Osaka Univ.*) replied: According to thermodynamic principles, the configurational changes of modifier such as DL → L or D and L → D or D → L are impossible. Therefore the conversions of negative to positive optical yield are not explained by the configurational change of modifier on the catalyst surface.

[1] Y. Izumi, M. Imaida, T. Harada, T. Tanabe, S. Yajima, and T. Ninomiya, *Bull. Chem. Soc. Japan*, 1969, **42**, 241.
[2] Y. Izumi, M. Imaida, H. Fukawa, and S. Akabori, *Bull. Chem. Soc. Japan*, 1963, **36**, 21.
[3] S. Tatsumi, *Bull. Chem. Soc. Japan*, 1968, **41**, 408.
[4] Y. Izumi, *Angew. Chem.*, 1971, **23**, 956.
[5] I. Yasumori, Y. Inoue, K. Okabe in *Catalysis, Heterogeneous and Homogeneous*, ed. B. Delmon and G. Jannes (Elsevier, Amsterdam, 1975), p. 41.
[6] T. Harada, *Bull. Chem. Soc. Japan*, 1975, **48**, 3236.

Synthesis of Asymmetric Hydroxy-compounds *via* Hydrosilylation

JAN BENEŠ and JIŘÍ HETFLEJŠ*

Institute of Chemical Process Fundamentals, Czechoslovak Academy of Sciences,
165 02 Prague 6-Suchdol, Czechoslovakia

ABSTRACT   A series of novel chiral phosphines derived from tartaric and malic acids, camphor and galactose have been tested as ligands in the rhodium-catalysed hydrosilylation of ketones by diphenylsilane. The catalytic effectiveness of the *in situ* prepared catalysts was compared with that of cationic rhodium(I) complexes and the hydrido-rhodium complexes $RhHP_2$ (P = a diphosphine) and the effect of the Rh: ligand, the catalyst: substrate, and the silicon hydride: ketone ratios on chemical yields and optical purity of the hydroxy-compounds formed by hydrolysis of the hydrosilylation products was determined. The complexes $RhHP_2$ were found to be superior to the other types tested in their asymmetric efficiency.

INTRODUCTION

Until now a limited number of chiral ligands have been used to induce asymmetry in the products of transition metal-catalysed synthesis of various organic substances. Of the phosphines, those containing chiral phosphorus atom have mainly been employed. Although more easily accessible synthetically, the phosphines with chirality at a carbon atom so far utilized in asymmetric synthesis have usually been DIOP,[1] menthyl-[2] and neomenthyl-diphenylphosphine.[3]

We have attempted to extend this class of compounds by preparing chiral phosphines derived from other optically active natural substances such as α-hydroxy-acids, terpenes, and monosaccharides.[4] In the present work we report on their efficiency in synthesis of optically active alcohols by hydrosilylation of ketones catalysed with rhodium(I) complexes of several different types.

EXPERIMENTAL

All manipulations with oxygen-sensitive compounds were carried out in an argon atmosphere using sealed reaction vessels and Schlenk's technique. Specific rotations were measured on Polamat and Perkin-Elmer 141 polarimeters. Reaction mixtures were analysed by gas chromatography [Chrom 31 instrument, a column (1·8 m × 4 mm) packed with 20% Carbowax 20 M on Chromosorb W (60/80 mesh)]. Melting points were determined with a Kofler hot stage microscope and are uncorrected.

Chemicals. Diphenylsilane,[5] [{RhCl(ethylene)$_2$}$_2$],[6] [{RhCl(cyclo-octene)$_2$}$_2$],[7] [{RhCl(cyclo-octa-1,5-diene)}$_2$],[8] and [{RhCl$_2$(2-methylallyl)}$_n$][9] were prepared by literature procedures as indicated. Carbonyl compounds used were commercial products (Fluka A.-G., Buchs) which were purified by distillation and stored over a molecular sieve.

Chiral Ligands. (—)-2,3-*O*-Isopropylidene-2,3-dihydroxy-1,4-bis(diphenylphosphino)-butane (I) was prepared according to ref. 1; m.p. 88·5—89·5°C, $[\alpha]_D^{20}$ —12·5 (*c* 4·12, benzene).

(—)-2,3-*O*-Cyclohexylidene-2,3-dihydroxy-1,4-bis(diphenylphosphino)butane   (II) was obtained by the following procedure: Diethyl (—)-2,3-*O*-cyclohexylidenetartrate,[10] b.p. 142°C/133·3 Pa, $[\alpha]_D^{20}$ — 35·2 (neat), n_D^{20} 1·4600, was converted by lithium aluminium hydride reduction into (—)-2,3-O-cyclohexylidene-L-threitol, m.p. 53—54·5°C (diisopropyl ether), $[\alpha]_D^{20}$ — 5·56° (c 3·56, chloroform), which in turn was treated with toluene-p-sulphonyl chloride to give 1,4-ditosyl-2,3-O-cyclohexylidene-L-threitol, m.p. 107·5—108·5°C (ethanol), $[\alpha]_D^{20}$ — 13·82° (c 6, chloroform). Its treatment with lithium diphenylphosphide afforded the diphosphine, m.p. 118·5—120°C, $[\alpha]_D^{20}$ —19·65° (c 0·6, benzene) (for details see ref. 4).

(—)-2,3-O-Cyclopentylidene-2,3-dihydroxy-1,4-bis(diphenylphosphino)butane (III) was prepared in a manner similar to that described for the cyclohexylidene derivative,[4] m.p. 90—91·5°C, $[\alpha]_{546}^{20}$ — 19·9° (c 0·5, benzene).

(—)-2,3-O-Cyclohexylidene-2,3-dihydroxy-1,4-bis(cyclohexylphenylphosphino)butane (IV) was obtained by treatment of the appropriate ditosyl derivative with lithium cyclohexylphenylphosphide[11] as an oily product which was purified by column chromatography on silica gel with acetone–hexane (2:8) as an eluent; $[\alpha]_{546}^{20}$ — 6·1° (c 10, benzene).

(—)-2-Benzyloxy-1,4-bis(diphenylphosphino)butane (V) was synthesized in the following way. The diethyl ester of L-(—)-malic acid, $[\alpha]_D^{20}$ — 10·3° (neat), was treated with benzyl bromide in the presence of silver(I) oxide to give diethyl (—)-2-benzyloxysuccinate, b.p. 138—143°C/266·6 Pa, $[\alpha]_{546}^{20}$ — 42·46° (neat), which in turn was reduced by lithium aluminium hydride to (—)-2-benzyloxybutane-1,4-diol, b.p. 147—150°C/399·9 Pa, $[\alpha]_{546}^{20}$ — 37·9° (c 3·35, 96% ethanol), n_D^{20} 1·5240. The diol was treated with toluene-p-sulphonyl chloride to give (—)-2-benzyloxy-1,4-ditosylbutane-1,4-diol, m. p. 93·0—93·5°C, $[\alpha]_{546}^{20}$ — 34·5 (c 3·14, chloroform), which was converted by treatment with lithium diphenylphosphide into the corresponding diphenylphosphinoderivative; an oily product, $[\alpha]_{546}^{20}$ — 29·4° (c 4·77, benzene) (for details see ref. 4).

(+)-2-Benzyloxy-1-(diphenylphosphino)propane (VI). Ethyl lactate was treated with benzyl bromide to give ethyl (—)-2-benzyloxypropionate, b.p. 109—113°C/266·6 Pa, $[\alpha]_{546}^{20}$ — 65·0° (neat), n_D^{20} 1·4993, which by a procedure similar to the preceding case was converted via (+)-2-benzyloxypropan-2-ol ($[\alpha]_D^{20}$ + 31·4° (neat), n_D^{20} 1·5151) and (+)-2-benzyloxy-1-chloropropane ($[\alpha]_D^{20}$ + 11·0° (c 1·41, chloroform), n_D^{20} 1·5088) into the phosphine. The oily product thus obtained was purified by molecular distillation at 150°C/0·001 Pa; $[\alpha]_{546}^{20}$ + 4·28° (c 5·83, benzene) for the t.l.c.-pure product [Silufol, acetone: hexane (1:8)] (the synthesis of this phosphine is reported in detail elsewhere[4]).

(+)-2,9-Dibenzyl-4,7-diphenyl-4,7-diphosphadecane (VII). 1,2-Bis(diphenylphosphino)ethane (11 g, 0·027 mol) in tetrahydrofuran (150 ml) was stirred at room temperature with lithium (3·5 g, 0·5 mol). A deep-red solution was filtered through glass wool, and t-butyl chloride (0·027 mol) was introduced into the filtrate. After boiling for 15 min under a reflux condenser, a solution of (+)-2-benzyloxy-1-chloropropane (15·4 g, 0·07 mol) in tetrahydrofuran (50 ml) was added, the reaction mixture was stirred at room temperature for another 2 h and then worked-up in the usual way. The resulting viscous oil was purified by molecular distillation (bath temperature 180°C, pressure 0·00066 Pa), $[\alpha]_{546}^{20}$ + 3·73° (c 3·49, benzene). Calc. for $C_{34}H_{40}I_2P_2$ (544·8): C, 74·95%; H, 7·40%. Found: C, 74·62%; H, 7·43%.

(+)-1,3-Bis(diphenylphosphinomethyl)-1,2,2-trimethylcyclopentane (VIII) was obtained by reaction of (+)-1,3-ditosyl-1,3-dihydroxymethyl-1,2,2-trimethylcyclopentane,[12] m.p. 149·0—149·5°C, $[\alpha]_D^{20}$ + 19·52° (c 0·6, chloroform), with lithium diphenylphosphide as an oily product which was purified by thin-layer chromatography on silica gel with acetone–hexane (1·9) as an eluent; $[\alpha]_{546}^{20}$ + 78·7° (c 5·16, benzene).

6-Deoxy-6-diphenylphosphino-1,2,3,4-di-isopropylidene-D-galactose (IX) was prepared by treatment of 6-tosyl-1,2,3,4-di-isopropylidene-D-galactose,[13] m.p. 89·5—91°C, $[\alpha]_{546}^{20}$ — 65° (c 3, 1,1,2,2-tetrachloroethane) with lithium diphenylphosphide. The product had m.p. 131·5–132·5°C, $[\alpha]_{546}^{20}$ — 101·4° (c 1·52, chloroform).

Hydrosilylation procedure. All hydrosilylation experiments were carried out at 20°C in an argon atmosphere. The catalyst was prepared *in situ* from the alkene–rhodium complexes and phosphines in benzene solution or, in the case of cationic complexes and the complexes $RhHP_4$ (P = a phosphine), it was weighed and dissolved in benzene. In a typical example 10 mmole of ketone was introduced into a glass ampoule, then the catalyst (5×10^{-5} mol), and diphenylsilane (10 mmol) were successively added. The ampoule was sealed, and shaken by means of a vibrator in a thermostatted bath. The course of the reaction was followed by gas chromatographic analysis of the reaction mixture. After disappearance of the ketone or after its concentration in the mixture attained a constant value, the products of the reaction were hydrolysed by aqueous-methanolic potassium hydroxide solution, the mixture was diluted with water, filtered, and extracted with three 30 ml-portions of diethyl ether. The organic layer was separated, washed with water (2×50 ml), and dried (K_2CO_3). After removal of the solvent by evaporation, the residue was distilled under reduced pressure.

RESULTS AND DISCUSSION

The results obtained with phosphines (I—IX) are summarized in Table 1. From inspection of the table it is evident that the derivatives of tartaric acid (I—IV) are exceptional in their asymmetric efficiency. The replacement of the isopropylidene by cyclohexylidene or cyclopentylidene groups (diphosphines II and III) has only a small effect on the asymmetric efficiency of the ligand, presumably on account of the great distance between the group and the reaction centre.

Another site which can be modified in derivatives of this type is phosphorus atom. During the reaction of a prochiral anion of the type $R^1R^2P^{(-)}$ with a chiral nonracemic

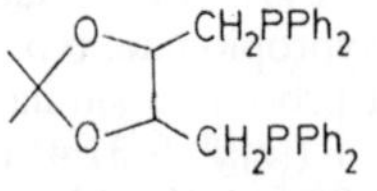

(I) (DIOP) 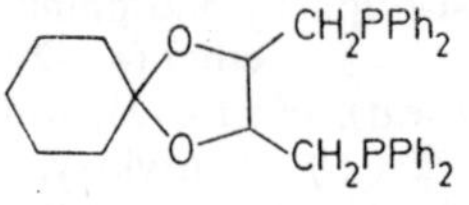 (II) 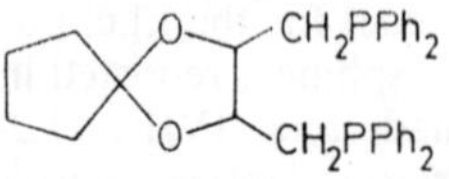 (III) 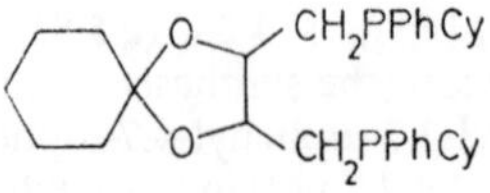 (IV) (Cy = cyclohexyl)

$Ph_2PCH_2CH(OCH_2Ph)CH_2CH_2PPh_2$

(V)

$CH_3CH(OCH_2Ph)CH_2PPh_2$

(VI)

$CH_3CH(OCH_2Ph)CH_2P(Ph)CH_2CH_2P(Ph)CH_2CH(OCH_2Ph)CH_3$

(VII)

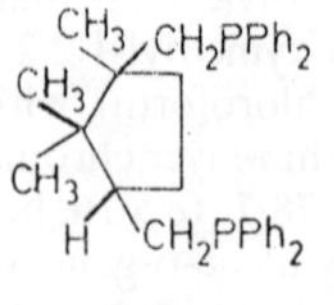

(VIII)

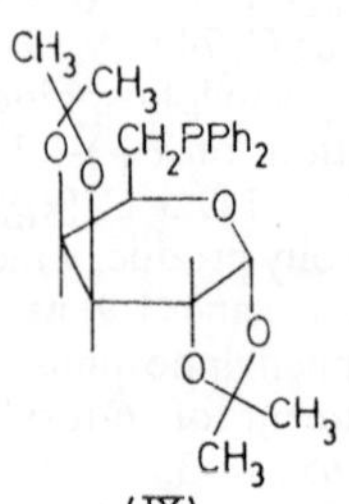

(IX)

Table I

Hydrosilylation of ketones by $PhSiH_2$ catalysed by rhodium complexes

Catalyst	Ketone	Chemical yield, %[a]	Alcohol Optical purity, %	Absolute configuration
[{RhCl(η-C_2H_4)$_2$}$_2$] + Phosphines (I–IX)[b]				
(I)	PhCOMe	72	22·5	(R)–(+)
	PhCOEt	85	22·5	(R)–(+)
	BzCOMe	84	18·8	(R)–(−)
	PhCH$_2$CH$_2$COMe	81	15·4	(R)–(−)
(II)	PhCOMe	80	24·8	(R)–(+)
	PhCOEt	82	26·0	(R)–(+)
	BzCOMe	93	16·0	(R)–(−)
	PhCH$_2$CH$_2$COMe	96	18·0	(R)–(−)
(III)	PhCOMe	78	25·2	(R)–(+)
	PhCOEt	80	25·7	(R)–(+)
	BzCOMe	81	16·8]	(R)–(−)
	PhCH$_2$CH$_2$COMe	85	18·0	(R)–(−)
(IV)	PhCOMe	79	44·0	(R)–(+)
	BzCOMe	87	18·0	(R)–(−)
(V)	PhCOMe	77	10·3	(S)–(−)
	PhCOEt	87	8·4	(S)–(−)
	BzCOMe	91	7·2	(S)–(+)
	PhCH$_2$CH$_2$COMe	88	8·6	(S)–(+)
VI)	PhCOMe	84	8·0	(R)–(+)
	BzCOMe	84	4·0	(R)–(−)
(VII)	PhCOMe	74	10·7	(R)–(+)
	BzCOMe	83	3·6	(R)–(−)
(VIII)	PhCOMe	69	4·0	(S)–(−)
	BzCOMe	56	6·0	(S)–(+)
(IX)	PhCOMe	56	2·4	(R)–(+)
	BzCOMe	87	8·7	(R)–(−)
Cationic and hydrido-rhodium complexes				
[L$_2$^{1}RhH][c,d]	PhCOMe	72	32·8	(R)–(+)
[L^{2}Rh(1,5-COD)]$^+$ BF$_4$$^-$ [e,f]	PhCOMe	79	27·6	(R)–(+)
(Rh-L^2-BPh$_4$-MeOH)[e,g]	PhCOMe	83	20·6	(R)–(+)
[L$_2$^{2}RhH][d,e]	PhCOMe	85	49·0	(R)–(+)
[L^{1}Rh(1,5-COD)]$^+$ BF$_4$$^-$ [c,f]	PhCOMe	77	22·0	(R)–(+)

[a] Determined by g.l.c.
[b] The ketone to silane mol. ratio = 1:1, the ketone to Rh mol. ratio = 200:1, the Rh to (I–V, VII–IX) mol. ratio = 1:1·05 (Rh: VI mol. ratio = 1:2·1), 20°C.
[c] L^1 = phosphine (II).
[d] Prepared according to ref. 15.
[e] L^2 = phosphine IV).
[f] Prepared according to ref. 14.
[g] Prepared according to ref. 16.
Bz = benzyl.

compound with formation of the tertiary phosphine $PR^1R^2R^3$, asymmetric synthesis is taking place on the phosphorus atom, resulting in formation of a mixture of diastereoisomers. These isomers could in favourable cases be separated and the substances with chirality at the phosphorus atom could be obtained, which otherwise are synthesized only with difficulty and by multistep procedures. Although we have so far been unable to realize this separation, data for diphosphine (IV) show that even the mixture of diastereoisomers produces the corresponding carbinol from acetophenone with optical purity nearly twice as high as that obtained with diphosphines (I—III). Separation of the aryl group from the C=O group by insertion of a methylene group results in a decrease of asymmetric induction to the level exhibited by diphosphine (I). We have obtained very similar results to those with the catalysts based on the rhodium–ethylene complex as a precursor, using [{RhCl(cyclo-octene)$_2$}$_2$],[{RhCl(cyclo-octa-1,5-diene)}$_2$], and [{RhCl$_2$(2-methylallyl)}$_n$].

Mechanistic interpretation of the absolute configurations obtained, as well as the observed variations of optical purity with the ligand used is at present difficult. By analogy with other asymmetric syntheses it seems likely that the rigidity of the system plays an important role. In the case of ligands (I—IV) this rigidity is ensured by the ring-closure of the OH functions *via* the alkylidene groups and also by co-ordination of the two phosphorus atoms to the rhodium.

Current theories of asymmetric synthesis are essentially based on the theory of the asymmetric rotator.[17-19] They do not, however, allow us to predict the absolute configuration of the alcohol formed in the case of bidentate ligands with the same, mutually dependent chiral centres [phosphines (I—IV)], nor for the ligands (VII—IX) containing several chiral centres. From the inspection of a Dreiding model of the ligand (VIII) it follows further that here the co-ordination of the two phosphorus atoms to the one rhodium atom is hardly likely.

The dependence of asymmetric yield on the reaction conditions is illustrated by data presented in Table 2. The strongest effect is exhibited by the change in the Rh:phosphine molar ratio, both on the optical purity of the product and on the amount of the carbinol present in the mixture after hydrolysis of the hydrosilylation product:

The observed re-formation of the starting ketone during hydrolysis has been explained by hydrolytic cleavage of the silylenol ether[20] arising from dehydrogenative condensation. Our results show that the extent of this reaction increases with the increasing concentration of the phosphine-nonstabilized rhodium complex which is an effective catalyst for this dehydrogenative condensation.

The use of cationic rhodium complexes, especially with tetraphenylborate anionic ligand (Table 1), decreases the stereoselectivity of the hydrosilylation, which effect is most pronounced for diphosphine (IV). This selectivity difference is surprising, since the *in situ* prepared rhodium complexes and cationic complexes showed identical behaviour in asymmetric hydrogenation of enamides.[16] With diphosphine (IV) it seems likely that this striking difference can be caused by a preferential crystallization of one of the diastereoisomers during preparation of this substance.

Also, the higher stereoselectivity of the rhodium–hydride complexes (Table 1) contrasts with their behaviour in hydrogenation of enamides.[15] The results obtained so far indicate that hydrosilylation of carbonyl compounds catalysed by rhodium

Table 2

The effect of experimental conditions on the yield and optical purity
of the carbinol obtained by hydrosilylation of PhCOMe by Ph_2SiH_2
catalysed by a $[\{RhCl(\eta-C_2H_4)_2\}_2] + (II)$ system[a]

Factor	Reacted ketone %	PhCH(OH)Me	
		Yield, %[b]	Optical purity, %
	Rh:II mol. ratio		
3:1	97	52	9·2
2:1	100	62	13·8
1:1	100	80	25·8
1:1·5	84	82	26·3
	Rh:PhCOMe mol. ratio		
4×10^{-2}	100	89	17·3
2×10^{-2}	100	86	19·7
1×10^{-2}	100	80·5	23·0
4×10^{-3}	95	78	26·4
2×10^{-3}	96	79	25·2
1×10^{-3}	81	69	25·9
	PhCOMe:Ph_2SiH_2 mol. ratio		
1:3	100	87	24·5
1:2	100	85	25·0
1:1	97	80	25·8
1:0·5	67	37	27·5

[a] For the other conditions see Table 1.
[b] Determined by g.l.c.

complexes of the above types might proceed with participation of different catalytically active species, the structure of which is at present unclear.

In connection with the results obtained for chiral phosphines described in the present work, it is worthy of note that hydrosilylation of ketones can also be effected in the presence of catalysts containing amines, or alkoxy- or amino-phosphines. This observation made in our laboratory has led us to examine the asymmetric efficiency of several chiral ligands of this type such as phenethylamine, menthylamine, α-pipecoline, *N*-methylephedrine, sparteine, nicotine, phenyldimenthyloxyphosphine, phenyldi-α-pipecolylphosphine, diphenyl-α-pipecolylphosphine, (diethylamino)menthylphosphine, and (1-ethoxycarbonylethoxy)diphenylphosphine. Experiments have, however, been discouraging. Although in all cases the corresponding alcohols were obtained in good chemical yields under conditions similar to those given in Table 1, the products were racemates on account of weak co-ordination of these ligands to the metal atom, resulting in their displacement by the silicon hydride.

REFERENCES

[1] H. B. Kagan and T. P. Dang, *J. Amer. Chem. Soc.*, 1972, **94**, 6429.
[2] Y. Kiso, K. Yamamoto, K. Tamao, and M. Kumada, *J. Amer. Chem. Soc.*, 1972, **94**, 4373.
[3] J. P. Morrison, R. E. Burnett, A. M. Aguiar, and C. J. Morrow, *J. Amer. Chem. Soc.*, 1971, **93**, 1308.

[4] J. Beneš and J. Hetflejš, *Coll. Czech. Chem. Comm.*, in the press.
[5] V. Bažant and M. Černý, *Coll. Czech. Chem. Comm.*, 1974, **39**, 1880.
[6] R. Cramer, *Inorg. Chem.*, 1962, **1**, 722.
[7] G. Winghaus and H. Singer, *Chem. Ber.*, 1966, **99**, 3602.
[8] J. Chatt and L. M. Venanzi, *J. Chem. Soc.*, 1957, 4738.
[9] F. Pruchnik, *Inorg. Nuclear Chem. Letters*, 1973, **9**, 1229.
[10] Y. Tsuzuki, *Bull. Chem. Soc. Japan*, 1937, **12**, 491.
[11] K. Issleib and H. Völker, *Chem. Ber.*, 1961, **99**, 392.
[12] M. Janczewski and T. Bartnik, *Roczniki Chem.*, 1962, **36**, 1243.
[13] A. L. Raymond and E. F. Schroeder, *J Amer. Chem. Soc.*, 1948, **70**, 2788.
[14] R. R. Schrock and J. A. Osborn, *J. Amer. Chem. Soc.*, 1964, **93**, 2402.
[15] W. R. Cullen, A. Fenster, and B. R. James, *Inorg. Nuclear Chem. Letters*, 1974, **10**, 167
[16] T. P. Dang, J. C. Paulin and H. B. Kagan, *J. Organometallic Chem.*, 1975, **91**, 105.
[17] L. Salem, *J. Amer. Chem. Soc.*, 1973, **95**, 94.
[18] D. J. Cram and Abd Elhafez, *J. Amer. Chem. Soc.*, 1952, **74**, 5828.
[19] I. Ojima, *Chem. Letters*, 1974, 223.
[20] I. Ojima, M. Nihonyanagi, T. Kogure, M. Kumagai, S. Horiuchi, and K. Nakatsugawa, *J. Organometallic Chem.*, 1975, **94**, 449.

DISCUSSION

A. Hoek (*Univ. of Leiden*) said: In your paper you show equations which describe the hydrolysis of the hydrosilylation product. Could reaction (1) also take place? This process would cause racemization and influence your optical yield.

$$\diagup \hspace{-0.3em}\diagdown CH - \overset{|}{C}HOSiR_3 \rightleftarrows \diagup \hspace{-0.3em}\diagdown C = \overset{|}{C} - OSiR_3 + H_2 \quad (1)$$

J. Hetflejš (*Inst. of Chem. Process Fundamentals, Prague*) replied: We have not detected this dehydrogenation. Under our conditions, using an open system, the reverse hydrogenation is also unlikely to occur. If the alkoxysilane formed is allowed to stand in the presence of the catalyst for several days at the reaction temperature the silyl enol ether is not formed. We did observe hydrogen evolution during hydrosilylation of ketones, but that hydrogen came from the organosilicon hydride used as reagent.

E. Wicke (*Univ. of Münster*) said: To what extent can your methods of synthesizing asymmetric compounds be heterogenized, for instance by fixing chiral phosphines as ligands in transition metal complexes at solid surfaces? It is a matter of general interest, as to whether asymmetric reactions may not be performed more advantageously in a heterogeneous system, because there the surface of the catalyst fixes a preferred direction in space (*i.e.* vertical to the surface) whereas in a homogeneous system the transition state is established in a cage between solvent molecules without a distinct direction in space.

J. Hetflejš (*Inst. of Chem. Process Fundamentals, Prague*) replied: The first partially successful attempt to produce a heterogenized catalyst of this type has been made by Kagan and co-workers.[1] Recently, we have prepared two chiral ligands bound to an inorganic surface. Concerning your general comment, the advantage of using a heterogeneous system has not yet been demonstrated. Kagan and co-workers[1] have reported that a polymer-supported rhodium-DIOP complex showed some catalytic activity for hydrogenation of some enamides. However, this catalyst was efficient for hydrosilylation of ketones, giving the corresponding alcohols in the same optical purity as obtained

---

[1] W. Dumont, J.-C. Pulin, T.-P. Dang, and H. B. Kagan, *J. Amer. Chem. Soc.*, 1973, **95**, 8295.

with a homogeneous analogue. We have used a rhodium-DIOP catalyst bound to an inorganic surface and obtained similar results. In our case, the heterogenized catalyst was less efficient compared to the homogeneous one, with no gain in optical purity of the product.

Reactions of Methylsilane at Gold and at Molybdenum Surfaces

DAVID I. BRADSHAW, RICHARD B. MOYES, and PETER B. WELLS*

Department of Chemistry, The University Hull, HU6 7RX

ABSTRACT   Chemisorption and reaction of methylsilane at gold and at molybdenum surfaces at 195 K and above has been investigated to determine how the metal-catalysed reactions of silanes differ from those of alkanes. Unlike alkanes, silanes are chemisorbed by gold which is superior to molybdenum and some other earlier transition elements as a catalyst for hydrogen isotope exchange in labelled methylsilane.

Methylsilane chemisorption is accompanied by the formation of hydrogen and methane. Mutual exchange between $MeSiH_3$ and $MeSiD_3$ is catalysed by both gold and molybdenum at 195 K. Exchange is confined to the silyl group, and reactions are kinetically well-behaved. Poisoning is absent (Au) or minimal (Mo). Exchange between $MeSiH_3$ and D_2 is catalysed by molybdenum but not by gold. Surfaces which become rapidly poisoned for this reaction retain activity for mutual exchange. Mechanisms of exchange are proposed and the origin of catalytic activity in gold discussed.

INTRODUCTION

Our understanding of catalysis by metals is firmly based on investigations of reactions of carbon compounds. In the last decade it has become evident that a wide variety of types of chemisorption bond may be established between carbon atoms of an adsorbate and metal atoms constituting sites for adsorption at the surface. By contrast, very few studies have been made of catalysed reactions of silicon compounds at metal surfaces. The reasons are two-fold, (*i*) the chemistry of the simple silanes is less extensive than that of carbon (*e.g.* compounds containing a silicon–silicon double bond are unknown) and (*ii*) mass spectrometric analysis of mixtures of deuteriated silanes presents severe difficulties in some cases.

Nevertheless, silane reactions catalysed at metal surfaces merit investigation simply because the electronic structure of silicon differs from that of carbon. First, the well-known ability of silicon to exceed four-co-ordination provides the possibility that silanes may chemisorb associatively, whereas alkanes cannot do so. Secondly, the availability of vacant *d*-levels in silicon may enable chemisorption bonds of the co-ordinate type to be established between metals having a full *d*-band and silane molecules, thus creating the possibility of catalysis by these metals which has no parallel in the catalysis of reactions of carbon compounds. Thus, metal-catalysed reactions of silanes have novel potentialities from the standpoint of the theory of metal catalysis.

Exchange of protium for deuterium in methylsilane has been catalysed by Au, Mo, W, Rh, and Ni;[1] reactions over Au and Mo are reported here, the remainder will be published elsewhere.

Methylsilane was chosen as adsorbate so that (*i*) the reactivity of protium bonded to silicon could be compared with that of protium bonded to carbon, and (*ii*) a comparison could be made in the case of the molybdenum-catalysed reaction, with exchange in ethane.[2] The chosen system permits the investigation of three types of exchange, (*a*) *self exchange* in which randomization of isotopes occurs in CH_3SiD_3 as sole reactant; (*b*) *mutual exchange* in which isotope exchange occurs between CH_3SiH_3 and CH_3SiD_3; and (*c*) *exchange with molecular deuterium*, the reaction between D_2 and CH_3SiH_3 or CH_3SiD_3.

EXPERIMENTAL

MeSiH$_3$ (Ralph Emanuel) was pure as supplied. CH$_3$SiD$_3$ was prepared by the reduction of methyltrichlorosilane with lithium aluminium deuteride in dibutyl ether solution at 273 K. Isotopic purity was >99%; the only chemical impurity was <1% Me$_2$SiD$_2$. Deuterium was purified by diffusion through a heated palladium–silver alloy thimble. Molybdenum wire (New Metals) was >99·9% Mo, and gold wire (Johnson, Matthey) was >99·99% Au.

Reactions were carried out using either a grease-free high-vacuum system (glass) employing mercury diffusion pumps and connected directly to a Vacuum Generators MM6 mass spectrometer, system (i), or an ion-pumped ultra-high-vacuum system (stainless steel) connected to a Vacuum Generators Q7 quadrupole mass spectrometer, system (ii). Each system was fitted with a Pyrex reaction vessel [0·30 l, system (i); 0·48 l, system (ii)].

Evaporated metal films were prepared by standard methods, molybdenum films by evaporation from a self-supporting filament, and gold films by evaporation from a thick layer of gold formed by the melting of a filament on to a thick tungsten loop. Pressures during evaporation were typically <0·7 mN m^{-2} in system (i) or <0·7 μN m^{-2} in system (ii).

When measurements were to be made using a gold film, great care was taken to ensure that the catalyst was not contaminated by tungsten from the support loop. The catalytic activities of the apparatus for the H$_2$–D$_2$ exchange reaction and for the MeSiH$_3$–MeSiD$_3$ exchange reaction were first measured under the conditions later to be used for the gold–catalysed reaction. A slow rate of exchange was observed in each of these experiments when the stainless steel apparatus was used. After deposition of the gold film, the activity for H$_2$–D$_2$ exchange was re-determined. Identical rates before and after film deposition demonstrated that tungsten contamination was absent.

The composition of the gas phase during adsorption experiments was determined from the magnitudes of (i) the ion currents in mass spectra attributable to hydrogen, methane, and methylsilane, and (ii) the variation of total pressure measured by a thermal conductivity gauge (LKB) calibrated for these gases.

Positive-ion mass spectrometry has not previously been used to determine the composition of a mixture of deuteriated methylsilanes. Analysis, which presents difficulties because of the absence of parent ions, can be achieved under certain restricted conditions of instrumental operation and of mixture composition. A full description of the analytical method is given elsewhere.[3]

Rate coefficients k_ϕ and k_0 for the exchange of MeSiH$_3$ with molecular deuterium have been determined in the standard graphical manner using Kemball's eqns. (1) and (3).[2,4]

$$\ln(\phi_\infty - \phi) = -k_\phi t/\phi_\infty + \ln(\phi_\infty - \phi_0) \tag{1}$$

where $\phi = [\%\text{MeSiH}_2\text{D}] + 2[\%\text{MeSiHD}_2] + 3[\%\text{MeSiD}_3]$ $\tag{2}$

and $\ln(d_0 - d_{0,\infty}) = -k_0 t/(100 - d_{0,\infty}) + \ln(100 - d_{0,\infty})$ $\tag{3}$

where $d_0 = [\%\text{MeSiH}_3]$. $\tag{4}$

A rate coefficient, k_{03} for the reaction of MeSiH$_3$ with an equal pressure of MeSiD$_3$ was obtained from eqn. (5).

$$\ln(d_{03} - d_{03,\infty}) = -k_{03} t/(100 - d_{03,\infty}) + \ln(100 - d_{03,\infty}) \tag{5}$$

where $d_{03} = [\%\text{MeSiH}_3] + [\%\text{MeSiD}_3]$.

Graphical values of these rate coefficients have units of percentage change in unit time, and are therefore dependent upon quantities of catalyst and of reactants used, and upon vessel volume. These graphical values have been converted to specific rates, having the units nM s^{-1} (mg catalyst)$^{-1}$ so that values obtained using vacuum systems (i) and (ii) may be compared directly.

RESULTS

Chemisorption of Methylsilane. The chemisorption of MeSiH$_3$ on gold at 273 K and on molybdenum at 195 K and 273 K has been investigated (Fig. 1). The first material admitted to a molybdenum film was completely adsorbed (*i.e.* total pressure in the vessel was <0·1 Nm^{-2}). With successive admissions of methylsilane to each film, hydrogen and methane were detected in the gas phase, the yield of hydrogen greatly exceeded that of methane. Later, methylsilane was detected in the gas phase and it is assumed that, at the point at which its presence first became detectable (point A, Fig. 1), a monolayer of methylsilane and/or its decomposition products had been formed. Gaseous silane was not formed. From a mass balance, the composition of the chemisorbed layer at point A was C$_{0.94}$Si$_{1.00}$H$_{3.10}$ for Au at 273 K, C$_{0.98}$Si$_{1.00}$H$_{3.97}$ for Mo at 273 K, and C$_{0.98}$Si$_{1.00}$H$_{4.95}$ for Mo at 195 K. Thus, the extent of dehydrogenation accompanying chemisorption increased with increasing temperature at the molybdenum surface, and was greater on gold than on molybdenum at 273K.

Self Exchange. Self exchange was not catalysed by gold in the range 195–273 K. On first admitting MeSiD$_3$ to a gold film prepared in ultra-high vacuum at 273 K, 1·4 moles of hydrogen of mean composition H$_{0.25}$D$_{1.75}$ were formed per mole of MeSiD$_3$ adsorbed, together with a trace of CH$_3$D.

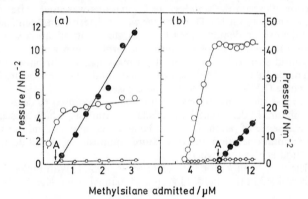

Methylsilane admitted / μM

FIGURE 1 Variation of component pressures in the gas phase when methylsilane was adsorbed at 273 K on films of (*a*) gold; (*b*) molybdenum. Film weights: 32·0 mg (Au); 25·0 mg (Mo). Vessel volume = 0·30 1. Large open circles = hydrogen; small open circles = methane; filled circles = methylsilane.

When 133 Nm^{-2} MeSiD$_3$ was admitted to (i) a 21·6 mg molybdenum film at 195 K, (ii) at 16·5 mg molybdenum film at 273 K, no exchange occurred, but decomposition took place as expected (Fig. 1) the methane having the composition: CH$_4$, 30%; CH$_3$D, 70% and the hydrogen, H$_2$, 12%; HD, 24%; D$_2$, 64%. Warming film (i) to 293 K and film (ii) to 350 K did not affect the hydrogen composition but caused the desorption of a little (CH$_3$)$_2$SiD$_2$ and of further and more highly exchanged methane so that the composition became: CH$_4$, 20%; CH$_3$D, 50%; CH$_2$D$_2$, 14%; CHD$_3$, 9%; CD$_4$, 7%.

Mutual Exchange. This reaction [eqn. (6)] was catalysed

$$CH_3SiH_3 + CH_3SiD_3 \rightarrow 2CH_3SiX_3 \; [X = H \text{ or } D] \tag{6}$$

by gold and by molybdenum at 195K. Graphs obtained by use of equation (5) were linear over the range 0–90% conversion, indicating that the progress of these exchange reactions did not poison the surfaces. Rate coefficients, k_{03}, (obtained from these graphs) remained unaltered when the pressure of a 1:1 reactant mixture was increased two fold, confirming that methylsilane was strongly adsorbed and indicating that its surface coverage was invariant with reactant pressure. Values of k_{03}^{Au} varied only slightly in successive reactions over a given film (Table 1) whereas k_{03}^{Mo} diminished slowly but steadily from one reaction to the next. For reactions at 243 K (Au) or 273 K (Mo) the function plots remained linear but values for k_{03} diminished slowly (Au) or rapidly (Mo) in successive reactions. Thus, processes occurring on evacuation of the vessel, and not the progress of the exchange reactions, were responsible for the poisoning of the catalysts.

Table 1

Rate coefficients for mutual exchange at 195 K

Metal	Vacuum system	Film wt/mg	Vessel vol/l	Initial pressures/ Nm^{-2}			k_{03}/nM s^{-1} mg^{-1}			
				MeSiH$_3$	MeSiD$_3$	Expt no	2	4	6	8
Au	h.v.	55·0	0·30	124	137		5·3	5·2	5·6	5·1
Au	u.h.v.	18·1	0·48	30	30		13·1	15·9	15·8	—
Mo	h.v.	14·5	0·30	117	116		18·8	18·5	15·4	10·1
Mo	u.h.v.	8·2	0·48	24	24		13·1*	—	—	—

* Expt. no. 1

Exchange with Molecular Deuterium. No exchange occurred in the range 195–300 K when (D$_2$ + CH$_3$SiD$_3$) mixtures were admitted to gold or molybdenum films, or when (D$_2$ + CH$_3$SiH$_3$) mixtures were admitted to gold films. However, reaction according to eqn. (7) was catalysed by molybdenum in the range 195–300 K, and at

$$CH_3SiH_3 + D_2 \rightarrow CH_3SiX_3 + X_2 \; [X = H \text{ or } D] \tag{7}$$

273 K exhibited orders measured by the initial rate method that were zero or negative in methylsilane and $0·6 \pm 0·1$ in deuterium. Linear function plots were obtained by

Table 2

Molybdenum-catalysed exchange of MeSiH$_3$ with D$_2$ and of MeSiH$_3$ with MeSiD$_3$ at 195K

Film weight = 18·8 mg   Vessel size = 0·30 1

Experiment No.	Initial pressures/Nm^{-2}			Rate coefficients/ nM s^{-1} mg^{-1}			$\dfrac{k_\phi}{k_0} = M$
	MeSiD$_3$	MeSiH$_3$	D$_2$	k_{03}	k_ϕ	k_0	
1	—	133	1320	—	10·2	5·1	2·0
2	—	133	1320	—	0·19	0·14	1·4
3	115	115	—	16·3	—	—	—
4	110	115	—	15·9	—	—	—
5	—	133	1320	—	0·15	0·12	1·3

application of eqns (1) and (3) to reactions at 195 K. Reaction over a clean film was fast at 195 K (Table 2), the multiplicity of exchange, M, ($= k_\phi/k_0$) being in the range 2·0–2·3. However, the second and subsequent reactions over a given film were slower by an order of magnitude, and for these reactions M approached unity. Remarkably, this deactivation of the $MeSiH_3$–D_2 reaction was *not* accompanied by a diminution in the rate of mutual exchange (compare values of k_{03} in Tables 1 and 2). As for mutual exchange, this catalyst poisoning at 195 K is attributed to processes occurring on evacuation and not to intermediates formed during exchange. However, function plots for reactions at 273 and 300 K were non-linear, indicating that at these more elevated temperatures species produced during exchange poisoned the surface.

DISCUSSION

Methylsilane is chemisorbed on both molybdenum and gold. The zero or negative orders in methylsilane shown by the exchange reactions, and the extent of the dehydrogenation that accompanies adsorption (Fig. 1), demonstrate that chemisorption is strong. Furthermore, in exchange with molecular deuterium catalysed by molybdenum, the order in deuterium was 0·6, indicating that deuterium is more weakly adsorbed than methylsilane. This behaviour is quite different from that observed in exchange reactions involving ethane and other simple alkanes. Alkanes are not chemisorbed at all at the gold surface at these temperatures, and are more weakly chemisorbed than hydrogen on earlier transition elements.[2,4]

The non-occurrence of self-exchange, and the absence of exchange when $(CH_3SiD_3 + D_2)$ mixtures were admitted to molybdenum, demonstrate that all intermediates containing both carbon and silicon that participate in exchange reactions are bonded to the surface only via silicon. Intermediates for consideration are thus:

MeSiH$_3$	MeSiH$_2$	MeSiH	MeSi
↑		‖	⫲
*	*	**	***
(I)	(II)	(III)	(IV)

Species (II), (III), and (IV) are products of the dissociative chemisorption of methylsilane. Formally analogous species may be written for chemisorbed alkanes, except that in (II), (III), and (IV) back-donation of electronic charge by means of $d\pi$–$d\pi$ interactions is expected to be present and to strengthen the metal–silicon bond(s). This $d\pi$–$d\pi$ bonding is probably crucial to the establishment of chemisorbed species on gold and we thus suggest that the associatively adsorbed species (I) may have an independent existence or that it may serve, as a weakly adsorbed state, to lower the activation energy for the chemisorption of methylsilane as (II).

Mechanisms of Exchange. The simplest mechanism for the interpretation of mutual exchange is given in eqn. (8). The

$$MeSiH_3(g) + MeSiD_3(g) \longrightarrow \begin{matrix} MeSiH_2 + H \\ | \quad | \\ * \quad * \\ MeSiD_2 + D \\ | \quad | \\ * \quad * \end{matrix} \longrightarrow MeSiH_2D(g) + MeSiHD_2(g)$$

(8)

multiplicity, M, cannot be determined for mutual exchange, and hence the participation of species (III) and (IV) cannot be adjudged from observations of mutual exchange alone (see below).

Exchange of methylsilane with molecular deuterium was not catalysed by gold because of the well-known inability of this metal to dissociate deuterium molecules at these temperatures.[5] However, this reaction was catalysed by clean molybdenum, two hydrogen atoms per molecule of methylsilane being exchanged for deuterium in the early stages of reaction (experiment 1, Table 2). Activity for this exchange was largely poisoned when the products were pumped from the vessel (compare experiments 1 and 2, Table 2) whereas activity for mutual exchange was not poisoned (compare Tables 1 and 2). Had the rates of the two types of exchange reaction been similarly diminished, then exchange with molecular deuterium via the interconversion of isotopically distinguishable (I) and (II) would have been postulated. The different behaviours of the two exchange reactions towards poisoning reveal that exchange with molecular deuterium involves the participation of a more highly dissociated species, probably (III) or perhaps (IV); eqns. (9–11) interpret the multiplicity $M = 2$. (The reaction shown

$$D_2(g) \; \overrightarrow{\longleftarrow} \; 2D \qquad\qquad (9)$$
$$\underset{*}{|}$$

$$\text{MeSiH}_3(g) \xrightarrow{-H} \underset{*}{\text{MeSiH}_2} \xrightarrow{-H} \underset{**}{\text{MeSiH}} \xrightarrow{+D} \underset{*}{\text{MeSiHD}} \xrightarrow{+D} \text{MeSiHD}_2(g) \; (10)$$

$$\underset{*}{\text{H}} + \underset{*}{\text{D}} \xrightarrow{\text{fast}} \text{HD}(g) \qquad\qquad (11)$$

in eqn. (11) is designated "fast" on the ground that the equilibration $H_2 + D_2 \rightleftharpoons 2HD$ proceeded rapidly at 195 K in the presence of methylsilane, faster by an order of magnitude than the rate of exchange of methylsilane with molecular deuterium).

It was suggested above (see Results) that catalyst poisoning at 195 K occurs during the evacuation of the vessel. We now suggest that, on evacuation, species (I) is desorbed, species (II) is desorbed after reaction with chemisorbed hydrogen, but that species (III) is retained and under vacuum conditions dissociates to give some unreactive residue or permanent poison. Thus, in the second and subsequent exchanges of methylsilane with deuterium over a given catalyst, the concentrations of sites available for the formation of (III) and of chemisorbed deuterium were much reduced, and the exchange rate diminished accordingly.

The multiplicity, M, of reactions at poisoned surfaces was reduced to a value close to unity; this reduction is not due to a change of mechanism but to a reduced chance of adsorbed-MeSiH and adsorbed-MeSiHD acquiring deuterium atoms [eqn. (10)]. The increased chance of protium acquisition arises because those sites which can accommodate only adsorbed-MeSiH$_2$, (and which are active for mutual exchange when the chosen reactants are MeSiH$_3$ and MeSiD$_3$) are not poisoned during evacuation of the vessel. These sites remain available for the dissociative chemisorption of MeSiH$_3$ to give species (II), and hence provide a rich source of chemisorbed protium.

This association of species (III) with surface deactivation provides a ground for dismissing it, and hence also (IV), as intermediates that participate in mutual exchange (see above).

Formation of Methane and Dimethylsilane. The complete absence of exchange in the methyl group of methylsilane is a remarkable feature of these reactions. Multiple exchange in ethane occurs at both carbon atoms and involves the interconversion of (A) with (B) or (C), but in methylsilane exchange the formation

$$H_2C-CH_3 \qquad H_2C-CH_2 \qquad H_2C=CH_2 \qquad H_2C-SiH_2 \qquad H_2C=SiH_2$$

(A)	(B)	(C)	(D)	(E)

of (D) or (E) from (II) clearly does not occur. Species (E) may be discounted on the ground that the carbon–silicon double bond is unknown in molecules in either the free or the complexed state. Species (D) is less easily dismissed, unless the bonding shown in (B) and (D) is simply a less satisfactory description than that in (C) or (E).

Thus the process of methane formation commences with the fission of the carbon–silicon bond in (II), (III), or (IV), and not with any preliminary dissociation of hydrogen from the methyl group. At 195 K, adsorbed-methyl acquired deuterium giving CH_3D as the only (Au) or major (Mo) product. Some dehydrogenation/hydrogenation of adsorbed-methyl occurred on molybdenum at 195 K, eqn. (12), yielding CH_4 and more highly dissociated intermediates which were converted to more highly deuteriated methane when the catalyst was

$$CH_3 \xrightarrow{-H} CH_2 \xrightarrow{-H} CH \qquad (12)$$
$$\big\downarrow_{+H} CH_4(g)$$

warmed. Warming also facilitated reaction between adsorbed-methyl and species (II), (III), or (IV) to give intermediates which, on further hydrogenation if necessary, desorbed as dimethylsilane.

Specific Activities of Gold and Molybdenum. Rate coefficients k_{03}^{Au} and k_{03}^{Mo} are compared in Table 1. These values do not provide a fair comparison of the relative specific activities of these metals because molybdenum (along with other metals of Groups VIA, VIIA, and VIII) gives porous films of relatively high surface area, whereas gold films are virtually non-porous[6] and only the geometric surface area is available for catalysis. Expression of k_{03} in the units $nM\ s^{-1}\ mg^{-1}$ makes no allowance for the higher surface area of molybdenum. An allowance can be made by conversion of the units to 'nanomoles methylsilane converted per second per micromole of methylsilane adsorbed', using point A on the adsorption isotherms (Fig. 1) as a measure of the number of micromoles adsorbed. This is equivalent to using points A as measures of surface area; surface areas so measured for films of Mo, W, Rh, and Ni vary in a manner similar to that shown by surface areas determined in this laboratory by benzene adsorption and by deuterium adsorption.[7] This conversion of units assumes that the proportion of the surface available for mutual exchange at 195 K is the same for gold and for molybdenum.

The converted values are:

$$k_{03}^{Au} = 420\ nM\ s^{-1}\ (\mu M\ adsorbed)^{-1}$$
$$k_{03}^{Mo} = 54\ nM\ s^{-1}\ (\mu M\ adsorbed)^{-1}$$

On this basis, the specific activity of gold at 195K greatly exceeds that of molybdenum. Provisional values for Ni, Rh, and W at the same temperature[1] are 36, 74, and 79 $nM\ s^{-1}\ (\mu M\ adsorbed)^{-1}$. This emphasizes further the exceptional catalytic activity of gold for this exchange reaction.

Bradshaw, Moyes, Wells

ACKNOWLEDGMENTS. We thank the Science Research Council and Imperial Chemical Industries for the grants for the purchase of mass spectrometers and the former for a maintenance award to D.I.B.

REFERENCES

[1] D. I. Bradshaw, R. B. Moyes, and P. B. Wells, *J.C.S. Chem. Comm.*, 1975, 137; D. I. Bradshaw, Ph.D. thesis, University of Hull, 1975.
[2] J. R. Anderson and C. Kemball, *Proc. Roy. Soc.*, 1954, **A223**, 361.
[3] D. I. Bradshaw, R. B. Moyes, and P. B. Wells, *Canad. J. Chem.*, 1976, **54**, 599.
[4] C. Kemball, *Adv. Catalysis*, 1959, **11**, 223.
[5] G. C. Bond, P. A. Sermon, G. Webb, D. A. Buchanan, and P. B. Wells, *J.C.S. Chem. Comm.*, 1973, 444; D. A. Buchanan, and G Webb. *J.C.S. Faraday I*, 1975, **71**, 134.
[6] G. C. Bond, *Catalysis by Metals* (Academic Press, London and New York, 1962), p. 37.
[7] K. Baron, R. James, and R. B. Moyes, unpublished work.

DISCUSSION

J. L. Garnett (*Univ. of New South Wales*) said: My question refers to species (I) and equation (10) in your mechanism of the methylsilane exchange. By analogy with the hydrocarbon–platinum system where M-values higher than 1 have also been observed under both homogeneous[1,2] and heterogeneous[3] conditions, is it necessary to invoke the formation of species such as MeSiH in equation (10) to explain your M-values?
An alternative possibility would be to consider the following steps:

$$CH_3SiH_3(g) \xrightarrow{(A)} CH_3SiH_3 \underset{+D}{\overset{-H}{\rightleftharpoons}} CH_3SiH_2$$
$$\text{(I)} \qquad\qquad \text{(II)}$$

where a large activation energy barrier (A) is associated with the initial formation of the charge-transfer species (I) followed by several exchange cycles of lower activation energy.
The analogy between ethane and methylsilane as reactants in their exchange reactions is interesting. As you have indicated, your data for methylsilane show strong interaction between silicon and the surface such that bonding between methylsilane and metal is exclusively *via* silicon. The silicon atom is thus an inbuilt marker for the molecule and would enable you readily to study electronic and related effects in chemisorption of methylsilane whereas in ethane the two carbons are identical. I would therefore like to ask whether any examination of electronic effects has been carried out in your system, *i.e.* have compounds such as XCH_2SiH_3 ($CH_3CH_2SiH_3$ would be the simplest) been examined?
R. B. Moyes (*Univ. of Hull*) replied: Prof. Garnett is correct in saying that multiple exchange with molecular deuterium can, if considered in isolation, be interpreted as an interconversion of (I) and (II). However, our proposed mechanism must interpret simultaneously both mutual exchange, which is not easily poisoned, and exchange with molecular deuterium, which is easily poisoned. Species (II) is required to participate in mutual exchange [see equation (8)] and hence the easily poisoned exchange with molecular deuterium must involve a more extensively dissociated species, *i.e.* species (III).
We envisage that species (I) may be formed as an intermediate during the dissociative

[1] J. L. Garnett and R. J. Hodges, *J. Amer. Chem. Soc.*, 1967, **89**, 4046.
[2] R. J. Hodges, D. E. Webster, and P. B. Wells, *JCS Dalton*, 1972, 2571.
[3] J. L. Garnett, *Catalysis Rev.*, 1971, **5** 199.

chemisorption of methylsilane; it was not included in equations (8) and (10) because its contribution is trivial.

Thank you for your comments concerning more complex silanes with which we agree. We have not yet examined the behaviour of such compounds.

A. Sárkány (*Inst. of Isotopes, Budapest*) said: The surface area of gold and molybdenum films was determined from point A in Figure 1 of your paper. I am not convinced, however, that the adsorbed amount (point A) corresponds to the same coverage on both metals. Did you measure the surface area of your samples by other independent methods? I wonder whether the low surface area of gold films might be the result of the low reactivity of methylsilane on gold?

R. B. Moyes (*Univ. of Hull*) replied: Surface areas of molybdenum have also been measured by hydrogen chemisorption and by benzene chemisorption. The values support our view that point A corresponds to full surface coverage by methylsilane.

We would point out that the activity of gold is certainly not low. On a weight for weight basis its activity for mutual exchange is equal to that of molybdenum (see Table 1) and when the surface area correction is applied its activity greatly exceeds that of molybdenum.

Kinetic Model for Selectivity Decline of Reforming Catalysts

G. J. M. van Keulen

Akzo Chemie Nederland bv., Research Centre Amsterdam, P.O. Box 15, Amsterdam, Netherlands

ABSTRACT Bimetallic reforming catalysts are characterized by a higher stability than a straight platinum catalyst. This holds especially for the product yield. However, the product yield depends on the octane number and is therefore not a simple parameter to describe the catalyst selectivity. In fact no single parameter has been found yet to describe adequately the selectivity decline of a reforming catalyst.

The present work shows that the overall paraffin conversion during reforming of a full range naphtha can be evaluated kinetically. The selectivity is defined by the ratio of aromatization and cracking rates, and depends only on the operating pressure and the catalyst age. Apparently aromatization of paraffins is more strongly influenced by the pressure than cracking, and has a higher decline rate. The order in pressure is -0.7 for the initial selectivity, and -1.65 for the selectivity decline rate. This model has been used successfully to estimate the selectivity of a bimetallic reforming catalyst during prolonged testing, as well as for accelerated deactivation tests.

INTRODUCTION

Catalytic reforming is a process to convert virgin naphtha into a high quality gasoline. This is achieved by raising the octane number. The conventional description of the process is based on the relation between octane number and product yield. This relation depends primarily on the operating pressure, but also on the quality of the feed.[1]

The level of product yield obtained at a specific RON depends further on the catalyst age. The intrinsic selectivity slowly declines during use. However the resulting decrease in product yield can only be established after a long time on oil. This is caused by the fact that the octane number also depends on the catalyst selectivity. Therefore the yield stability is a long term effect, requiring tests during a prolonged period under accurately controlled conditions.

Catalyst evaluations

A straight Pt catalyst (monometallic) deactivates rapidly at severe reforming conditions, e.g. high space velocity and low operating pressure. These conditions are favourable for higher gasoline yields and aromatics.

The deactivation is caused mainly by coke deposition on the catalyst surface. Sintering of the metal crystallites may be neglected during regular reforming operation, although it has a strong influence on the catalyst performance.[2]

A bimetallic catalyst containing Pt and Re shows a lower deactivation rate. Moreover the product yield drops off less rapidly as compared with Pt alone. In other words the bimetallic also shows a lower selectivity decline which is decisive for its commercial application.[3]

A proper evaluation of a bimetallic on pilot plant scale requires accurate determination of the yield stability. However the conditions in such a unit may be quite different from commercial reformers. This especially holds for the hydrogen purity and the temperature profile within the reactor.

Generally the reformer feed contains 20—40% of naphthenic hydrocarbons. These are dehydrogenated very rapidly in the front part of the catalyst bed. The reaction gives an estimated heat loss of 5 cal/g feed for each percent of naphthenes.[4] An adiabatic fall in temperature of 50—100°C is therefore possible in reforming. When operating nonadiabatically the heat loss results in both longitudinal and radial temperature differences. The relatively high outer surface of the catalyst bed in a small pilot plant is beneficial for achieving isothermal conditions. Temperature differences in the bed are levelled out through radial heat transport. This depends on the heat conductivity of catalyst bed and gas atmosphere, and the superficial velocity of the gas.[5] The overall heat conductivity of the bed is very low because of a low Reynolds number, and can hardly be increased by diluting the catalyst with conductive material.

However, dilution of the catalyst results in a better distribution of heat loss over the reactor, so that equalization of the temperature can take place more readily. In this way the accuracy of the measured gas temperature can be improved.

The average catalyst temperature may still be considerably lower[6] than the measured gas temperature. Therefore the activity of a reforming catalyst at a specific, or even local (*e.g.* bed inlet) temperature is difficult to determine accurately. The same holds inevitably for the activity stability, because this depends on the temperature increase requirement for a constant octane level. It is thus essential to evaluate a selectivity parameter, that is insensitive to the temperature and hardly depends on the activity level as well.

Such a parameter may be used more properly for deactivation studies than the product yield at constant octane number, which is commonly applied.

Reforming kinetics

A detailed matrix of relative reaction rates, based on pseudo first order kinetics, was published in 1959.[7] It seems rather difficult to incorporate the catalyst parameters into such a complicated system. Simplifications can be made by assuming that each hydrocarbon class is represented by a single compound having the average properties of that class.[8] But even then it is too difficult to evaluate the separate deactivation rates. Therefore we suggest the use of a simplified pattern[9] as illustrated in Table 1 for C_7 hydrocarbons. This table also shows some elementary kinetic data.[10] Obviously the reaction rate for naphthene conversion is one or two orders of magnitude higher than for the conversion of paraffins. Therefore the major reaction steps that are kinetically controlled during reforming are cyclization and cracking of the paraffin molecule. We propose to determine the ratio of the reaction rate constants as defined by the selectivity parameter:

$$S = k_1/k_2 \tag{1}$$

with the numbers 1 and 2 standing for cyclization and cracking, respectively. Both reactions are competitive and supposed to be first order in paraffin concentration. Therefore the selectivity does not depend on the concentration and can be estimated from integral data by means of:

$$S = \ln (1 - x_1)/\ln (1 - x_2) \tag{2}$$

with x_1 and x_2 the conversion of paraffins into aromatics and gas, respectively. This parameter is strongly correlated with both octane number and yield. The influence of temperature is small, due to the fact that the activation energies (Table 1) are nearly equal.

The effect of carbon atom number is indicated in Table 2.[7] Apparently the differences in S are large, going from C_6 to C_8; but small when dealing with hydrocarbons in the end boiling point range usual for reforming (C_9—C_{12}).

1052

van Keulen

Table 1

Reforming of C_7 hydrocarbons

Reaction pattern

$$G \longleftarrow C_7 \rightleftarrows DMCP \rightleftarrows MCH \rightleftarrows T$$
$$-H_2 \qquad\qquad +H_2 \qquad\qquad +3H_2$$

Kinetics

Reaction	Activation energy, kcal/mole	1st order rate constant, $\dfrac{\text{g moles}}{\text{g cat} \times \text{h}}$
Dehydrogenation MCH → T	14·4	11·8
Isomerization MCH → DMCP	17·3	4·9
Dehydrocyclization C_7 → T	32·7	0·14
Hydrocracking C_7 → G	31·1	0·19

Conditions: 496°C, 18 kg/cm², $H_2/HC = 6$, Ref. 10.

Table 2

Selectivity of paraffin conversion

Reaction pattern

$$\text{Paraffins} \xrightarrow{k_1} \text{Aromatics}$$
$$\Big\downarrow k_2$$
$$\text{Gas}$$

Selectivity: $S = \dfrac{k_1}{k_2}$

Effect of carbon atom number (relative rates)

Paraffins	C_6	C_7	C_8	C_9	C_{10}
k_1	0·0	0·6	1·3	1·8	2·5
k_2	0·3	0·3	0·4	0·7	1·1
$S = k_1/k_2$	0·0	2·0	3·2	2·6	2·3

Conditions: 496°C, $H_2/HC = 5$. Ref. 7.

1053

We investigated the combined effects of pressure and time on the selectivity of a bimetallic catalyst, to estimate the selectivity decline rate with data obtained from accelerated deactivation tests. Test runs were made at low and high severity in a small pilot plant, to evaluate the selectivity decline in one particular reactor configuration.

EXPERIMENTAL

A commercial bimetallic catalyst, CK-433, was tested in a small pilot plant without hydrogen recycle. The catalyst contained 0·3 % w Pt and 0·3 % w Re on a high purity γ-Al$_2$O$_3$. The unit consists of two stainless-steel reactor tubes (20 mm i.d.) within one bronze block furnace. Each reactor was charged with 20 g catalyst calcined for 1 h at

Table 3
Feed properties

Feed	A	B
Origin	Qatar Marine	Mixed
Distillation (ASTM)		
IBP, °C	92	93
10%	106	111
30%	113	124
50%	120	138
70%	129	152
90%	144	170
FBP	164	189
Density (15°C), g/ml	0·7432	0·7510
Sulphur content, ppm w	1·4	0·4
Composition (g.l.c.)		
Paraffins, % w	61	63
Naphthenes	21	23
Aromatics	18	14

450°C in a muffle furnace. The catalyst was reduced overnight and presulphided at 480°C. The feed was introduced afterwards. It was dried over activated molecular sieve 3A. This material was also used for drying hydrogen after oxygen removal with a conventional deoxo unit. Because of the once-through operation of the reactors, a low water level was obtained easily after the start-up.

Two different feeds have been applied during our investigations. The properties are listed in Table 3. Feed A is a low boiling paraffinic naphtha *ex* Qatar Marine, and feed B has a higher end boiling point. Experiments with feed A were performed without substantial dilution of the catalyst. A steep decrease of temperature in the first part of the bed was observed. Therefore, the average temperature could not be measured adequately. The temperature profile was improved at high space velocity by a twofold dilution of the catalyst.

Isothermal experiments were made with feed B. The maximum temperature difference in the bed could be maintained within 5°C by a progressive dilution. The catalyst concentration at the top was about 60 times lower than at the bottom of the reactor. However the average dilution factor was only 3. This value is supposed to be low enough to avoid backmixing in the reactor. The dilution material was crushed carborundum (16 mesh). Experiments with crushed catalyst (20—60 mesh) showed that paraffin conversion was not limited by pore diffusion.

Testing procedures

An accelerated deactivation procedure was applied with feed A, see Table 4. The duration was only 60 h. The operating pressure was only 8 kg/cm². This was necessary for a sufficient activity and selectivity decline within such a short period of time. Testing

Table 4

Test procedures

Feed	A	B	
Duration, h	60	1000	
Conditions			
LHSV, h^{-1}	2·5–5·0	2·0–3·0	
WABT, °C	variable	495–510–525	
pressure, kg/cm²	8	15–20	
H$_2$/HC, mole/mole	4	4	
Sampling			
time, h	10	16–24	
frequency/period	1	8	
Periods	(RON-O)	(LHSV)	(WABT)
1	>102	3	495
2	101	2	495
3	99	3	510
4	97	2	510
5	95	3	525
6	93	2	525

was started at a very high RON, to speed up the stabilization of the catalyst by coke formation. At each following period, the activity was set at a lower level by on-line control of the refractive index.

Obviously the temperature corrections necessary to maintain activity were rather small, due to the short time of deactivation. Data from the first 10 h have been disregarded for our calculations on the selectivity decline. Obviously the applied selectivity parameter should be independent of the activity or temperature level. To investigate the deactivation during a much longer time on oil, experiments were made with feed B during almost 1000 h. The pressure was set at 15 and 20 kg/cm². To maintain sufficient deactivation of the catalyst during the run, the reactor temperature was increased in three stages from 495 to 525°C. The space velocity was changed once at each level of temperature to control the activity at a reasonable level, *e.g.* 100—95 RON (Table 4).

Reformate samples have been collected over each period in a high pressure product receiver at 12°C. The remaining C_4^- cracking products were removed by batch distillation afterwards. The product yield was corrected for the C_5^+ in reactor off-gas. The stabilized reformate was analysed on paraffins, naphthenes, and aromatics by capillary g.l.c. Data obtained in this way were corrected at % w on feed.

The g.l.c. analyses of feed and product were performed on DC 550 Silicone oil at a temperature from 25—150°C. The evaluation of the chromatograms with up to 300 peaks was done by on-line dedicated computer.

RESULTS AND DISCUSSION

During several initial experiments it was found that the following equations could be applied satisfactorily on feed A:

$$k = \text{WHSV} \ln \frac{P_o}{P} \tag{3}$$

$$k = k_o p^n e^{-Ea/RT} e^{-\alpha t} \tag{4}$$

to describe the separate effects of pressure (p), temperature (T) and time (t) on the overall paraffin conversion. The order in pressure (n) varied from 0 to —1, and the average activation energy was about 30 kcal/mole. This is·in line with data mentioned above (Table 1). The effect of temperature is, however, more complicated due to variations in the profile over the reactor at different WHSV.

The determination of average reactor temperature was also inaccurate. New experiments were done with feed B at more isothermal reactor conditions, by diluting the catalyst particles as explained in the previous section.

Activity decline

Simultaneous variation of time and temperature is often applied on testing the activity decline of a reforming catalyst. This may be represented by the following differential equation:

$$\frac{d \ln k}{dt} = \left(\frac{\partial \ln k}{\partial T}\right)\frac{dT}{dt} + \left(\frac{\partial \ln k}{\partial t}\right) = \left(E_a/RT^2 - \frac{\partial \alpha}{\partial T}t\right)\frac{dT}{dt} - \left(\alpha + \frac{\partial \alpha}{\partial t}t\right) \tag{5}$$

The effect of temperature on $\alpha(\partial \alpha/\partial T)$ can be described with an activation energy, E_d. The effect of time ($\partial \alpha/\partial t$) is negligible, if the catalyst structure and/or environment are not changing during the run. For a constant activity (d ln $k/dt = 0$), the temperature can be increased, so that:

$$\alpha = \frac{\beta E_a/RT^2}{1 + \beta E_d/RT^2 t}, \text{ with } \beta = \frac{dT}{dt} \tag{6}$$

Although this method resembles the commercial practice, it is much better to correct the WHSV. The deactivation rate α should be determined at isothermal conditions.

It can be made independent of throughput by defining the aging rate:

$$\alpha^* = \frac{\alpha}{\text{WHSV}} \text{ (dimensionless)} \tag{7}$$

According to Myers, Lang, and Weisz[12] the aging rate (α^*) depends on the pressure with an order equal to or below —1 at a constant hydrogen to hydrocarbon molar ratio. This was confirmed by our experiments on feed B, but optimum correlations between overall paraffin conversion or ln k and the conditions (4) could not be established for isothermal experiments on feed B. The variations of octane number with time at different conditions are illustrated in Figure 1. It is shown very clearly that the catalyst activity declines more rapidly at a lower operating pressure (*e.g.*, 15 kg/cm² compared to 20). The effect of space velocity did not completely agree with the first order kinetic description given before (eqn. 3). This can be explained with complex adsorption kinetics. The presence of increasing levels of aromatics through the reactor may result in growing competition on the catalytic sites. It is obvious that this will be equally noticeable on both cyclization and cracking reactions.

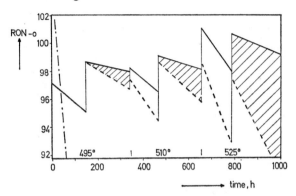

FIGURE 1 Effect of pressure on octane number decline
— · — 8 kg/cm², feed A, LHSV = 2·5–5·0
(accelerated test procedure, 60 h).
⎯⎯⎯ 20 kg/cm²⎫
- - - - 15 kg/cm² ⎬ feed B, LHSV = 2·0–3·0
(shaded area: lower space velocity).

Selectivity decline

Without speculating on the real mechanism it may be suggested that cyclization and cracking of paraffins are both steps in a coupled reaction system,[13] *e.g.*, several reactions proceed simultaneously on the same catalyst. The rate of each reactions is a function of the local concentrations of all compounds present in the system, and of the corresponding number of adsorption constants. This can be symbolized with:

$$\frac{dx_1}{d\tau} = k_1 K_1 C_1 / (1 + \Sigma K_1 C_1) \tag{8}$$

The time variable is eliminated by:

$$\frac{dx_1}{dx_2} = \frac{k_1}{k_2} \times K_{rel} \tag{9}$$

This demonstrates clearly that the selectivity parameter defined before (eqn 1) is still valid in the case of a more complex mechanism. Beranek[13] demonstrated that the relative reactivity data are equally well obtained from integral data by means of the known relation (eqn 2). For estimating the partial conversions of cyclization and cracking, we applied the following equations:

$$1 - x_1 = \frac{P + G}{P_0} \qquad 1 - x_2 = \frac{P_0 - G}{P_0} \tag{10}$$

with P, P_0: paraffins in % w on feed
G: gas production in % w on feed.

The decrease in molecular weight was not taken into account. The selectivity was evaluated as a function of time at each of three levels of pressure, separately. The initial selectivity differed only twofold from 8—20 kg/cm², whereas the selectivity decline rate varied by a factor of 5. This can be explained with a different order in pressure for the two parameters, as demonstrated in Table 5. Obviously the effect of pressure on initial

Table 5

Test results on CK-433

Feed	Pressure kg/cm²	S_o[a]		$\alpha^* \times 10^3$	
		experimental	calculated	experimental	calculated
A	8	2·77(0·08)[d]	2·80[b]	1·0(0·2)	0·97[c]
B	15	1·93(0·07)	1·80	0·35(0·02)	0·34
	20	1·32(0·05)	1·47	0·20(0·02)	0·21

[a] $\ln S = \ln S_o - \alpha^* \times \text{WHSV} \times t$
[b] $S_o = 12 \times p^{-0\cdot7}$
[c] $\alpha^* = 0\cdot03 \times p^{-1\cdot65}$
[d] Standard deviations.

selectivity can be expressed with an order of —0·70. The aging rate however shows an order of —1·65.

The standard deviation of the initial selectivity was about 3%. The accuracy of the selectivity decline rate is much lower and can be given by a standard error of 10—20% in α^*. This is caused by the experimental error when determining the product yield and paraffin conversion. It may be improved by on-line stabilization or g.l.c. analyses. We assume however that the above kinetic evaluation of the selectivity is still reliable because

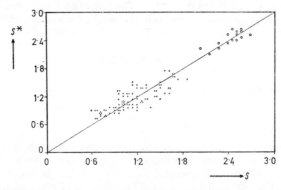

FIGURE 2   Correlation between experimental and calculated selectivity.
$S = \ln (1 - x_1)/\ln (1 - x_2) = k_1/k_2$
$S^* = \exp (2\cdot48 - 0\cdot70 \ln p - 0\cdot03 \times p^{-1\cdot65} \times \text{WHSV} \times t)$
Standard error of the estimate S^*: 0·16
The straight line represents: $S = S^*$
.: 15 and 20 kg/cm², 0–1000 h
○: 8 kg/cm², accelerated 0–60 h.

of the big range of time and pressure that has been taken into account. This is confirmed by the correlation between experimental and calculated values of the selectivity illustrated in Figure 2. The standard error of the estimate appeared to be 0·16 in selectivity.

CONCLUDING REMARKS

We have investigated the selectivity stability of a bimetallic reforming catalyst at various conditions. The experiments were made in a small pilot plant without hydrogen recycle. The intrinsic catalyst selectivity and selectivity decline rate could be equally well determined for low and high severity operations.

Calculations were based on paraffin conversion, showing that the selectivity mainly depends on pressure and time on oil. The cyclization of paraffins is more strongly influenced by the pressure than is cracking. The difference is even bigger when comparing the deactivation rates. This results in a sharp increase of the selectivity decline rate, when lowering the pressure.

In general a hydrogen recycle is applied during commercial reforming. The hydrogen purity gradually decreases with catalyst age. This will most likely influence the selectivity decline of the catalyst.

LIST OF SYMBOLS

k first order reaction rate constant (h^{-1})
k_1 rate constant for paraffin aromatization
k_2 rate constant for paraffin cracking
k_o frequency factor
E_a activation energy
ln   natural logarithm
P_o paraffin content of feed (% w).
P Paraffin content of product in % w on feed
WHSV   weight hourly space velocity (h^{-1})
LHSV   liquid hourly space velocity (h^{-1})
p absolute pressure (kg/cm^2)
n order in pressure for activity
T absolute temperature (K)
R gas constant (kcal/mole K)
α first order deactivation rate constant (h^{-1})
β temperature increase requirement to correct for deactivation (°C/h)
α^* catalyst aging rate (dimensionless)
E_d activation energy for deactivation (kcal/mole K)
m order in pressure for deactivation
t catalyst age (h)
x_1 paraffin conversion into aromatics (w/w)
x_2 paraffin conversion into gas products (w/w)
G gas production in % w on feed
S_o initial selectivity
DMCP   dimethylcyclopentane
MCH   methylcyclohexane
T   toluene
WABT   Weight Average Bed Temperature
H_2/HC   Hydrogen to hydrocarbon ratio
RON-O Research Octane Number

REFERENCES

[1] W. L. Nelson, *Petroleum & Petrochemical Int.*, February, 1972, **12**, 2, 44.
[2] H. J. Maat and L. Moscou, *Proc. 3rd Internat. Congr. Catalysis*, Amsterdam, 1964, p. 1277.
[3] G. D. Gould, A. E. Ravicz, A. I. Salka, and J. W. Wasson, *Nat. Petr. & Ref. Ass. 72nd*, April 1974.

⁴ A. M. Kugelman, *Hydrocarbon Processing*, December 1973, p. 67.
⁵ S. Yagi and N. Wakao, *A.I.Ch.E.J.*, 1959, **5**, 79.
⁶ C. N. Satterfield, *Mass Transfer in Heterogeneous Catalysis*, M.I.T. Press, Massachusetts, 1970.
⁷ H. G. Krane, A. B. Groh, B. L. Schulman, and J. H. Sinfelt, *Proc. 5th World Petr. Congr.*, New York, 1959, II-4, p. 39.
⁸ R. B. Smith, *Chem. Eng. Prog.*, 1959, **55**, 76.
⁹ Yu. M. Zhorov, G. M. Panchenkov, S. P. Zel'tser, and Yu. A. Tirak'yan, *Kinetics and Catalysis*, 1965, **6**, 986.
¹⁰ R. L. Burnett, H. L. Steinmetz, E. M. Blue, and J. J. Noble, *A.C.S. Div. Petr. Chem.*, Detroit, April 1965.
¹¹ O. Levenspiel, *Chemical Reaction Engineering*, 2nd edn., Wiley, New York, 1972, p. 542.
¹² C. G. Myers, W. H. Lang, and P. B. Weisz, *Ind. and Eng. Chem.*, April 1961, **53**, 4, 299.
¹³ L. Beránek, *Adv. Catalysis*, 1975.

DISCUSSION

W. J. Thomas (*Univ. of Bath*) said: I fear that there may be difficulties concerning the quantitative interpretation of your interesting results unless you took into account the dispersion effects which must surely have been present during non-isothermal operation of your packed reactor. Did you take into account such effects and, if so, how were they estimated?

G. J. M. van Keulen (*Akzo Chemie, Amsterdam*) replied: We did not take into account dispersion effects of heat transfer in the catalyst bed. These are difficult to estimate because the main temperature effect occurs in a small region of the reactor, where little paraffin conversion takes place. Dispersion effects of mass transfer are small because of adequate dimensions of the catalyst bed. However, we avoided difficulties associated with dispersion effects by investigating the selectivity parameter.

J. Bousquet (*Elf Erap, Solaize*) said: You do not mention injection of chloride, which is frequently practised to maintain constant acidity during a run. Did your procedure include chloride injection?

G. J. M. van Keulen (*Akzo Chemie, Amsterdam*) replied: We added 2 p.p.m.w. of Cl as dichloroethane to the feed, to maintain the chloride level of the catalyst. The chloride level also depends on the concentration of water vapour present in the reactor and the type of catalyst used.

N. N. Bakhshi (*Univ. of Saskatchewan, Saskatoon*) said: For feed A, a two-fold dilution of the catalyst improved the temperature profile at high space velocity. Did you optimize the diluent:catalyst ratio in order to obtain isothermal conditions?

With feed B, the maximum temperature difference in the bed could be maintained to within five Kelvin degrees by progressive dilution. Please define 'progressive dilution'? Were the catalyst particle size and the diluent particle size the same?

G. J. M. van Keulen (*Akzo Chemie, Amsterdam*) replied: The diluent:catalyst ratio was varied over a wide range with different procedures, but precisely isothermal conditions have never been obtained. Progressive dilution gives the best results, because the front part of the bed needs a higher dilution than the bottom part. Progressive dilution means a logarithmic increase in catalyst concentration along the reactor. Experiments were made with catalyst particles bigger, as well as smaller, than the diluent material. The regular catalyst particles were $\frac{1}{16}$ inch extrudates.

The Role of Sulphur in Reforming Reactions on Pt–Al$_2$O$_3$ and Pt–Re–Al$_2$O$_3$ Catalysts

P. G. MENON and J. PRASAD*

Research and Development Centre, Indian Petrochemicals Corporation Limited,
Baroda 391320, India

ABSTRACT To study the exact role of sulphur in catalytic reforming reactions and the lower sulphur tolerance of Pt–Re–Al$_2$O$_3$(I) as compared with Pt–Al$_2$O$_3$(II) reforming catalyst, microcatalytic pulse reactor investigations and metal dispersion measurements were undertaken. On exposure to thiophen at 500°C, all surface platinum atoms on both catalysts are fully sulphided irreversibly, but the activity for dehydrogenation, dehydroisomerization, dehydrocyclization, and hydrocracking is affected only marginally for (II) but drastically for (I). Further interaction of sulphur with platinum results in reversible adsorption, which leads to poisoning of the two catalysts to different degrees depending on the partial pressure of sulphur. The suppression of hydrocracking and consequent enhancement of aromatization is achieved for (II) by the sulphur held reversibly over and above that held irreversibly, whereas for (I) irreversibly held sulphur alone does this. Hence, the lower sulphur tolerance limit of < 1 p.p.m. for (I), compared with 20 p.p.m. for (II) in commercial reforming processes.

INTRODUCTION

The poisoning effect of sulphur on platinum catalysts has long been known. In the case of catalytic reforming of naphthas to increase their octane rating or to produce aromatic feedstocks, however, the usually dreaded harmful effect of sulphur on Pt–Al$_2$O$_3$ catalysts has been found to be beneficial to some extent: it minimizes excessive hydrocracking. For this reason processors even presulphide the catalyst before naphtha is brought in contact with it. To obtain optimum activity of the catalyst in relation to the quality and yield of product and cycle life, the sulphur in the feedstock is maintained below certain levels. Thus for commercial catalytic reforming on Pt–Al$_2$O$_3$ catalysts the sulphur concentration should generally be below 20 p.p.m., while for the newer bimetallic reforming catalysts like Chevron's Pt–Re–Al$_2$O$_3$ it should be preferably below 1 p.p.m.[1]

The manner in which sulphur influences Pt–Al$_2$O$_3$ reforming catalysts has been a subject of several investigations. In 1957 Engel[2] reported that presulphiding prolongs the life of the catalyst. Haensel[3] reported that sulphur in the feed has a stabilizing effect on catalyst activity for naphtha reforming. Minachev and co-workers studied the effect of thiophen on dehydrogenation of cyclohexane[4] and dehydroisomerization of methylcyclopentane[5] on a platinum–alumina catalyst under reforming conditions. They also reported studies on the amount of sulphur on the catalyst under equilibrium conditions with a sulphur-containing feed as well as on a catalyst "regenerated" with thiophen-free cyclohexane.[5-8] Chang and Kalechits[9] and Lin and Cheng[10] investigated the poisoning effect of sulphur on dehydrogenation and isomerization reactions encountered in platforming. The effect of varying concentrations of sulphur on the activity and selectivity of platinum reforming catalysts was reported by Bursian and Maslyanskii.[11] Sulphiding of platinum–alumina was shown by Maslyanskii et al.[12] to increase its thermal stability but decrease its activity as tested for dehydrogenation of cyclohexane. Pfefferle[13] demonstrated that there should be an optimum partial pressure of sulphur over the catalyst in order to achieve the best activity. Quite recently, Hayes et al.[14] conducted some novel studies and obtained data which indicate that sulphur can be

1061

used as a tool to enhance catalyst stability, particularly in applications involving high-severity processing conditions.

The role played by sulphur by its interaction with supported platinum catalysts used in hydrocarbon transformations, summarized from literature reported so far, is: (1) it attenuates the activity of platinum and thus tends to suppress dehydrogenation, dehydrocyclization, and hydrocracking reactions and (2) it induces considerable catalyst stability both towards selectivity as well as activity. In order to explain this behaviour of sulphur, Hayes et al.[14] suggested that either (a) essentially all of the platinum interacts with sulphur to form a platinum–sulphur complex of lower reactivity than the original platinum species such that coke-forming excessive dehydrogenation does not occur, or (b) the platinum–sulphur interaction is simply an adsorption phenomenon wherein only a fraction of the available platinum sites are deactivated by sulphur adsorption at any given time, but the identity of such deactivated sites changes constantly because of the dynamic nature of adsorption. According to the latter premise the platinum–sulphur species is catalytically inactive for all practical purposes. The authors have not been able to conclude specifically which of the above two hypotheses is correct. The nature of the sulphur species which influences the selectivity in catalytic reforming and also imparts stability to the catalyst is thus not known from the literature.

In the case of the Pt–Re–Al$_2$O$_3$ there is no published information on the influence of sulphur on this catalyst except that the naphtha should preferably contain less than 1 p.p.m. of it.[1] At the same time this catalyst too is sometimes subjected to sulphur treatment before the "break-in" of the naphtha. Thus, in addition to the fact that the role of sulphur in reforming is not clear, there is also no explanation as to the different sulphur tolerance limits of these two types of reforming catalysts. To throw more light on these problems, micro-catalytic pulse reactor studies and hydrogen chemisorption measurements were undertaken. Thiophen was used as the model sulphur compound and the reactions studied were dehydrogenation, dehydroisomerization, dehydrocyclization, and hydrocracking, which constitute the main types of reactions occurring in catalytic reforming.

EXPERIMENTAL

The Pt–Al$_2$O$_3$ catalyst, supplied by Ketjen Catalysts, Amsterdam, was the commerical grade CK-306 (0·6% Pt on CK-300 γ-Al$_2$O$_3$ extrudates). The Pt–Re–Al$_2$O$_3$ catalyst was prepared by us by co-impregnation of 0·6% Pt and 0·6% Re on the same CK-300 carrier. Both catalysts had the same surface area (182 m^{2}g^{-1}) and pore volume (0·53 cm^{3}g^{-1}). The dried catalysts were crushed and sieved and the fraction containing 0·3—0·6 mm particles was used in the experiments.

Pulse Tests. A typical microcatalytic pulse reactor system connected to a gas chromatograph equipped with a TC detector was employed. The catalyst (0·2 g), predried at 550°C for 1 h in air, was taken in the reactor and reduced *in situ* in a stream of hydrogen, 75 cm^3 STP min^{-1}, emerging from an Elhygin hydrogen generator (Milton Roy make) for 1 h at 500°C. For the reduction, the temperature was raised slowly at about 5°C min^{-1}. This was the pretreatment given to each catalyst charge taken. The following tests were then made:

1. *With cyclohexane (CH)*

After reduction of a fresh catalyst charge, the temperature was lowered to 315°C in flowing hydrogen. The catalyst was tested for its activity with 2 μl pulses of cyclohexane. The temperature of the catalyst was then raised to 500°C. A few 2 μl pulses of thiophen were injected to the catalyst. The catalyst was maintained at 500°C for 2 h while hydrogen continued to pass during which time unconverted thiophen and free hydrogen sulphide formed from it would be purged. The temperature was then lowered to 315°C and the activity of the catalyst towards dehydrogenation of cyclohexane was again determined. Then, at 315°C, one 2 μl pulse of thiophen was injected to the catalyst, followed by pulses of cyclohexane at different time intervals after the introduction of thiophen to observe the dehydrogenation activity. This testing was done for 60 min.

2. *With methylcyclopentane (MCP)*

A fresh catalyst charge was pretreated as earlier and after reduction at 500°C for 1 h, 2μl injections of MCP were given to see the activity of the catalyst. A few 2 μl pulses of thiophen were injected to the catalyst which was then purged with hydrogen as earlier for 2 h. Catalyst activity towards MCP was again determined. Then 2 μl of thiophen was injected to the catalyst followed by MCP injection at different time intervals up to 60 min.

3. *With n-Hexane (n-H)*

Same procedure as with MCP above.

Hydrogen chemisorption measurements. This was done by the gas-chromatographic pulse technique adopted for separate determination of Pt and Re dispersion on a composite Pt–Re–Al$_2$O$_3$ catalyst.[15] The predried catalyst (1 g) was reduced *in situ* in the reactor at 500°C for 1 h with pure dry hydrogen, 75 cm^3 STP min^{-1}, the temperature being raised slowly at about 5°C min^{-1}. The catalyst was then cooled to room temperature in flowing hydrogen. Hydrogen was then replaced by oxygen-free dry nitrogen, 40 cm^3 STP min^{-1}, and the catalyst was purged for 30 min. The amount of hydrogen that could be chemisorbed on Pt in both catalysts was then determined.[15] The nitrogen was replaced by hydrogen, 75 cm^3 STP min^{-1}, and the catalyst temperature was again raised slowly to 500°C. A few 2 μl pulses of thiophen were then injected to the catalyst. The flow of hydrogen was continued for 4 h at 500°C. The catalyst was then cooled to room temperature and once again hydrogen chemisorption on it was determined.

Table 1

Influence of exposure of catalysts to sulphur on activity and hydrogen chemisorption capacity

Experiment	0·6% Pt–Al$_2$O$_3$ Before/After exposure to Sulphur		0·6% Pt–0·6% Re–Al$_2$O$_3$ Before/After exposure to Sulphur	
	Before	After	Before	After
Hydrogen chemisorbed on Pt only (cm^3 STP g^{-1})	0·27	0	0·34	0·01
CH Test at 315°C (% Benzene in liquid product)	23·0	19·8	28·5	19·3
MCP Test at 500°C Dehydroisomerization to benzene (%)	22·5	18·0	25·9	65·7
Conversion into hydrocracked products (%)	74·0	77·0	74·1	31·3
n-H Test at 500°C Dehydrocyclization to benzene (%)	20·9	18·1	3·1	11·8
Conversion into hydrocracked products (%)	76·6	79·4	96·9	80·7

All the above experiments were carried out with both types of catalysts. Wherever a few 2 μl pulses of thiophen were injected to the catalyst, it was observed that actually one 2 μl pulse was sufficient to bring about the desired effect. However more than one pulse were injected just to make sure that all the catalyst taken was exposed to sulphur.

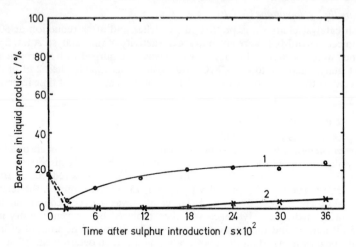

FIGURE 1   Effect of reversible sulphur chemisorption on dehydrogenation of cyclo-hexane at 315°C on sulphided 0·6% Pt–Al₂O₃ (1) and 0·6% Pt–0·6% Re–Al₂O₃ (2).

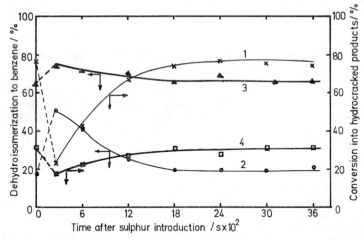

FIGURE 2   Effect of reversible sulphur chemisorption on conversion of MCP into benzene and hydrocracked products at 500°C on sulphided 0·6% Pt–Al₂O₃ (1,2) and 0·6% Pt–0·6% Re–Al₂O₃ (3,4).

RESULTS

The data on measurements of hydrogen chemisorption on both types of catalysts before and after exposure to sulphur are given in Table 1. The stable activities of both catalysts, again before and after sulphur treatment, as determined by the pulse tests, are also given.

The results of pulse tests with cyclohexane, methylcyclopentane, and n-hexane at different time intervals up to 60 min after introduction of sulphur to the catalysts, which had earlier already been exposed to sulphur, are given in Figures 1—3. The activity at zero time on the Figures corresponds to the stable activity before commencement of the 60 min activity cycle determination. In the case of cyclohexane, only benzene was obtained as reaction product. Dehydrogenation activity towards cyclohexane is therefore expressed as percent benzene in liquid product. In activity tests with methylcyclopentane and n-hexane, benzene was obtained on dehydroisomerization and dehydrocyclization of the two reactants, respectively. In addition to benzene and unconverted reactants, several other components were present in the products. For convenience, all non-benzene products are grouped as hydrocracked products. Hence, catalyst activities towards methylcyclopentane and n-hexane are expressed as their capacities to produce benzene and hydrocracked products.

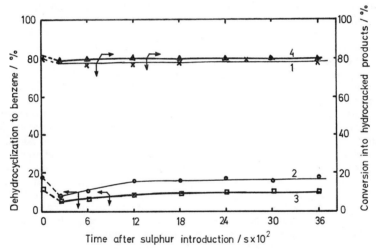

FIGURE 3   Effect of reversible sulphur chemisorption on conversion of n-hexane into benzene and hydrocracked products at 500°C on sulphided 0·6% Pt–Al$_2$O$_3$ (1,2) and 0·6% Pt–0·6% Re–Al$_2$O$_3$ (3,4)

DISCUSSION

The amount of hydrogen chemisorbed at room temperature on the freshly reduced catalyst surface was found to be 0·27 and 0·34 cm³ STP g^{-1} for the Pt–Al$_2$O$_3$ and Pt–Re–Al$_2$O$_3$ catalysts, respectively (Table 1). However, after exposure of the same catalyst samples to thiophen at 500°C and purging for 4 h with hydrogen at that temperature, and then repeating the chemisorption measurements at room temperature, very little or no measurable amount of hydrogen could be chemisorbed on these catalyst surfaces. The conclusion therefore is that all or most of the platinum in the two catalysts has been converted into a platinum–sulphur species on exposure to sulphur. The reaction between sulphur (usually represented as H$_2$S) and platinum may be written as:

$$Pt + H_2S \rightarrow PtS_i + H_2$$

where PtS$_i$ signifies sulphur in an irreversibly chemisorbed state.

During industrial catalytic reforming of sulphur-containing petroleum fractions with Pt–Al$_2$O$_3$, and Pt–Re–Al$_2$O$_3$, the temperature of operation being in the vicinity of

500°C, the interaction of sulphur with the platinum will result in the conversion of all or most of the latter into a stable sulphided species. The presence of hydrogen over the catalyst at 500°C cannot remove this sulphur, as seen in the present chemisorption determination experiment where the sulphur-treated catalyst was even purged for a few hours in the absence of the sulphur source. Moreover, in actual reforming, the constant interaction of sulphur (at p.p.m. levels) from the feed with the platinum will only doubly ensure that the platinum exists mostly or wholly as a sulphided species, PtS_i, and not as free Pt. This would be the situation whether or not the catalyst is presulphided, as in some processes.

Whether this PtS_i species is catalytically active or not is the next question. Data in Table 1 clearly show that the sulphided catalyst is essentially active towards dehydro-genation, dehydroisomerization, dehydrocyclization, and hydrocracking. Activities of both catalysts are no doubt somewhat attenuated as is already very well known from the literature. But attenuation up to a permissible level is beneficial to the catalyst and its selectivity, as is known from commercial practice. Attenuated dehydrogenation is believed to be helpful in decreasing excessive dehydrogenation which may lead to increased coke formation rates. A lowering of hydrocracking would result in better selectivity towards aromatization and increased liquid yields. These features are seen in Table 1. Presulphiding of the catalyst before naphtha is brought in contact with it thus helps attain an attenuated catalyst activity in less time than if this level of activity were to be achieved only by the sulphur in the feedstock when presulphiding is not carried out. Hence, some processors prefer to presulphide the catalyst.

On the basis of the above discussion, in Figure 1, 2, and 3 the state of the catalysts at zero time will be in a sulphided form with sulphur irreversibly chemisorbed as soon as platinum "sees" sulphur, as the catalysts had been exposed to sulphur at 500°C in the pretreatment. The activities at zero time would therefore correspond to the stable activities of the sulphided catalysts. Soon after a thiophen pulse is injected to the catalysts, the activities decline but again rise with passage of time and, in most cases, return to their original levels. This shows that: (1) the sulphur from the thiophen pulse becomes adsorbed on the catalyst surfaces but can gradually be removed with hydrogen; in other words sulphur now becomes *reversibly* chemisorbed on an already sulphided catalyst surface; this interaction can be represented as

$$PtS_i + H_2S \rightleftarrows PtS_iS_r + H_2$$

where S_r signifies reversibly chemisorbed sulphur on platinum. (2) the presence of reversibly chemisorbed sulphur on the catalyst surface inhibits catalytic activity and its gradual removal with time restores the original activity. The observed influence of reversible sulphur chemisorption is seen to affect both catalysts and all the model reactions, though to varying degrees.

The reversible chemisorption of sulphur on sulphided platinum of both catalysts therefore results in the creation of a PtS_iS_r species which is catalytically inactive for all practical purposes. In industrial reforming of sulphur-containing feeds, the inter-action of sulphur with the platinum would result in the formation of PtS_i and PtS_iS_r species on the catalyst surface. The latter type being the result of a dynamic interaction, its concentration depends on the partial pressure of sulphur over the catalysts and also its location on the surface would be constantly changing. Thus, while the primary irreversible sulphiding is actually beneficial to the catalyst, it is the extent of the secondary interaction producing the deactivated PtS_iS_r sites that determines the actual poisoning role of sulphur. The concentration of sulphur in the feed would therefore be permissible only to a level at which the undesirable reactions like hydrocracking and excessive dehydrogenation (leading to coke formation) are suppressed sufficiently by the second-ary interaction. An increase in sulphur concentration beyond this level would result in a further suppression of catalytic activity leading to an overall decrease in catalyst performance. These results are in accordance with those of Weisz[16] who has shown that progressive deactivation of Pt–Al$_2$O$_3$ results in a decrease in hydrocracking activity with an increase in aromatization but later aromatization also decreases ultimately giving a totally deactivated catalyst. The results of the present studies also explain the "hump" that Pfefferle[13] observed in his studies with naphtha contaminated with sulphur which led him to suggest that there should be an optimum for sulphur concentration in the feed.

The results of Table 1 can also explain at least qualitatively some of the well-known superiorities of Pt–Re–Al$_2$O$_3$ catalyst over Pt–Al$_2$O$_3$ catalyst in commercial catalytic reforming. The substantial increase in dehydroisomerization of five-membered naphthenes and the suppression of their hydrocracking will increase the aromatics yield and decrease the coke formation. The reduced hydrocracking of five-membered naphthenes and to some extent of paraffins enables the use of higher temperatures and lower pressures, thereby favouring thermodynamically the formation of still more aromatics.

The suppression of hydrocracking and consequent enhancement of aromatization is achieved for Pt–Al$_2$O$_3$ catalyst mainly by the sulphur held reversibly over and above that held irreversibly, whereas for Pt–Re–Al$_2$O$_3$ irreversibly held sulphur alone readily does this. Hence, further addition of reversible sulphur to the latter may only suppress the overall activity. This perhaps explains the lower sulphur tolerance limit of < 1 p.p.m. for the latter, compared with about 20 p.p.m. for the former, in commercial reforming processes.

ACKNOWLEDGEMENTS. We thank Ketjen Catalysts, Amsterdam, for gifts of CK-306 Pt–Al$_2$O$_3$ catalyst and CK-300 alumina extrudates. The experimental help of Mr. Y. D. Bhatt and Mr. K. N. Jhala is also acknowledged.

REFERENCES

1. H. E. Kluksdahl, *U.S. Patent* 3,415,737 (1968).
2. W. F. Engel, *Dutch Patent* 84,714 (1957).
3. V. Haensel, *U.S. Patent* 3,006,841 (1961).
4. Kh. M. Minachev and D. A. Kondrat'ev, *Izvest. Akad. Nauk S.S.S.R., Otdel. Khim. Nauk.*, 1961, 877; [*Chem. Abs.*, 1961, **55**, 27145d].
5. Kh. M. Minachev and D. A. Kondrat'ev, *Izvest. Akad. Nauk S.S.S.R., Ser. Khim.*, 1965, **7**, 1169; [*Chem. Abs.*, 1965, **63**, 11375c].
6. Kh. M. Minachev and D. A. Kondrat'ev, *Izvest. Akad. Nauk S.S.S.R., Otdel. Khim. Nauk.*, 1960, 300; [*Chem. Abs.*, 1960, **54**, 2091i].
7. Kh. M. Minachev, G. V. Isagulyants, and D. A. Kondrat'ev, *Izvest. Akad. Nauk S.S.S.R., Otdel. Khim. Nauk.*, 1960, 702; [*Chem. Abr.*, 1960, **54**, 23671d].
8. Kh. M. Minachev and G. V. Isagulyants, *Proc. Third Internat. Congr. Catalysis* (North Holland, Amsterdam, 1965) p. 308.
9. Yen-Ching Chang and I. V. Kalechits, *K'o Hsüeh T'ung Pao*, 1958, **15**, 478; [*Chem. Abs.*, 1959, **53**, 10718g].
10. Tao-Yu Lin and Yen-Ch'in Cheng, *Wu Han Ta Hsüeh, Tzu Jan K'o Hsüeh Hsüeh Pao*, 1959, **5**, 32; [*Chem. Abs.*, 1960, **54**, 55987e].
11. N. R. Bursian and G. N. Maslyanskii, *Khim. i Technol. Topliv i Masel*, 1961, **6**, 6; [*Chem. Abs.*, 1962, **56**, 7582d].
12. G. N. Maslyanskii, B. B. Zharkov, A. Z. Rubinov, and T. M. Klimenko, *Kinetika i Kataliz*, 1971, **12**, 1060.
13. W. C. Pfefferle, *Preprints Division of Petrol. Chem., Amer. Chem. Soc.*, 1970, **15**, No. 1, AZI.
14. J. C. Hayes, R. T. Mitsche, E. L. Pollitzer, and E. H. Homeier, *Preprints 167th National Meeting, Amer. Chem. Soc., Los Angeles*, 1974.
15. P. G. Menon, J. Sieders, F. J. Streefkerk, and G. J. M. van Keulen, *J. Catalysis*, 1973, **29**, 188.
16. P. B. Weisz, *Proc. Second Internat. Congr. Catalysis, Paris*, 1960, p. 937.

DISCUSSION

W. F. de Vleesschauwer (*Texaco, Ghent*) said: You conclude that, with an all-platinum catalyst, the sulphur mainly held reversibly acts as promoter. We progressively sulphided a monometallic catalyst. At completion of the sulphidation, MCP aromatization rose sharply and was accompanied by a further reduction of the extent of hydrocracking. Such behaviour was not observed for n-hexane. Subsequent flushing of the catalyst with hydrogen at 873 K for 1 h to remove the loosely held sulphur increased the hydrocracking and reduced the MCP aromatization (Table below). This further

Table
Effects of sulphur content on aromatization and hydrocracking

Catalyst	Sulphur (wt %)	Aromatization (mole %)		Hydrocracking (mole %)	
		MCP	n-hexane	MCP	n-hexane
1[a]	0·12	40·0	38·3	51·8	56·5
	0·20	41·8	39·1	40·2	52·0
	0·21	52·5	40·6	24·4	47·9
	0·21[b]	44·6	41·4	47·5	54·5
2[c]	0·26	73·0	32·1	7·1	31·4
	0·61	64·9	30·7	9·1	31·1

[a] 0·6 % platinum supported on γ-Al$_2$O$_3$.
[b] Catalyst flushed with hydrogen at 873 K for 1 h.
[c] 0·3 % platinum, 0·3 % rhenium supported on γ-Al$_2$O$_3$.

supports your conclusions. For a catalyst containing 0·6 wt % platinum and having a dispersion of 0·70, the critical sulphur level was at 0·21 wt %. This corresponds to a sulphur to surface-platinum atom ratio of 3 to 1. Would you comment on this?

The Table also shows two results for a commercial rhenium-promoted catalyst which exhibits an activity edge that can be effectively removed by presulphidation. The sulphur appetite of this catalyst is much greater than of an all-platinum catalyst, and the real danger of oversulphidation is demonstrated.

P. G. Menon (*Indian Petrochem. Corp., Baroda*) replied: Dr. de Vleesschauwer's results agree with ours. The distinct roles of sulphur held (i) reversibly and (ii) irreversibly, and the difference in behaviour between Pt–Al$_2$O$_3$ and Pt–Re–Al$_2$O$_3$ on sulphidation are also confirmed.

We have not determined the stoicheiometric ratio of sulphur to surface platinum. However, a value as high as 3 is very interesting, particularly since it is believed that sulphur poisons only the metal sites and not the acidic sites of the carrier.[1] This aspect deserves further investigation.

H. Charcosset (*CNRS, Villeurbanne*) said: You have published (ref. 15 of the paper) a method of separate titration of the exposed platinum and rhenium atoms in Pt–Re–Al$_2$O$_3$ bimetallic catalysts based on successive oxygen–hydrogen titrations at room temperature. Did you apply that method in the present study, and what value did you obtain for the dispersion of rhenium? From our work it appears that in most cases, if the titration of all surface platinum and rhenium atoms is possible, then the

[1] F. G. Ciapetta and D. N. Wallace, *Catalysis Rev.*, 1971, **5**, 107.

separate titration of the surface metal atoms of each kind is hardly possible. We found that the presence of platinum allowed the reduction at room temperature of most of the oxygen chemisorbed on rhenium. On the other hand, oxygen chemisorbed on pure Re–Al$_2$O$_3$ was not reduced by molecular hydrogen under the same conditions (in agreement with ref. 15 of the paper and other publications).

P. G. Menon (*Indian Petrochem. Corp., Baroda*) replied: We used the gas chromatographic pulse technique, described in ref. 15, for titration of platinum both in Pt–Al$_2$O$_3$ and in Pt–Re–Al$_2$O$_3$. In the examination of the bimetallic catalyst successive oxygen–hydrogen titrations give a reliable chemisorption value for platinum only, and not for rhenium (I have pointed out this in ref. 15). That is why in the present work we have given the gas titre values for platinum, but not for rhenium.

Recently, we have found that the choice of arbitrary experimental conditions (*e.g.* quantity of catalyst, pulse volume) in the above method can give unreliable and non-reproducible titres even for monometallic Pt–Al$_2$O$_3$. Hence, we have suggested[2] unique conditions under which the gas chromatographic pulse technique should be used for the determination of metal dispersion in Pt–Al$_2$O$_3$.

P. H. Emmett (*Portland State Univ., Portland, Oregon*) said: The author has pointed out that pure platinum and platinum covered with a tightly bound layer of sulphur have the same activity toward the dehydrogenation of cyclohexane. I believe this is the first time that a catalyst has been found to have identical activity for a reaction before and after it has been covered with a layer of a foreign substance. It brings to mind a recent contention by Prof. Merrill to the effect that carbon dimers in a more or less complete layer on platinum are active in the hydrogenation of ethylene, and also the observation by Prof. Somorjai and his associates that a template of ordered carbon on a face of a single crystal of platinum may be involved in the cyclodehydrogenation of heptane to toluene.

P. G. Menon (*Indian Petrochem. Corp., Baroda*) replied: I am grateful to Prof. Emmett for pointing out the parallelism between our results and those of Merrill and Somorjai. It seems to be a fairly general phenomenon in catalysis that the active sites are not those present initially in the fresh catalyst, but those formed *in situ* during reaction. An outstanding example of this has been illustrated by Prof. Emmett, who showed[3] that the 'coke' deposited on a silica–alumina cracking catalyst could take part in subsequent cracking reactions on the catalyst, the coke even contributing to the formation of gaseous products such as propylene. All these observations underline the need for us to study more equilibrated catalysts and catalysts under actual reaction conditions, rather than catalysts in the virgin state.

J. R. Katzer (*Univ. of Delaware, Newark, Del.*) said: Your comment that a platinum catalyst which has been contacted with thiophen can still almost as effectively catalyse cyclohexane dehydrogenation (and such more complex reactions as methylcyclopentane dehydroisomerization to benzene, and n-hexane dehydrocyclization to benzene) is most interesting in view of the proposed absence of hydrogen chemisorption on the metal. My questions relate more to the reaction network involved.

(*a*) Would you comment on the almost complete absence of sulphur poisoning of cyclohexane dehydrogenation?

(*b*) At higher pressures methylcyclopentane dehydroisomerization and n-hexane dehydrocyclization occur to almost equal extents as (i) a monofunctional reaction on the metal and (ii) a bifunctional reaction on the metal–acid function. I do not know how this division occurs at atmospheric pressures. Would you state the chloride content of your catalysts and estimate the extent to which the reactions are occurring

[2] J. Prasad and P. G. Menon, *J. Catalysis*, in press.
[3] P. H. Emmett and W. A. van Hook, *J. Amer. Chem. Soc.*, 1962, **84**, 4421.

by a bifunctional reaction network? Hexane dehydrocyclization is particularly important. It has been shown that sulphur poisoning stops the more complex reactions on platinum, but does not poison simple dehydrogenation; thus n-hexane could be dehydrogenated to hexatriene, which is favoured by equilibrium considerations at low hydrogen pressure and undergoes non-catalytic thermal cyclization to benzene. This would not occur at higher hydrogen pressures, and thus the poisoning effect could be very different under the higher hydrogen pressures used in actual commercial operations. Have you investigated the effect of pressure? Please comment on these possibilities.

(*c*) Have you carried out any experiments with hydrogen sulphide which might confirm the results obtained using triophen? Are the results you observe with thiophen due solely to sulphur?

(*d*) Many of us use pulse reactors; have you carried out any continuous-flow experiments to support your results and to establish equilibrium poisoning effects?

(*e*) Please comment on the mechanism of sulphur poisoning of the metal surface.

P. G. Menon (*Indian Petrochem. Corp., Baroda*) replied: (*a*) The activities of unsulphided and sulphided catalysts towards cyclohexane dehydrogenation are given in Table 1. The effect of sulphur towards both types of catalyst is appreciable. Perhaps a more marked effect may be seen in continuous-flow studies. Sulphidation of the catalysts does indeed produce an attenuation of dehydrogenation activity. We wish to emphasize that, although almost all the surface platinum atoms are in a sulphided state (irreversibly adsorbed sulphur), the catalysts still possess a good measure of their original activity and hence cannot be considered to be poisoned in the conventional sense. The real poisoning occurs with reversibly adsorbed sulphur when the catalyst activity can be totally inhibited (see Fig. 1).

(*b*) We have started high-pressure studies, but it is too early to comment on our results. We believe that, under commercial reforming conditions and using sulphur-containing naphtha, all surface platinum atoms will be in the irreversibly sulphided form. The chloride content of the $Pt-Al_2O_3$ was 0.067%, and that of the $Pt-Re-Al_2O_3$ catalyst 0.65% by wt.

(*c*) We have not done experiments with hydrogen sulphide, but we will do some in the future in view of your comments and those of Prof. Emmett.

(*d*) We also are aware of the limitations of pulse-reactor results. Continuous-flow experiments are in progress in our laboratory.

(*e*) Our results contradict the popular belief that sulphur is a poison for platinum catalysts. Irreversibly sulphided platinum is a reactive catalyst in its own right, whereas further reversible adsorption of sulphur attenuates activity. Any attempt to comment on the mechanism of these platinum–sulphur interactions, in the absence of results obtained by other techniques, would be too speculative.

Conversion–Selectivity Trends in Fluid Catalytic Cracking of Hydrotreated Heavy Vacuum Gas Oil Over Zeolite Catalyst

Hartley Owen, Paul W. Snyder, and Paul B. Venuto*

Mobil Research and Development Corporation, Research Department,
Paulsboro, New Jersey 08066, U.S.A.

ABSTRACT Cracking patterns of hydrotreated (HDT) heavy vacuum gas oil (HVGO) at 482—593°C and short contact times have been investigated in a riser pilot plant using commercial fluid zeolite catalyst. Trends in conversion, selectivity, and product quality have been analysed.

Below 80% conversion, crackability, temperature dependency ($E_A = 12 \cdot 9$ kcal/mole) and overall selectivity characteristics of HDT HVGO were similar to those of non-HDT gas oil of comparable hydrogen content. Gasoline paraffins and aromatics increased, while olefins and naphthenes decreased with increasing conversion, indicating intermolecular hydrogen-transfer reactions catalysed with great facility by zeolites. Octane numbers varied only slightly with conversion, reflecting compensating effects by molecules of different classes.

At 593°C and very high conversions, the residual concentration of polynuclear aromatics (remaining in the cycle oil after more reactive species had been cracked) appeared to have limited accessibility to zeolite pores.

Introduction

With the advent of the energy crisis and escalation in the price of imported crude, petroleum refiners have been forced to process increasingly heavy fractions to meet continuing product demand. Many naturally-occurring impurities, which are potential cracking catalyst poisons and environmental pollutants, are concentrated in the high-molecular-weight, hydrogen-deficient "bottom of the barrel". There has been considerable interest recently in hydrotreating feedstocks prior to catalytic cracking to improve activity/yield performance and as a means to remove sulphur, nitrogen, and organo-metallic contaminants.[1-3]

We now report detailed crackability/selectivity patterns in the catalytic cracking of HDT HVGO at 482—593°C and short contact times in a riser FCC pilot plant, using zeolite catalyst. Data on yield structure, and product quality, and some mechanistic interpretation are included.

Experimental

Catalyst. Equilibrium commercial fluid catalyst containing 15% wt. rare-earth Y-type zeolite and burned carbon-free was employed: S.A., 172 m²/g; packed density, 0·74 g/cm³; pore volume, 0·64 cm³/g; Ni/V/Fe (p.p.m.) = 118/244/1204. This catalyst had a Fluid Activity Index of 67·5 (defined as the conversion obtained with a 180°C at 90% A.S.T.M. gasoline product when cracking a Light East Texas Gas Oil at 2 C/O, 454°C, 6 W.H.S.V. for 5 min on stream time in a fixed fluidized bed.

Charge. Feedstock was prepared by hydrogenation of raw Arab light gas oil. Inspections for this HDT HVGO and for a distillate gas oil (straight run gas oil, SRGO, not hydrotreated) are shown in Table 1.

Reactor System. Experiments at 482—538°C were run in a 9·23 m, 5-loop, bench-scale riser FCC pilot unit, while those at 593°C were conducted in a 4·86 m, 3-loop riser. Hot catalyst was displaced vertically downward by nitrogen pressure from a storage

Owen, Snyder, Venuto

Table 1

Inspections of Gas Oil Feedstocks

	HDT[a] HVGO	SRGO
Sp. Gravity at 16°C, g/cm³ (°API)	0·8849 (28·4)	0·8822 (28·9)
Refractive Index, 70°C	1·47716	
Molecular Weight (V.P.)	405	354
Hydrogen, wt %	13·02	12·97
Sulphur, wt %	0·14	0·37
Total Nitrogen, wt %	0·037	0·062
Basic Nitrogen, p.p.m.	33	192
Total Metals (Ni, V, Fe), p.p.m.	<0·3	<0·3
Molecular Type, wt % (Mass Spec.)		
Paraffins	25·7	23·7
Naphthenes	30·5	44·2
Aromatics	43·8	32·1
Distillation, A.S.T.M., °C		
IBP	325	214
10 vol %	381	296
30 vol %	401	353
50 vol %	431	399
70 vol %	466	449
90 vol %	512	516

[a] Raw charge (343°C+); Sp. Gr. = 0.9153 g/cm³ (23.1 °API), n_D 70°C 1.49495, % wt Hydrogen 12.07, % wt Sulphur 2.24, % Paraffins 25.2, % Naphthenes 22.5, % Aromatics 52.3.

hopper through a calibrated orifice to a mixing zone, where preheated oil (421°C) was injected into the riser inlet (pressure, 210 kN/m²). Reactor loop temperatures were controlled by salt baths. Reaction effluent was passed through a steam-stripper/cyclone, where vapour was disengaged from spent catalyst, and processed in a conventional separation/distillation/analysis sequence. Total inerts (as steam equivalents) and recoveries-on-feed averaged 3·0 wt % and 97·4 wt %, respectively. Operating conditions and yield structure for all runs are shown in Table 2. Conversion is defined as 100 minus vol % cycle oil.

RESULTS

Crackability. Catalyst/oil (C/O) responses (effect of C/O on conversion) at 482—593°C are shown in Figure 1. At 482°C, the response was +8 vol % conversion for doubling the C/O in the 60—80 vol % conversion range; at 583°C, the response for doubling the C/O was +10 vol % in the conversion range of 75—80 vol %, which compares well with that for SRGO (+8 vol %) at identical conditions. Thus, HDT HVGO behaves like a distillate gas oil in crackability. However, at 593°C, and in the high conversion range of 85—90 vol %, the response was much lower (+2·2 vol %/doubling C/O).

Data from runs at 482—593°C with catalyst residence times (τ) near 4·5 s were adjusted to determine C/O ratios required to give 79 vol % conversion at 4·5 s τ. Reciprocals of C/O ratios were assumed to be proportional to relative rate constants, k. Apparent activation energies, E_A, for the intervals 482—538°C and 538—593°C were calculated to be 12·9 and 1·8 kcal/mole, respectively (Fig. 2).

1072

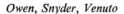

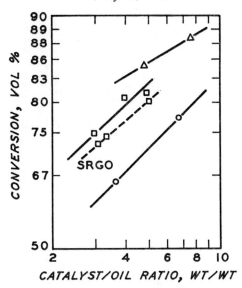

FIGURE 1 Conversion, vol % against catalyst/oil ratio, wt./wt. in riser cracking of HDT HVGO: $\tau = ca.$ 4·5 s; $\bigcirc$, $\square$, $\triangle$ = 482,538, and 593°C., respectively; —— = HDT HVGO, — — — = SRGO.

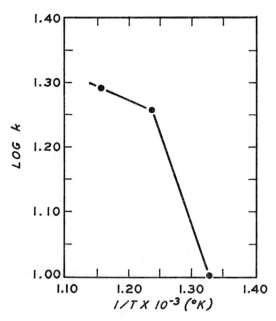

FIGURE 2 Arrhenius plot, log k against $(T/K)^{-1} \times 10^{-3}$, for riser cracking of HDT HVGO at $\tau = ca.$ 4·5 s.

Table 2

Operating Conditions and Yield Structure for Riser Fluid Catalytic Cracking of HDT HVGO

Run Number	1	2	3	4	5	6	7	8	9	10
Conditions										
Reactor Inlet Temperature, °C	577	←——— 482 ———————→							←— 593 —→	
Catalyst Inlet Temperature, °C	538	538	532	736	666	638	671	632	727	688
Catalyst/Oil Ratio, wt/wt	3·7	6·8	7·3	3·0	4·0	5·0	4·0	5·6	4·8	7·7
Catalyst Residence Time, s	5·3	4·6	9·8	4·6	4·4	4·4	2·3	8·4	3·6	3·9
Oil Partial Pressure, Inlet, p.s.i.a.	30·8	25·8	23·2	32·1	30·0	29·0	32·7	24·6	22·6	27·6
Carbon on Spent Catalyst, wt %	0·629	0·594	0·616	0·569	0·595	0·635	0·563	0·750	0·860	0·812
Yields (No Loss Basis on Fresh Feed)										
Conversion, vol %[a]	65·2	77·41	79·41	74·48	80·57	81·34	76·91	83·75	84·87	87·97
Hydrogen, wt %	0·02	0·02	0·02	0·05	0·05	0·04	0·03	0·05	0·10	0·08
Methane, wt %	0·18	0·26	0·31	0·63	0·71	0·73	0·54	1·04	1·78	2·01
Ethylene, wt %	0·26	0·34	0·42	0·59	0·59	0·59	0·48	0·78	1·26	1·36
Ethane, wt %	0·22	0·29	0·34	0·56	0·56	0·54	0·43	0·73	1·23	1·32
Propylene, vol %	3·29	4·48	4·60	6·93	8·14	7·52	6·50	8·54	12·20	13·39
Propane, vol %	1·18	1·81	2·22	1·44	1·99	2·09	1·62	3·03	3·28	4·75
Total C$_3$'s, vol %	4·47	6·29	6·82	8·37	10·13	9·61	8·12	11·57	15·48	18·14

Butenes, vol %	4·63	5·40	5·11	7·51	9·83	7·38	8·46	8·77	13·26	11·47
Iso-butane, vol %	4·80	7·29	8·11	4·69	6·06	6·67	5·76	9·67	7·91	10·57
n-Butane, vol %	0·98	1·59	1·89	1·07	1·68	1·61	1·37	2·35	2·45	3·37
Total C₄'s, vol %	10·41	14·28	15·11	13·27	17·57	15·66	15·59	20·79	23·62	25·41
Pentenes, vol %	3·30	2·87	1·99	4·50	4·83	3·67	5·85	4·15	8·23	5·35
Iso-pentane, vol %	5·28	7·54	8·65	3·53	4·89	5·40	6·07	9·53	7·54	8·98
n-Pentane, vol %	1·16	1·55	1·86	0·61	1·17	1·24	0·98	2·08	1·77	1·90
Total C₅'s, vol %	9·74	11·96	12·50	8·64	10·89	10·31	12·90	15·76	17·54	16·23
C₆+ Gasoline, vol %ᵇ	51·42	56·16	56·64	57·33	55·55	58·81	53·46	48·75	40·40	38·08
Cycle Oil, vol %	34·77	22·59	20·59	25·52	19·43	18·66	23·09	16·25	15·13	12·03
Coke, wt %	2·48	4·39	4·81	1·86	2·58	3·44	2·49	4·49	4·36	6·56
Gasoline Efficiency, vol %	93·6	88·0	86·9	88·6	82·46	84·9	86·2	77·0	68·0	61·7
Gasoline Octane Number, R + O°	80·7	<82	<82	87·4	87·8	85·8	87·1	86·9	91·7	93·0

ᵃ Based on C_5–180°C at 90% A.S.T.M. cut-point on small, one-piece, non-refluxing still.

ᵇ 180°C at 90% A.S.T.M. cut-point.

ᶜ C_5+ R + O Mini-Micro octane test; based on measured value for raw gasoline with correction for C_5+ in gas and C_4— in gasoline using blending values as follows: C_6+ 85; n-C_5 61·5; i-C_5 92·5; C_5 = 98; n-C_4 94; i-C_4 101; C_4 = 103; C_3 112; C_3=102.

Yield Structure. Both total dry gas (C_3-) and total C_4's increased with temperature as favoured by both kinetics and thermodynamics;[4] and, for a given temperature, with conversion. Molar expansions, observed moles of product/mole of feed (ex coke) extrapolated to 100% conversion, for runs at 482, 538, and 593°C were 4·45, 4·95, and 5·50, respectively. Ethylene/ethane ratios were insensitive to temperature or conversion, but C_3H_6/C_3H_8, C_4H_8/C_4H_{10}, and C_5H_{10}/C_5H_{12} ratios increased sharply with temperature in accord with thermodynamic constraints,[5] and decreased with conversion (secondary hydrogen transfer), especially at higher temperatures. Kinetically-favoured high iso-/n-butane ratios decreased with both conversion (constant temperature) and increasing temperature, reflecting thermodynamic influences.[4]

At each temperature, the characteristic[6,7] increase of coke with conversion was observed. Coke decreased sharply as temperature increased from 482 to 538°C. At 593°C, however, coke levels were relatively high. Coke sulphur levels were all < 0.1 wt % on coke. The effect of catalyst residence time, τ, on coke formation at 538°C was estimated *via* the Voorhies Equation, carbon-on-catalyst $= A\tau^n$, after adjusting data at 2, 4 and 8 s. τ to 4·0 C/O. The Voorhies slope, n, for this HDT stock was 0·16, which is lower than values near 0·28 obtained[8] under comparable conditions for non-HDT gas oils.

C_5+ gasoline yields and efficiencies (Table 2), at a given conversion, decline with increasing temperature. A maximum gasoline yield of 68—69 vol % was observed at 538°C near 79—80 vol % conversion. The highest gasoline efficiency (*ca.* 94% at 65 vol % conversion) was observed at 482°C (mild conditions), while severe secondary cracking of gasoline occurred at 593°C.

Product quality. As seen in Table 2, $C_5+ R + O$ octane number increased with temperature (averaging 81·3, 86·7, and 92·4 at 482, 538, and 593°C, respectively). At 538°C, over the range of 74—84 vol % conversion, octane number remained essentially constant, while at 593°C, a slight uptrend was observed. Variations with temperature for gasoline properties are listed in Table 3. Dominant C-numbers in all gasolines were C_5—C_9, with overall range of C_5—C_{13}.

Table 3

Average Properties of Cracked Gasolines

	482°C	538°C	593°C
Sp. Grav., g/cm³, 16°C, C_5+	0·7358	0·7385	0·7554
Molec. Wt., C_6+	114	112	106
Hydrogen, Wt. %, C_6+	13·5	13·0	12·5
Sulphur, Wt. %, C_5+	← <0·005 →		0·02

Trends in molecular composition (P,O,N,A-analysis) against conversion for C_6+ gasolines are shown in Fig. 3. Over the temperature range 482–593°C, paraffins and aromatics generally increased; conversely, olefins and naphthenes decreased, particularly at the higher temperatures. Within the class of aromatics, the ratio of naphthalenes/tetralins increased greatly with temperature (Fig. 4), but the *total* concentration of two-ring species remained nearly constant (*ca.* 4 vol %).

In cycle oils, density increased smoothly with conversion (range 0·9256—1·0413 g/cm³), with points from all temperatures falling on the same correlating line. Trends in cycle oil composition against conversion are shown in Fig. 5, where aromatics increase, and paraffins and naphthenes decline. Paraffin and aromatic trends seem to correlate points from all temperatures, while with naphthenes, low and high temperature lines

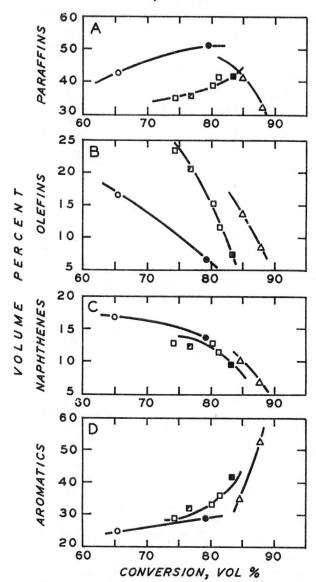

FIGURE 3 Molecular composition, vol % paraffins, olefins, naphthenes, and aromatics against conversion, vol % for depentanized gasolines; 482°C: ○ = 4·5 s τ, ● = 10 s τ; 538°C: □ = 4·5 s τ, ◪ = 2 s τ, ■ = 8 s τ; 593°C: △ = 4 s τ.

appear. The P,O,N,A and density data reflect loss of the lighter, more crackable components as conversion increases, with consequent accumulation of denser, more refractory species. Aromatic ring demethylation probably accounts for the larger amount of methane formed at 593°C.

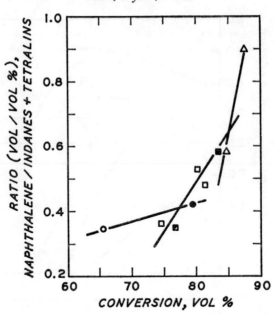

FIGURE 4   Trends in ratio of naphthalenes/indanes + tetralins, against conversion, vol % for depentanized gasolines; 482°C: ○ = 4·5 s τ; ● = 10 s τ; □ = 4·5 s τ, ◪ = 2 s τ, ■ = 8 s τ; 593°C: △ = 4 s τ.

DISCUSSION

Below 80 vol % conversion, the C/O response for HDT HVGO (where H_2 was added catalytically) and overall selectivity/molar expansion patterns were similar to those of good quality non-HDT SRGO of comparable hydrogen content. A similar association has been made for the cracking of HDT Kuwait stock.[1] Residence time showed little apparent influence on selectivity.

The changes in gasoline P,O,N,A-analyses with increasing conversion (Fig. 3) reflect carbonium ion-type intermolecular hydrogen transfer processes that are well known in catalytic cracking,[4,9–11] where olefin + naphthene or olefin + olefin form paraffin + aromatic (and/or polymer/coke). Nace[10] suggested that $k_{\text{hydrogen transfer}}/k_{\beta\text{-scission}}$ is greater with zeolites than with amorphous catalysts, and such hydrogen redistribution is particularly simple with HDT stocks.[11] Similar chemistry may contribute to trends in cycle oil composition with increasing conversion (Fig. 5). The data in Fig. 3 suggest that the rate of hydrogen transfer (but not necessarily the rate of hydrogen transfer relative to cracking) is increasing with temperature.[10]

The striking increase of +18 vol % aromatics (Fig. 3-D)—with corresponding decrease in paraffins and naphthenes at 593°C—also reflects, in part, an "aromatic concentration effect", as saturated species are cracked away at higher severity. The hydrogen transfer process is particularly visible in the 2-ring aromatic fraction at the upper end of the gasoline boiling range (Fig. 4), where reactions of the type $C_{10}H_{12} \rightarrow C_{10}H_8 + 4''H''$ can be inferred. Thus, the net result of increased cracking severity is more aromatic, more hydrogen-deficient, and lower-molecular-weight gasoline.

At 538°C, despite the marked changes in relative proportions of gasoline P,O,N,A components over the conversion range 74—84 vol %, measured octanes vary only slightly. This infers the operation of compensation effects, as one class of high-octane compounds (*i.e.*, olefins) is replaced by another (*i.e.*, aromatics).

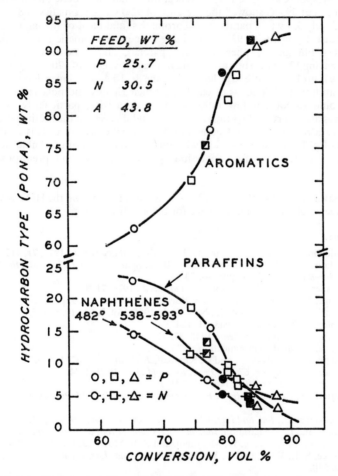

FIGURE 5 Hydrocarbon Type (P,O,N,A), wt % against conversion, vol % for cycle oils in riser cracking of HDT HVGO; 482°C: O = 4·5 s τ; ● = 10 s τ; □ = 4·5 s τ, ◨ = 2 s τ, ◼ = 8 s τ; 593°C: △ = 4 s τ; signs indicated with N in the figure represent naphthenes only.

A range of 10—35 kcal/mole has been reported for E_A by Voge,[4] and Nace[8] has calculated E_A values of 14—17 kcal/mole for the cracking of the SRGO in Table 2 at conversions less than 80 vol % at 510—566°C; our value of 12·9 kcal/mole for the interval 482—538°C, while relatively low, seems reasonable. The value of 1·8 kcal/mole for the interval 538—593°C, however, represents a very low temperature response, in the direction of the sharply lower E_A values encountered in severely diffusion-limited catalytic regimes.[12]

Owen, Snyder, Venuto

In gas oil cracking, reactivity is itself a function of conversion,[13] because the complex mixtures of reactant species crack at different rates, and the more chemically resistant classes of hydrocarbon persist longer than the more reactive ones. The data in Fig. 5 and the flat C/O-response at 593°C (Fig. 1) reflect this. Analysis of cycle oil from the highest conversion run at 593°C showed the presence of 92·0 wt % aromatics, 4·8 wt % naphthenes, and only 3·3 wt % paraffins. All aromatics (and most of the naphthenes) contained at least 2 rings, with 93% containing 2 or more aromatic rings (a mixture of substituted naphthalenes, phenanthrenes, pyrenes, and some heterocyclic sulphur analogues having 13—17 carbon atoms and molecular weights of 170—225). The average size of these is probably close to or larger than the 8—10 Å pore dimension of faujasite.[14] Assuming an intrazeolitic locus of catalysis,[15] their rigidity and polarizable π-electron systems would foreordain low intracrystalline diffusivity, particularly at the high temperatures and short contact times involved, and if access to the zeolite pore-channel system were already somewhat constrained from coke laydown. This residual concentration of polynuclear aromatics (with minimal retention of crackable side-chains) would show little tendency for conversion by β-scission,[4] but high predisposition[16] to form coke.

ACKNOWLEDGEMENTS. We thank Dr. R. T. Pavlica for preparing the HDT HVGO, and Mr. E. J. Demmel and Dr. D. M. Nace for helpful discussions.

REFERENCES

[1] R. E. Hildebrand, G. P. Huling, and G. F. Ondish, *Oil and Gas J.*, 1973, **71**, 112.
[2] W. M. Haunschild, D. O. Chessmore, and B. G. Spars, *79th A.I.Ch.E. Meeting, Houston, Texas, March 16–20*, 1975, Paper No. 70–D.
[3] C. L. Hemler and W. L. Vermillion. *Oil and Gas J.*, 1973, **71**, 88.
[4] H. H. Voge, *Catalysis* (Reinhold, New York, 1958) Vol. 6, Ch. 5, p. 407.
[5] K. K. Kearby, *Catalysis* (Reinhold, New York, 1955) Vol. 3, Ch. 10, p. 455.
[6] J. B. Pohlenz, *Oil and Gas J.*, 1963, **61**, 124.
[7] A. Voorhies, *Ind. and Eng. Chem.*, 1945, **37**, 318.
[8] D. M. Nace, personal communication, 1973.
[9] F. E. Shephard, J. J. Rooney, and C. Kemball, *J. Catalysis*, 1962, **1**, 379.
[10] D. M. Nace, *Ind. and Eng. Chem., Prod. Res, and Dev.*, 1969, **8**, 24.
[11] P. B. Weisz, *U.S. Patent*, 1968, No. 3,413,212.
[12] C. N. Satterfield, *Mass Transfer in Heterogeneous Catalysis* (MIT Press, Cambridge, Mass., 1970), Ch. 1, p. 4.
[13] V. W. Weekman, Jr., *Ind. and Eng. Chem., Proc. Res. and Dev.*, 1968, **7**, 90.
[14] P. B. Venuto and L. A. Hamilton, *Ind. and Eng. Chem., Prod. Res. and Dev.*, 1967, **6**, 190.
[15] P. B. Weisz, V. J. Frilette, R. W. Maatman, and E. B. Mower, *J. Catalysis*, 1962, **1**, 307.
[16] D. M. Nace, S. E. Voltz, and V. W. Weekman, Jr., *Ind. Eng. Chem., Proc. Res. and Dev.*, 1971, **10**, 530.

DISCUSSION

A. H. Weiss (*Worcester Polytechnic, Worcester, Mass.*) said: Prof. Antoshin (U.S.S.R. Inst. of Org. Chem.) informs me that rare-earth-Y zeolites treated at 973 K show bulk crystallinity by *X*-ray analysis but surface disorder by ESCA. The conclusion thus emerges that very large gas oil molecules do not enter the zeolite cage, but rather react on the zeolite-like 'bowls' on extensions of the microcrystals. What do you believe about the mechanism of cracking of very large molecules? Does it occur inside the zeolite cage?

P. B. Venuto (*Mobil, Paulsboro*) replied: Prof. Antoshin's work is interesting, but it would be unwise to make interpretative comments without seeing his results. Concerning the mechanism of cracking of very large molecules, the internal geometry of zeolites permits facile sorption (rapid rates, rectangular isotherms) of potential reactant molecules with critical dimensions less than those of the pore diameters. Such molecules can include some C_{18}–C_{25} (or higher) mono-, di- and tri-nuclear aromatics present in heavy gas oil,[1] although exclusion by size or shape will, of course, occur with very large or bulky molecules (this is an extension of the molecular size–selectivity principle enunciated by Weisz and co-workers). For those molecules with critical dimensions less than those of the zeolite pores, we believe that primary as well as secondary cracking reactions can occur within zeolite pores. This conclusion would seem to be well documented in the literature.

J. Bousquet (*Elf Erap, Solaize*) said: The exponent of the Voorhies equation normally has the value 0·5, whereas you report values of 0·1 to 0·2. Could it be that the higher values relate to amorphous catalysts and the lower values to zeolite catalysts?

P. B. Venuto (*Mobil, Paulsboro*) replied: Yes. Higher values have been found for amorphous catalysts: as cited in our paper we have observed values of n in the Voorheis equation near 0·28 for non-hydrotreated gas oils in cracking over *zeolite* catalysts. Our value of 0·16 for cracking of the hydrotreated stock is somewhat lower, but not unreasonable.

H. J. Lovink (*Akzo Chemie, Amersfoort*) asked: What was the nature of the coke on the equilibrated catalyst? What thermal cracking takes place at 4–10 s contact time at 811 K and at 866 K? Is the catalytically cracked naphtha made at 811 K still sufficiently stable?

P. B. Venuto (*Mobil, Paulsboro*) replied: By 'coke' we mean the hydrogen-deficient polymeric residue left on the catalyst after the relatively efficient steam-stripping of the spent catalyst; we did not make detailed studies on its chemical properties.

While slight thermal cracking may have occurred at 866 K, we believe that the overall selectivity patterns and product quality [particularly at the short *oil* residence times (2·9–3·1 s) and low pressures (210 kN m^{-2})] reflect zeolitic cracking. The dominance of intermolecular H-transfer reactions is noteworthy in this respect. Also, the increase in methane yield at 866 K probably arises from aromatic ring demethylation which ultimately emerges at the very high conversions.

Although we conducted no special stability tests on the catalytically cracked naphtha from runs at 866 K, we noticed no sediment or precipitate in the vials after storage in an icebox for several days in the presence of a trace of free radical inhibitor (our standard storage procedure prior to testing for octane number). Also, the gasolines from runs at 866 K contained relatively little olefin and were highly aromatic; this is the consequence of intermolecular H-transfer reactions (notably of olefins) which presumably would have preferentially removed dienes and other reactive unsaturated hydrocarbons.

[1] B. J. Mair and M. Shamaiengar, *Analyt. Chem.*, 1958, **30**, 276.

Catalytic Steam Gasification of Carbons: Effect of Ni and K on Specific Rates

KLAUS OTTO and MORDECAI SHELEF*

Research Staff, Ford Motor Company, Dearborn, Michigan 48121 U.S.A.

ABSTRACT   The object of the present work was to establish on the basis of a model system whether small additions of catalyst can enhance markedly the gasification rate of coal chars, and to differentiate their effect from that of the vast array of impurities already present in the coal chars. Gasification rates of characterized coal chars, obtained from coal of different ranks, were measured and compared with the gasification rates of graphite. The specific rate of the non-catalytic gasification, considered on a weight basis, is considerably faster in the coal char than in pure graphite, but rate differences examined on a unit area basis are small. The effect of the added catalysts is only evident in the pre-exponential factor of the rate equation; the activation energy remains the same in catalysed and non-catalysed graphite or chars, provided the process is chemically controlled. Inherent difficulties to be overcome for practical implementation are emphasized.

INTRODUCTION

Steam-gasification of carbonaceous materials may become an important industrial chemical process if imminent shortages of natural gas are to be alleviated by coal-to-gas conversion. Since the projected industry could be huge, even small improvements in processing or energy utilization provide an economic incentive. Catalysis of the gasification can be useful in lowering the operating temperature and increasing the methane yield, thereby enhancing the energy recovery.[1] It can be used to even out the reactivity of various coals in the gasifier feed and to permit the coupling of the endothermic gasifier with lower temperature nuclear or solar heat sources.

Carbon gasification is a relatively mature branch of chemical engineering.[1,2] The catalysis of this process is less well understood, although significant progress is being made.[3,4] The process is even more complicated than ordinary heterogeneous catalysis as the "support"-carbon is being gasified away and its relation to the catalyst continuously altered.

In practical applications the use of small amounts of a disposable catalyst would be preferable, since recovery does not appear to be economically feasible. Acid mine effluents or natural brines could be considered. Since the carbonaceous materials contain varying amounts of potential catalysts as ash, two questions immediately arise: how much do these contribute to the reactivity, and can an external catalyst enhance this inherent effect?

In this work an attempt has been made to answer these questions for steam-gasification. The approach differs from previous investigations in that emphasis has been placed on the specific reactivity referred to the surface area. Catalysts of both main classes, alkali and transition metals, were employed. A comparison was also made between the specific rates of graphite gasification and well-characterized coal chars.

EXPERIMENTAL
Materials

Six coals of different rank, employed before in non-catalytic gasification studies,[5] were the subject of this investigation. Proximate analyses are given in this reference. Fig. 1 gives the relative amounts of the major ash impurities, together with the absolute

1082

amount of each impurity in the coal sample in which it is most abundant. The absolute values were obtained by wet chemical analysis; the relative values were measured by programmed multi-element *X*-ray fluorescence.

Single crystal SP-1 graphite (Union Carbide) with 6 ppm total impurities was used as reference material. The graphite particle size was 75 % below 72 μm.

Cylindrical specimens of 9 mm diameter were pressed from the graphite, or from the coals ground to —200 mesh in an agate mortar, at 5·5 × 10⁵ kPa. The coal pellets were

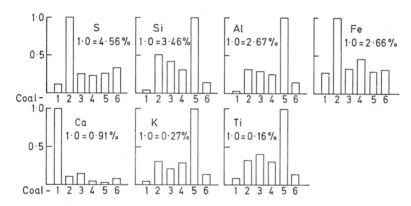

FIGURE 1 Relative major impurity levels in the investigated coals. 1, Braunkohle Lignite; 2, Pittsburgh No. 8; 3, Pittsburgh Seam; 4, Hagen; 5, Zollverein; 6, Geitling.

devolatilized to chars by heating in a flow of nitrogen at a rate of 25°C min⁻¹, up to a temperature of 1000°C. The bulk of the devolatilization occurred below 450°C. Addition of Ni to the Specpure graphite was done by blending-in before pelletizing Specpure Ni particles (Johnson-Matthey, av. size 9·3 μm) dispersed in benzene, and evaporation of the solvent. Since no catalytic effect could be observed by the use of particles in the case of char gasification, the catalysts, nickel (as nitrate) and potassium (as carbonate), were impregnated into the pellet either before pressing or after charring. Pellet impregnation was effected by stepwise saturation, using a miniature syringe, and drying until a pre-measured amount of solution had been introduced.

Apparatus and measurement procedure

Gasification rates were measured gravimetrically using a Cahn Electrobalance. The sample was suspended from the balance beam by a quartz fibre. The gas was circulated by a pump which removed the reaction products from below the pellet to the other end of the system. The gas volume was recirculated once every 5 min. The balance was attached to an Hg-free pumping system containing oil-diffusion and VacIon pumps. A Pt/Pt–Rh thermocouple placed below the sample was used for temperature measurement. The experiment began with overnight evacuation at 800°C, followed by area measurement, evacuation, temperature adjustment and admission of degassed water to a pressure of 2·33 kPa (17·5 Torr). The balance was filled with 66·7 kPa Ar carrier gas and circulation started. After a constant isothermal gasification rate has been recorded, the temperature was changed. For each sample the gasification rates were measured at five temperatures, alternatively lowering and raising the temperature.

Surface area measurements were made using as adsorbate Ar at —196°C or CO_2 at —79°C. Depending on the sample and burn-off there were considerable differences between the surface areas as measured by the two methods (see below). The cross-sectional areas of Ar and CO_2 were taken as 16·9 and 24·4 Å^2, respectively.[6]

RESULTS

Measurements of surface areas

Measurement of the internal areas of the chars, a pre-requisite for the determination of specific rates, presents considerable difficulties.[7] It was decided initially to characterize the areas by Ar adsorption at —196°C using the BET equation. Ar adsorption at this temperature in coals and non-gasified chars is slow and activated, and therefore measurements were also made using CO_2 adsorption at —79°C. The use of inert gases measures the surface area of pores with unrestricted access, while CO_2 adsorption is representative of the area associated with both micro- and macro-porous systems. Control experiments with CO_2 adsorption at 0°C gave values in agreement with those at —79°C.

A comparison of the two measures and their changes in the course of gasification was carried out with one non-catalytically and one catalytically gasified coal sample. The results are given in Table 1. After devolatilization the area measured by CO_2

Table 1

Changes in surface area measurement with gasification

Non-catalytic gasification of char (Zollverein)				Catalytic gasification[a] of char (Pittsburgh Seam Coal)			
Wt. of char/mg	% Gasified	BET area/m^{2}g^{-1} Ar	CO$_2$ Ratio CO$_2$/Ar	Wt. of char/mg	% Gasified	BET area/m^{2}g^{-1} Ar	CO$_2$ Ratio CO$_2$/Ar
281	0	32·1	197	6·1	310	0	41·2	318	7·7
275	2·1	86·5	236	2·7	293	5·5	110	321	2·9
265	5·7	151	311	2·1	270	12·9	183	348	1·9
251	10·7	308	390	1·3	251	19·0	242	366	1·5
240	14·6	396	442	1·1	171	44·8	462	476	1·03

[a] Impregnated after charring with 1·9 % Ni from nitrate solution; BET area before impregnation: Ar 38·2; CO_2 347 m^2 g^{-1}.

adsorption was 6—8 times larger than that measured by Ar. As the gasification proceeds along the path of the diffusing reactants, the pore openings are enlarged and differences between the two measures are gradually narrowed. Due to the high gasification temperatures, limitations associated with activated H_2O adsorption are unlikely and the higher CO_2 area is a better measure for expressing the specific rate. The rates calculated on the basis of the lower Ar area may be considered as upper bound values.

Non-catalytic gasification of chars and graphite

Table 2 summarizes the data on the specific gasification rates of pure SP-1 graphite and the six investigated chars. The burn-off range of the measurements was between 15 and 25 %. In all cases where measurements were made, the area measured by Ar approximated that measured by CO_2 adsorption. The activation energy was measured over a range of ± 50°C around 850°C. The values given in Table 2 represent an average of at least two measurements and the possible average error is ± 5 kcal mol^{-1}. The possibility of retardation by products was minimal as only 4 mg run^{-1} was gasified which corresponded to 10% conversion of the water vapour in the gas phase. The E_a value for graphite corresponds closely to the accepted value and indicates the absence of diffusional limitations. The char having the slowest rate shows the same value of E_a.

The slightly lower E_a values and associated higher specific rates, as expected, in the gasification of the other chars do indicate that the utilization factors of the internal area deviate from unity because of mass-transport limitations.[2] Due to such limitations the specific rate based on CO_2 area may be somewhat underestimated. The specific reactivity of five of the six chars exceeds that of the graphite by a factor of only 2—3. This is surprising in view of the large amounts of ash and the differences in structural

Table 2

Specific gasification rates of chars and graphite

Gasified material	% Gasified	BET area/m²g⁻¹ Ar	BET area/m²g⁻¹ CO_2	Gasification rate at 850°C $\mu g\ m^{-2}\ min^{-1}$ Ar area	Gasification rate at 850°C $\mu g\ m^{-2}\ min^{-1}$ CO_2 area	Activation energy kcal mol⁻¹
Braunkohle char	23·6	417	436	4·96	4·74	59·3
Pittsburgh No. 8 char	22·3	244	n.m.	0·67	n.m.	82·0
Pittsburgh seam char	18·6	291	407	0·95	0·68	66·3
Hagen char	24·5	214	n.m.	1·01	n.m.	71·1
Zollverein char	14·6	396	442	1·11	0·99	65·0
Geitling char	15·2	290	n.m.	1·04	n.m.	69·7
Graphite	13·4	4·2	3·6	0·29	0·34	80·5

n.m. = not measured.

perfection. Only in the case of the char derived from the highly volatile lignite coal does the specific reactivity exceed that of graphite by an order of magnitude. It is worth noting that this char has relatively little ash. Even in the case of lignite where the rate, referred to unit weight of carbon, exceeds the graphite gasification rate by 4 orders of magnitude, the larger part of this difference must be ascribed to the difference in the available surface area, and only a small part can be associated with catalytic activity of the ash and structural imperfections.

Table 3 shows the changes in rates and E_a with burn-off for a representative coal (the corresponding changes in S.A. measurements are given in Table 1). The specific rate (based on the CO_2 area) remains fairly constant as does the activation energy. This shows that changes in the pore structure, as evidenced by changes in low temperature adsorption of Ar, are not sensed by the mass-transport of H_2O molecules at the reaction temperature. This lends support to the choice of the CO_2-measured area as that on which the reaction takes place.

Catalysed gasification

(a) Graphite

The catalytic gasification of graphite is shown in Fig. 2. For comparison the non-catalytic gasification is also included (curve A). Since this run was separate from that shown in Table 2, the agreement attests to satisfactory reproducibility. Curves B and C

show the catalytic effect of particulate Ni added in small amounts, 303 ppm and 1·4%
respectively. Although iron is more interesting from practical considerations, we were
precluded from using it because under our conditions the change of the added iron from
active metallic state to non-active oxide[4,8] introduces complications. Addition of

Table 3

Changes in non-catalysed gasification rate
with burn-off for Zollverein char (No. 5)

% Gasified	Gasification rate at 850°C $\mu g\ m^{-2}\ min^{-1}$ Ar area	CO$_2$ area	Activation energy kcal mol^{-1}
0	4·40	0·72	70·5
2·1	1·65	0·60	72·6
5·7	1·03	0·50	70·5
10·7	1·43	1·13	n.m.
14·6	1·11	0·99	65·0

n.m. = not measured.

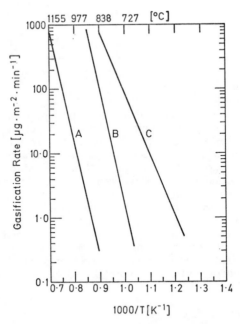

FIGURE 2 Arrhenius plots for non-catalysed and catalysed steam gasification of
graphite.
A: 0·0% Ni, $E_a = 80·6 \pm 6·7$ kcal mol^{-1}; B: 303 ppm Ni, $E_a = 80·4 \pm 4·5$ kcal mol^{-1};
C: 1·4% Ni, $E_a = 43·3 \pm 4·5$ kcal mol^{-1}.

61 μg Ni m^{-2} as relatively coarse particles (curve B, Fig. 2) enhances the rate by 2·5 orders. The reproducible value of E_a allows two deductions: (i) the reaction is still chemically controlled and (ii) the rate-determining step in the catalysed and non-catalysed reaction is apparently similar.

Further increase in the catalyst loading leads to another substantial rise in the rate and to the halving of E_a. This is taken as direct evidence that now the rate is limited by mass-transport within the porous system.[2] The strong catalysis by Ni particles of the gasification of pure SP-1 graphite by H_2O is similar to that observed for the same carbon in the C–CO_2 reaction[4] and also to that noted for the same reaction using a different graphite.[8]

Table 4

Catalytic gasification of Pittsburgh seam char

Wt. of char/mg	BET area/m² g⁻¹ (Ar)	Gasification rate at 850°C/μg m⁻² min⁻¹ based on		Activation energy kcal mol⁻¹
		Ar area	400 m² g⁻¹	
(a)				
248	226	4·53	2·55	42·4
244	226	3·58	2·02	56·9
236	382	3·43	2·75	59·2
234	398	1·47	1·46	61·2
231	416	1·52	1·58	63·2
230	416	1·43	1·48	67·2
Catalyst: Ni 2·4%, added before charring.				
(b)				
230	109	6·41	1·74	58·6
227	187	3·81	1·77	63·9
217	266	1·50	1·00	61·3
208	290	1·76	1·28	70·0
203	319	1·44	1·15	70·2
201	341	1·11	0·95	73·7
Catalyst: K 2·9%, added after charring.				

(b) *Chars*

An arbitrary choice was made to use the Pittsburgh seam coal char. Addition of particulate Ni has shown no enhancement of the rate, probably due to minimal contact with the char surface. Table 4, parts (a) and (b), indicate the catalytic effect of Ni and K, respectively. The amounts added were scaled up with respect to the graphite in proportion to the internal area of the char.

Based on 400 m² BET area the Ni-catalysed rate increases by a factor of 3—4 as compared with the char gasification without added Ni; the potassium only doubles the rate. The enhancement of the rate introduces a diffusional limitation which manifests itself in the decrease of the activation energy. As the gasification progresses, there is a drop in the specific rate, the values tending towards the non-catalysed rate. The sharper decrease, when based on Ar area, reflects the pore opening during burn-off and a wider access for Ar at low temperature.

The rate enhancement is small compared with the catalytic effect on graphite gasification, and the only plausible explanation is that impregnation of porous chars from aqueous solutions under ambient conditions does not ensure contact with the whole available surface area. Most of this contact is probably confined to the restrictions of the pore openings and is associated with a small fraction of the area. During gasification the carbon of the restrictions is gasified away and the contact decreases, depressing the catalytic effect even further. Re-impregnation can restore this contact and with it the catalytic effect. Thus, after the rate decayed, as shown in Table 4, with successive gasifications, a second re-impregnation has restored the rate at 850°C to 3·82 μg m^{-2} min^{-1}; when this rate has decayed a third re-impregnation has restored it again to 3·89 μg m^{-2} min^{-1}.

DISCUSSION

The E_a of 80·5 kcal mol^{-1} for the steam–carbon reaction in the chemically controlled region is quoted as the accepted value.[2] Since conditions were such that retardation by product was minimal, E_a is associated with the ratio k_1/k_3 in the rate equation which is: $r = k_1 P_{H_2O}/(1 + k_2 P_{H_2} + k_3 P_{H_2O})$. It also falls within the range observed by Montet and Myers for single crystals of graphite.[9] On the other hand, comparison of the absolute rates is difficult, in particular because the gasification rate is so sensitive to catalysis and to the uncertainties in area measurement. For instance, McKee[8] did not notice any reaction in uncatalysed steam-gasification at 23 Torr H$_2$O up to 1000°C while the rates reported by Giberson and Walker[10] and Overholser and Blakely[11] at lower H$_2$O pressures are higher than those measured here, when referred to the areas quoted.

Whether the E_a changes upon catalysis, was a question still unresolved a decade ago[2] and appears still to be so. Only recently Feates *et al.*[12] elaborated a new model for the "compensation effect" in catalysed carbon gasifications, whose underlying assumption is the effect of catalysis on the activation energy and therefore also of a basic difference between the catalysed and non-catalysed sequence of primary reaction steps. In contrast, the evidence obtained here is that E_a variations are all towards smaller values and can be explained by mass-transport. If diffusional interference is absent, the Arrhenius plots for the catalysed and non-catalysed steam gasification are parallel. Numerous observations of such a nature have been observed experimentally.[4]

It should be noted that the constancy of E_a on passage from catalysed to non-catalysed reaction supports the "oxygen-transfer" mechanism of catalysis, in which the role of the catalyst is to provide a reduced site for the abstraction of an oxygen atom from the oxidizing molecule, for subsequent transfer to a carbon site in the immediate proximity. All the other elementary gasification steps then proceed as in the absence of a catalyst. Hence, the catalytic effect will manifest itself only in an increased number of loci where such transfer events may occur *i.e.*, an increased pre-exponential factor. Changes in E_a, on the other hand, strongly imply the effect of the catalyst on the C–C bonds to be broken, and therefore support an "electronic" mechanism. At present the "oxygen-transfer" mechanism is able to explain the experimental facts more consistently.[4]

The specific reactivity in steam gasification of char surfaces was found to exceed that of pure and ordered graphite to a lesser extent than anticipated. The catalytic effect of natural impurities in the chars, although present, is limited by the segregation of the impurities into large clusters. In reactivity studies of chars with oxygen[13] and CO$_2$[14] no correlation was found between reactivity and alkali or transition metal impurities. A correlation, though, was established with the Ca content. In our case the more active lignite char had a high Ca content but a causal relationship has yet to be established.

The narrow spread in specific oxidation rates between graphite and chars is still more surprising considering the vast difference in structural perfection. It is commonly accepted that only the edge planes of graphite (usually a few percent of the area) are reactive in the non-catalysed gasification. Obviously, in chars heated to 1000°C the extent of this "active area" is not vastly different. It is well known that the reactivity of chars decreases rapidly with carbonization temperature[13] and it is quite possible that the annealing of the "active area" of chars occurs at temperatures considerably below graphitization temperature.

An important conclusion is that the catalytic enhancement of graphite gasification is much easier than that of chars. This is due to the relatively easy access to the surface of graphite and the maintenance of close contact between catalyst and the carbon substrate by catalyst particle motion on the basal planes[8] or by penetration from the edge planes.[15] If any practical accomplishment of the catalytic gasification of coals (chars) is to be achieved, future efforts have to be directed towards methods for providing and maintaining such contact under reaction conditions.

ACKNOWLEDGEMENT. We thank Dr. K. H. Van Heek and Dr. H. Jüntgen of the Bergbau-Forschung Institut in Essen, for the coal samples.

REFERENCES

[1] C. G. von Fredersdorff and M. A. Elliott in *Chemistry of Coal Utilization*, ed. H. H. Lowry, (Wiley, N.Y., 1963) Supplementary Vol., chap. 20, p. 892.
[2] P. L. Walker, Jr., F. Rusinko, Jr., and L. G. Austin, *Adv. Catalysis*, 1959, **11**, 133.
[3] C. Kröger, *Z. angew. Chem.*, 1939, **52**, 129.
[4] P. L. Walker, Jr., M. Shelef, and R. A. Anderson in *Physics and Chemistry of Carbon*, ed. P. L. Walker, Jr. (Marcel Dekker, N.Y., 1968) Vol. 4, p. 287.
[5] K. H. Van Heek, H. Jüntgen, and W. Peters, *J. Inst. Fuel*, 1973, 249.
[6] H. L. Pickering and H. C. Eckstrom, *J. Amer. Chem. Soc.*, 1952, **74**, 4775.
[7] H. Marsh, *Fuel*, 1965, **44**, 253.
[8] D. W. McKee, *Carbon*, 1974, **12**, 433.
[9] G. L. Montet and G. E. Myers, *Carbon*, 1968, **6**, 627.
[10] R. C. Giberson and J, P. Walker, *Carbon*, 1966, **3**, 521.
[11] L. G. Overholser and J. P. Blackely, *Carbon*, 1965, **2**, 385.
[12] F. S. Feates, P. S. Harris, and B. G. Reuben, *J. C. S. Faraday I*, 1974, **11**, 2011.
[13] R. G. Jenkins, S. P. Nandi, and P. L. Walker, Jr., *Fuel*, 1973, **52**, 288.
[14] E. Hippo and P. L. Walker, Jr., In press.
[15] R. T. K. Baker and P. S. Harris, *Carbon*, 1973, **11**, 25.

DISCUSSION

H. Bosch (*Twente Univ. of Technology, Enschede*) said: In your calculation of specific reactivities, you refer to total surface area. Your Table 1 shows that the total surface area increases with increasing burn-off, even when the surface areas are expressed per gram of starting material. Walker et al.[1] have shown that such a newly formed area of surface does not contribute to the gasification rate. There exists an active surface area, estimated by chemisorption of oxygen, and this appears to be less than 4% of the total surface area. In our opinion their ratio is not constant. Our work[2] on the gasification of carbon from saccharose shows that the activity of the newly formed surface is lower by at least one order of magnitude. How do you explain that, in the case of your materials, there is not such a marked difference between the surface originally present and the new surface formed?

M. Shelef (*Ford, Dearborn*) replied: You are right in saying that the active surface area is a fraction of the total. So it is in all surface reactions. Its extent is notoriously

[1] P. L. Walker, Jr. and Y. Yanov, *J. Colloid Interface Sci.*, 1968, **28**, 449.
[2] H. Bosch, unpublished work.

difficult to measure and it also changes from sample to sample and with the operating conditions. The newly formed area may not be active in catalytic gasification, simply because the catalyst cannot make contact with it, in the case of externally added catalysts. In the non-catalytic case, the new area should not *a priori* be less reactive unless associated with a less accessible pore system or blocked by non-catalytic ash from the burned-off material.

H. Bosch (*Twente Univ. of Technology, Enschede*) said: You conclude, from evidence based on argon isotherms, that pore-opening occurs during burn-off. Marsh and Rand[3] have shown that a sharp rise of the low-temperature isotherm with increasing burn-off does not permit the conclusion that closed pores are opened. Additional evidence is necessary, and they have proposed an experimental method for the investigation of this matter. They report that several materials (*e.g.* active carbon from coconut shell, Australian Coke) do not show an appreciable contribution to the total pore volume by opening of closed pores, although their low-temperature isotherms show a sharp rise with increasing burn-off. This is confirmed by our work which shows that, at least in the case of carbon from saccharose, the starting material is composed of small regions of different density.[4] During gasification the less dense material is removed from the inner part of the grains, leaving a micropore structure which is accessible to adsorbate molecules at low temperatures. So in this case the process is not an opening of pores already present, but the creation of new pores which were not present before gasification. Since there are at least two mechanisms, and no distinct choice can be made from adsorption measurements, I wish to ask whether you have additional information to prove that, in your work, the opening of closed pores really occurs, or do you think that our mechanism of new pore creation would interpret your results?

M. Shelef (*Ford, Dearborn*) replied: From our results no choice can be made as to whether the surface created during burn-off is in new pores, or in old pores opened up by gasification. We chose the latter because it was a simpler model.

J. B. Butt (*Northwestern Univ., Evanston*) asked: Were the samples devolatilized *in situ*, or were they exposed to the atmosphere following devolatilization before the rate measurements were made? Our results, using Western lignite at higher water partial pressures, indicate that much higher char gasification rates are obtained in the former case, although the general kinetics appear similar to those reported here. What was the product distribution and how did it vary with reaction time?

M. Shelef (*Ford, Dearborn*) replied: The samples were not devolatilized *in situ*. They were devolatilized in a tube furance, cooled, and reinserted into the Cahn balance. This was done to avoid contamination of the balance by devolatilization products. The products were only carbon monoxide and hydrogen; there was no methane formation.

N. D. Parkyns (*British Gas, London*) said: Which of your methods (if either) gives the more realistic values of the geometric surface area accessible to the reacting molecules?

In existing commercial processes for steam gasification of coal, the endothermicity of the reaction has to be compensated by the simultaneous admission of oxygen. Besides maintaining the reaction temperature at acceptably high levels, it is possible that oxygen has a further activating effect. It could work either by maintaining a high surface area in the coal by a rapid burn-off of small pores, or by increasing the concentration of oxygen-containing species at the coal surface. Have you any evidence, either from the present work or from other studies, as to whether either of these hypotheses is tenable?

[3] H. Marsh and B. Rand, *Carbon*, 1971, **9**, 47.
[4] H. Bosch, unpublished work.

M. Shelef (*Ford, Dearborn*) replied: The adsorption of carbon dioxide at 194 K is preferable to that of argon at 77 K as a measure of the char area (see Table 1). If the area is accessible to carbon dioxide at this low temperature, it should be accessible also to water molecules at about 1070 K. There is a growing body of evidence that the area measured by carbon dioxide can be correlated with char reactivity under various conditions and in different gasification reactions.[5] Regretfully, I cannot say anything about the effects of the simultaneous presence of oxygen and steam in the reacting mixture.

J. L. C. C. Figueiredo (*Univ. of Porto*) said: In recently published work on the gasification of carbon deposits obtained by hydrocarbon decomposition over nickel catalysts[6] it was suggested that a process of diffusion of carbon through nickel particles might be involved. Have you any indication that such a process might occur in your systems?

M. Shelef (*Ford, Dearborn*) replied: I have no evidence that the diffusion of carbon through nickel may (or may not) be involved in catalytic gasification. I am aware of your work and can only remark that the suggestion that carbon dissolution in the catalyst is a prerequisite for catalytic gasification is new and interesting and, I think, amenable to experimental test.

N. N. Bakhshi (*Univ. of Saskatchewan, Saskatoon*) asked: How were nickel and potassium added to the char? Did you impregnate the char or was it added in some other form? From the standpoint of gasification, which is the better method of addition?

M. Shelef (*Ford, Dearborn*) replied: As stated in the Experimental section, particulate nickel was added to the graphite while the chars were impregnated either before volatilization (*i.e.* as coal) or after. Due to the porous nature of the coal char and its very tortuous structure, there is very little access of particulate additives to the major part of the surface within the pores. It is difficult to determine how well the impregnation distributed the catalyst on the surface of the pores, but such impregnation is nevertheless the preferred method.

T. R. Phillips (*British Gas, Solihull*) said: The conspicuous exception among your results is the low activation energy and the high gasification rate attributed to Braunkohle char. It is also conspicuous that this contains the most calcium. Have you conducted an experiment in which you added calcium salts to a char?

M. Shelef (*Ford, Dearborn*) replied: There may indeed be a causal relationship between the calcium content and the reactivity of Braunkohle char. We have not investigated this thoroughly, but have noticed that ash extracted from Braunkohle catalyses the gasification of graphite, whereas the ash from Pittsburgh seam coal has no catalytic effect.

W. C. Conner (*Allied Chemical, Morristown*) asked: Since there is a movement in the position of the $C-CO-CO_2$ equilibrium over your temperature range do you have any evidence of selectivity or reactivity changes relative to the equilibrium as the temperature is varied? And do you have any evidence for a change in mechanism over this temperature range?

M. Shelef (*Ford, Dearborn*) replied: This question is related to results in Figure 2. There is no reason to assume any effect of the equilibrium between the carbon oxides on the mechanism by which carbon atoms are detached from the solid. There may be an effect, though, on the product distribution, especially in the lower temperature range.[7] Below 1073 K the ratio of carbon monoxide to hydrogen will deviate from

[5] J. L. Johnson, *Preprints of The Division of Fuel Chemistry* (The American Chemical Society), Vol. 20, No. 4, pp. 85–101.
[6] J. L. C. C. Figueiredo and D. L. Trimm, *J. Catalysis*, 1975, **40**, 154.
[7] P. L. Walker, Jr., F. Rusinko, Jr., and L. G. Austen, *Adv. Catalysis*, 1959, **11**, 133 (see especially Figure 3).

unity, some of the carbon monoxide being converted into carbon dioxide, perhaps on the graphite surface itself, by the steam in the gas phase. We did not measure the extent of this shift.

E. Wicke (*Univ. of Münster*) said: How does the effectiveness of your gasification catalysts change with increasing burn-off? Actually, the carbon around the catalyst particles is burnt off preferentially, and thereby the catalyst becomes separated from the carbon by a layer of ash.

M. Shelef (*Ford, Dearborn*) replied: Indeed, the loss of contact between the catalyst and the carbon is the main reason for the loss of the catalytic effect with burn-off and for practical applications one must strive to minimize this separation.

Oxygen Intermediates at Flash-illuminated Metal Oxide Surfaces studied by Dynamic Mass Spectrometry

Joseph Cunningham,* Brendan Doyle, Denis J. Morrissey, and Nicolas Samman

Chemistry Department, University College, Cork, Ireland

ABSTRACT   Information is presented on short-lived oxygen intermediates at surfaces of metal oxide catalysts. Application of a dynamic mass spectrometer technique yielded time-profiles of mass-resolved changes in pressure or composition of gases over the catalyst surfaces following illumination with a u.v.-pulse of 50 μs duration. Flash-illumination of Cr_2O_3, Fe_3O_4, or ZnO in the presence of 2×10^{-4} N m^{-2} of $^{18}O_2$ resulted in appearance of isotopically scrambled oxygen with post-flash rise times of 0·1 s. Evidence is presented that scrambling stems from oxygen-16 intermediates produced by flash photolysis of the metal oxide. Relative efficiencies for four flash-initiated processes involving oxygen are presented for $^{18}O_2$ in contact with oxides of first-row transition metals. Results of experiments carried out with low pressures of N_2O and/or aliphatic alcohols present at flash-illuminated zinc oxide surfaces are shown to be consistent with formation and reaction of $^{16}O^-$ at the gas/solid interface.

INTRODUCTION

Surfaces of zinc oxide or titanium dioxide have been reported to exhibit additional catalytic activity, relative to the non-illuminated system, when simultaneously exposed to u.v.-illumination, molecular oxygen, and an oxidizable reactant.[1-4] Prominent examples of such photocatalytic activity include oxidation of hydrogen or carbon monoxide,[5,6] partial oxidation of C_3—C_{10} alkanes to corresponding aldehydes or ketones[3,7] over TiO_2 and the photosynthesis of H_2O_2 in u.v.-illuminated oxygenated aqueous suspensions of ZnO. A mechanism recently suggested for this latter process by Dixon and Healy[8] typifies one general hypothesis concerning these photocatalysed oxidations, since it involves both the formation of an active oxygen intermediate on the metal oxide surface (O_2^- in that case) and product formation by reaction between oxidizable reactant and the oxygen intermediate on the surface. The present study attempts, by application of fast detection techniques, to develop information on the reactivity, lifetime, and identity of any such oxygen intermediates produced at metal oxide surfaces by intense pulses of u.v. photons.

EXPERIMENTAL

Metal oxides were used as finely divided powders of highest purity available commercially, *viz.* oxides of first-row transition metals as spectroscopically pure standards from Spex Industries, and research-grade TiO_2 (Code TiO_2-MR 128) or ZnO (Code ZnO-SP 500) obtained through the courtesy of New Jersey Zinc Co. Surface areas were determined from N_2 adsorption at 77 K using a Sartorius vacuum microbalance. Thin layers of metal oxide on a quartz substrate were prepared by coating it with a thick paste made from 0·2—1·0 g of metal oxide in triply distilled water and subsequent vacuum-evacuation, first at 380 K and finally at 623 K for 16 h at 10^{-6} N m^{-2}.

Nitrous oxide enriched to $>95\%$ in nitrogen-15 at the central atom was used as obtained from Stohler Isotopes. Oxygen enriched to $>98\%$ in oxygen-18 was obtained from Miles Laboratories. Reagent grade alcohols were dried, distilled from molecular sieve, and purified by trap-to-trap distillations prior to use.

Procedures. The great majority of experiments here reported were carried out in *dynamic* conditions with gas at reduced pressure of 2×10^{-4} N m^{-2} flowing continuously over a metal oxide catalyst to an ion-pump of 80 l s^{-1} pumping speed. Full details have been published elsewhere[9,10] of the high-vacuum system, of the equipment utilized to deliver 50 μs light pulses to these dynamic gas/metal oxide interfaces and of the dynamic mass spectrometer (DMS), used to monitor time-profiles of flash-initiated changes in gas phase pressure or composition.

Use of an asterisk thus, gas/metal oxide*, denotes a u.v.-illuminated interface and more conventional studies of photoeffects were made in *static* conditions at such interfaces by contacting reactant gas(es) at pressures 10—1000 N m^{-2} with vacuum-activated metal oxide surfaces at room temperature and then continuously illuminating them by photons of $\lambda = 254$ nm. Pressure measurements with an Edwards type GC 52 Pirani gauge and analysis of samples by a CEC 021–620A mass-spectrometer were employed to identify products and their increase with time.

Relative efficiencies of photoassisted processes

A potassium ferrioxalate actinometer located in the elliptical cavity of the 50 μs flash-tube was used to obtain values of "photons incident per flash". Relative efficiencies, ϕ, of flash-initiated changes in pressure of $^{18}O_2$, $^{16}O_2$, or $^{16}O–^{18}O$ were derived by dividing measured increases (or decreases) of gas phase oxygen by the appropriate value of photons per flash.

RESULTS AND DISCUSSION

No evidence for isotopic exchange between flowing $^{18}O_2$ and oxygen-16 of metal oxide surface was obtained at room temperature in the dark at 2×10^{-4} N m^{-2} since the isotopic composition of the $^{18}O_2$ gas remained unaffected. Exposing dark-equilibrated $^{18}O_2$/metal oxide interfaces to the output of the flash-tube, produced readily measurable changes in signal level at $m/e = 36, 34$, or 32. Time-profiles of representative changes at $m/e = 36$ and 34 are illustrated in Figure 1 for the flash-illuminated $^{18}O_2/Cr_2O_3*$ interface. The two upper traces of Figure 1 were measured with light incident

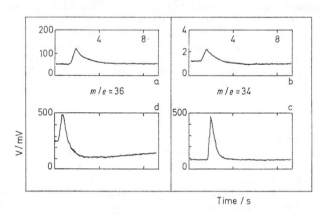

FIGURE 1   Oscilloscope traces showing time profiles of flash-initiated changes in signal level (V/mV) monitored at $m/e = 36$ or 34 during and after flash-illumination of the $^{18}O_2/Cr_2O_3$ interface. *Top*: Changes initiated by flashes incident through a 38 A filter monitored at $m/e = 36$ (trace a) and $m/e = 34$ (trace b) at indicated sensitivities. *Bottom*: Change initiated by flashes incident *via* quartz monitored at $m/e = 36$ (trace d) and $m/e = 34$ (trace c).

via a 38A filter and show very similar time profiles for appearance of additional $^{18}O_2$ or $^{16}O-^{18}O$ in the gas phase. The relative magnitudes of the maximum post-flash increases (Δ^+_{max}) are 50 mV and 2 mV, respectively, for $m/e = 36$ and $m/e = 34$, which is the same as the relative abundance of $^{18}O_2$ and $^{16}O-^{18}O$ in the gas phase. Furthermore, both rise to 50% of Δ^+_{max} in half times, $t_{\frac{1}{2}}^+$, of 0·1 s. Photoeffects for $^{18}O_2/Cr_2O_3*$ *via* a 38A filter are thus consistent with flash-assisted desorption of molecular oxygen having the same isotopic composition as the gas phase and show no isotopic scrambling. Similar observations were made and the same conclusion drawn for all the metal oxides flash-illuminated *via* a 38A filter in this study.

The lower traces of Figure 1 were measured with the full output of the flash tube incident on to $^{18}O_2/Cr_2O_3$ *via* a quartz envelope. Time profiles measured at $m/e = 36$ in these conditions reveal, not only a rapid flash-initiated desorption of $^{18}O_2$ similar to that in the upper trace of Figure 1, but also a slower process which caused $P(^{18}O_2)$ to decrease at times 0·2 to 1·5 s after a flash and to remain below the pre-flash steady-state value for post-flash times up to 28 s. This latter process is not of central concern in this paper but can reasonably be interpreted[10] as an uptake of molecular oxygen occurring *via* collisonal encounter of $^{18}O_2(g)$ with relatively long-lived ($t_{\frac{1}{2}} \geqslant 15$ s) reactive centres produced by the flash at the O_2/Cr_2O_3* interface. As such it serves to illustrate the slow time-profiles expected for processes requiring post-flash collision of gas phase molecules with the flash-activated surface.

If the signal level at $m/e = 34$ were influenced only by the two flash-initiated processes which affected $m/e = 36$, then the time-profile at $m/e = 34$ should correlate closely with that measured at $m/e = 36$ in similar conditions, but with features reduced to 4%. Comparison of traces c and d of Figure 1 reveals instead that the time profiles are very different and that $m/e = 34$ consists mainly of a rapid rise with $t_{\frac{1}{2}}^+ \sim 0·1$ to $(\Delta^+_{max}) \sim$ 320 mV. This is disproportionately large relative either to Δ^+_{max} for $m/e = 36$ in another flash or to the decrease (Δ^-_{max}) (*cf.* trace d of Figure 1). The discrepancy in size, together with the "fast" profile of this transient, are taken as evidence for a flash-initiated oxygen-scrambling process involving preadsorbed $^{18}O_x$ and an oxygen-16 species on the surface of flash-activated Cr_2O_3. No evidence for release of ^{16}O to the gas phase was obtained, but measurements were made on flash-initiated changes at $m/e = 32$ as an indication of extent of formation and combination of oxygen-16 species on the surface of flash-activated Cr_2O_3. Such measurements demonstrated much-enhanced "fast" release of $^{16}O_2$ from $^{18}O_2/Cr_2O_3$ under flash-illumination through quartz ($\Delta_{max} = 230$ mV at $m/e = 32$) than through the 38A filter ($\Delta_{max} \sim 0·3$ mV). When taken together with our observation that significant oxygen-scrambling occurred for flash-illumination incident through quartz but not through the 38A filter (*cf.* traces b and c of Figure 1), these data point strongly to probable involvement of oxygen-16 fragments produced by flash photolysis of the Cr_2O_3 surface in the flash-initiated oxygen-scrambling.

The main characteristics of the four flash-initiated processes distinguished at the $^{18}O_2/Cr_2O_3*$ interface by the DMS technique may be summarized as follows:

Process I: Increase with $t_{\frac{1}{2}}^+ \sim 0·1$ s of signal $m/e = 36$, attributable to flash-initiated desorption of preadsorbed $^{18}O_2$.

Process II: Slow decrease with $t_{\frac{1}{2}}^- \sim 1$ s of signal at $m/e = 36$ attributable to uptake of $^{18}O_2$ from the gas phase by interaction with active long-lived centres produced on the metal oxide by a flash.

Process III: Increase with $t_{\frac{1}{2}}^+ \sim 0·1$ s of signal at $m/e = 32$, attributable to formation of $^{16}O_2$ by photolysis of the metal oxide.

Process IV: Increase with $t_{\frac{1}{2}}^+ \sim 0·1$ s of signal at $m/e = 34$ attributable to desorption of $^{18}O-^{16}O$ produced by interaction of preadsorbed $^{18}O_x$ with oxygen-16 species from lattice photolysis.

In order to test whether these processes were of general occurrence, thin layers of the stable oxides of other first-row transition metals were prepared, vacuum-activated and flash-illuminated in identical manner to $^{18}O_2/Cr_2O_3*$. Results of this survey are presented in Table 1 as relative efficiencies (*cf.* Experimental Section). Absence of any entry for a process indicates that flash-initiated changes in signal-level were too small for certain detection with the prevailing noise level. This effectively limited observable processes to those having relative efficiencies $\geqslant 3 \times 10^{-7}$. Inspection of Table 1 shows that, with $2 \times 10^{-4}\,N\,m^{-2}$ pressure of $^{18}O_2$ present over the various metal oxides, one or more of processes I—IV was readily measurable for all oxides except NiO. Entries in Table 1 have been corrected for any small changes measured in blank experiments carried out with light incident on to the $^{18}O_2$/quartz substrate interface. Entries in parentheses show, for comparison, the relative efficiency, if any, of the same flash-initiated process when light was incident on to the metal oxide *in vacuo* prior to its exposure to $^{18}O_2$. The following trends emerge:

Process I: Presence of $^{18}O_2$ caused a significant increase in photodesorption of $^{18}O_2$ for all except the *p*-type semiconducting metal oxides NiO, CuO, and CoO. This trend appears fully consistent with arguments presented elsewhere by the authors[9,10] that availability of conduction-band electrons (*n*-type semiconductivity), influences extent of O_2 photodesorption, because it controls extent of $O_2^-(s)$ formation at the O_2/metal oxide interfaces prior to the flash.

Process II: This slow long-persisting uptake of $^{18}O_2$ likewise did not occur for the *p*-type metal oxides but was readily measurable for other metal oxides, except V_2O_5, with efficiency increasing in the sequence $ZnO < TiO_2 < Cr_2O_3 \approx Fe_3O_4$. This trend would be consistent with post-flash reaction of gas-phase O_2 with long-persisting surface centres bearing excess negative charge. Such centres could exist at surfaces of *n*-type semiconducting oxides, after a flash, either as lower valency states of the cations or as surface-trapped electrons. Present results do not suffice to distinguish between these possibilities.

Process III: Presence of $^{18}O_2$ at $2 \times 10^{-4}\,N\,m^{-2}$ either decreased the extent of lattice breakdown to $^{16}O_2$ (as occurred for CoO, ZnO, and to lesser extent for V_2O_5 and Fe_3O_4), or increased it (as occurred strongly for Cr_2O_3 and to lesser extent for CuO). Comparison of the efficiencies of Process I with Process III for these various metal oxides reveals no apparent correlation between these processes, such as has been noted elsewhere[9b] for various zinc oxides.

Process IV: Oxygen scrambling with apparent efficiency $>10^{-6}$ was detected only for $^{18}O_2/Cr_2O_3*$, $^{18}O_2/Fe_3O_4*$, $^{18}O_2/ZnO*$, and $^{18}O_2/CoO*$ interfaces and these correspond to interfaces which simultaneously underwent photolysis to $^{16}O_2$ with larger or comparable efficiencies. Other $^{18}O_2$/metal oxide* interfaces, which were shown not to photolyse to $^{16}O_2$ with appreciable efficiency also did not produce significant flash-initiated oxygen scrambling. Such correlation between extent of flash-initiated oxygen scrambling and lattice photolysis could be understood if photolysis produced activated surface ^{16}O intermediates capable of interacting with preadsorbed ^{18}O to yield $^{16}O_2-^{18}O_2$ product. Our observations that extent of Processes I or IV appeared independent of ^{18}O pressures in the range 10^{-4} to $10^{-3}\,N\,m^{-2}$ pointed to involvement of chemisorbed, rather than physisorbed, ^{18}O species, i.e. $^{18}O_x{}^{n-}$ with $x = 1$ or 2 and $n = 0$ or 1. These observations did not definitively distinguish between O^-, O, O_2 or O_2^- as the activated ^{16}O intermediates. Improved definition was therefore sought from experiments chosen for their ability to form $^{16}O^-$ selectively on the illuminated interface.

Use of N_2O *to enhance* O^- *formation on illuminated zinc oxide.* A technique employed by previous workers[11] to enhance selectively formation of O^- on metal oxide surfaces, and so facilitate study of its reactivity and e.s.r. spectrum[12,13] involves transfer of an

Table 1

Relative Efficiencies of Oxygen Photoeffects Flash-Initiated *via* Quartz

System	Efficiency of Process[a]			
	I $m/e = 36$	II $m/e = 36$	III $m/e = 32$	IV $m/e = 34$
$^{18}O_2/TiO_2$*	4.4×10^{-6}	-5.6×10^{-5}	4.0×10^{-7}	—
vac/TiO$_2$*	(2.0×10^{-7})	—		—
$^{18}O_2/V_2O_5$*	3.5×10^{-6}	—	1.2×10^{-5}	—
vac/V$_2$O$_5$*	(5.0×10^{-7})		(2.0×10^{-5})	—
$^{18}O_2/Cr_2O_3$*	2.5×10^{-5}	-1.0×10^{-4}	2.3×10^{-5}	3.0×10^{-5}
vac/Cr$_2$O$_3$*	—	—	(4.0×10^{-7})	—
$^{18}O_2/Fe_3O_4$*	6.3×10^{-6}	-1.3×10^{-4}	6.6×10^{-5}	4.8×10^{-6}
vac/Fe$_3$O$_4$*	—	—	(9.3×10^{-5})	—
$^{18}O_2/CoO$*	—	-3.4×10^{-6}	7.4×10^{-5}	1.2×10^{-6}
vac/CoO*	—	—	(2.7×10^{-4})	—
$^{18}O_2/NiO$*	—	-3.4×10^{-6}	4.2×10^{-7}	—
vac/NiO*	—	—	(3.0×10^{-7})	—
$^{18}O_2/CuO$*	—	—	2.9×10^{-5}	3.5×10^{-7}
vac/CuO*	—	—	(1.0×10^{-6})	—
$^{18}O_2/ZnO$*	1.5×10^{-6}	-1.7×10^{-5}	7.0×10^{-6}	1.0×10^{-6}
vac/ZnO*	—	—	(4.1×10^{-5})	—

[a] See text for detailed designation of each process.

electron to N_2O from the metal oxide and dissociation according to the overall scheme, (1). Results in this section relate to our attempts to use this process to generate high

$$N_2O_{(g)} \rightleftharpoons N_2O_{(s)} \xrightarrow{e^-/MO} N_2O^-_{(s)} \rightarrow N_{2(g)} + O^-_{(s)} \qquad (1)$$

transient concentrations of O^- at the interface between gaseous N_2O and a flash-illuminated metal oxide and thence to study reactivity of O^-. The feasibility of thus employing gaseous N_2O as a source of additional O^- at a flash-illuminated interface was first checked by observing the effect of $N_2{}^{16}O$ upon the oxygen scrambling process noted above for $^{18}O_2/ZnO$* interfaces. Introduction of $N_2{}^{16}O$ at pressure 3×10^{-4} N m^{-2} caused a four-fold increase in the amount of ^{16}O–^{18}O released from the $(N_2{}^{16}O + {}^{18}O_2)/ZnO$* interface relative to that measured from the $^{18}O_2/ZnO$* interface but only for a fully oxidized surface. Such an increase was fully consistent with flash-initiated production of additional $^{16}O^-(s)$ species from N_2O *via* (1) and their contribution to rapid oxygen scrambling *via* scheme (2).

$$^{16}O^-(s) + {}^{18}O_x(s) \rightarrow ({}^{16}O-{}^{18}O_x)^-(s) \rightarrow {}^{16}O-{}^{18}O(g) + {}^{18}O^-_{x-1}(s) \qquad (2)$$

Aliphatic alcohols were used as additional tests of O^- involvement, in view of reports that $O^-(g)$ species underwent secondary reaction with alcohols in the presence of nitrous oxide.[14] Warman attributed the appearance of additional product nitrogen to schemes (3a) and (3b).

$$ROH + O^- \rightarrow X + Y^- \; (\equiv RO^- + OH) \; (or \; RO + OH^-) \qquad (3a)$$

and

$$X + Y^- + 2N_2O \rightarrow 2N_2 + Products \qquad (3b)$$

Warman argued that absence of additional N_2 product from tertiary alcohols originated because process (3a), involving abstraction of H or H^+ from the alcohol, could not occur in the absence of an α-hydrogen.

Results obtained in this study, on the extent of N_2 product formation at N_2O/metal oxide interface in the presence of various alcohols are summarized in Figure 2 and are in

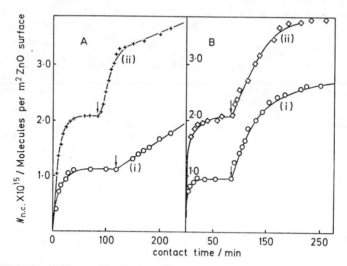

FIGURE 2 Growth of nitrogen product with contact time (minutes) from inter-actions of zinc oxide with N_2O alone or premixed with aliphatic alcohols. U.v -illumination of the dark-equilibrated interface by photons (254 nm) was commenced after contact times indicated by an arrow. 2A Trace (i) Gas phase consisting initially only of N_2O. Trace (ii) Gas phase consisting initially of $(N_2O + C_2H_5OH)$. 2B Trace (i) Gas phase consisting initially of $(N_2O + Bu^tOH)$. Trace (ii) Gas phase consisting initially of $(N_2O + Pr^iOH)$.

good agreement with Warman's explanation. Curve (i) of Figure 2A shows an initial rapid production of N_2 product at the N_2O/ZnO interface in the dark which is in accordance with scheme (1) and published work. Comparison of the N_2 product formed when ethanol (curve ii of Figure 2A) or isopropyl alcohol (curve ii of Figure 2B) were simultaneously admitted to the ZnO interface together with N_2O reveals additional formation of N_2 product—as expected from reaction of O^- with these alcohols in the dark via schemes (3a) and (3b). That $O^-(s)$ produced via scheme (2) on the ZnO interface retained selectivity similar to gas-phase O^- and did not react with t-butyl alcohol via schemes (3a) plus (3b) to produce additional nitrogen, is illustrated by the similarity of the plots for $(N_2O + t$-butyl alcohol)/ZnO [plot (i) of Figure 2B] and N_2O/ZnO [plot (i) of Figure 2A].

Additional type (1) processes at N_2O/ZnO interfaces had previously been reported under continuous u.v.-illumination and attributed to migration of photogenerated species to the negatively charged N_2O/ZnO interface.[11b] Additional N_2 product from extra type (3a) and (3b) events were therefore expected, on Warman's mechanism, from illuminated mixtures of N_2O with primary or secondary alcohol. The section of curve ii of Figure 2A which lies immediately to the right of the arrow denoting start of u.v. illuminations, demonstrates an initial rapid rise consistent with u.v. illumination initiating additional production of N_2, probably via (1) → (3a) → (3b) at the

(N_2O + C_2H_5OH)/ZnO* interfaces. This system was examined under flash-illumination on the DMS system in the hope that individual steps of this 3-step illumination-induced production of N_2 might thereby be resolved. Time profiles in Figure 3A and 3B allow detailed comparison to be made of fast changes in gas phase pressure of

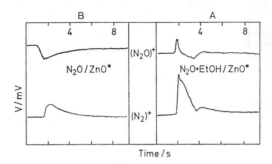

Time / s

FIGURE 3 Comparison of time-profiles for changes in signal level (V/mV) for N_2O^+ and N_2^+ flash-initiated at an N_2O/ZnO* or an (N_2O + EtOH)/ZnO$^+$ interface. 3A Upper trace: Profile at $m/e = 45$ ($\equiv$ $^{14}N^{15}N^{16}O^+$) from mixture of $^{14}N^{15}N^{16}O$ with C_2D_5OD flowing over a flash-illuminated ZnO surface; 3A Lower trace: Profile for flash-initiated production of N_2^+ from (N_2O + EtOH)/ZnO*. 3B Upper trace: Profile of flash-initiated depletion of $^{14}N^{15}N^{16}O$ from the gas phase over N_2O/ZnO*. 3B Lower trace: Profile of flash-initiated growth of N_2 in gas phase above N_2O/ZnO*.

N_2O and N_2 over the (N_2O + EtOH)/ZnO* and N_2O/ZnO* interfaces following identical flash-illumination through quartz. The profile for N_2O^+ over N_2O/ZnO* shows a decrease in gas-phase N_2O attributable to process (1) at the flash-activated surface and subsequent restoration of the pre-flash steady-state condition within 3s (upper trace Fig. 3B). On the other hand, the trace for N_2O^+ over (N_2O + EtOH)/ZnO* (upper trace Fig. 3A) indicates an initial rapid desorption of N_2O and superimposed upon that a secondary process which depleted N_2O from the gas phase and resulted in an almost linear section of the profile at times 1—2·5 s after the flash. Our preliminary interpretation of this section is that it arises from step (3b) involving reaction of N_2O from the gas phase with intermediates $(X + Y)^-$ rapidly produced *via* (3a) and involving interaction of O^- with preadsorbed ethanol. As required by this interpretation, the time-profile for production of N_2^+ over flash-illuminated (N_2O + EtOH)/ZnO* (lower trace in Figure 3A) likewise shows a linear section corresponding to a secondary process yielding additional N_2 (g) after fast initial N_2 production (*cf.* lower trace of Figure 3B and note that no linear secondary process was apparent in the time profile for N_2^+ from N_2O/ZnO*).

ACKNOWLEDGEMENTS. Financial support from the National Science Council of Ireland, from the Department of Education of the Irish Government, and from USAF through European Office of Aerospace Research under Grant AFOSR 71–2148, is gratefully acknowledged.

REFERENCES

[1] F. F. Volkenstein, *Adv. Catalysis*, 1973, **23**, 157.
[2] G. Heiland, E. Mollwo, and F. Stockman, *Solid State Phys.*, 1959, **8**, 191.
[3] N. Djeghri, M. Formenti, F. Juillet, and S. J. Teichner, *Faraday Discuss.*, 1975, **58**, 177.
[4] (a) R. I. Bickley and F. S. Stone, *J. Catalysis*, 1973, **31**, 389; (b R. I. Bickley, G. Munuera, and F. S. Stone, *J. Catalysis*, 1973, **31**, 398; (c R. I. Bickley and R. K. M. Jayanty, *Faraday Discuss.*, 1975, **58**, 186.

[5] F. Steinbach, *Electronic Phenomena in Chemisorption and Catalysis on Semiconductors* (Walter de Gruyter, Berlin, 1969), p. 196.

[6] K. Tanaka and G. Blyholder, *J. Phys. Chem.*, 1972, **76**, 1807.

[7] M. Formenti, F. Juillet, P. Meriaudeau, and S. J. Teichner, *Chemical Technology*, 1971, **1**, 680.

[8] D. R. Dixon and T. W. Healy, *Austral. J. Chem.*, 1971, **24**, 1193.

[9] (a) J. Cunningham, E. Finn, and N. Samman, *Faraday Discuss.*, 1975, **58**, 160; (b) J. Cunningham and B. Doyle, *JCS Faraday I* (submitted for publication 1975).

[10] J. Cunningham and N. Samman, *Dynamic Mass Spectrometry* (Heyden and Son, London 1970), Vol. 4, Ch. 17.

[11] J. Cunningham, J. J. Kelly, and A. L. Penny, (a) *J. Phys. Chem.*, 1970, **74**, 1992; (b) *ibid.*, 1971, **75**, 617.

[12] N. Wong, Y. Ben-Taarit, and J. H. Lunsford, *J. Phys. Chem.*, 1974, **60**, 2148.

[13] (a) N. B. Wong and J. H. Lunsford, *J. Chem. Phys.*, 1971, **55**, 3007; (b) Y. Ben-Taarit, and J. H. Lunsford, *Chem. Phys. Letters*, 1973, **19**, 348.

[14] (a) O. Warman, *J. Phys. Chem.*, 1968, **72**, 52; (b) O. Warman, *Nature*, 1967, **213**, 381.

DISCUSSION

T. A. Egerton (*Tioxide International, Stockton-on-Tees*) asked: To what extent do the efficiencies listed in Table 1 of your paper depend on the sample pretreatment conditions? In particular, do the efficiencies vary with the degree of surface hydroxylation of the oxides?

G. Munuera (*Univ. of Seville*) asked: Can you distinguish between the actual photo-effect of the flash and the likely thermal effect in your experiments?

M. Deane (*AERE, Harwell*) said: Process II is an order of magnitude larger than Process I. You do not specify whether you cleaned the surface between flashes; if not, do the amounts of $^{18}O_2$ adsorbing and desorbing converge as the surface is saturated with $^{18}O^{2-}$? Alternatively, do you see the scrambling process extending into the bulk and offering a sink for $^{18}O^{2-}$?

J. Cunningham (*University College, Cork*), in reply to the three previous speakers, said: If Mr. Deane is asking whether we *thermally* outgassed the metal oxide surface between flashes, the answer is 'no'. Rather, we left sufficient time between flashes for the gas/metal oxide interfaces to return towards a 'dark equilibrium' condition. When this precaution was observed, flash-initiated photo-effects involving oxygen were generally reproducible within $\pm 50\%$ for metal oxide surfaces pretreated by the standard method described in the paper. (Greater variability was observed for manganese dioxide and to lesser extent with chromia.) At present we do not know why Process II, where it occurs, is an order of magnitude more efficient than Process I but, in answer to Dr. Egerton, I can report that attempts to rehydroxylate titania and zinc oxide by exposure to water vapour at room temperature did not markedly affect the efficiency of either process. A partial answer to Prof. Munuera's questions is possible for flash-assisted desorption of $^{18}O_2$. The most likely thermal effect would be desorption of physically adsorbed oxygen and the extent of such a process should vary with oxygen pressure. However, the extent of $^{18}O_2$ desorption, as initiated by photons incident *via* the 38A filter, did not vary with oxygen pressure, thereby indicating that the photo-effect was real. I feel that this is part of a wider problem, namely that of defining the efficiencies of all channels for dissipation of radiant energy by metal oxide surfaces in particular conditions. Already at this conference we have seen that the radiative channel for energy dissipation (luminescence) at ultraviolet-illuminated magnesium oxide was markedly affected by the presence of nitrous oxide or oxygen. I believe that other effects of surface condition upon efficiencies of radiative and radiationless processes are to be expected.

F. S. Stone (*Univ. of Bath*) said: Following up Dr. Egerton's question, I too have a query about the possible role of hydroxyl groups, although I realize that such remarks may seem rather uncharitable in view of the care and effort which Prof. Cunningham and his colleagues have put into designing and commissioning this complex u.h.v. system. The assumption is presumably made that surface hydroxyl groups are absent and hence not playing a role here, but could this be positively tested by raising the outgassing temperature above 623 K? It would also be interesting to include magnesium oxide and calcium oxide in the series in view of the information on electron-donor properties now coming from luminescence and reflectance studies.[1,2]

J. Cunningham (*University College, Cork*) replied: In relation to these questions concerning the possible roles of surface hydroxyls, we have made the following experimental observations and preliminary deductions for our partially dehydroxylated titania surfaces. First, there was no significant enhancement of the extent of dissociation of nitrous oxide either at the dark or the ultraviolet-illuminated N_2O–TiO_2 interface when the surface was treated with deuterium oxide at 623 K for several hours before admitting nitrous oxide at room temperature. We deduce from this that any surface hydroxyls re-introduced by the treatment with deuterium oxide vapour at 623 K did not represent 'active sites' for nitrous oxide dissociation on titania. Secondly, the introduction of 5 mole % D_2O into the vapour of perdeuterioethanol flowing over the titania surface (previously outgassed for 16 h at 623 K) at room temperature and 2×10^{-4} N m^{-2} pressure produced an approximate doubling of the extent of photo-assisted formation of acetaldehyde detected down-stream from the ethanol–TiO_2 interface during three hours exposure to the output of a 15 W low-pressure mercury arc lamp. These results (to be published) would be consistent with the reintroduction of surface hydroxyls at the $(C_2D_5OD + D_2O)$–TiO_2 interface and their activation towards alcohol dehydrogenation by hole-trapping at OH$^-$ to yield OH· radicals. However, since the partial rate constant for abstraction of α-hydrogen from an alcohol by OH· is reportedly[3] only twice as great as for O$^-$, and since O$^-$ would be produced by hole-trapping at co-ordinatively unsaturated O^{2-} on the dehydroxylated TiO_2 surfaces,[4] it seems to us that, contrary to an implication of Prof. Stone's remarks, photo assisted dehydrogenation of alcohol would be expected even for a fully dehydroxylated titania surface.

P. Pichat (*CNRS, Villeurbanne*) said: This comment concerns the formation of oxygen species on the surface of metal oxides irradiated by u.v. light in the presence of oxygen. By use of electrical photoconductivity measurements, we have shown recently that O$^-$ and O_2^- species are formed on ultraviolet-irradiated anatase (TiO_2). We also found that isotopic exchange of oxygen takes place at room temperature via a mechanism involving one surface oxygen atom at a time. These results will be presented in part at the forthcoming Fifth Ibero-American Symposium on Catalysis. With regard to surface oxygen species, these results are in agreement with those of Prof. Cunningham.

J. Cunningham (*University College, Cork*) replied: I am pleased that your results corroborate ours. We shall present other results at the forthcoming Ibero-American Symposium which will emphasize the important role of the monatomic oxygen anion radical, O$^-$, in photo-assisted elimination reactions from aliphatic alcohols on rutile (TiO_2) and zinc oxide surfaces under conditions of u.v. illumination.

[1] S. Coluccia, M. Deane, and A. J. Tench, this Congress, p. 171
[2] A. Zecchina, M. G. Lofthouse, and F. S. Stone, *JCS Faraday I*, 1975, **71**, 1476: F. S. Stone and A. Zecchina, this Congress, p. 162
[3] P. Neta and R. H. Schuler, *J. Phys. Chem.*, 1975, **79**, 1.
[4] M. Che, C. Naccache, and B. Imelik, *J. Catalysis*, 1972, **24**, 328.

Infrared Studies of Oxidation of Olefins Adsorbed on Zinc Oxide

Kazuo Hata, Shoji Kawasaki, Yutaka Kubokawa,* and Hisashi Miyata

Department of Applied Chemistry, University of Osaka Prefecture, Sakai, Osaka, 591 Japan

ABSTRACT The oxidation of olefins adsorbed on zinc oxide has been investigated by infrared spectroscopy. On oxidation of the π-allyl species formed from adsorbed propene or *trans*-but-2-ene, formate and acetate ions or formate and propionate ions, respectively, were formed simultaneously. This suggests that the π-allyls on zinc oxide are not oxidized *via* allyl peroxide. In the case of oxidation by nitrous oxide only propionate ions were formed from adsorbed propene. From the spectral change on increasing temperature together with the amounts of oxygen taken up during oxidation it is concluded that before the π-allyl species were oxidized to the carboxylate species a precursor composed of olefin and oxygen had been formed in the case of oxygen as well as nitrous oxide.

Introduction

In view of the industrial importance of the reaction, the oxidation of olefins over oxide catalysts has been investigated by many workers. The mechanism of the oxidation and the nature of its reaction intermediates have been studied by a number of techniques.[1] Existing data seem insufficient to clarify these questions. The information on the mechanism of the complete oxidation seems necessary for understanding the selectivity of the oxidation. In order to clarify the mechanism of the oxidation we have investigated the interaction of oxygen with olefins adsorbed on zinc oxide by analysis of reaction products in the gas phase.[2] Kokes and co-workers[3] have recently studied the i.r. spectra of olefins adsorbed on zinc oxide. We have already carried out i.r. studies of the interaction of oxygen with propan-2-ol and acetone on MgO and NiO.[4] It seemed of interest to extend such i.r. studies to the olefin–zinc oxide system. Some of the results have already been published as a preliminary communication.[5]

Experimental

Materials. The zinc oxide used in the present work was Kadox-25 (New Jersey Zinc Co.). Propene, *trans*-but-2-ene, and nitrous oxide, (99·9% pure) were obtained from the Takachiho Shoji Co. and used without further purification. Propene-d_6 was prepared by deuteriation of propyne-d_4 (obtained by the reaction of magnesium carbide with deuterium oxide at 473 K and converted into propene-d_6 by deuteriation on Pd black). Magnesium carbide was prepared by heating magnesium powder to 873 K in a stream of pentane.[6] Finally, propene-d_6 was purified by chromatographic separation with a 5 m DMS column operated at room temperature. Its isotopic purity determined by mass spectrometry was 90%. Oxygen was obtained from a cylinder and purified by passing through a molecular sieve immersed in liquid nitrogen.

Apparatus and Procedures. The i.r. cell was the same as that used previously.[4] All spectra were measured at room temperature using a Hitachi EPI-G2 grating i.r. spectrometer with a spectral slit width 2·8 cm^{-1} at 1000 cm^{-1}. The zinc oxide sample (0·5—0·7 g) was pressed into a disc of 20 mm dimater. After the disc had been placed in the cell, the disc temperature was slowly increased from room temperature to 723 K under evacuation and kept at that temperature for 10 h. Then, oxygen at *ca.* 6·7 kN m^{-2} was

1102

introduced into the cell. Heating at 723 K for 2 h, followed by degassing at the same temperature for 2 h, was repeated several times. Finally the disc was cooled to room temperature in oxygen and briefly degassed.

The uptake of oxygen and the amounts of reaction products formed on oxidation of the adsorbed olefins were determined manometrically or by analysis of products in the gas phase using a closed circulation system. For these studies 5—10 g of the ZnO was used.

RESULTS AND DISCUSSION

Reactions of Oxygen with Adsorbed Propene. The i.r. spectra of propene adsorbed on zinc oxide which we have obtained are essentially the same as those reported by

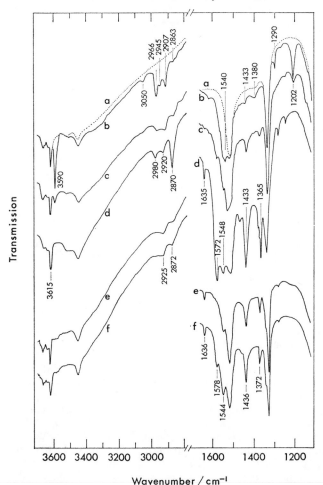

FIGURE 1   Interaction of oxygen with propene (or acraldehyde) adsorbed on ZnO. (a) Background (0·5 g); (b) after 1 h adsorption of propene followed by 5 min evacuation at 298 K; (c) followed by 1 h at 323 K in oxygen (2·0 kN m⁻²); (d) 1 h at 423 K in oxygen; (e) after 1 h contact of the ZnO containing acraldehyde with oxygen at 323 K; (f) followed by 1 h at 423 K in oxygen.

1103

Dent and Kokes[3] (Fig. 1). After oxygen was admitted to the ZnO plus adsorbed propene at room temperature, the temperature of the ZnO was raised in stages. At room temperature no spectral change was observed. At 323 K the band at 3590 cm^{-1} due to OH groups formed on dissociative adsorption of propene, as well as the bands at 3050, 2966, 2945, 2907, 2863, and 1202 cm^{-1}, which are attributable to π-allyl species, began to decrease in intensity. These bands disappeared completely at 353 K. Simultaneously, new bands appeared *ca.* 3500, 2900, and 1500—1300 cm^{-1}. The broad band at *ca.* 3500 cm^{-1} is attributable to surface OH groups since a small amount of water

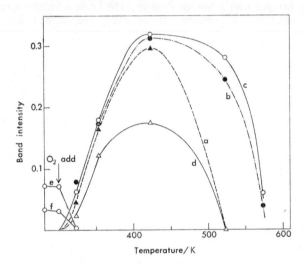

FIGURE 2 Change in the band intensities with increasing temperature of the ZnO. (a) 1572 cm^{-1}; (b) 1548; (c) 1433; (d) 1365; (e) 3590; (f) 1202.

was detected at 353 K. Figure 2 shows the change in the intensity of these bands with increasing temperature of the ZnO. On raising the temperature to 423 K, the new bands increased in intensity. At 523 K the bands at 2870, 1572, and 1365 cm^{-1} disappeared, while the intensity of the bands at 2980, 1548, and 1433 cm^{-1} hardly changed. This suggests that these bands form two groups, which can be attributed to different adsorbed species.

As regards the bands at 1572 and 1365 cm^{-1}, the ratio of the intensity of the former to the latter was always constant in the temperature range 323—423 K. In addition, the corresponding bands were observed at 1573 and 1338 cm^{-1} with propene-d$_6$. According to the work of Ueno, Onishi, and Tamaru,[7] formate ions on ZnO show bands at 2870, 1572, and 1369 cm^{-1}. Consequently, the bands at 2870, 1572, and 1365 cm^{-1} are attributable to $\nu(CH)$, $\nu_{as}(CO_2^-)$, and $\nu_s(CO_2^-)$ of formate ions, respectively. The bands at 2980, 1548, and 1433 cm^{-1} are in agreement with those of acetate ions on ZnO and also with those of zinc acetate (Table 1). In addition, the ratio of the intensity of the band at 1548 to that at 1433 cm^{-1} was always constant. This suggests that those bands are due to $\nu_{as}(CO_2^-)$ and $\nu_s(CO_2^-)$ of acetate ions, respectively. A summary of the band frequencies for these carboxylate species is shown in Table 1.

The band at 1635, 1507, and 1342 cm^{-1} may be attributed to adsorbed carbon dioxide or carbonate ions produced by the reactions of surface carboxylate ions. The positions of these bands are similar to those reported for such species.[8]

Table 1

Wavenumbers ($\bar{\nu}$/cm^{-1}) of the bands for the carboxylate species formed from propene on ZnO

This work				Ueno et al.[b]		Vibrational
Propene $CH_3CO_2^-$ HCO_2^-	Propene-d_6 $CD_3CO_2^-$ DCO_2^-	$CH_3CO_2^-$ on ZnO[a]	$Zn(CH_3CO_2)_2$	HCO_2H	DCO_2D	mode[c]
2980			2983			$\nu_{as}(CH_3)(A)$
2920			2925			$\nu_s(CH_3)(A)$
1548	1540	1550	1556			$\nu_{as}(CO_2^-)(A)$
1433	1430	1430	1420			$\nu_s(CO_2^-)(A)$
2870				2870	2190	$\nu(CH$ or $CD)(F)$
1572	1573			1572	1572	$\nu_{as}(CO_2^-)(F)$
1365	1338			1369	1342	$\nu_s(CO_2^-)(F)$

[a] This was prepared by the adsorption of C_2H_5OH on ZnO.
[b] Ref. 7.
[c] ν_{as}, asymmetric stretch; ν_s, symmetric stretch; A, acetate; F, formate ion.

In a separate experiment the reaction of oxygen with π-allyl species on ZnO was investigated by analysis of the reaction products in the gas phase. Oxygen was admitted to the ZnO at 273 K, and its temperature was raised in stages under circulation of oxygen. With increasing temperature the amount of oxygen uptake increased. At 353—423 K where no formation of CO_2 occurred, the ratio of the amount of oxygen uptake to that of adsorbed propene was about 2·0 as described in a previous paper.[2] From the i.r. studies described above, it is concluded that such an oxygen uptake is attributable to formation of surface formate and acetate ions as well as OH groups. Furthermore, the ratio of the oxygen uptake to the propene adsorbed was about 1·0 at ca. 323 K where no bands due to the carboxylate species were observed, while the bands at 3590 and 1202 cm^{-1} arising from π-allyl species disappeared. This suggests that the carboxylate species are not formed directly from π-allyl species but via a precursor composed of propene and oxygen.

Reactions of Oxygen with Adsorbed trans-But-2-ene. Similar experiments were carried out for trans-but-2-ene. The results are shown in Fig. 3. The spectra of adsorbed trans-but-2-ene are similar to those obtained by Kokes and co-workers.[3] Since the bands at 1620 and 1574 cm^{-1} are due to $\nu(C = C)$ of the π-complex and of π-allyl in the anti-form, respectively, the spectra in Fig. 3 suggest that most of the trans-but-2-ene forms π-allyl species on adsorption. Oxygen was admitted to the ZnO containing π-allyl species at room temperature. After 1 h the band at 3590 cm^{-1} arising from formation of π-allyl species disappeared while new bands appeared at 2974, 1575, 1540, 1432, and 1361 cm^{-1}. On raising the temperature to 323 K, the band at 2974 cm^{-1} shifted to 2967 cm^{-1} and a new band appeared at 2865 cm^{-1}. All the bands increased in intensity up to 423 K. At 523 K the bands at 2865, 1575, and 1362 cm^{-1} disappeared while the bands at 2967, 1540, and 1432 cm^{-1} still remained. This suggests that these bands form two groups arising from different adsorbed species. The bands at 2865, 1575, and 1362 cm^{-1} can be attributed to formate ions on ZnO, in a similar manner to that for propene on ZnO. The bands at 2967, 1540, and 1432 cm^{-1} are in agreement with those of the propionate on ZnO prepared by the adsorption of propan-1-ol, which was found to show absorption at 1542 and 1430 cm^{-1}. In addition, the ratio of the intensity of the band at 1540 to that at 1430 cm^{-1} was always constant. Thus, it is concluded that the bands at 2967, 1540, and 1432 cm^{-1} are attributable to $\nu_{as}(CH_3)$, $\nu_{as}(CO_2^-)$, and $\nu_s(CO_2^-)$ of propionate ions.

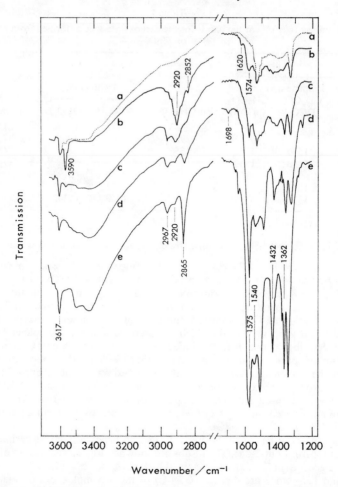

FIGURE 3. Interaction of oxygen with *trans*-but-2-ene adsorbed on ZnO. (a) Back-ground (0·6 g); (b) after 1 h adsorption of *trans*-but-2-ene followed by 5 min evacuation at 298 K; (c) after 1 h exposure to oxgyen (1·3 kN m⁻²) at 298 K; (d) followed by 1 h at 323 K in oxygen; (e) 1 h at 423 K in oxygen.

In a separate experiment it was found that on admission of oxygen to the ZnO plus adsorbed *trans*-but-2-ene the uptake of oxygen occurred to an appreciable extent at room temperature as described previously.[2] The amount of oxygen uptake was almost equal to that of π-allyl species after 1 h contact with oxygen at 298 K, where the π-allyl species disappeared while the amount of the carboxylate species was very small. Thus, essentially the same conclusion can be drawn: the π-allyl species formed from *trans*-but-2-ene is also oxidized to the carboxylate species *via* a precursor.

Intermediates in Oxidation of π-Allyls. Kugler and Kokes[9] have carried out i.r. studies of the interaction of oxygen with propene adsorbed on ZnO and obtained essentially the same spectra as those in Fig. 1. They have attributed the bands in their spectra to

1106

adsorbed acraldehyde plus CO_2 and H_2O from their similarity to the spectra of acraldehyde adsorbed on ZnO. For acraldehyde on ZnO we have obtained spectra almost identical with theirs (Fig. 1). However, in the case of propene on ZnO the intensity of the bands in the 1500—1300 cm^{-1} as well as the 2900—2800 cm^{-1} region increases markedly with increasing temperature while the corresponding increase observed with acraldehyde on ZnO is much smaller (Fig. 1). This suggests that their assignment is incorrect, *i.e.* acraldehyde is not formed in the course of the oxidation of propene on ZnO. Our assignment, *i.e.* that these bands are due to the carboxylate species, has been checked by deuteration. Thus, it is concluded that π-allyl species on ZnO is not oxidized *via* allyl peroxide.

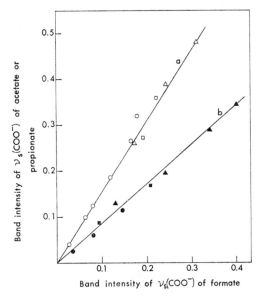

FIGURE 4 The correlation between the band intensities for formate and acetate (or propionate) species.
(a) Propene (1433 and 1365 cm^{-1}); (b) *trans*-but-2-ene (1432 and 1362 cm^{-1}).

Figure 4 shows that the ratio of the intensity of $\nu_s(CO_2^-)$ of formate species to that of acetate (or propionate) species is always constant. This suggests that both carboxylate ions are formed simultaneously on the scission of the carbon–carbon bond in the π-allyl species by the attack of oxygen.

As described above, the carboxylates are not formed directly from π-allyl species but *via* a precursor composed of olefin and oxygen. As regards the nature of the precursor, the following tentative conclusion may be drawn: the precursor may be $C_3H_6O_2$ (ads) formed by addition of two oxygen atoms to the carbon–carbon double bond in propene. Some information is available from the inspection of the i.r. spectra (Fig. 5). It is seen that the intensity of the band at 1147 cm^{-1} passes through a maximum at *ca.* 323 K, then decreases at higher temperatures. Considering that the C–O stretching vibration shows absorption at *ca.* 1100 cm^{-1}, such behaviour would be expected if the precursor just described is actually formed.

As to the nature of adsorbed oxygen species, it has been suggested by various workers that O_2^-(ads) is a reaction intermediate in the oxidation of propene on ZnO[9] and on

$V_2O_5–SiO_2{}^{10}$ on the basis of e.s.r. results. In previous work[2] it was found that molecular adsorbed oxygen participates in the oxidation of propene. Although it seems too early to conclude that $O_2{}^-$(ads) reacts directly with the π-allyl species, there seems little or no doubt that molecular adsorbed oxygen plays a significant role in the oxidation.

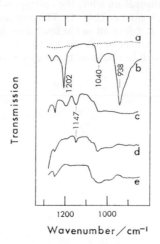

FIGURE 5   Spectra of propene adsorbed on ZnO in the range below 1300 cm^{-1}. (a) Background (0·7 g); (b) adsorbed propene; (c) followed by 1 h at 323 K in oxygen (2·0 kN m^{-2}); (d) 1 h at 353 K in oxygen; (e) 1 h at 423 K in oxygen. The reference sample was used for compensation.

On the oxidation of adsorbed *trans*-but-2-ene both formate and propionate species are formed, suggesting that the attack of oxygen occurs at the 1- and 2-positions of the π-allyl species. The oxidation of adsorbed *trans*-but-2-ene takes place even at room temperature, compared to *ca.* 323 K in the case of propene. This indicates that the reactivity of π–allyl species for oxygen increases with increasing electron concentration of its double bond.

Reaction of Nitrous Oxide with Adsorbed Propene. For propene adsorbed on ZnO similar experiments were carried out using nitrous oxide instead of oxygen. On admission of nitrous oxide to the ZnO containing π-allyl species at room temperature, no spectral change was observed except for appearance of the bands at 2236 and 1278 cm^{-1} which are attributable to adsorbed nitrous oxide.[11] The temperature of the ZnO was raised in stages. At 353 K the bands arising from formation of the π-allyl species such as 3590 and 1202 cm^{-1} began to decrease in intensity. At 423 K these bands became much weaker and simultaneously new bands appeared at 1548, 1466, and 1430 cm^{-1}. With an increase of the temperature to 473 K these new bands increased in intensity. In a separate experiment it was found that oxygen was incorporated in adsorbed propene in the course of the reaction with nitrous oxide, *e.g.* at 423 K the number of incorporated oxygen atoms was almost equal to that of adsorbed propene molecules. Such behaviour suggests that, apart from the band at 1466 cm^{-1}, a pair of bands at 1548 and 1430 cm^{-1} may be attributable to the carboxylate species. It is to be noted that no bands due to formate species appeared in the COO stretching region (1600—1300 cm^{-1}) as well as the CH stretching region (2900—2800 cm^{-1}), suggesting that no scission of the carbon–carbon bond in propene takes place. Consequently, the bands at 1548 and 1430 cm^{-1} appear to be due to propionate ions. In fact, the observed frequencies are in agreement with those of propionate ions just described above.

1108

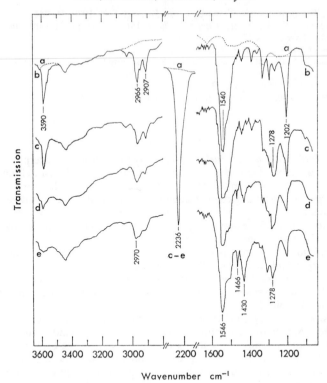

Wavenumber cm⁻¹

FIGURE 6   Interaction of nitrous oxide with propene adsorbed on ZnO. (a) Background (0·7 g); (b) after adsorption of propene followed by 5 min evacuation at 298 K; (c) followed by 2 h at 353 K in nitrous oxide (2·3 KN m⁻²); (d) 2 h at 423 K in nitrous oxide; (e) 2 h at 473 K in nitrous oxide. The reference sample was used for compensation.

The fact that the concentration of the π-allyl species had decreased appreciably before the carboxylate species were formed suggests that, by analogy with the case of oxygen, the carboxylate species are not formed directly from the π-allyl species, but probably *via* a precursor, the nature of which is unclear at present.

Since nitrous oxide is a well-known source of dissociated adsorbed oxygen,[12] such a marked difference in the spectral behaviour between the oxidation by oxygen and by nitrous oxide may be explicable in terms of the concept that O^-(ads) participates in the oxidation by nitrous oxide while molecular adsorbed oxygen such as O_2^-(ads) in the oxidation by oxygen.

REFERENCES

1 *e.g.* H. H. Voge, and C. R. Adams, *Adv. Catalysis*, 1967, **17**, 151; L. Ya. Margolis, *Catalysis Rev.*, 1974, **8**, 241.
2 Y. Kubokawa, T. Ono, and N. Yano, *J. Catalysis*, 1973, **28**, 471.
3 A. L. Dent, and R. J. Kokes, *J. Amer. Chem. Soc.*, 1970, **92**, 6709, 6718; *J. Phys. Chem.*, 1971, **75**, 487; C. C. Chang, W. C. Conner, and R. J. Kokes, *ibid.*, 1973, **77**, 1967.
4 H. Miyata, M. Wakamiya, and Y. Kubokawa, *J. Catalysis*, 1974, **34**, 117.
5 Y. Kubokawa, H. Miyata, T. Ono, and S. Kawasaki, *J.C.S. Chem. Comm.*, 1974, 655.
6 C. Leitch and R. Renaud, *Canad. J. Chem.*, 1952, **30**, 79.

[7] A. Ueno, T. Onishi, and K. Tamaru, *Trans. Faraday Soc.*, 1970, **66**, 756.
[8] J. H. Taylor and C. H. Amberg, *Canad. J. Chem.*, 1961, **39**, 535; L. H. Little, *Infrared Spectra of Adsorbed Species* (Academic Press, 1966), p. 47.
[9] E. L. Kugler and R. J. Kokes, *J. Catalysis*, 1974, **32**, 170.
[10] S. Yoshida, T. Matsuzaki, T. Kashiwazaki, K. Mori, and K. Tarama, *Bull. Chem. Soc. Japan*, 1974, **47**, 1564.
[11] A. Zecchina, L. Cerruti, and E. Borello, *J. Catalysis*, 1972, **25**, 55.
[12] *e.g.*, V. W. Herzog, *Ber. Bunsenges. phys. Chem.*, 1972, **76**, 64.

DISCUSSION

F. Bozon-Verduraz (*Univ. de Paris VI*) said: My first question concerns your interpretations based upon band intensities (Figures 2 and 4). Zinc oxide is a well-known *n*-type semiconductor and its transmittance depends on the nature of the surrounding atmosphere. Oxygen acts as an electron acceptor and this leads to a considerable increase in transparency. Thus, any comparison of intensities appears to be delicate. Do you take this into account, for example, by studying the change of transmittance following adsorption of oxygen alone?

My second question concerns the range below 1100 cm^{-1}. We have shown that the spectrum of zinc oxide contains rather strong bands in this region which we attribute to harmonics of Zn–O vibrations.[1] These bands, which make difficult the observation of specific bands due to adsorbed species, are not mentioned on your paper. Would you comment on this point?

Y. Kubokawa (*Univ. of Osaka*) replied: The increase in transparency of zinc oxide following addition of oxygen causes a shift in the background. Our measurements of the optical density always took into account this shift in the background.

In order to offset the background absorption of zinc oxide a matched disc was inserted in the reference beam, which made it possible to observe the weak bands below 1100 cm^{-1}.

W. C. Conner (*Allied Chemical, Morristown*) said: I have several comments to make on the interaction of butene with oxygen over zinc oxide. First, it is unsafe to assign the band at 1572 cm^{-1} to a 'formate ion'. When butene is adsorbed in the absence of oxygen a strong band at 1574 cm^{-1} dominates the spectrum which is attributed to anti-π-allyl. Secondly, it is hard to understand a shift of 27 cm^{-1} for one band of the proposed formate ion on deuteriation and no shift for the other band. This and the previous point bring into question the 1575–1572, 1365–1362 couple as a formate species. Next, there is no reason to suppose that a 250° rise in temperature will result in the similar behaviour of the bands attributed to two different forms of a surface species.

The mechanism

$$O_2 + C_nH_{2n} \rightarrow HCO_2(\text{formate}) + H + C_{n-1}H_{2(n-1)} \; etc.$$

needs to be proved by further work. If this occurs it is most unusual and interesting.

Lastly, the authors observed weak adsorption bands for butene adsorption alone, whereas we have observed strong bands. This is difficult to understand. The conclusions of the paper need to be supported by experiments involving $^{18}O_2$; such labelling would have the effect of moving the desired bands away from those already assigned, and should allow these questions to be answered.

Y. Kubokawa (*Univ. of Osaka*) replied: The band at 1574 cm^{-1} in Figure 3b is attributable to anti-π-allyl. However, such an assignment is not valid in Figures 3c, 3d, and 3e, since it would be very difficult to explain the spectral change with increasing

[1] F. Bozon-Verduraz, *J. Catalysis*, 1970, **18**, 12.

temperature, including the increase in the intensity of the band at 1575 cm^{-1} with temperature.

A marked difference in the shifts on deuteriation between antisymmetric and symmetric O—C—O stretching vibrations for formate ions is not unusual; Ueno, Ohnishi, and Tamaru (see ref. 7 of the paper) and Greenler[2] have found similar phenomena. It should be noted that such an increase in the intensity of the bands at 1572 and 1365 cm^{-1} with increasing temperature is always accompanied by oxygen uptake by zinc oxide, and that a band pair at essentially the same positions also appears in the case of other systems such as oxygen with acetone, propan-2-ol, or 2-methyloxiran on zinc oxide. Also, Budneva *et al.*[3] have reported the formation of formate and acetate species on oxidation of propene on chromia and copper(II) oxide. On this basis, we believe that these bands can be assigned to formate species. The experiments with $^{18}O_2$ suggested by Dr. Conner would, of course, be very useful for confirmation of the assignments.

Your third and fourth points show that you have misunderstood our paper. Our mechanism is as follows:

$$CH_2 \cdots CH \cdots CH_2 + \underline{O}H \xrightarrow{O_2} C_3H_6O_2 \xrightarrow{O_2} CH_3COO + HCOO + 2\underline{O}H$$

where all carbon-containing species are adsorbed, and $\underline{O}$ represents lattice oxygen. On increasing the temperature to 523 K, the formate species is removed by oxidation, but the acetate species remains.

Comparison of band intensities should be made after detailed examination of the experimental conditions in both cases.

E. L. Kugler (*Stanford Univ.*) said: Prof. Kubokawa, you assert that the assignment of the reaction product spectrum by Kugler and Kokes (your ref. 9) to adsorbed acraldehyde is incorrect. I wish to make four comments concerning your conclusion.

First, the results of Kugler and Kokes are of experiments carried out in the absence of excess oxygen in the range room temperature to 313 K. They assigned the spectrum of the reaction product to adsorbed acraldehyde because there is a one-to-one correspondence between the spectrum of the reaction product and that of adsorbed acraldehyde. I should point out that there is definitely a band at 1635 cm^{-1} in the spectrum of adsorbed acraldehyde obtained at room temperature with no oxygen present. This band cannot be assigned to either formate or acetate and since, it is obtained with oxygen absent, your assignment of it to carbon dioxide or carbonate species seems highly unlikely.

Secondly, the observation that some bands decrease in intensity at 523 K while others do not does not necessarily mean that two species are present at room temperature.

Thirdly, there are no frequencies listed in the carbon–deuterium stretching region in Table 1, column 2, which correspond to the reaction of perdeuteriopropene with oxygen. However, in the work of Ueno, the carbon–hydrogen stretching frequency at 2870 cm^{-1} for adsorbed formate shifts to 2190 cm^{-1} for the deuteriated species. This discrepancy requires an explanation.

Lastly, the observation that the band intensities do not increase as much for acraldehyde and oxygen as for propene and oxygen upon heating is not inconsistent with the assignment of the reaction product spectrum to acraldehyde. In fact, this behaviour might be expected, since there is no reason whereby propene and acraldehyde should react with oxygen at the same rate.

Thus, the behaviour which you have observed in reactions in the presence of excess

[2] R. G. Greenler, *J. Chem. Phys.*, 1962, **37**, 2094.
[3] A. A. Budneva, A. A. Davydov, and V. G. Mikhalchenko, *Kinetika i Kataliz*, 1975, **16**, 480, 486.

oxygen and at high temperatures is quite consistent with assignment of the reaction product spectrum to acraldehyde.

You suggest an intermediate in the oxidation of the form $C_3H_6O_2$. My recent work[4] has produced evidence that this intermediate is glycidaldehyde, and the paper contains further evidence for the original assignment of the product spectrum to acraldehyde.

Y. Kubokawa (*Univ. of Osaka*) replied: My answers to your comments are as follows. First, the assignment of the band at 1635 cm^{-1} is not conclusive. At 323 K where the precursor ($C_3H_6O_2$) is formed this band is not observed (Figure 1c); this contrasts with its appearance in the case of acraldehyde (Figure 1e). On increasing the temperature to 423 K (Figure 1d) this band appears. Such behaviour suggests that this band is not necessarily associated with the precursor.

Secondly, neither formate nor acetate species is present at room temperature. The precursor is formed at around 323 K as described above. Appreciable formation of the carboxylate species occurs at higher temperatures, as seen in Figure 2.

Thirdly, impurities in perdeuteriopropene made the position of the band unclear. Recent experiments with pure C_3D_6 have shown the presence of bands at 2228 and 2170 cm^{-1}.

Lastly, of course different adsorbed species react with oxygen at different rates. However, if acraldehyde were the intermediate in propene oxidation, at 323 K (Figure 1c) most of the propene would be oxidized to acraldehyde, *i.e.* in both cases ($C_3H_6 + O_2$ and $C_3H_4O + O_2$) the same adsorbed species would be formed, and the same spectral change with increasing temperature would be expected. Accordingly, such a marked difference in the temperature dependence of the spectral changes suggests that the similarity of the spectra (Figure 1c and 1e) is only apparent, and does not mean that the same adsorbed species is formed.

Y. Kubokawa (communicated): Following your remark we have made an i.r. study of glycidol on zinc oxide. On the basis of preliminary results, I doubt that the precursor is glycidaldehyde.

J. Haber (*Inst. Catalysis, Krakow*) said: The formation of formate and acetate ions in propene oxidation may indicate that the reaction intermediate is a perepoxy-complex formed by electrophilic attack of oxygen radicals on the double bond, not necessarily involving π-allyl species as precursors. One would expect the formation of formate and propionate complexes from analogous reaction with but-1-ene, which might have been formed by isomerization of but-2-ene. Did you observe any isomerization of but-2-ene? Have you determined also the spectra after outgassing at higher temperatures without oxygen addition to check whether the disappearance of certain lines is not simply due to desorption, rather than to interaction with oxygen?

Y. Kubokawa (*Univ. of Osaka*) replied: The reaction probably proceeds *via* the intermediate suggested by Prof. Haber. We have found that the rate of oxygen uptake by zinc oxide on which mono-olefins are adsorbed increases with increasing carbon number of the olefin (ref. 2 of the paper). To be certain it is necessary to deduce the structure of the precursor of the carboxylate species. Studies to achieve this are now in progress.

It seems unlikely that isomerization of *trans*-but-2-ene takes place under our experimental conditions. There appear to be several pathways leading to the formation of formate and propionate species; at present it is difficult to decide between the possibilities.

We have not investigated the spectra after outgassing at high temperatures without oxygen addition. Experiments with other systems (*e.g.* acetone on magnesium oxide, ref. 4 of the paper), suggest that spectral changes with increasing temperature arise mainly from reaction with oxygen.

[4] E. L. Kugler and J. W. Gryder, *J. Catalysis*, in press.

Kinetic Manifestations of Reversible Changes Induced by Ammonia Synthesis upon Iron Catalysts

H. Amariglio* and G. Rambeau

Laboratoires de Catalyse et Cinétique Hétérogènes, Université de Nancy I, Centre de 1er Cycle, Case Officielle No. 140, 54037-Nancy-Cédex, France

ABSTRACT   Progressive changes of catalytic activity have been observed during ammonia synthesis upon different iron catalysts. Between 250 and 500°C, every increase in ammonia production on the catalyst increases the activity of the surface, whereas every decrease of this production is followed by a deactivation which also occurs in nitrogen, hydrogen, or in an equilibrated mixture of nitrogen, hydrogen, and ammonia. It is concluded that the working iron surface acquires some sort of mobility leading to a non-equilibrium surface, more defective than its initial form. The enhancement of the number of surface defects, in which the activation originates, is induced by the reaction itself whereas the deactivation is the kinetic result of spontaneous surface relaxation. The autocatalytic character of the ammonia synthesis, so far unknown, opens new perspectives for the understanding of this reaction.

Introduction

Our initial aim was to investigate the effect of the dispersion of iron upon its catalytic activity in NH_3 synthesis. By the choice of this reaction, we meant to avoid any contaminant effect often reported in studies devoted to organic reactions. Moreover the very important kinetic work already accumulated about this system led us to think that correct activity determinations should be easy. In fact, when submitting the catalyst to any set of experimental conditions we always observed transient activities and long periods were necessary before steady rates could be measured. It thus appeared difficult to ascribe a well defined activity to a given set of experimental conditions if the reason for the transient behaviour was ignored. So the modified purpose of our work has been the elucidation of that point. It may seem surprising that such phenomena are not referred to by classical and important reviews[1-3] on NH_3 synthesis. However they have sometimes been reported[4,5] though without any explanation, perhaps because the closed flowing systems most often used were not appropriate for rate determinations other than steady ones.

Experimental

The apparatus was designed to allow precise and very rapid measurements needed by this sort of study. The reactor is of the flow type and fed with alternate flows of N_2, H_2 or $N_2 + 3H_2$. The gases are purified through Cu filings (400°C) and a trap immersed in liquid air. The NH_3 produced reacts with a sulphuric acid solution contained in a beaker through which the exit gas bubbles, and in which a conductivity cell is immersed. By choosing appropriate volume and concentration of the sulphuric acid solution it is possible to determine the production rate of ammonia, R_s (mole/min), very quickly and precisely. The lowest measurable rate is 10^{-8} mole/min; the corresponding volume decrease which would have allowed the measurement of the reaction rate in the apparatus previously described[4,5] would have been as low as 10^{-4} cm^3/min and so hardly attainable.

Catalyst samples.

1 g of the well known triply-promoted KMI catalyst, powdered into 0·2—0·25 mm particles and reduced at 500°C.

1 g of very pure iron powder (Johnson-Matthey) after reduction at 300°C.

0·1 g of iron supported on 5 g of graphitized carbon black (Spheron 6, Cabot) after reduction at 400°C.

Reduction of the catalyst was carried out with a stoicheiometric mixture of N_2 and H_2 at a flow rate of 900 cm³/min (Standard mixture).

RESULTS AND COMMENTS

Among the different types of transient activities, some show maxima and minima and so reveal probable superimposition of two distinct effects which can be disentangled by appropriate operations. Fig. 1, in which R_s is plotted against time, illustrates the main results obtained on KMI catalyst at 253°C.

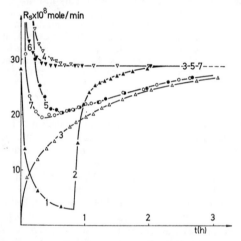

FIGURE 1   Rate evolutions in standard mixture ($N_2 + 3H_2$, 900 cm³/min) at 253°C on 1 g KMI following short or long exposures to H_2 (2 and 3), N_2 (4 and 5) or 30 cm³/min $N_2 + 3H_2$ (6 and 7).

(a) *Transient Response after H_2 treatments*

At first the catalyst is fed with the standard mixture until it reaches its steady activity, which is attained at the end of the curves 2, 3, 4, 5, 6, 7 of Fig. 1. The standard mixture is then replaced by pure H_2. The ammonia production decreases as shown on curve 1 while H_2 uses up the nitrogenated species left on the surface by its prevous stationary working. After about 1 h the N_2 stored on the surface reaches a negligible value, as revealed by the production level. The standard mixture is then re-established, which makes the product rate rise (curve 2) while the composition of the superficial film tends towards its stationary value again. This transient response lasts approximately 1 h.

After that, the same procedure is repeated, but the H_2 treatment lasts 15 h instead of 1 h. Following this longer stop in the production, the rate in the standard mixture rises much slower (curve 3) than in the previous run (curve 2) and the steady state is not reached in less than 5 h.

Were the phenomena merely to consist of the consumption or replenishment of the

1114

chemisorbed film, curves 2 and 3 would be superimposable, in that the same NH_3 production, once attained, would have revealed the same surface composition. We therefore conclude that the catalyst progressively deactivates in H_2 and reactivates in the reacting mixture, and we are led to examine whether the reactivation is due to N_2.

(b) *Transient response after N_2 treatments*

In a new set of experiments a flow of pure N_2 is substituted for the standard mixture after the steady state is established. NH_3 production stops immediately. In the first run the catalyst is exposed to N_2 for 15 min and the standard mixture is then admitted again. The rate decreases as indicated on curve 4, till the steady state operates again. This decrease may be explained as the progressive disappearance of the N_2 which was chemisorbed over its stationary value.

The catalyst is then exposed to N_2 for 15 h, after which the gas is replaced by the standard mixture. The production rate is plotted along curve 5 and we can see that it decreases to a value lower than the final stationary value. The standard mixture first removes an excess of chemisorbed N_2 upon a surface which was deactived by N_2 treatment. The increase which follows can no longer result from the kinetics of chemisorbed film formation and so must be wholly related to an activation of the catalyst.

Therefore, the catalyst behaviour following N_2 treatment greatly depends upon the length of such treatment, and so cannot result from the mere consumption of the preadsorbed N_2. Here again we must conclude that a sufficiently long N_2 treatment results in a deactivation of the catalyst which progressively reactivates in the reacting conditions.

The deactivation being observed in N_2 as well as in H_2, whereas the reactivation operates under the reacting mixture, it remains to examine what may be the effect of the NH_3 produced.

(c) *Transient response after production of higher NH_3 contents.*

Once the steady state has been re-established, the flow-rate of the standard mixture is diminished to a low value (30 cm^3/min) either for 15 min or 15 h. It is then restored to its standard value again (900 cm^3/min). With a reactor of the integral type, any decrease in the flow rate results in a decrease in the production of NH_3, whereas its pressure increases as well as the concentration of nitrogenated species in equilibrium with NH_3.[4]

After a short exposure to a higher NH_3 pressure the production rate follows curve 6 while the standard mixture eliminates (as for curve 4) the excess of N_2 chemisorbed on a surface which does not appear to be deactivated. However, after longer exposure the rate decreases as shown by curve 7: the decrease below the steady value demonstrates a new deactivation, whereas the rising part of the curve is exactly superimposable upon that of curve 5, and does not differ much from that of curve 3. On the other hand, since the surface has more N_2 adsorbed on it when it has been pretreated with N_2 alone, we can easily explain why curves 6 and 7 decrease more quickly than curves 4 and 5. The fact that these successive deactivations and reactivations end in the same steady state proves the reversibility of the changes in the active surface.

Thus we can see that the same activation is observed after an increase (curves 3 and 5) or a decrease (curve 7) of the NH_3 pressure occurring when the standard mixture is re-established. This activation always appears when the NH_3 production takes up agaiu after the catalyst has ceased to work (or has worked less well) for a long enough time. What appears to be the decisive factor is the reaction itself, which tends to activate the surface when operating, whereas the same surface deactivates when it is left at rest.

(d) *Activation and ammonia production*

A simple way to affect the production is to vary the temperature. From 283°C, chosen as a reference temperature, the catalyst is made to work at higher temperatures ($<$ 600°C) until the establishment of the stationary activity in the standard mixture. The temperature is then lowered to 283°C and the catalyst appears to be activated as shown by curve 1 on Fig. 2, where the initial activity in the standard mixture at 283°C following thermal treatment is plotted against the treatment temperature. The activation thus

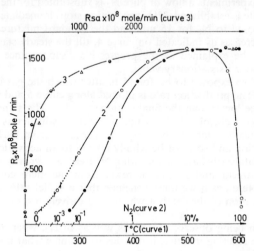

FIGURE 2   Rate of NH_3 production in the standard mixture on 1 g KMI at 283°C following activations by: the standard mixture, against temperature (curve 1); 900 cm³/min of N_2, H_2 mixtures at 509°C, against N_2 content (curve 2) or against the steady rate during the activation period (curve 3 ◖); stoicheiometric mixture at various flow rates at 509°C, against the steady rate during the activation period (curve 3 △).

obtained is at a maximum for a treatment temperature of about 500°C, the temperature at which equilibrium is reached for NH_3 synthesis, so that no further increase in production can be made. The catalyst activation appears to be well related to the ammonia production.

At the optimum temperature so determined, a second way to affect the production is to change the composition of the gas at a constant flow rate (900 cm³/min). Starting with pure H_2, the N_2 content is increased up to 100%. Once the steady activity relating to each gaseous composition has been reached at 509°C, the temperature is lowered to 283°C. The initially observed rate for the standard conditions (283°C, $N_2 + 3H_2$, 900 cm³/min) is plotted on curve 2 (Fig. 2) against the N_2 content of the mixture, operating at 509°C.

The catalyst activity is at a minimum for pure H_2 but rises sharply with the N_2 content in the 0—1% range. It is worth stressing that the pressure and flow rate of H_2 do not vary significantly under these conditions, so that the activation obtained can be related neither to H_2 nor any impurity carried along with it. The activation should be ascribed to N_2 or any promoting impurity it might contain. However, the activation remains practically constant for N_2 contents from 10 to 25% and then decreases sharply at above 50%. It becomes very low in pure N_2 and still lower if N_2 is diluted with He. It might be inferred that the promoting effect of N_2 or any hypothetical impurity would reverse when the N_2 content is high enough, but the dilution with He should restore the promoting role; this is not the case. Now let us plot on curve 3 (◕) of Fig. 2 the initial rate at 283°C against the steady rate R_{sa} at 509°C relating to the various mixture compositions. The activation which first rises and then decreases with N_2 pressure (curve 2) rises continuously with R_{sa}.

Since in these experiments carried out at a constant flow rate, R_{sa} is proportional to NH_3 pressure, it becomes necessary to discriminate between the two factors. At 509°C, varying the flow rate affects neither the N_2 pressure nor the H_2 pressure, since the extent of reaction does not exceed 10^{-3}. When the flow rate is increased from 0 to 900 cm^3/min the NH_3 exit pressure remains constant and the production rate, R_{sa}, is proportional to the flow rate. But at a lower flow rate the equilibrium of the reacting mixture is reached nearer to the top of the catalyst bed and a larger part of the catalyst is subjected to the equilibrium NH_3 pressure.

Once the steady state relating to each of the various flow rates has been achieved at 509°C, the catalyst temperature is reduced to 283°C. The initial rate at 283°C in the standard mixture is plotted against the steady rate R_{sa} at 509°C. Curve 3 (△) in Fig. 2 is thus obtained and superimposes curve 3 (◕) referring to activating mixtures of various compositions. We therefore conclude that the activation essentially depends on the value of R_{sa}. The catalyst activity is at a minimum when the flow rate is nil, *i.e.* when the catalyst is wholly subjected to the equilibrium conditions. The increase in the flow rate from 0 to 500 cm^3/min results in a sharp increase of the activation, though the NH_3 pressure inside the catalytic bed continuously decreases. Therefore the activation is a consequence of the NH_3 production itself.

Before examining whether the phenomena we are discussing are also displayed by other kinds of iron catalysts, let us stress that the slow deactivation which can be observed at 283°C has nothing to do with a mere consumption of the nitrogenated species formed on the surface at 509°C: 20 min would certainly be sufficient to eliminate a complete N_2 layer over the 10 m^2 or so of catalyst[6] with a rate of 10^{-5} mole/min, whereas the N_2 which might be present inside the catalyst[7] can be neglected (10^{-6} N/g at 500°C). During 20 min the rate decreases by about 2%.

(e) *Pure iron*

The changes of catalytic activity following exposure to the various gases are studied at 274°C (Fig. 3) on the supported iron sample and at 283°C (Fig. 4) on pure iron powder. The curves of Fig. 3 and 4 are numbered in a way similar to that in Fig. 1.

On the supported sample (Fig. 3) the elimination (curve 1) and the formation (curve 2) of the stationary N_2 layer as well as the surface activation (curves 3, 5, and 7) are slower than on the promoted catalyst. Also the deactivation is slower.

On the powdered iron (Fig. 4) the reverse is true. All the changes occur much more rapidly and some deactivation appears after a short H_2 or N_2 treatment (curves 2 and 4 respectively). The fast deactivation of bulk iron causes the progressive disappearance of the minimum displayed by curve 4 when the N_2 treatment is more prolonged (30 min, curve 9; 3 h, curve 8; 15 h, curve 5).

We can thus say that the three kinds of iron catalysts display similar activations and deactivations, induced by the occurrence of the reaction itself, and which differ only in their rates and extents.

DISCUSSION AND CONCLUSION

We have just described the various reversible changes of the catalytic activity of an iron surface and we have shown that they do not originate from changes in the composition or in the extent of the chemisorbed layer, or any impurity which might have exerted a promoting or poisoning effect.

We must also discard the notion that the catalyst may become nitrided under our experimental conditions[8] when we take into account the pressures of N_2 (0·75 Atm) and NH_3 ($\leqslant 10^{-3}$ Atm). Besides, such a hypothesis would lead to inconsistencies since an

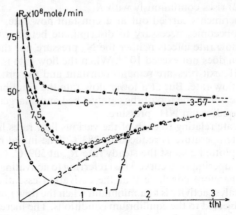

FIGURE 3 Rate evolutions in standard mixture at 274°C on 0·1 g supported iron following short or long exposures to H_2 (2 or 3), N_2 (4 or 5) or 30 cm³/min $N_2 + 3H_2$ (6 or 7).

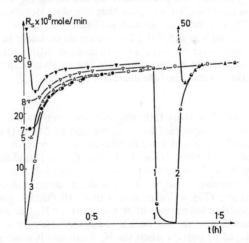

FIGURE 4 Rate evolutions in standard mixture at 283°C on 1 g pure iron powder following:

10 min exposures to H_2 or N_2 (2 or 4);
15 h exposures to H_2, N_2 or 30 cm³/min $N_2 + 3H_2$ (3, 5 or 7);
3h or 30 min exposure to N_2 (8 or 9).

increase in the flow rate or an H_2 treatment would both favour the decomposition of the nitride and thus result in the same kinetic effect, which is not the case.

We are therefore led to suggest that the reaction itself induces some mobility among the superficial iron atoms and so causes the formation of some special configurations

1118

which are especially active towards this reaction. We recently[9] reported a related case in which the active sites generated by one reaction (H_2–O_2) may catalyse another one (H_2–C_2H_4). Such a chemical generation of active sites causes the surface to be out of equilibrium as to its content of defects, with the result that the active sites will spontaneously tend to disappear. In such circumstances the steady-state activity relates to the case of the reaction occurring on the normal deactivated surface and producing as many active sites as their spontaneous decay causes to disappear. So we can explain the fact that any increase in the production of NH_3 activates the catalyst, though when the reaction stops or becomes slower the catalyst deactivates. All the changes we have reported are then easily accounted for.

Of course, displacing the catalyst surface out of its equilibrium state needs an expenditure of energy which is furnished by the reaction. By this means some part of the chemical energy is accommodated by the metal surface rather than being wasted as heat. Now if we take into consideration the fact that the rate of this reaction is determined by the N_2 adsorption[4,5] we conclude that it is the only step that may yield energy to the surface. Such a conclusion is not at variance with the deactivation previously observed after an N_2 treatment. We can indeed easily understand that during a sufficiently long exposure of the catalyst to N_2 or H_2 or any equilibrated mixture of N_2, H_2, and NH_3 a deactivated surface finally appears.

Unfortunately the literature does not give direct evidence of a reconstruction of the iron surface during N_2 chemisorption. But such a hypothesis has been put forward in the recent work from Boudart *et al.*[10] Also, Brill[11] *et al.* have reported that N_2 favours the formation of 111 faces on iron. It is also worth recalling that Roginskii *et al.*[12] have observed catalytic corrosion caused by the decomposition of NH_3 occurring on Fe, Pt, and Cu surfaces whereas no modification could be observed in pure H_2 or N_2.

If our interpretation is adopted, the system studied offers an example of a catalyst surface which changes progressively and fits the conditions to which it is subjected. Rozovskii *et al.*[13] have suggested the same conclusion in the case of synthesis of alcohols from CO and H_2 on iron and called this phenomenon "self-regulation in catalysis".

From a kinetic point of view we arrive at the conclusion that the simple and usual model of a catalyst surface which tacitly contains a constant number of active sites must be given up in this case at least. All the kinetic determinations should be effected in such a manner that the active surface remains constant, which may be approximated in this case on account of the slow deactivations. We shall report on that later.

REFERENCES

1 W. G. Frankenburg, *Catalysis* (Reinhold, New York, 1955), Vol. 3, Ch. 6, p. 171.
2 C. Bokhoven, C. van Heerden, R. Westrik, and P. Zwietering, *Catalysis* (Reinhold, New York, 1955), Vol. 3, Ch. 7, p. 265.
3 A. Nielsen, *Catalysis Rev.*, 1971, **4**, 1.
4 A. Ozaki, H. S. Taylor, and M. Boudart, *Proc. Roy. Soc.*, 1960, *A*, **258**, 47.
5 K. Aika and A. Ozaki, *J. Catalysis*, 1969, **13**, 232.
6 A. Nielsen and H. Bohlbro, *J. Amer. Chem. Soc.*, 1952, **74**, 963,
7 J. D. Fast, *Chem. Weekblad*, 1955, **51**, 427.
8 S. R. Logan, R. L. Moss, and C. Kemball, *Trans. Faraday. Soc.*, 1958, **54**, 922.
9 P. Paréja, A. Amariglio, and H. Amariglio, *J. Chim. phys.*, 1974, **71**, 999; *J. Catalysis* 1975, **36**, 379.
10 J. A. Dumesic, H. Topsöe, and M. Boudart, *J. Catalysis*, 1975, **37**, 513.
11 R. Brill, E. L. Richter, and E. Ruch. *Angew. Chem.*, 1967, **79**, 905.
12 S. Z. Roginskii, I. I. Tretyakov, and A. B. Shekhter, *Doklady Akad. Nauk. S.S.S.R.* 1963, **91**, 881.
13 A. Ya. Rozovskii and Yu. B. Kagan, *Proc. 4th Internat. Congr. on Catalysis*, Moscow, 1968 (Akademiai Kiado, Budapest 1971), Vol. 2, p. 347.

DISCUSSION

P. H. Emmett (*Portland State Univ., Portland*) said: As one who has been following the work on ammonia synthesis for fifty years I would like to make a few comments on this interesting paper. It is thrilling to think that some new factors are being discovered in regard to the activity of ammonia catalysts that we may have overlooked. I would like to point out two things. In the first place, as the authors point out, Brill claims that nitrogen atmospheres tend to produce a transformation of the (100) and (110) planes of iron into active (111) planes at 670 K. He further claims that the (111) planes are much more active than the (110) and (100) planes. Possibly some of the effects being reported here may be due to a readjustment of the relative exposures of these planes as a function of the type of gas to which the catalyst is exposed. The other point is that these effects should be checked in the ammonia synthesis region of 620–670 K and high pressures. Great care should be taken, too, to be sure they are not due to impurities in the gases.

H. Amariglio (*Univ. de Nancy*) replied: It has already been postulated (refs. 10 and 11 of the paper) that the crystallographic nature of the exposed faces of iron might change upon exposure to one or another species involved in this reaction. This point of view would allow an alternative explanation for the transient responses of the catalyst and the progressive adjustment of the surface to operational conditions.

However, we have observed that mere exposures of the catalyst to each reactant or to product causes a deactivation if it is long enough, whereas the surface reactivates when reaction conditions are restored. Moreover, we have shown that the activation effect produced during pretreatments correlates precisely with the rate of production prevailing during these pretreatments, regardless of the mode of activation (Figure 2) which may relate to different compositions and result in the same activating effect.

We think that the experimental facts are best rationalized by the suggestion that the chemical energy can be used to dislodge atoms and produce active configurations at some of the points where it is released. Dislodged atoms might then be involved in rearrangements resulting in the formation of new surface planes, but that would be a macroscopic by-product of the primary effect which occurs at the atomic scale and is directly revealed by the catalytic behaviour.

N. Pernicone (*Montedison, Novara*) said: I wonder if the interesting results you have presented are relevant under the actual operating conditions of ammonia synthesis (200–300 atm, 670–770 K). During our experiments with various industrial catalysts (including KM1) in these ranges of pressure and temperature, we have found no evidence of transient phenomena. We have used differential reactors and gas chromatographic analysis, so we should have had no difficulty in detecting phenomena such as those you have presented here. The faster transient phenomena you have found for pure iron could be explained, if your interpretation based on mobility of iron atoms is correct, by taking into account the long-range disorder that we and Hosemann have demonstrated to exist in alumina-containing iron catalysts.[1] The introduction of $FeAl_2O_4$ or γ-Al_2O_3 groups in the lattice of α-Fe could well lower the mobility of iron atoms.

H. Amariglio (*Univ. de Nancy*) replied: Figures 1, 3, and 4 describe the transient responses of different iron catalysts under the conditions which are referred to in the paper. For each of the three samples it is obvious that the length of the response to some sudden modification of one or another kinetic parameter depends greatly upon the experimental conditions. For sufficiently high temperatures and/or pressures steady states may be expected to be reached quickly. Such an occurrence would not mean

[1] N. Pernicone G, Fagherazzi, F. Galante, F. Garbassi, F. Lazzerin, and A. Mattera, *Proc. 5th Internat. Congr. Catalysis*, Miami Beach, 1972.

that the number of active sites would remain constant but merely that the surface would adjust quickly to any changes in the operating conditions.

Our experiments do not reveal the factors that determine iron mobility, so I cannot comment on any proposal concerning a higher or lower mobility induced by alumina.

P. Mars (*Twente Univ. of Technology, Enschede*) said: Because ammonia synthesis is affected by concentrations of less than 0·1 p.p.m. of oxygen-containing species in the feed, sophisticated purification trains have been used.[2] These may be more efficient than that used in this investigation. For example, in an empty cold trap such as you used, ice or solid carbon dioxide may be formed if cooling occurs quickly. Moreover, in your experiments back-diffusion of water vapour from the sulphuric acid solution to your catalyst may occur. This may cause the poisoning of the catalyst that you observe after using a low flow-rate of gas. Some further experiments should be carried out to determine whether or not the observed phenomena have been caused by impurities.

H. Amariglio (*Univ. de Nancy*) replied: Our purification train contains 200 g of copper filings and is maintained at 673 K; the cold trap, which is a coiled pipe 6 m in length and of 6 mm internal diameter is not empty but is filled with glass beads. We checked that the oxygen and water contents were lower than 10^{-2} and 10^{-1} p.p.m., respectively. Furthermore, by-passing the purification did not modify qualitatively the existence and features of the transient responses.

Concerning your proposed back-diffusion of water vapour, I make the following remarks. First, a sufficiently long pipe (50 cm in length, 0·2 cm internal diameter) separates the reactor from the gas exist. Secondly, the gas bubbles through the solution only during the rate measurements. Thirdly, when a high flow-rate of $(N_2 + 3H_2)$ is substituted for pure nitrogen the transient response is exactly the same regardless of the value (high or low) of the nitrogen flow rate.

Our interpretation accounts for all the transients observed, and this would not be the case with an explanation based upon the influence of some impurity. For instance (Figure 2, curve 2) the back-diffusion hypothesis would not explain why the greater activation occurs when the composition change of the gas is very small (0–1 % nitrogen) whereas a large composition change (10–50%) does not alter the activity. Furthermore, the catalyst deactivation is strongest when the reactor is closed.

A. V. Krylova (*Mendeleyev Inst. of Chem. Eng., Moscow*) asked: Is there a difference between the action of hydrogen and of nitrogen on fresh catalyst when there are no special configurations on the surface and on the samples with steady activity?

H. Amariglio (*Univ. de Nancy*) replied: All experiments we have described are reproducible, and we have found no difference between the first and subsequent runs.

R. W. Coughiln (*Lehigh Univ., Bethlehem, Pa.*) (communicated): The fact that you have also found similar behaviour with pure iron powder as catalyst reminds me of the story, perhaps apocryphal, that Haber's success in fixing nitrogen hinged on the fact that he carried out his reactions in the bores of cannon which provided the catalytic surface. This suggests that massive iron might be investigated for microscopically observable surface transformations resulting from catalysis of this reaction. Such observations are well known for platinum catalysts in ammonia oxidation and for silver catalysts in ethylene oxidation.

---

[2] C. Bokhoven and W. van Raayen, *J. Phys. Chem.*, 1954, **58**, 471.

Activity and Selectivity of Supported Noble Metals for Steam Dealkylation of Toluene

KAREL KOCHLOEFL*

Süd-Chemie A.G., Girdler Katalysatoren, München, G.F.R.

ABSTRACT The effect of various noble metals (Pt, Pd, Rh) and carriers (α-Al_2O_3, γ-Al_2O_3, SiO_2, Al_2O_3-SiO_2, CaO, MgO, ZnO, α-Cr_2O_3) on activity and selectivity of catalysts for toluene steam dealkylation (TSD) was studied in the temperature range 375–625°C and at pressures between 1 and 22 atm. It was found that the activity of supported noble metals for TSD increases in the order Pd, Pt, Rh. Oxides known as insulators are associated with high activity of Rh for TSD as well as for toluene hydrodealkylation (THD) as a consecutive reaction. CaO, MgO, and ZnO diminish the activity of Rh for TSD and THD. A Rh–Cr_2O_3 catalyst was found to have the most suitable catalytic properties for TSD. The concentration of Rh and basic additives affects the activity and selectivity of the catalyst for TSD with respect to toluene steam reforming (TSR). On the basis of deuterium isotope effects, and kinetic and adsorption measurements, mechanisms for TSD and THD on Rh catalysts are suggested.

INTRODUCTION

Toluene steam dealkylation (TSD), eq. (1), represents a very interesting hetero-

$$C_6H_5CH_3 + 2H_2O \rightarrow C_6H_6 + CO_2 + 3H_2 \tag{1}$$

geneously catalysed reaction, from the theoretical as well as from the practical point of view. Industrial interest can be expected, only if benzene and pure hydrogen will be produced. The idea of using water instead of hydrogen in dealkylation of toluene, eq. (2), is not new. In 1948 Haensel[1] described different types of solid catalysts for

$$C_6H_5CH_3 + H_2 \rightarrow C_6H_6 + CH_4 \tag{2}$$

TSD. Later on Russian authors[2–7] in particular focused their interest on the study of this reaction. They found that Rh supported on γ-Al_2O_3 possesses high activity for TSD. However, this catalyst also accelerates some undesirable parallel and consecutive reactions, e.g. toluene steam reforming (TSR), eq. (3), and toluene hydrodealkylation

$$C_6H_5CH_3 + 14H_2O \rightarrow 7CO_2 + 18H_2 \tag{3}$$

(THD), eq. (2). Reaction (3) lowers the yield of benzene, reaction (2) lowers the purity of the hydrogen produced. Recently much effort has been devoted to the development of an industrial catalyst[8–13] and to the elucidation of a mechanism[3] for TSD.

EXPERIMENTAL

Chemicals. The toluene was "purum" grade (Fluka, Switzerland), $D_2O \geqslant 99.5\%$ (Roth, G.F.R.) and $D_2 \geqslant 99.5\%$ (J. T. Baker, Chemical Co., USA). $RhCl_3$, $PdCl_2$, and H_2PtCl_6 p.A. grade, commercial chemicals of Heraeus (G.F.R.).

Carriers. γ-Al_2O_3, α-Al_2O_3, and Al_2O_3-SiO_2 (Süd-Chemie, G.F.R.), α-Cr_2O_3 (Bayer, G.F.R.), porous glass (GS–1, PPG Industries, USA), and SiO_2 (Hermann, G.F.R.), were all commercial products. ZnO, MgO, and CaO were prepared by thermal decomposition of their carbonates (Riedel de Haen, G.F.R.) at 1000°C for 3 h. Their surface basicity or acidity was determined by C_6H_5COOH and $C_4H_9NH_2$ titration in benzene using standard procedures[14] (Table 1). The preparation of a carrier for industrial purposes (pellets 4 × 4 mm) containing Cr_2O_3, hydraulic binder, and promoters has been described elsewhere[13].

Table 1

Physical Properties of Catalysts Investigated

Catalyst code	Metal (0·6 wt. %)	$S_T{}^a$ (m^2/g)	S_M (m^2/g)	Carrier	Basicity (B) or Acidity (A) of carrier	
					mmol/g	pK_A
Cat.–1	Rh	192	3·1	γ-Al_2O_3		
Cat.–2	Pt	200	1·9	γ-Al_2O_3	$3·1 \times 10^{-1}A$	+3·3
Cat.–3	Pd	202	3·0	γ-Al_2O_3		
Cat.–4	Rh	5·6	2·5	α-Al_2O_3	$2·0 \times 10^{-2}A$	+3·3
Cat.–5	Rh	199	3·4	porous glass	$1·0 \times 10^{-3}A$	+4·8
Cat.–6	Rh	327	3·6	SiO_2	$5·5 \times 10^{-3}A$	+4·8
Cat.–7	Rh	114	3·0	Al_2O_3(5·0 wt. %) –SiO_2	$4·5 \times 10^{-2}A$	+1·5
Cat.–8	Rh	2·0	<0·1	CaO	$5·5 \times 10^{-1}B$	+9·3
Cat.–9	Rh	28·0	1·3	MgO	$7·0 \times 10^{-3}B$	+9·3
Cat.–10	Rh	3·4	0·1	ZnO	$1·0 \times 10^{-2}A$	+6·8
Cat.–11	Rh	4·6	2·1	α-Cr_2O_3		
Cat.–12	Pt	4·8	1·1	α-Cr_2O_3	$1·5 \times 10^{-2}A$	+3·3
Cat.–13	Pd	6·0	2·0	α-Cr_2O_3		

[a] Total surface area measured by BET method.

Catalysts. All catalysts investigated were prepared by impregnation of granulated (0·2–0·4 mm or 1·0–2·5 mm) or pelletized carriers with an aqueous solution of the corresponding noble metal salt. Subsequently these were dried (120°C, 14 h) and calcined (550°C, 3 h). The catalysts were activated in a hydrogen stream at 250°C for 2 h in the reactor. They were characterized by their surface area (S_T), metallic surface area (S_M) measured by chemisorption of CO, X-ray diffraction, and chemical analysis (Table 1).

Apparatus and Analysis. The kinetic measurements and performance tests of catalysts were conducted in a microreactor or in a pressure flow lab-scale unit. For adsorption measurements the pulse reactor was used. The micro- and pulse-reactors were connected with a gas chromatographic column filled with Fluoropak coated with 10 wt. % poly-(ethylene glycol) 1500, on which benzene, toluene, and water were separated at 95°C using H_2 as carrier gas. In experiments with D_2, D_2O or on deuteriated catalysts, reaction products were trapped at —78°C for i.r. analysis. Biphenyl, fluorene, and anthracene were determined in liquid reaction products by means of gas chromatography (Apiezon L, 2·5 m column, 250°C). The gaseous reaction products were measured and analysed gas chromatographically by standard procedure. The experimental parameters such as temperature (T) [375—625°C]; space velocity (F/W) [4·0–60·0 mol T/h kg_{cat}] and in some cases also total pressure (P) [1·0—22·0 atm]; molar ratio water/toluene R (H_2O/T) [4·0—22·0] were varied in the range given. The initial reaction rate $r_0 = d(x)/d(W \cdot F^{-1})$ [where x is the conversion of toluene into benzene (mol), F feed of toluene (mol/h), W weight of catalyst (kg)], was determined graphically from the plot x against W/F. The apparent activation energy E_a (kcal/mol), catalyst selectivity S (%) for TSD in view of TSR and THD given by following formula

$$S = \frac{r_0(\text{TSD}) \times 100}{r_0(\text{TSR}) + r_0(\text{THD}) + r_0(\text{TSD})}$$

and deuterium isotope effects $\alpha_{D/H}$ [$= r_0(\text{THD})/r_0(\text{TDD})$], α_{D_2O/H_2O} [$= r_0(\text{TH}_2\text{OD})/r_0(\text{TD}_2\text{OD})$] were calculated from experimental data using the Hewlett–Packard computer (Model 9810).

RESULTS AND DISCUSSION

Preliminary experiments performed with Rh–γ-Al$_2$O$_3$–catalyst (Cat.–1) showed that the CH$_4$ content in gaseous products of TSD increases considerably with increasing total pressure. For example, at 450°C, F/W = 9·1 mol T/h kg$_{cat}$, R (H$_2$O/T) = 7·0, the methane content increases from 2·1 vol % at atmospheric pressure to 11·4 vol % at 11·0 atm and to 34·8 vol % at 21·0 atm. These data are in a good agreement with recent results of Rabinovich and co-workers[7] who investigated TSD at pressures ranging between 1·0 and 16·0 atm. Such a high hydrodealkylation activity makes Rh–γ-Al$_2$O$_3$ catalyst uneconomic for its application on an industrial scale, because of the high expenses for purification of hydrogen. In view of these facts we studied the effect of other noble metals and carriers on the activity and selectivity of catalysts for TSD.

Effect of Different Noble Metals on Characteristic Properties of Catalysts for Steam Dealkylation of Toluene. As can be seen from Table 2, the activity of the catalyst containing Rh was found to be approximately 5·8 times higher than that of a Pd and 4·5 times higher than of a Pt catalyst at the same weight concentration of the metals. Using other carriers, *e.g.* α-Cr$_2$O$_3$, the same order in activity was found for the noble metals investigated. The apparent activation energies of TSD increase also for noble metals supported on α-Cr$_2$O$_3$ in the same order as for noble metals supported on γ-Al$_2$O$_3$. The methane content in gaseous products of TSD was higher on Pd and Pt than on Rh catalysts. In view of these results it can be concluded that Rh possesses more suitable catalytic properties for TSD than Pt or Pd. This conclusion is in a good agreement

Table 2

Effect of Various Noble Metals on Activity and Apparent
Activation Energy (E_a) of Catalyst for TSD

Catalyst code	Metal (0·6 wt. %)	Activity $r_0{}^a$		E_a (kcal/mol)	CH$_4$ Content[b] (vol %)
		(mol/h kg$_{cat}$)	(mol/h m$^2{}_M$) $\times 10^{-3}$		
Cat.–3	Pd	1·4	0·45	18·8	4·8
Cat.–13	Pd	0·8	0·40	21·0	0·6
Cat.–2	Pt	1·8	0·95	15·5	3·2
Cat.–12	Pt	3·0	2·73	17·6	0·3
Cat.–1	Rh	8·1	2·70	24·3	2·1
Cat.–11	Rh	6·7	2·86	27·5	0·1

[a] At 450°C, R (H$_2$O/T) = 7, P = 1 atm;
[b] In gaseous reaction products.

with results given in some patents[15,16] in which Rh, Rh–Pt, or Rh–Pd catalysts are preferred. The application of supported pure Pt as a catalyst for TSD has rarely been described,[17] obviously because of its low activity.

Effect of Carriers on Catalytical Properties of Supported Rh for Steam- and Hydrodealkylation of Toluene. Having regard to the suitable catalytic properties of Rh in TSD, the carrier effect was studied on this metal only. Since the predominant part of the methane in the gaseous products of TSD is formed by consecutive hydrodemethylation of toluene, eq. (2), we studied simultaneously the effect of different oxides on the

Table 3

Effect of Various Carriers on Catalytic Properties of Supported
Rhodium in Toluene Steam- and Hydro-dealkylation

Catalyst[a]	TSD		THD		R (TSD/THD)	Catalyst life (24 h)
	Activity[b] r_0 (475°C)	E_a (kcal/mol)	Activity[c] r_0 (475°C)	E_a (kcal/mol)		
Cat.–5	2·4	26·1	25·7	13·3	0·09	e
Cat.–6	3·8	22·5	10·2	12·0	0·37	f
Cat.–1	14·8	24·3	15·7	12·8	0·94	e
Cat.–4	5·7	21·0	5·6	11·0	1·01	e
Cat.–7	2·3	21·3	2·3	10·3	1·00	b
Cat.–8	0·1	—	—	—	—	
Cat.–9	0·6	20·6	0·03	31·0	20·00	e
Cat.–10	0·3	22·0	0·08	38·0	3·75	e
Cat.–11	13·4	27·5	6·9	19·2	1·94	d

[a] Concentration of Rh 0·6 wt. %;
[b] R (H_2O/T) = 14·3, P = 1 atm;
[c] R (H_2/T) = 10, P = 1 atm;
[d] stable;
[e] slight decrease of activity in TSD;
[f] very sharp drop of activity in TSD and THD.

catalytic properties of Rh for TSD and THD. From the data given in Table 3 it is evident that different oxides affect very strongly the activity of Rh for both TSD and THD. Oxides known as insulators are associated with high activities of Rh for both reactions. Alkaline-earth and semiconducting oxides (with the exception of Cr_2O_3) on the other hand diminish the activity of Rh, especially for THD. Taylor[18] also found that carriers reduced the activity of supported Ni for hydrogenolysis of hydrocarbons in the order: SiO_2, Al_2O_3, Al_2O_3–SiO_2. Particularly interesting are the values of the ratio R (TSD/THD) which increase smoothly from 0·09 to 10·1 within the group of insulators and then very sharply for the semiconductors (Cr_2O_3 < ZnO < MgO). The carrier effect on the apparent activation energies of different Rh-catalysts in TSD is not as pronounced as in THD. For Rh supported on SiO_2, Al_2O_3 or Al_2O_3–SiO_2 lower activation energies of THD were found than for Rh on MgO, ZnO, or Cr_2O_3. The same qualitative order in activation energies was found by Schwab[19] for Ni supported on semiconductors such as ZnO, MgO, and Cr_2O_3 for the decomposition of formic acid. In view of this fact it is reasonable to assume that THD proceeds with a different mechanism on Cat.–1, –4, –5, –6, and –7 from that on Cat.–9, –10, and –11.

A very important criterion in catalyst screening is the life time (stability) of catalyst or its regeneration. As observed, Rh supported on acidic carriers *e.g.* Al_2O_3–SiO_2 loses its initial activity both for TSD and THD very rapidly. The cause of this effect was found in the strong adsorption of polynuclear aromatic hydrocarbons formed by dehydrocondensation, eq. (4), on the acidic sites of the carrier. The drop in activity can be explained by indirect blocking of Rh active sites by bulky aromatic molecules adsorbed in the vicinity.

$$C_6H_5CH_3 + C_6H_6 \rightarrow C_6H_4CH_2C_6H_4 + 2H_2 \qquad (4)$$

The extent of reaction (4) and similar processes depends on catalyst, temperature, and on molar ratio R (H_2/T) or R (H_2O/T). At T > 525°C and R < 7 the conversion of toluene or benzene into biphenyl, fluorene, and anthracene reached a maximum 1·6 wt. %. However, on Rh catalysts containing oxides of basic character, the adsorption of these polynuclear hydrocarbons is considerably weaker and so the rate of TSD or THD is not affected. These conclusions result from separate experiments in which biphenyl, fluorene, or anthracene were added to the toluene feed in amounts of 0·1–5·0 mol %.

Taking into account all the criteria to be set for catalyst screening, Cr_2O_3 seems to have the best properties as a carrier for Rh and therefore the system Rh–Cr_2O_3 was chosen as starting point for the development of an industrial catalyst.

Effect of Rh and K Concentration on Activity and Selectivity of Catalysts for TSD. The data summarized in Table 4 demonstrate clearly the pronounced effect of Rh concentration on the rate of TSD and TSR. However, with decreasing content of Rh, the r_0 of TSR decreases more sharply than that of TSD as can be seen from the values of the ratio R (TSD/TSR).

Table 4

Effect of Rh Concentration on Activity of Catalysts[a] for TSD

Rh (wt. %)	S_{Rh} (m^2/g)	r_0[b] (mol T/h kg_{cat})		R (TSD/TSR)	S (%)	CH_4 Content (vol %)
		TSD	TSR			
0·9	2·7	15·5	1·53	10·1	91·0	0·25
0·6	2·1	13·4	1·17	11·5	92·0	0·20
0·3	1·1	10·0	0·63	15·9	94·1	0·15
0·15	0·4	6·5	0·24	27·1	96·5	0·15

[a] Carrier α-Cr_2O_3;
[b] T 475°C, R (H_2O/T) 14·3, P = 1 atm.

The addition of alkali to Rh supports, *e.g.* α-Al_2O_3 or SiO_2, has been recommended by many authors.[9,10,12] Our experiments, performed with different amounts of K_2CO_3 (0·5–6·0 wt. %) added to the Cr_2O_3, have shown that with increasing concentration of K the selectivity of catalyst for TSD (92·0–96·0%) increases. However, at higher concentrations of K_2CO_3 (4·0—6·0%) the rate of TSD falls considerably. In view of these facts the addition of 0·5–1·0 wt. % of K_2CO_3 led to the most satisfactory results (S = 94·7%; x = 38·1 mol %; at 475°C; R (H_2O/T) = 14·3; P = 1 atm; F/W = 0·284 mol T/h kg_{cat}).

Table 5

Behaviour of Rh–Cr$_2$O$_3$ Catalyst[a] at Conditions Close to
Industrial Performance of TSD
(T 625 °C, F/W 7·8 mol T/h kg$_{cat}$)

Reaction conditions		x^b (mol %)	S (%)	CH$_4$ Content[c] (vol %)	Content of polynuclear aromatic hydrocarbons[d] (wt. %)
R (H$_2$O/T)	P (atm)				
4·0	1·0	30·2	98·6	0·7	0·6
7·0	1·0	40·0	96·0	0·4	0·3
14·0	1·0	55·2	90·2	0·2	0·1
21·0	1·0	60·0	78·8	<0·1	<0·1
14·0	11·0	56·5	96·3	1·0	0·5
14·0	21·0	48·5	98·2	1·7	0·8

[a] Catalyst containing 0·3 wt. % Rh, *ca.* 20 wt. % hydraulic binder, and 0·3 wt. % K;
[b] Conversion of toluene into benzene;
[c] In gaseous reactions products;
[d] In liquid reaction products.

Behaviour of Rh–Cr$_2$O$_3$ *Catalyst at Conditions Close to the Industrial Performance of TSD.* As can be seen from Table 5, the conversion of toluene into benzene increases considerably with growing molar ratio R (H$_2$O/T). On the other hand, the selectivity decreases because the TSR is accelerated more with increasing partial pressure of water than TSD. The total pressure affects mainly the selectivity and the formation of methane. However, the quantity of CH$_4$ found at a pressure of 21 atm is surprisingly low in comparison with the amount of CH$_4$ formed (34·8 vol %) on the Rh–γ-Al$_2$O$_3$ catalyst. In view of this fact the Rh–Cr$_2$O$_3$ catalyst[13] is applicable for the industrial performance of TSD. The reason for the formation of such small quantity of CH$_4$ on a Rh–Cr$_2$O$_3$ catalyst is the strong retardation of the consecutive THD by water, which is contrary to the behaviour of a Rh–γ-Al$_2$O$_3$ catalyst. These conclusions resulted from kinetic and adsorption measurements made on both catalysts.

Using Cat.–1 we found for the THD at 475 °C ($p_T = 0·1$ atm, $p_{H_2} = 0·9$ atm) $r_0 = 15·7$ mol T/h kg$_{cat}$ (Table 3). This value decreases to 6·1 mol T/h kg$_{cat}$ when the feed contains water (p_{H_2O} 0·3 atm, p_{H_2} 0·6 atm, and p_T 0·1 atm). On the other hand on Cat.–11 the r_0 (THD) falls from 6·9 to 0·63 mol T/h kg$_{cat}$ when water is added to the feed in the same amount.

Adsorption measurements in the pulse reactor at 300 °C revealed large differences in the adsorptivity of toluene, H$_2$, and H$_2$O on both catalysts. On Cat.–1 the following relative adsorption coefficients were determined: ($K_T = 1$) $K_{H_2} = 1·4$; $K_{H_2O} = 1·6$. On the other hand, on Cat.–11 a much smaller value for H$_2$ ($K_{H_2} = 0·07$) and a higher value for H$_2$O ($K_{H_2O} = 3·6$) were found. These results are in a good agreement with the conclusions drawn on the basis of kinetic measurements.

Kinetic course of TSD on Rh–Cr$_2$O$_3$ *catalyst.* In experiments with different space velocities (estimation of r_0) at constant temperature and molar ratio R (H$_2$O/T) we observed that at very high values of F/W (> 20 mol T/h kg$_{cat}$) and low values R (H$_2$O/T) < 10, the CO content in gaseous reaction products of TSD increases considerably (> 4·0 vol %).

This fact indicates that TSD, eq. (1), consists of two steps, eq. (5, 6).

$$C_6H_5CH_3 + H_2O \rightarrow C_6H_6 + 2H_2 + CO \tag{5}$$

$$CO + H_2O \rightleftharpoons H_2 + CO_2 \tag{6}$$

However, reaction (6) was found to be much faster than reaction (5) under our standard experimental conditions, and reached equilibrium on all catalysts investigated. Reaction (5) was the rate-determining step in our kinetic measurements.

The kinetic data for TSD obtained using Cat.–1 and Cat.–11 at 475°C treated with different types of Langmuir–Hinshelwood equations for bimolecular reactions[20] have shown that r_0 can be described by eq. (7)

$$r_0 = \frac{kK_TK_{H2O}p_Tp_{H2O}}{(1 + K_Tp_T + K_{H2O}p_{H2O})^2}. \tag{7}$$

Mechanistic Considerations. To obtain further information concerning the mechanisms of TSD and THD on different Rh catalysts, the kinetic isotope effects α_{D_2O/H_2O} and $\alpha_{D/H}$ were measured under conditions given in Table 3. For the TSD on Rh–γ-Al$_2$O$_3$ (Cat.–1) α_{D_2O/H_2O} was 1·55 and on Rh–Cr$_2$O$_3$ (Cat.–11) α_{D_2O/H_2O} was 1·62. On the other hand, for the THD on Cat.–1 we obtained for $\alpha_{D/H}$ the value of 1·38, and on Cat.–11 the value of 1·05.

In all experiments carried out with D_2 or D_2O the incorporation of deuterium in benzene or toluene molecules caused by dealkylation or H–D exchange was observed. An especially high content of $C_6H_5CD_3$ (*ca.* 70%) was detected in unreacted toluene. The experiments carried out in a pulse reactor with H_2 and toluene on Rh–γ-Al$_2$O$_3$ catalyst which was pretreated with D_2O have shown that benzene formed in the initial stage of reaction contained a detectable amount of C_6H_5D species. This fact can be explained by the participation of surface OD groups of Al$_2$O$_3$ in the hydrodealkylation process. The interpretation of isotope effects in this case is complicated and does not lead to unambiguous conclusions, expecially in the case of TSD, in which very close values were determined for both catalysts (Cat.–1 and Cat.–11). However, the large difference in values of $\alpha_{D/H}$ found for Rh–γ-Al$_2$O$_3$ and Rh–Cr$_2$O$_3$ catalysts could be further evidence that the THD proceeds on both catalysts by different mechanisms.

On catalysts in which Rh was supported on oxides such as Al$_2$O$_3$, SiO$_2$, or Al$_2$O$_3$-SiO$_2$ we can assume that the surface OH groups of carriers take part in the THD, obviously in the form of "spillover" effects. In this case not only the concentration of OH groups but primarily their acid strength, seems to play an important role (Table 1). On Rh supported on MgO, ZnO, and Cr$_2$O$_3$ another mechanism operates in which adsorbed hydrogen directly attacks molecules of adsorbed toluene.

On the other hand, the results already mentioned seem to indicate that TSD proceeds with the same mechanism on all Rh catalysts studied. Concerning Cat.–1, –4, –5, –6, –7 and –11, adsorbed water and toluene molecules obviously participate in the surface reaction.

However, further experiments are necessary to obtain a detailed mechanistic picture of the course of TSD and some other parallel and consecutive reaction occurring on Rh catalysts.

ACKNOWLEDGEMENTS. The author is indebted to Dr. P. Günther, Dr. W. Rohm (Süd-Chemie) and Prof. Dr. H. Knözinger (Physikalisch-chemisches Institut, Universitat München) for valuable discussions and Mr. O. Bock and Mr. H. Rudolph for their technical assistance.

REFERENCES

1 V. Haensel, *US Pat.* 2,436,923 (1948).
2 G. V. Dydikina, G. L. Rabinovich, G. N. Maslaynaskii, and M. I. Dementeva, *Kinetika i Kataliz*, 1969, **10**, 497.
3 G. L. Rabinovich, G.N. Maslyanskii, and L. M. Treiger, *Kinetika i Kataliz*, 1971, **12**, 1567.
4 L. M. Treiger, G. L. Rabinovich, and G. N. Maslyanskii, *Neftekhimiya*, 1972, **12**, 29.
5 G. L. Rabinovich, G. N. Maslyanskii, V. S. Vorob'ev, L. M. Biryukova, and I. I. Ioffe, *Neftekhimiya*, 1973, **13**, 518.
6 L. M. Treiger, G. L. Rabinovich, and G. N. Maslyanskii, *Kinetika i Kataliz*, 1973, **14**, 1582.
7 G. L. Rabinovich, V. S. Vorob'ev, I. A. Vasil'ev, and L. M. Biryukova, *Zhur. priklad. Khim.*, 1973, **46**, 1798.
8 G. N. Maslyanskii, G. L. Rabinovich, and co-workers, *Brit. Pat.* 1,174,879 (1969).
9 G. R. Lester, *US Pat.* 3,436,433 (1969).
10 G. R. Lester, *US Pat.* 3,436,433 (1969).
11 G. R. Lester, *US Pat.* 3,649,706 (1972).
12 G. L. Rabinovich, and G. N. Maslyanskii, *Khim. Tekhnol. Topl. Masel*, 1973, **2**, 1.
13 K. Kochloefl, *Ger. OS* 2,357,405 (1974).
14 H. A. Benesi, *J. Amer. Chem. Soc.*, 1956, **78**, 5490.
15 G. L. Rabinovich, and G. N. Maslyanskii, *Zhur. priklad. Khim.*, 1974, **45**, 1351.
16 G. N. Maslyanskii, and G. L. Rabinovich, *French. Pat.* 1,588,876 (1970).
17 R. J. Sampson, C. B. Spencer, and J. G. Cheroweth, *Ger. OS* 2,214,512 (1972).
18 W. F. Taylor, D. J. C. Yates, and J. H. Sinfelt, *J. Phys. Chem.*, 1964, **68**, 2962.
19 G. M. Schwab, J. Block, J. H. Müller, and D. Schultze, *Naturwissenschaften*, 1957, **44**, 582.
20 J. Šimoniková, L. Hillaire, J. Pánek, and K. Kochloefl, *Z. phys. Chem.* (*Frankfurt*,) 1973, **83**, 287.

DISCUSSION

T. R. Phillips (*British Gas, Solihull*) said: Steam dealkylation can be regarded as the removal of a methyl group from the benzene nucleus and subsequent steam reforming. The final products, if the catalyst is active for this reaction, will depend on the pressure and temperature: high pressure will favour methane manufacture, whereas high temperature and lower pressure will favour hydrogen production. Your results obtained with rhodium supported on γ-alumina follow this pattern and this would also explain the much lower methane content that you found with the chromia-supported rhodium at 898 K. If pure hydrogen is required a relatively inactive catalyst used at high temperature is probably desirable. Unless this is taken into account erroneous conclusions could be drawn concerning the relative rates of steam dealkylation and hydrodealkylation of toluene. Have you allowed for these effects?

M. Taniewski (*Silesian Polytechnic Univ., Gliwice*) said: Steam dealkylation of toluene and its hydrodealkylation are important ways of transforming excess toluene into benzene. Such transformations will be used until we are able, industrially, to produce from toluene those aromatic derivatives which at present we make from benzene. Steam dealkylation of toluene belongs to the large group of reactions of organic compounds with steam in which one type of conversion is desirable whereas some other is undesirable. Here the steam conversion of a side-chain is desirable but that of the benzene ring is undesirable. Fortunately, the relatively high stability of the benzene ring makes it possible to develop selective catalysts. But quite often we face a much more difficult problem, namely how to suppress completely the steam conversion of alkylbenzene (*e.g.* ethylbenzene) during its reaction carried out in the presence of water (*e.g.* dehydrogenation) while retaining the self-regenerative functions of the catalyst bed. Self-regeneration in such systems consists in the continuous removal of hydrocarbon deposits from the catalyst surface by a steam conversion similar in its nature to steam conversion of hydrocarbons. We have demonstrated in recent years that even in the presence of a highly selective catalyst developed for dehydrogenation of ethylbenzene (contact PO-12) some proportion of the by-products (benzene, toluene) undoubtedly originated from the steam conversion of chemisorbed alkylaromatics.

You seem to regard the extent of methane formation as a measure of the extent of the consecutive toluene hydrodealkylation. What is your opinion as to the possibility that some part of the methane may be formed also in another way, namely by a secondary hydrogenation (methanation) of carbon oxides? Did you try to hydrogenate carbon oxides on your catalysts? The kinetics of these two methods of methane formation would be different. As we know, catalysts containing platinum, ruthenium, *etc.* (but not nickel) are also used for methanation. Methanation of carbon oxides, being a pressure-dependent reversible reaction, would also be favoured by an increase of pressure, as you find in your experiments.

R. W. Coughlin (*Lehigh Univ., Bethlehem, Pa.*) (communicated): Have you excluded a mechanism which would include steam reforming of methane produced by hydrogenolysis of toluene?

K. Kochloefl (*Girdler-Südchemie, Munich*), in reply to the three previous contributors, said: As has been shown, toluene steam dealkylation consists of two steps [equations (5) and (6)] and can be accompanied by different consecutive reactions [equations (2), (3), and (4)]. To these should be added, *e.g.*

$$CO + 3H_2 \rightleftharpoons CH_4 + H_2O$$

$$CO + H_2 \rightleftharpoons C + H_2O$$

$$CH_4 + CO_2 \rightleftharpoons 2CO + 2H_2$$

and the like. The thermodynamic possibility of these consecutive reactions accompanying steam dealkylation has been verified for a broad range of conditions.[1,2] However, the extent to which these reactions really participate depends not only on reaction conditions (thermodynamic control) but also on the properties of the catalyst (kinetic control). For example, at 698 K ($P = 1$ atm; $F/W = 30.9$ mol T/h kg$_{cat}$ $R(H_2O/T) = 11$) using Cat.-1 we found 14.8 vol % carbon monoxide in the gaseous reaction products of toluene steam dealkylation. On the other hand, under the same conditions but using Cat.-11, the carbon monoxide content was only 1.9 vol %. This indicates that the shift reaction [equation (6)] proceeds on Cat.-11 faster than on Cat.-1. This conclusion was confirmed recently by direct kinetic measurements of the carbon monoxide shift reaction on these catalysts. At still higher space velocities of toluene the carbon monoxide content in gaseous reaction products increases and the methane content decreases. If methane was an intermediate in steam dealkylation, as Dr. Phillips suggests, then the opposite would have been observed.

Complementary experiments carried out with addition of hydrogen to toluene steam dealkylation ($T = 748$ K, $P = 1$ atm, $p_T = 0.05$ atm, $p_{H_2O} = 0.51$ atm, $p_{H_2} = 0.44$ atm) revealed that, on Cat.-1, the addition of hydrogen increased toluene conversion by about one third over the initial value without hydrogen (to keep p_T constant, helium was added) but that carbon monoxide conversion into methane remained practically constant. Over Cat.-11 the conversion of toluene decreased on addition of hydrogen by about 25%) and carbon monoxide conversion slightly increased.

These facts indicate that carbon monoxide methanation on Cat.-1 is more strongly retarded by water than is toluene hydrodealkylation. On the other hand, the opposite is true on Cat.-11. Consequently, methane produced during toluene steam dealkylation on Cat.-1 is formed mainly by hydrodealkylation. On Cat.-11 carbon monoxide

[1] P. Beltrame, L. Forni, P. Schwarz, and S. Torrazza, *Chimica e Industria*, 1972, **54**, 1072.
[2] S. Kasaoka, M. Omoto, T. Watanabe, and K. Takamatsu, *Nippon Kagaku Kaishi*, 1975, **8**, 1418.

hydrogenation is the main process for methane formation. However, on this catalyst the methane content in the gaseous products of toluene steam dealkylation is very low because the carbon monoxide shift reaction is very rapid and carbon monoxide methanation is retarded by water.

In reply to Prof. Coughlin, I would say that, when studying n-hexane and methane steam-reforming on both catalysts at 748 K we found that the activity of Cat.-1 for this reaction was lower than that of Cat.-11. With respect to all results observed in these complementary experiments, the steam-reforming of methane need not be considered as a consecutive reaction of toluene with steam dealkylation.

R. Maatman (*Dordt College, Sioux Center*) said: Earlier in the meeting, in connection with the paper by Beyer *et al.*, I commented that calculation of a site density for a reaction, assuming a certain step of a mechanism to be rate-determining, provides a criterion for determining whether or not that step can be the rate-determining step. Calculation of the site density for each of 14 possible rate-determining steps in the steam dealkylation of toluene for your rhodium catalysts, numbered 1 and 11, reveals that only three of these steps could be a rate-determining step provided that only one step is rate-determining. Three other steps are outside possibilities. Of the three most likely, it seems (taking into account chemical intuition and certain of your results) that the rate-determining step is one in which water is strongly adsorbed, compared to toluene, and in which reaction is between two adsorbed molecules. I realize you have already made some conclusions concerning mechanism. Is reaction between strongly adsorbed water and weakly adsorbed toluene consistent with what you know of this reaction?

K. Kochloefl (*Girdler-Südchemie, Munich*) replied: The adsorption coefficients for toluene (K_T) and for water (K_{H_2O}) have been calculated from equation (7). For both catalysts the adsorption coefficient for water was found to be higher than that for toluene. For example the ratio K_{H_2O}/K_T was 1·4–2·5 for Cat.-1 and 3·2–9·5 for Cat.-11, the range showing the variation with reaction temperature. These values are in good agreement with the results of adsorption measurements. The lower value of K_{H_2O} for Cat.-1 can be explained by assuming that the water molecules are bonded to surface hydroxyl groups by relatively weak hydrogen bonds. On the other hand on Cat.-11 where the higher value of K_{H_2O} was found, the water molecules are obviously strongly held by incompletely co-ordinated chromium ions that have Lewis acid character.

Y. Ogino and A. Igarashi (*Tohoku Univ., Sendai*) said: We have found that a nickel–beryllia catalyst is very active and selective for the steam dealkylation of alkylbenzenes. We have determined the structure of the catalyst as well as the kinetics of reaction, and have proposed the model[3] shown in the Figure (see p. 1132). This model has proved useful in designing and preparing effective catalysts.

Would you agree that the use of a carrier prepared from an acidic oxide and a basic oxide would yield interesting results?

K. Kochloefl (*Girdler-Südchemie, Munich*) replied: Hitherto the experimental information available has not been sufficient to enable us to determine detailed mechanisms for hydrodealkylation, steam dealkylation, and steam reforming of toluene over acid-supported noble metal catalysts. However, it seems probable that the geometrical orientation of toluene molecules in the activated state plays an important role. In hydrodealkylation or steam dealkylation the removal of the methyl group will be facilitated by a perpendicular or cross-orientation of the benzene ring with respect to the catalyst surface, whereas in steam reforming adsorption of toluene parallel with the surface will clearly be important.

In answer to your second point, a rapid loss of initial activity for steam dealkylation

---

[3] Y. Ogino *et al.*, *J. Japan Petroleum Inst.*, 1975, **18**, 108.

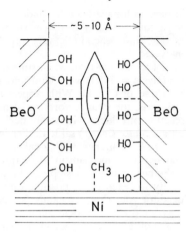

FIGURE 1   A model for the adsorption of toluene on nickel–beryllia.

of toluene over rhodium has been observed when the metal was supported on a mixture of oxides such as (MoO_3 + MgO) or (Fe_2O_3 + ZnO).

Z. M. George (*Alberta Res. Council, Alberta*) (communicated): To what extent was adsorption of toluene on your catalysts reversible, and did coking complicate your kinetic measurements?

K. Kochloefl (*Girdler-Südchemie, Munich*) replied: The adsorption of toluene on Rh–γ-Al_2O_3 and on Rh–α-Cr_2O_3 between 523 and 623 K was studied in helium using the gas chromatographic pulse method.[4] Under these conditions only a very small proportion of the injected toluene (the difference between the first and the other pulses) was irreversibly adsorbed.

The kinetic measurements were carried out between 648 and 798 K where no coke deposits were found. The observed decrease of activity (in the initial period of the kinetic experiments especially) was caused by the adsorption of polynuclear hydrocarbons (*e.g.* anthracene) on the acid sites of the carrier. The real coking of rhodium catalysts was observed above 900 K and was a function of the steam:toluene ratio. It is not clear whether the adsorbed polynuclear aromatic hydrocarbons are the coke precursors or whether the coke is formed through other reactions such as:

$$CH_4 + CO \rightleftharpoons H_2O + H_2 + 2C.$$

G. Y. Gati (*High Pressure Research Inst., Budapest*) (communicated): The degree of surface hydroxylation and the excess oxygen content (in the case of chromia) can be quite different during acidity and basicity determinations from that which obtains under reaction conditions, because of interaction with the reactants at the reaction temperature. You have divided the carriers into two groups with respect to the mechanism by which toluene hydrodealkylation is achieved. The acidic carriers such as alumina and silica–alumina belong to the first group. In the second group you have the basic magnesium oxide together with chromia, which was acidic by your titration method. Have you any explanation for this? I believe that chromia reduced at the higher reaction temperatures exposed only basic oxygen ions and co-ordinatively unsaturated chromium ions as the Lewis acid sites, and that there are no acidic hydroxyl

[4] J. Svoboda, E. Ročková, and K. Kochloefl, *Coll. Czech. Chem. Comm.*, 1970, **35**, 1671.

groups which could take part in the hydrogen spillover which you suggest to be important in the case of the acidic carriers.

K. Kochloefl (*Girdler-Südchemie, Munich*) replied: As indicated in my paper, the hydrogenolysis of toluene over rhodium supported on α-chromia, zinc oxide, or magnesium oxide proceeds by one mechanism, whereas reaction over rhodium supported on silica, γ-alumina, or silica–alumina proceeds by another mechanism, in which hydroxyl groups take part in hydrodemethylation. Using Benesi's method, we found α-chromia precalcined at 923 K to be acidic ($pK_A = 3\cdot3$). However, the concentration of hydroxyl groups which are responsible for this acidity is very small. The gas chromatographic 'titration' with the complex $(CH_3)_2Zn(tetrahydrofuran)$ confirmed the previous results. By this method the following concentrations of hydroxyl groups were found: α-chromia, $0\cdot05$; α-alumina, $0\cdot9$; silica, $1\cdot24$; γ-alumina, $2\cdot68$ mmol g^{-1}. This demonstrates why the mechanism involving spillover is irrelevant to toluene hydrogenolysis over α-chromia-supported rhodium.

H. Noller (*Technical Univ., Vienna*) (communicated): Instead of classifying the carriers as semiconductors and insulators, I would suggest classifying them according to their basicity or electron-pair donor (EPD) strength (at the oxygen, of course). The O ($1s$) binding energy obtained by X-ray photoelectron spectroscopy can be used as an indicator; the lower the values the higher the EPD strength. The values[5] are: MgO, $530\cdot2$ eV; ZnO, $530\cdot4$ eV; Cr_2O_3, $530\cdot7$ eV; Al_2O_3, $531\cdot8$ eV; SiO_2, $533\cdot1$ eV; a value for CaO is not available but it is certainly lower than that for MgO. This is, in general, the order of activity for both steam dealkylation and hydrodealkylation of toluene, with a maximum in activity with γ-Al_2O_3 (not including porous glass). Consequently, the most important feature determining the interaction must be one in which at least one of the reactants acts as an electron-pair donor. Selectivity varies in the opposite sense; it is highest with the most basic carrier (strongest EPD), which agrees with the rule that high selectivity is usually not associated with high activity.

With the above concept in mind it is possible to predict the effect of potassium carbonate on chromia. K_2CO_3 increases the basicity (EPD strength) of chromia. This should make the catalyst less active and more selective, as is actually found.

K. Kochloefl (*Girdler-Südchemie, Munich*) replied: I initially classified oxides used as carriers for noble metals as insulators and semiconductors. However, this classification has only a formal meaning. The important criterion for the classification of oxides with respect to their hydrogenolytic activity in the combination with noble metals is the acid strength of the surface hydroxyl groups and their concentration (see my reply to Dr. Gati). However, the values of EPD-strength that you mention can be a further valuable criterion for classification.

[5] N. P. Sergushim, V. I. Nefedov, B. F. Djurinsky, D. Gati, and Ya. V. Salym, *ESCA Symposium* (Reimhartsbrunn, DDR, 1975).

Author Index

This index has been compiled from the names of the authors of the papers, and the contributors to the discussions.

Author Index

Subject Index

This subject index has been compiled from key-words that appear in the title and abstract of each paper. Normally, reaction processes are not indexed: thus, for 'Hydrogenation of benzene' see 'Benzene, hydrogenation of'.

Date Due